Metal Speciation in the Environment

NATO ASI Series

Advanced Science Institutes Series

A series presenting the results of activities sponsored by the NATO Science Committee, which aims at the dissemination of advanced scientific and technological knowledge, with a view to strengthening links between scientific communities.

The Series is published by an international board of publishers in conjunction with the NATO Scientific Affairs Division

A Life Sciences B Physics	Plenum Publishing Corporation London and New York
C Mathematical and Physical Sciences D Behavioural and Social Sciences E Applied Sciences	Kluwer Academic Publishers Dordrecht, Boston and London
F Computer and Systems Sciences G Ecological Sciences H Cell Biology	Springer-Verlag Berlin Heidelberg New York London Paris Tokyo Hong Kong Barcelona

Series G: Ecological Sciences Vol. 23

Metal Speciation in the Environment

Edited by

J. A. C. Broekaert

Institut für Spektrochemie und angewandte Spektroskopie (ISAS), Bunsen-Kirchhoff-Straße 11, 4600 Dortmund, FRG

Ş. Güçer

Department of Chemistry, Inönü University, Art & Science Faculty, 44069 Malatya, Turkey

F. Adams

Universiteit van Antwerpen (UIA), Departement Scheikunde, Universiteitsplein 1, 2610 Wilrijk, Belgium

Springer-Verlag Berlin Heidelberg New York
London Paris Tokyo Hong Kong Barcelona
Published in cooperation with NATO Scientific Affairs Division

Proceedings of the NATO Advanced Study Institute on Metal Speciation in the Environment held in Cesme, Turkey, October 9–20, 1989

ISBN-13: 978-3-642-74208-8 e-ISBN-13: 978-3-642-74206-4
DOI: 10.1007/ 978-3-642-74206-4

Softcover reprint of the hardcover 1st edition 1990

2131/3140-543210 – Printed on acid-free-paper

Contents

EDITORIAL

The environmental persistence, fate and interactive effects with living organisms - beneficial or toxic - of trace elements are directly related to the physico-chemical forms in which they occur. An increasing awareness of this species-specific behaviour, over the course of the past few decades, has triggered the development of new analytical methods, commonly referred to as "speciation". In its broadly accepted sense, speciation may be defined as the qualitative identification and the quantitative determination of the individual chemical forms that comprise the total concentration of a given trace element in a sample.

Knowledge on the essential character of chemical compounds up to certain concentration levels and on their toxicity at other concentration levels as well as information about the circulation of chemical compounds through the different compartments of the environment are indispensable for the progress of life sciences. It is now well-known that the interactions of all the partly still unidentified species with plants, animals and man are complex. But their detailed understanding is a most exciting challenge to science, as this will be required for decisions influencing the evolution of life on one side and for the quality of life on the other.

Progress in environmental sciences can only be achieved by interdisciplinary research efforts where geologists, biologists, chemists and engineers closely cooperate and define and use one common language. It was the aim of the NATO ASI "Metal Speciation in the Environment" to foster these research endeavours by reporting on the state-of-the-art of the knowledge on the occurence and formation of relevant species of the elements and on their mobility in the different compartments of the environment. For making progress in this field, analytical techniques with which these species can be identified and determined down to the extremely low trace level, are required. For a series of these tasks, analytical tools only now become readily available. Accordingly, it is now of importance to arrive at an exact formulation of the environmental analytical problems to be solved on one hand and to be able to report on progress in environmental analytical techniques enabling the determination of species on the other hand.

In the NATO Advanced Study Institute 180/88, which was held from October 9 -

20, 1989 in Cesme (Turkey), participants from 18 countries took active part in the lectures, posters and discussion sessions. The scientific program was organized around three main topics. A first series of problems concerning the dynamic behaviour of metal species in the environment was extensively discussed. A second part of the meeting dealt with processes of metals complexation in the environment while a third part concerned analytical techniques for metal speciation in environmental analysis.

These proceedings of the meeting should reflect the state-of-the-art of the knowledge and the methodology available for the speciation of the elements in the environment today. The editors especially want to stress their gratitude to two members of the organization committee, namely to professor D. Klockow from the Institut für Spektrochemie und angewandte Spektroskopie in Dortmund for his efforts in the preparation of the programme and in the organization of the ASI, as well as to professor Erol Izdar from the University of Izmir. Their active collaboration during the meeting were instrumental for the quality level of the discussions and presentations.

ASSESSMENT OF METAL MOBILITY IN SLUDGES AND SOLID WASTES

U. Förstner, W. Ahlf, W. Calmano, M. Kersten, and J. Schoer
Department of Environmental Science and Engineering
Technical University of Hamburg-Harburg
Eißendorferstr. 40
D-2100 Hamburg 90
West Germany

1. Introduction

Among the criteria to assess which element or elemental species, beside its toxic potential, may be of major concern in ecological evaluations, one question deserves primary attention (Nriagu, 1984): Is the element mobile in geochemical processes because of either its volatility or its solubility in natural water, so that the effect of geochemical perturbations can propagate through the environment? In this context, "mobility" mostly deserves a negative aspect of environmental "speciation" - here in terms of "describing the distribution and transformation of metal species in various media" (Bernhard et al., 1986) - in that faster transfer from one environmental medium into another generally involves greater reactivity and bioavailibility of potentially toxic elements.

Problems with solid "speciation" - now with the second meaning of "operational procedures for determining typical metal species in environmental samples" (Bernhard et al., 1986) - are connected to the complexity of *heterogenous systems*, e.g. of soils, sediments and aerosol particles, and to disequilibria between dissolved and particulate fractions. It has been stressed, that due to the important role of kinetically controlled processes in these systems, the actual speciation is often different from what can be expected (Andreae et al., 1984). Particularly in polluted solid systems, an even greater increase of entropy will induce a concominant increase in instability in both physical and chemical context; this can mainly be seen in the difficulties in sample handling and storage prior to analysis (Wood et al., 1986). On the other hand, it is just these systems, where action is immediately needed

NATO ASI Series, Vol. G 23
Metal Speciation in the Environment
Edited by J. A. C. Broekaert, Ş. Güçer, and F. Adams

and where for an assessment or prognosis of possible adverse effects the species and the transformations of pollutants have to be evaluated.

The present review on metal mobility in relation to "speciation" in typical examples of solid waste materials is a follow-up study to our earlier compilations, e.g. from the NATO Advanced Research Workshop on Trace Element Speciation held at Nervi/Italy in 1981 (Förstner & Salomons, 1983), the Dahlem Konferenzen on Chemical Speciation in Environmental Processes held at Berlin in 1984 (Förstner, 1986), the NATO/-ISSS Workshop on Soil Colloid/Solution Interface held at Gent in 1986 (Förstner, 1989a), and from an international workshop on Speciation of Metals in Water, Sediment and Soil Systems held at Sunne/Sweden in October 1986 (Förstner, 1987). Here we will focus on the progress, which - to our opinion - has been achieved since the latter conference with respect to the conceptual aspects and operational procedures in the subject area of distribution and transformation of metal species in solid waste materials.

2. Dynamic Behaviour of Solid Metal Species in the Environment

In addition to the manifold procedural aspects of speciation, which form the major focus of the present symposium, we consider it as an important challange in this new field of environmental research that the findings established on the molecular level are transferred into a macroscopic scale, where practical action can take place.

2.1 The Mobility Concept in Relation to Metal Speciation

Typically for systems involving solution/solid interactions, "*mobility*" reflects the flux of metal species in a certain medium, which contains both accelerating and inhibiting factors and processes. The former influences comprise effects of pH-lowering, redox changes, inorganic and organic complexation, and microbially mediated species transformations such as biomethylation. Among the spectrum of "barriers", physical processes include adsorption, sedimentation, and filtration;

chemical barriers comprise mechanisms such as complexation and precipitation; biological barriers are often associated with membrane processes, which can limit translocation of metals, for example, from plant roots to the shoots and fruits. "Complexation" in its various forms can both inhibit and accelerate metal fluxes, particularly in biological systems consisting of different types of membranes. Four different ways can be envisaged for the assessment of metal mobility in typical environmental compartments:

1 The most comprehensive approach involves evaluation of all species changes along the pathway of the element through the system ("full-system speciation").

2 For many objectives, e.g. for studying effects on biota, determination of the elemental species distribution at a critical section within the system or at its efflux may be sufficient ("part-system speciation").

3 Model considerations, which are aimed to generalize typical distribution or transformation processes, need data on dominant species combined with typical mobilizing/inhibiting parameters such as pH, and pH_2S in the dissolved phase and surface properties of the solid phase ("model species/mobility approach").

4 Long-term prognosis of the behaviour of metals at critical sites requires both the knowledge of interactions of element species in solid matter and solution, and an estimation of the future borderline conditions in a dynamically evolving medium ("prognostic mobility/species approach").

In the present contribution we will mainly discuss the two latter approaches, which typically include solid/solution interactions of metal species. As will be shown later, long-term prognosis of such interactions (approach 4) at this stage is still based mainly on characteristic changes of total metal fluxes, i.e. on mobility variations derived from laboratory experiments, rather than on informations on real "speciation".

2.2 Metal Mobility in Dynamic Environments

Under the aspect "dynamic behaviour of metal species in the environment" (Session 1 of the Symposium) typical "macro-environments" can be considered, where the assessment of metal mobility is of particular relevance for estimating both the actual and future effects of critical components. Such "dynamic" sites, which have been described during the present symposium, include: (1) Excessive treatment of *soil* with organic-rich agricultural waste materials (Del Castilho); (2) transformation of metal-rich components in *mining residues* in tropical regions (Salomons); (3) behaviour of polluted particulate matter in the *estuarine mixing zone* (Calmano); (4) long-term development of *municipal and industrial waste materials* (this contribution). In these dynamic environments characteristic factors and processes are acting as "driving forces":

- *High-energy mechanical processes*, such as resuspension of particles by wave action, bioturbation, and dredging activities;
- *Strong chemical gradients*, e.g. of salinity, as well as of redox conditions, pH-values, and organic ligand concentrations, the latter three variables mainly induced by the degradation of organic matter;
- *Distinct temporal variations* and developments, such as seasonal changes of biological activity in aquatic systems and typical successions in metabolic transformations, e.g., reductive processes in waste materials.

From a practical view, for example, for application in waste-water treatment and solid waste disposal techniques, knowledge on "speciation" of characteristic elements may be useful; however, it seems that with the concept of mobility assessment, the often complex information on interactions between solid and dissolved species can be narrowed down to more simple parameters, which can both be measured under field conditions and used for generalizations in models. The questions arising in this context are:

1 Can the *actual metal species* in both solid and dissolved phase of a given system satisfactorily be described?

2 What are the *typical variables*, apart from metal species distribution, which can be measured in a *dynamic system*?

If these informations are available, experimental and/or model designs can be developed, from where - under expected boundary conditions - interrelations between metal species and typical variables can be studied and extrapolated for a long-term prognosis of metal behaviour in critical cases.

3. Modeling Solid/Solution Interactions of Metal Species

One of the greatest challenges of environmental chemistry has been to describe the behaviour of trace elements in natural aquatic systems based solely on knowledge of their fundamental physical-chemical properties. Initial efforts for applying quantitative models have been undertaken for the prediction of metal speciation in solution (Baham, 1984). The theoretical foundations for solving the problem of chemical speciation, which is usually not solvable by use of experimental analysis, are based upon a model that relates the equilibrium activities of metals and ligands to the formation of complexes in solution (see contributions by Buffle and other authors in this volume). These models are also being used increasingly to predict the mobilizable fraction of metals in the sediment (Baes & Sharp, 1983) and the transport of toxic metals and radionuclides in ground waters (Lewis et al., 1987), to assess pollution potential to the ecosystem in general (Luoma & Davis, 1983), and to develop sediment quality criteria (Jenne et al., 1986). Quantification of competing effects is difficult, and thus the equilibrium partition approach for sediment quality assessment of metals is still limited (Honeyman & Santschi, 1988; Shea, 1988; see contribution by Calmano et al. in this volume).

3.1 Factors Affecting Distribution Coefficients (K_D-Values)

Mobility of an element in the terrestrial and aquatic environment is reflected by the ratio of dissolved and solid fractions. Evaluation of the current literature indicates at least three major factors affecting the distribution of trace metals between solution and particulates: (i) the chemical form of dissolved metals originating both from natural and human-caused sources (Förstner & Salomons, 1983); (ii) the type of interactive processes, such as sorption/desorption- or precipitation-controlled mechanisms (Salomons, 1985); and (iii) concentration and composition of particulate matter, mainly with respect to surface-active phases.

Typical curves for the adsorption of metals onto inorganic substrates, such as iron oxyhydrate, increase from almost nothing to near 100% as pH increases through a critical range 1-2 units wide (Benjamin et al., 1982). It is important to note, that the location of the pH-adsorption-"edge" depends on adsorbent concentration. This effect is due to the existence of a range of specific site-binding energies. High-energy adsorption sites, since they are fewer in number than lower energy sites, become limiting first. As lower energy sites are gradually filled, the overall binding constant decreases. In the few cases, where kinetics of sorption have been investigated, surface reactions were not found to be a single step reaction (Anderson, 1981). Experiments performed by Benjamin & Leckie (1981) showed an initial, rapid and almost complete metal uptake process perhaps lasting no more than a few minutes to hours, followed by a second, slower uptake process requiring from a few days to a few months. The first effect was thought to be true adsorption, and the second to be slow adsorbate diffusion into the solid substrate or coagulation of colloidal to filterable particles (Santschi & Honeyman, 1989). Extrapolations to geologic time scales have been made by Förstner & Schoer (1984) by a comparison of the extractibility of stable metal isotopes and their unstable counterparts - the latter supplied from radioactive emissions of nuclear power and reprocessing plants; using reducing extraction agents, for example, only 15% of the natural stable manganese was mobilized from a river sediment sample, while the artificial, more recently (during the past 20-30 years) induced isotope Mn-54 from the reprocessing plant is extracted by more than 80% during the same treatment.

For systems rich in organic matter, metal adsorption curves cover a wider pH-range than is observed for inorganic substrata. Typically, a reduced reversibity of metal sorption has been observed in these organic systems (Lion et al., 1982). Such effects may be important restrictions for using distribution coefficients in the assessment of metal mobility in rapidly changing environments, such as rivers, where equilibria between the solution and the solid phases can often not be achieved completely due to the short residence times. In practice, applicability of distribution coefficients may find further limitation from methodological problems. Sample pretreatment (e.g., dry or wet conditions), solid/liquid separation technique (filtration or centrifugation), and grain size distribution of solid material is strongly affecting K_D-factors of metals (e.g., Calmano, 1979; Duursma, 1984; Table 1). Such effects also have to be considered for the interpretations of in-situ processes, where the above-mentioned influences of reversibility usually are playing a minor role than in the case of open-water conditions.

With respect to the possible differences of metal sorption between natural and polluted systems it can be argued from the preceding information that the higher concentrations of metals expected in contaminated systems will tend (see also Brümmer et al., 1986):

- to be less strongly bound to the substrate,
- to form precipitates rather than to interact via adsorption, and
- to be subject to more intensive transformations, including competition between various substrates, than do the natural constituents in aquatic and terrestrial systems.

In addition, mobility of metals is influenced by the chemical form of the *inputs*. Direct emissions, for example of Cd, into the environment from waste materials are approximately 10-fold greater from solid materials such as pigments, phosphate

Table 1 Factors and Mechanisms Influencing Distribution of Elements between Solid and Dissolved Phases

Factor/mechanism	Reference
Sample preparation (e.g. drying)	Duursma (1984)*
Separation (filtration/centrifugation)	Calmano (1979)*
Grain size distribution	Duursma (1984)*
Suspended matter concentration	Salomons (1980)
Kinetics of sorption/desorption	Schoer & Förstner (1985)*
Non-reversibily of sorption	Lion et al. (1982)
Colloids	Santschi & Honeyman (1989)*

* Experiments with artificial radionuclides

fertilizers, sewage sludge, municipal and mining wastes, and smelting residues, than from dissolved inputs such as lead-zinc mines, sewage treatment plant, effluents from battery factories, and electroplating plants. For atmospheric precipitation the percentage of dry deposition (by which aerosols or gaseous compounds are deposited on surfaces such as soil particles and plant leaves) has been observed between 10% and 90% of bulk cadmium deposition, depending upon the emission sources, climatic conditions and - in particular - upon pH (Nriagu, 1980; Nürnberg et al., 1983).

3.2 Equilibrium Models for Interstitial Waters

The use of equilibrium models has been to some extent successful for anoxic environments. "Buffering" of metal ion mobility not only involves interactions with numerous inorganic and organic *ligands*, but also with inorganic and organic (biological) *surfaces* - via coordinative bonding -, and *heterogenous equilibria* between metal precipitates and dissolved metal species (Stumm & Morgan, 1981). In sediment pore water and landfill leachates precipitation as sulfides is considered as the dominant mechanism limiting the solubility of many trace elements (Boulegue et al., 1982; Peiffer, 1989). Sulfide coordination is particularly strong for metals exhibiting so-called "B-character" (see contribution by Buffle in this book), such as Cu(I), Ag, Hg, Cd, Pb, and Zn; it is an important mechanism for transition elements in decreasing order of the Irving-Williams-series Cu(II) > Ni(II) > Co(II) > Fe(II) > Mn(II). There is strong direct (Luther et al., 1980; Lee and Kittrick, 1984) and indirect (Lu and Chen, 1977) evidence that the concentrations of copper, zinc and cadmium in sulfidic pore waters are determined by precipitation-dissolution processes, and in this case the metal concentrations in the pore waters are independent of the concentrations in the solid phase; on the other hand, the concentrations of arsenic and chromium in anoxic pore waters are probably controlled by adsorption-desorption processes, and mainly depend on the concentrations in the solid phase (Salomons, 1985).

According to Davies-Colley et al. (1985) two situations can be distinguished in natural systems, either the existence of a certain *sulfide precipitation capacity* (SPC), or - when exceeding the SPC - the accumulation of free sulfide (as H_2S or HS^-) in the water phase. Concentrations of sulfate and reducible iron seem to be the principle factors influencing different evolutionary sequences in anoxic fresh and marine waters (Salomons et al., 1987; Kersten, 1988). At excess sulfide concentrations (e.g., in Fe-poor environments), metal solubility can be increased by the formation of thio-complexes. There is still a discussion about the stability of thio-complexes, which possibly has been underestimated in most equilibrium models (Peiffer, 1989).

There is a remarkable agreement that complexation by natural organic ligands is not important for most metals except for copper, mainly due to the competition by ions such as Ca^{2+}, Mg^{2+}, Fe^{2+}, and Mn^{2+} (Stumm & Bilinski, 1973; Chian & De Walle, 1977; Tauchnitz et al., 1983; Van den Berg & Dharmvanij 1984; Salomons et al., 1987). Even at high concentrations of organic substances (and high complexing capacity) in solutions from digested sewage sludge only copper and lead seem to be slightly competitive with Ca- and Mg-ions (Fletcher & Beckett, 1987). It has been stressed by Peiffer (1989) that only in the presence of selective, basic complexing agents exhibiting small molar masses - such as organic sulfur and nitrogen compounds - significant oversaturation of Cu and Pb can be expected. With respect to copper speciation, the formation of more soluble polysulfides as a result of oxidation of H_2S by waste-borne oxidants may play a certain, not yet quantified, role (Peiffer, 1989).

4. Metal Mobilization from Aquatic Sediments and Solid Wastes

Regarding the potential release of contaminants from sediment, soil and solid waste material changing the pH and redox conditions are of prime importance (Table 2). It can be expected that *changes from reducing to oxidizing conditions*, which involve transformations of sulfides and a shift to more acid conditions, increase the mobility of typical "B-" or "chalcophilic" elements, such as Hg, Zn, Pb, Cu, and Cd. On the other hand, the mobility is characteristically lowered for Mn and Fe under oxidizing conditions. Elements exhibiting anionic species, such as S, As, Se, Cr, and Mo are solubilized, e.g., from fly ash sluicing/ponding systems at neutral to alkaline pH-conditions (Dreesen et al. 1977, Turner et al. 1982.)

Table 2 Relative Mobilities of Elements in Wastes and Soils as a Function of Eh and pH. (After Plant & Raiswell, 1983)

Relative	Electron activity		Proton activity	
Mobility	*Reducing*	*Oxidizing*	*Neutral-alkal.*	*Acid*
Very low mobility	Al,Cr,Mo,V, U,Se,S,B, Hg,Cu,Cd,Pb	Al,Cr,Fe,Mn	Al,Cr,Hg,Cu, Ni,Co	Si
Low mobility	Si,K,P,Ni, Zn,Co,Fe	Si,K,P,Pb	Si,K,P,Pb, Fe,Zn,Cd	K,Fe(III)
Medium mobility	Mn	Co,Ni,Hg,Cu, Zn,Cd	Mn	Al,Pb,Cu, Cr,V
High mobility	Ca,Na,Mg,Sr	Ca,Na,Mg,Sr, Mo,V,U,Se	Ca,Na,Mg,Cr	Ca,Na,Mg, Zn,Cd,Hg, Co,(Mn)
Very high mobility	Cl,I,Br	Cl,I,Br,B	Cl,I,Br,S,B, Mo,V,U,Se	Cl,I,Br,B

4.1 Metal Mobilisation from Aquatic Sediments

Release of potentially toxic metals from contaminated sediments pose problems both in aquatic systems and subsequent to land deposition of dredged materials (Förstner, 1989). Examples are given by various authors, indicating the major factors, processes and rates of metal mobilisation:

- Field evidence for changing cadmium mobilities was given by Holmes et al. (1974) from Corpus Christi Bay Harbor: During the summer period when the harbor water was stagnant cadmium was precipitated as CdS at the sediment/-water interface; in the winter months, however, the increased flow of oxygen-rich water into the bay resulted in a release of the precipitated metal.

- In the St. Lawrence Estuary, Gendron et al. (1986) found evidence for different release mechanisms near the sediment-water interface: The profiles for cobalt ressemble those for manganese and iron with increased levels downwards, suggesting a mobilization of these elements in the reducing zone and a reprecipitation at the surface of the sediment profile. On the other hand, cadmium appears to be released at the surface, probably as a result of the aerobic remobilization of organically-bound cadmium.

- Biological activities are typically involved in these processes: Remobilization of trace metals has been explained by the removal of sulfide from pore waters via ventilation of the upper sediment layer with oxic overlying water, allowing the enrichment of dissolved cadmium that would otherwise exhibit very low concentrations due to the of formation of insoluble sulfides in reduced, H_2S-containing sediments (Emerson et al., 1984); the authors suggest a significant enhancement of metal fluxes to the bottom waters by these mechanisms. It was evidenced by Hines et al. (1984) from tracer experiments that biological activity in surface sediments greatly enhances remobilization of metals by the input of oxidized water; these processes are more effective during spring and summer than during the winter months.

- From enclosure experiments in Narragansett Bay it has been estimated by Hunt and Smith (1983) that by mechanisms such as oxidation of organic and sulfidic material, the anthropogenic proportion of cadmium in marine sediments is released to the water within approximately 3 years; for remobilization of copper and lead, approximately 40 and 400 years, respectively, is needed, according to these extrapolations.

Typical early diagenetic geochemical changes and subsequent element mobilization via the porewater result from dredging activities. A study performed by Darby et al. (1986) in a human-made estuarine marsh demonstrates characteristic effects of oxidation (Table 3): Compared to the river water concentration, the channel sediment porewater is enriched by a factor of 200 for iron and manganese, 30-50 for nickel and lead, approx. 10 for cadmium and mercury, and 2-3 for copper and zinc. When the expected concentration of metals following hydraulic dredging, which were calculated from a rate of porewater to river water of about 1:4, were compared with the actual measurements at the pipe exiting the dredging device, negative deviations were found for iron and manganese, suggesting reprecipitation of Fe/Mn-oxide minerals; the positive deviations of zinc (factor 80), copper, lead and cadmium (factors 7-8) indicate, that during dumping of the sludge-water mixture significant proportions of these elements were mobilized and transferred into the effluent water.

Table 3 Mobilization of metals and nutrients during dredging (after Darby et al., 1986). Concentrations in mg/l

Metal	Channel sediment porewater (a)	River water concn. (b)	a/b	Effluent at man-made marsh: Expected concn.	Measured concn.	% Change
Mn	6.94	0.03	230	1.34	1.19	- 11
Fe	57.3	0.26	220	11.12	6.01	- 46
Ni	0.054	0.001	54	0.011	0.035	+ 218
Pb	0.077	0.002	38	0.016	0.142	+ 788
Hg (μg/l)	3.2	0.26	12	0.82	2.0	+ 144
Cd	0.009	0.001	9	0.0025	0.019	+ 660
Cu	0.012	0.004	3	0.0055	0.051	+ 827
Zn	0.12	0.052	2	0.065	5.30	+ 8069

These results demonstrate the problematic effect of dispersing anoxic waste materials in ecologically productive, high-energy nearshore, estuarine, and inlet zones (Khalid, 1980). Similar effects may also pertain to procedures such as "sludge-harrowing", which is occassionally performed in the cold season in some sections of Hamburg harbor. By application of these techniques highly contaminated sediments are transferred into zones of lower pollution intensity; oxygen-consuming substances, such as ammonia, are released from the pore water; increased turbidity affects "light climate" and thus the ecosystem in the lower reaches of the estuary.

4.2 Metal Mobility in Solid Wastes

Subsequent to landfilling, the raw waste compounds undergo a variety of *early diagenetic processes* accompanying microbially mediated degradation of the organic compounds (Aragno, 1989). The metabolic intermediates of organic matter decay (e.g. HCO_3^-, HPO_4^{2-}, carbohydrates and other low molecular organic acids) and those of the coupled inorganic reduction processes (e.g. Fe^{2+}, Mn^{2+}, S^{2-}, NH_4^+) accumulate in the interstitial water until concentrations are limited by physical convection-/dispersion, by subsequent microbial utilization, or by diagenetic formation of secondary ("authigenic") minerals such as metal sulfides. This secondary inventory of a reactor landfill is critical both in buffering leachate water chemistry (Stumm & Morgan, 1981) and in affecting transport of pollutants to underlying groundwater aquifers.

In municipal solid waste landfills initial conditions are characterized by the presence of oxygen and pH-values between 7 and 8. During the subsequent "acetic phase", pH-values up to 5 were measured due to the formation of organic acids in a more and more reducing milieu; concentrations of organic substances in the leachate are high. In a transition time of 1 to 2 years chemistry of landfill changes from acetic to methanogenic conditions; the methanogenic phase is characterized by higher pH-values and a significant drop of BOD_5 (biochemical oxygen demand)-values from more than 5.000-40.000 mg/l in the acetic phase to 20-500 mg/l. By comparison of the long-term evolution of a sewage sludge landfill to similar natural sediments (peat, organic soils) and their diagenesis it has been suggested by Lichtensteiger and Brunner (1987), that the transformation of organic material will last for geological time scales (10^3 to 10^7 years).

Particular problems occur when leachate collection pipes are plugging during the acidic decomposition period (Ham et al., in Baccini, 1989). Typically increased concentrations of metals have been found for iron, manganese and zinc in leachates during this phase compared with the "methanogenic phase" (Table 4). The lack of significant differences for other trace elements may be related to difficulties in sampling and chemical analysis since similar effects, for example, of pH on zinc mobility can be expected for other related elements such as Cd, Ni, Pb, and Cu. Comparison of inorganic groundwater constituents upstream and downstream of 33 waste disposal sites in West Germany (Arneth et al., 1989) indicates characteristic differences in pollutant mobilities, which may partly be related to releases during the acidic phase of the landfill development. High contamination factors (contaminated mean/uncontaminated mean) have been found for boron (>63), ammonia (62), and arsenic (CF = 34; this element may pose particular problems during initial phases of landfill operations; Blakey, 1984); heavy metals such as cadmium (CF >6.5), chromium (>5), lead (5.0), copper (4.7) and nickel (3.0) are significantly enriched in the leachates as well.

Table 4 Concentrations of trace elements (µg/l) in leachates from municipal solid waste landfills (review by Ehrig, 1989)

	"Acetic Phase"				"Methanogenic Phase"	
Element	average	range	average	range	average	range
Iron	780	20 - 2100			15	3 - 280
Manganese	25	0.3 - 65			0.7	0.03 - 45
Zinc	5	0.1- 120			0.6	0.03 - 4
Arsenic			160	5 - 1600		
Cadmium			6	0.5 - 140		
Chromium			300	30 - 1600		
Copper			80	4 - 1400		
Lead			90	8 - 1020		
Mercury			10	0.2 - 50		
Nickel			200	20 - 2050		

It has been inferred that oxidation of sulfidic minerals by intruding rainwater may mobilize trace metals from landfills subsequent to the methanogenic phase, and the impact on the underlying groundwater could be even higher if a chromatographic-like process, involving continuous dissolution and reprecipitation during passage of oxidized water through the deposit, would preconcentrate critical elements prior to final release with the leachate (Förstner et al., 1989). Experimental investigations performed by Peiffer (1989) on long-term development of sewage sludge materials provide detailed insight into the sequence of processes taking place in the post-methanogenic stage of such deposits: Transition from anoxic to oxic conditions involves a pH-decrease from 6.7 to 6.4, an accumulation of sulfate on the expense of sulfide, and a release of Mn- and Ca-ions, either by cation exchange (by protons) or by an indirect redox effect via oxidation of Fe^{2+} to Fe^{3+}; organic substances act as an acid buffer. Time-dependent release of zinc and cadmium is similar to as calcium and manganese, whereas lead and copper are not remobilized under these conditions. It has been inferred by Peiffer (1989), that due to slow oxidation kinetics of the sparingly soluble metal sulfides ZnS, CdS, PbS and CuS ion exchange will become the rate-determining mechanism in this system. From the current pH-decrease it can be expected that zinc and cadmium are being exchanged for protons, whereas lead and copper do not, because of their stronger bonding to the solid substrate. Because of their eminent practical significance, these initial findings need further confirmation. The same is valid for the effect of residual organic carbon in municipal solid waste incinerator slag as a potential proton producer due to microbial degradation to CO_2 (Krebs et al., 1988).

5. Assessment of Metal Mobility in Solid Materials

Since adsorption of pollutants onto particles is a primary factor in determining the transport, deposition, reactivity, and potential toxicity of these materials, analytical methods should be related to the chemistry of the particle's surface and/or to the metal species highly enriched on the surface. However, as it has been stressed before, the greater the instability of the polluted system the more difficulty is in sample handling and storage prior to analysis. Many of the analytical techniques are handicapped by disruptive preparation techniques which may alter the chemical speciation of inorganic components or lead to loss of analyte before analysis, e.g. freezing, lyophilization, evaporation, oxidation, changes in pH, light catalyzed reactions, reactions with the sample container, time delays before analysis with biologically active samples, and sample contamination.

5.1 Leaching Procedures

Various laboratory techniques have been reported for the characterization of reactivity and mobility of pollutants generating leachates from hazardous waste, and are generally grouped into batch and column extraction methods. The batch extraction method offers advantages through its greater reproducibility and simplistic design, while the column method is more realistic in simulating leaching processes which occur under field conditions. It has been noted by Jackson et al. (1984) that relative standard deviation of experimental data was over twice as great for the column method than for the batch method; this result was attributed to varying degrees of "channeling" that may have occurred during leaching of waste constituents using the column method.

Apart from the procedures aimed for the study of metal species distribution and transformation, other parameters controlling the mobility of elements have to be measured, preferentially in a continuous way during the time of the experiment. In many cases, the variables to be determined are pH, electric conductivity, oxygen potential, and sulfide activity. For the latter parameter, a pH_2S (glass/Ag^o, Ag_2S) electrode can be used, which has been introduced by Frevert & Galster (1978) and successfully applied for geochemical studies in anoxic systems, even at very low H_2S-concentrations (Peiffer & Frevert, 1987).

Initial estimation of potential release of metal from contaminated solids is mostly based on elutriate tests, which - apart from the characterization of the reactivity of specific metals - can provide information on the behavior of metal pollutants under typical environmental conditions. Common single reagent leachate tests, e.g. U.S. EPA, ASTM, IAEA, ICES, and DIN use either distilled water or acetic acid (Theis & Padgett, 1983). A large number of test procedures have been designed particularly for soil studies. These partly used organic chelators such as EDTA and DTPA, both as single extractants or in sequential procedures (Sauerbeck & Styperek, 1985). Physiological solutions have been applied, e.g. for the estimation of organ-specific effects of atmospheric particulates (Harris & Silberman, 1988).

A "mobility test" procedure for soils based on variations of pH-values has been proposed by Kiekens & Cottenie (1985). Application to a large number of polluted and non-polluted soils indicate that typical mobilisation patterns are obtained for the different heavy metals. Beside the nature of the element, the pH-curves reveal typical textural features of different soil substrates. Best results with respect to the estimation of middle-term effects can be attained by "cascade" test procedures at variable solid/solution ratios: A procedure of the U.S. EPA (Ham et al., 1979) designed for studies on the leachability of waste products consists of a mixture of sodium acetate, acetic acid, glycine, pyrogallol, and iron sulfate. For the study of combustion residues a standard leaching test has been developed by the Netherland Energy Research Centre (Van der Sloot et al., 1984). In the column test the material under investigation is percolated by acidified demineralized water (pH = 4; for evaluating most relevant effects of acid precipitation) to assess short-term leaching (< 50 years). In the cascade test the same quantity of material is extracted several times with fresh demineralized water (pH = 4) to get an impression of medium-term leaching behavior (50-500 years). As a time scale the liquid/solid ratio (L/S) is used (De Groot et al., 1987).

Single reagent leaching tests, in particular short-term batch procedures, can be used for easily soluble components such as halides or sulfates, but in most cases are not adequate for assessing mobility of trace metals. For example, cascade test procedures demonstrated much high increases in the percent elution of elements such as arsenic than can be expected from the increase of the solution/solid ratio (Van der Sloot et al., 1985). Another example, from the time-dependent leachability of cadmium and lead in filter-dust samples, indicated much greater percentages of released metals after short-time treatment (0.5 and 5 hours) than after longer contact (50 hours) between dissolved and solid metal species; a reverse time-dependency was observed for lead in dust samples solidified with cement; both developments can be interpreted as resulting from pH changes (Brunner & Baccini, 1987). Most important is the effect of time-dependent change in metal release from waste materials containing oxidizable components: In a low buffered industrial waste containing less than 1% sulfide minerals, proton activity decreased from initial pH 10.2 to pH

5.5 after 50 days, whereas E_h increased from -200 mV to +500; it is obvious, that application of standard elutriate procedures would have provided totally wrong results, also with respect to the design of remedial measures (see section 6).

In connection with the problems arising from the disposal of solid wastes, particularly of dredged materials, chemical extraction sequences have been applied which are designed to differentiate between the exchangeable, carbonatic, reducible (hydrous Fe/Mn oxides), oxidizable (sulfides and organic phases) and residual fractions. The undisputed advantage of this approach with respect to the estimation of long-term effects on metal mobilities lies in the fact, that rearrangements of specific solid "phases" can be evaluated prior to the actual remobilisation of certain proportions of an element into the dissolved phase (Förstner, 1985). One of the more widely applied extraction sequences of Tessier and co-workers (1979) has been modified by various authors (Table 5).

Table 5 Sequential Extraction Scheme for Partitioning Sediment Samples (Kersten & Förstner 1986, 1987a)

Fraction	Extractant	Operational Phase
Exchangeable	1 *M* NH_4OAc, pH 7	Exchangeable ions
Carbonatic	1 *M* NaOAc, pH 5 HOAc	Carbonates
Easily reducible	0.01 *M* NH_2OH HCl 0.01 *M* HNO_3	Mn-oxides
Moderately reducible	0.1 *M* oxalate buffer pH 3	Amorphous Fe-oxides
Sulfidic/ organic	30% H_2H_2 pH 2, 0.02 *M* HNO_3 extracted 1 *M* NH_4OAc-6% HNO_3	Sulfides together with organic matter
Residual	hot HNO_3 conc.	Lithogenic Material

The widely used extraction sequence of Tessier and collaborators (1979) consists of five steps. Applications of these methods have been given for atmospheric particulates (e.g., Lum et al., 1982), street dust and roadside soils (Harrison et al., 1981), sewage sludge (Rudd et al., 1986), incinerated sludge ash (Fraser et al. 1982), and in a review by Kersten and Förstner (1989) for sediment-related studies. It has been stressed that - providing the use of a proper sampling and preparation protocol - the more sophisticated extraction schemes such as Tessier's, significantly improves the specificity and efficiency of extraction.

Despite the advantages of a differential analysis over investigations of total metal concentrations and the fact that sequential chemical extraction is probably the most useful tool for predicting long-term adverse effects from contamined solid material, it is obvious that there are still many questions and uncertainties associated with these procedures (Kersten & Förstner 1986; Rapin *et al.* 1986), for example:

- Reactions are *not selective* and are influenced by the duration of the experiment and by the ratio of solid matter to volume of extractants. An excessive solid content, together with an increased buffer capacity may cause the system to overload; such an effect is reflected, for example, by changes of pH-values in time-dependent tests.
- Labile phases could be *transformed* during sample preparation, which can occur especially for samples from reducing sites.

In this respect, earlier warnings have been made by various authors, not to forget changes of the sample matrix during *recovery* and *treatment* of the material. This, in particular, relates to the *anoxic sediment* material, where changes are quite obvious.

5.2 Applications

From a practical view the following questions indicate the significance of informations on the mobility of critical elements in aquatic and terrestrial ecosystems (Förstner, 1987):

- How reactive are the metal compounds introduced with solid materials from anthropogenic activities (hazardous waste, sewage sludge, atmospheric fallout) in comparison to their natural existence?
- Are the mechanisms and kinetics of interaction between dissolved and solid metal phases comparable for natural and contaminated systems (this particularly pertains to the question, how fast a quasi-stable state is reached for the latter systems)?
- What are the factors and processes of remobilization to become particularly effective, when either the solid inputs or the solid/solution interactions lead to weaker bonding of critical metal species in contaminated compared to natural systems?

In the following examples emphasis will be given to the latter aspect - the assessment of remobilization of potentially toxic elements from polluted solid materials in both aquatic systems and land deposits.

5.2.1 Evaluation of Threshold Values for Soil Pollutants

Among the various factors affecting the transfer of metal pollutants to plant, lowering of pH - locally from oxidation of sulfidic components and regionally by acid precipitation - is the most important. Generally the lowering of pH by one unit will increase metal solubility by a factor 10. For the example of *zinc*, Figure 1 (after Herms & Brümmer, 1980) indicates how the permissible limit of total metal concentration in the soil is affected by pH as the dominant factor with respect to metal solubility. At pH 7, the limit of 1 mg Zn per liter soil equilibrium solution, which already may lead to slight depressions in yield for cabbage (Hara & Sonoda, 1979), would be attained at approx. 1200 mg Zn per kg soil. However, at pH 6, the maximum permissible Zn concentration in soil equilibrium solution would be reached at approx. 100 mg Zn per kg soil, at pH 5 even at approx. 40 mg Zn per soil. Under the latter conditions (pH 5), adverse effects can be found even in unpolluted soils.

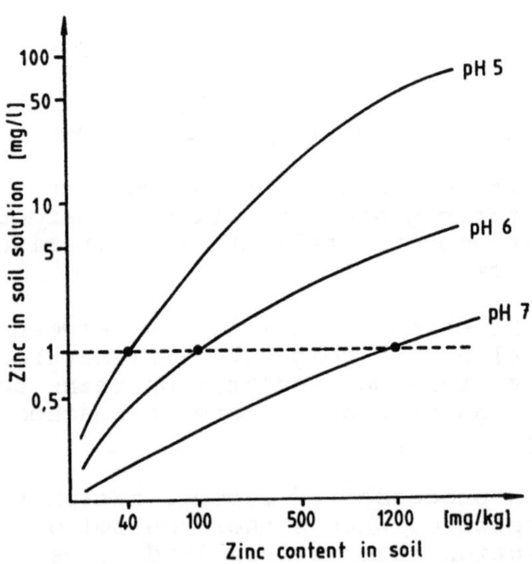

Figure 1

Solubility of Zn in Relation to pH and Total Zn-Content of Soils Not or to A Different Degree Polluted with Zn (Herms & Brümmer, 1980)

5.2.2 Applications for Sediment Quality Criteria

New objectives regarding the improvement of water quality as well as problems with the resuspension and land deposition of dredged materials require a standardized assessment of sediment quality. Numerical approaches are based on (1) accumulation, (2) pore water concentrations, (3) solid/liquid equilibrium partition (see above), and (4) elution properties (e.g., Förstner et al., 1990). In Table 6, an example is given for the possibilities of standardizing the data from elution experiments with respect to numerical evaluation. An "elution index" for sediment samples from various rivers in West Germany is based on the metal concentrations exchangeable with 1 N ammonium acetate at pH 7. These metal fractions are considered to be remobilizable from polluted sediments at a relative short term under more saline conditions, for example, in the estuarine mixing zone. Comparison of the release rates from oxic and anoxic sediments clearly indicates, that the oxidation of samples gives rise to a very significant increase in the mobilization of the metals studied. This effect was particularly important for Cd. When proceeding further in the extraction sequence, more long-term effects can be estimated, but generally with a reduction of prognostic accuracy.

Table 6 Elution-Index for Selected River Sediment, as Determined from Exchangeable Proportions (1 M Ammonium-Acetate). Calculated Relative to Background Data from Elbe River Sediments. These Values are Multiplied by a Factor of 100

	Neckar	Main	Elbe	Weser
Copper	<1	-	1	-
Lead	1	1	1	1
Zinc	7	10	40	10
Cadmium	22	22	25	-
Total oxic	**30**	**33**	**67**	**11**
(Anoxic	0.5	0.3	>4	4)

5.2.3 Metal Partitioning in Combustion Residues and Their Transfer to Other Environmental Compartments

Waste incineration ashes usually exhibit relative high concentrations of trace metals - Zn and Pb up to the percent range - and particularly strong enrichment factors compared to natural contents have been observed for these elements and for cadmium and silver (Brunner and Zobrist, 1983). Sequential extractions performed by Wadge and Hutton (1987) indicate (Figure 2), that about 20% of total Cd and 1% of total Pb in coal fly ash was in exchangeable fraction; in contrast, the single largest fractions of Cd and Pb in refuse ash, at 72% and 41%, respectively, were present in the exchangeable form. It is suggested that this effect is mainly due to higher concentrations of metal associations with chlorides (Baccini and Brunner, 1985).

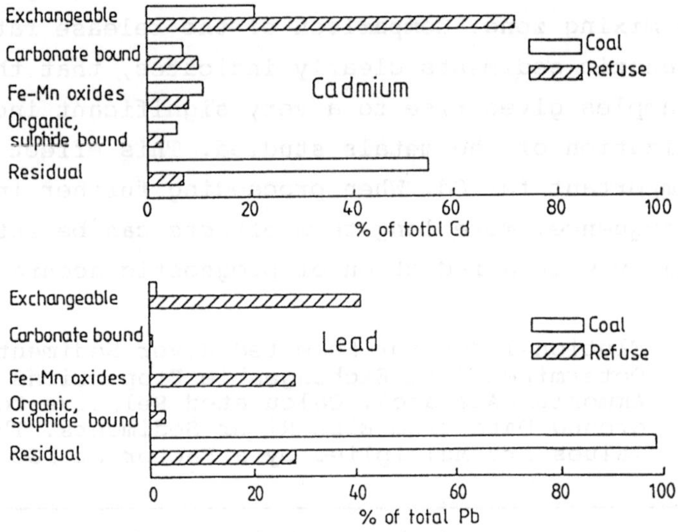

Figure 2 The Chemical Associations of Cadmium and Lead in Coal Fly Ash and Refuse Fly Ash (Wadge and Hutton, 1987). Extraction Sequence After Tessier et al. (1979)

The release of cadmium and other metals from particulate matter during interactions with saline waters is an important effect in the estuarine mixing zone and at the seawater interface. Such mobilization of cadmium in estuaries has been observed in several examples (Salomons & Förstner 1984); it seems that microbially mediated oxidation of particulate organic carbon is a prerequisite for that effect (Prause et al. 1985). According to Chester et al. (1986), the latter mechanism could be the major factor in the transfer of metals from aerosols to seawater. Figure 3 indicates, that a certain proportion of metals associated with contaminated aerosols, which is related to the more labile fractions, may be solubilized at the seawater interface. Percentage of released metals is greatest for cadmium, followed by zinc, lead and copper, whereas components such as chromium, aluminium and iron are only solubilized to a very small extent. Via selective accumulation in organic solids, released metals can be transferred into the marine food chain. It has been suggested that with such pathways the enrichment of Cd, Pb, and Hg, can be explained in both pelagic and benthic organisms of the northern and central part of the North Sea (Kersten et al., 1988).

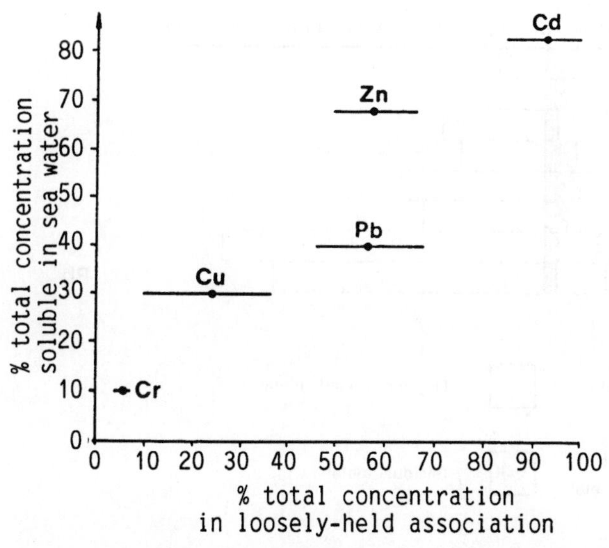

Figure 3

Relation between Water Solubility of Trace Metals from Polluted Atmospheric Aerosols and the Percentage Associated with Loosely Bound Fractions (Chester et al., 1986)

5.2.4 Study of Diagenetic Processes in Sediment Core Profiles

Partitioning studies of sediment from core profiles are particularly useful, since they provide information on the relative changes of elemental phases irrespective of the method applied, and thereby an insight into diagenetic processes taking place after deposition of the sedimentary components. Two examples are presented here, both indicating significant changes in the *partition of zinc and cadmium* during a relative short period of time.

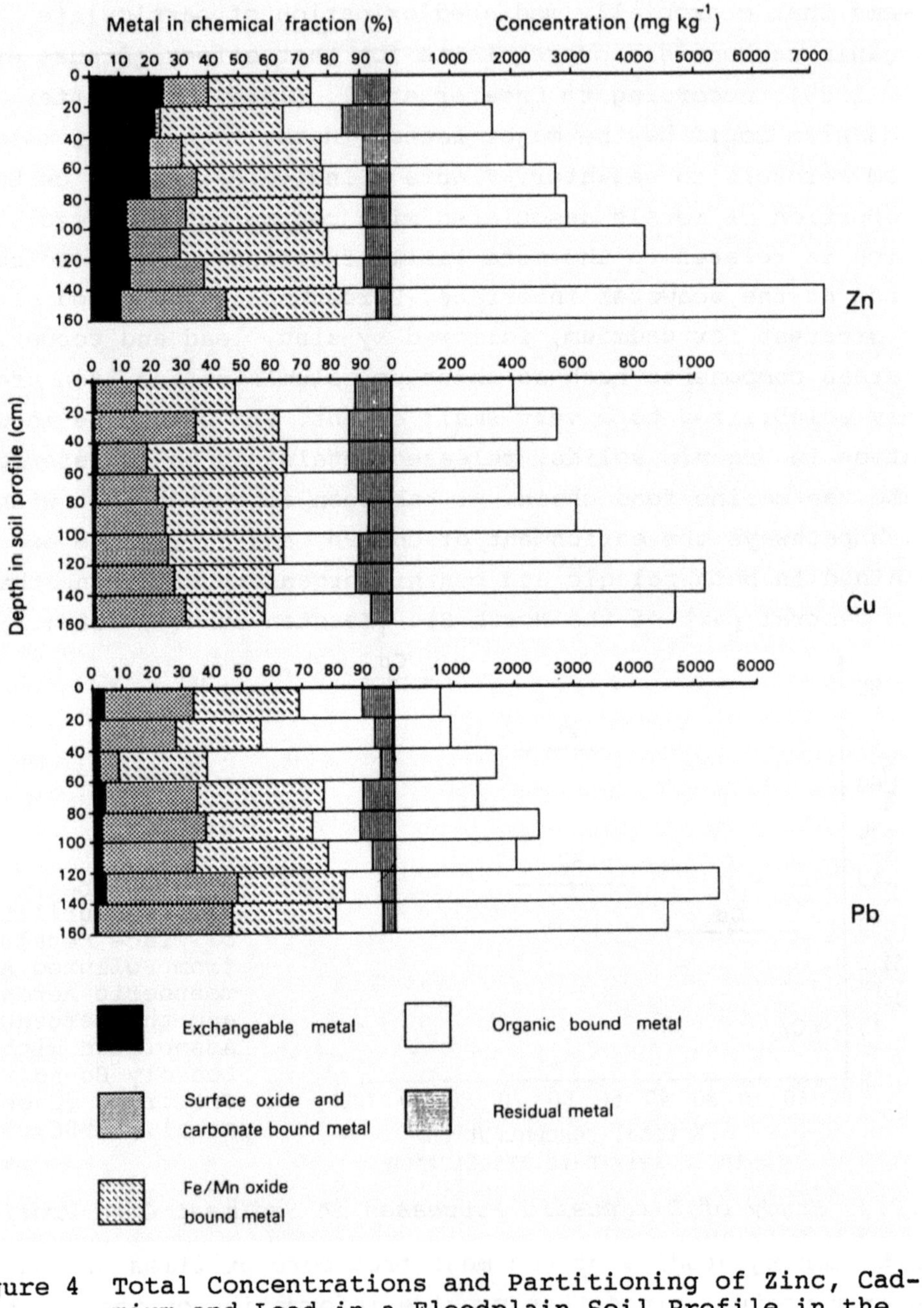

Figure 4 Total Concentrations and Partitioning of Zinc, Cadmium and Lead in a Floodplain Soil Profile in the Manifold Catchment (Bradley, 1988)

Floodplains on the River Manifold contain sediments with large metal concentrations (Bradley & Cox, 1986, 1987). The sedimentary units described were deposited during concerted mining, and as mining ceased in 1870, pedogenic processes have influenced these sediments for over 100 years. A general decrease of total Zn concentrations over time, which can primarily be interpreted as the result of improvements in the processing of ores, is given here (Fig. 4, right bars). Phase differentiation (Fig. 4, left) indicate that there has been a significant temporal increase of the percentage of exchangeable Zn. These changes can be interpreted from pedogenic processes. Since these transformations will result in a reduction of the bonding strength, it cannot be excluded, that the decrease of total metal concentrations is partly due to a diagenetic remobilisation of metals from the floodplain deposit. However, Pb concentrations have decreased similar to Cd and Zn, without the respective changes of chemical fractions. Thus, it is suggested that the process of diagenetic release generally is not a significant factor in this environment.

Metal release from *tidal Elbe River sediments* by a process of "*oxidative remobilization*" has been described by Kersten (1989). Short (30 cm) sediment cores were taken from a site, where diurnal inundation of the fine-grained fluvial deposits takes place; subsamples were analyzed with sequential extraction according to the methods described in section 5.1. In the upper part of the sediment column, total particulate cadmium content was approximately 10 mg kg^{-1}, whereas in the deeper anoxic zone the total particulate concentration of Cd was 20 mg kg^{-1}. The results of the sequential extractions of the core sediment samples separated at 2-cm levels (Figure 5) indicate, that in the anoxic zone 60-80% of the Cd was associated with the sulfidic/organic fraction. In the upper - oxic and transition - zone the association of Cd in the carbonatic and exchangeable fractions simultaneously increase up to 40% of total Cd. Thus, high proportions of mobile cadmium forms correlate with the reduction in total cadmium contents. This distribution of total and partitioned Cd in the "Heukenlock" sediment profile suggest that the release of metals from particulate phases into the pore water and further transfer into biota is controlled by the frequent downward flux of oxygenated surface water. In the oxic zone, Cd is leached from the labile particulate-binding sites, where the predominant mechanism controlling the availability of Cd is adsorption/desorption. With the downward flux of pore water, the mobilized metal moves into the anoxic environment, where Cd is reprecipitated by the formation of sulfidic associations. From the observed concentrations, it would be expected that long-term transfer of up to 50% of the Cd from the sediment subsurface would take place in the anoxic zone located further below the sediment-water interface. By this process of "*oxidative pumping*" (Kersten & Förstner 1987b) a maximum downward flux of 0.4 g Cd/m^2 per year can be estimated in the Heukenlock sediments.

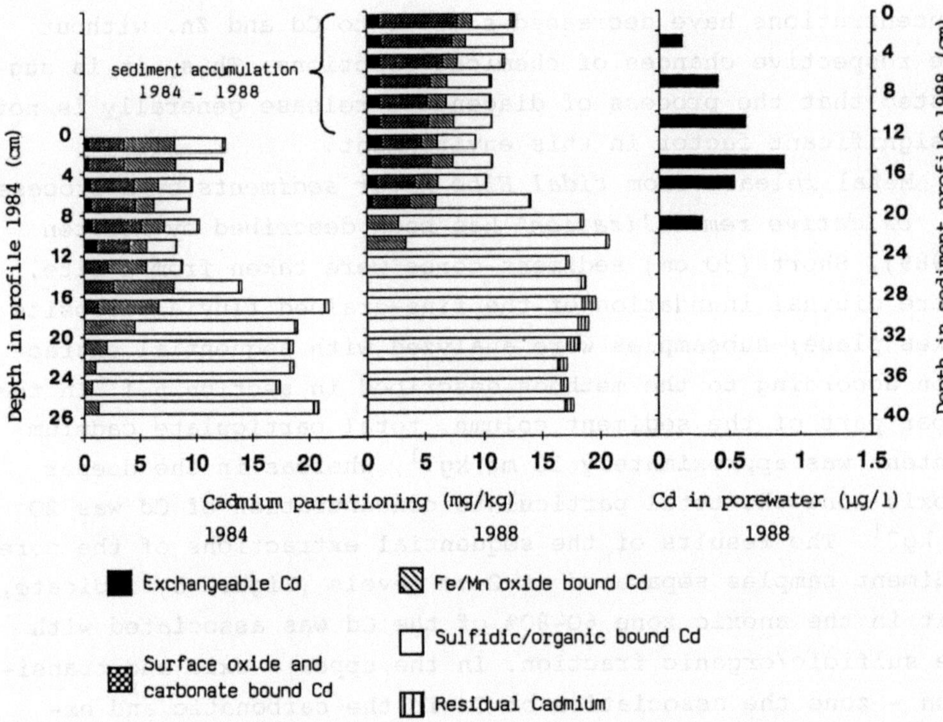

Figure 5 Total Concentrations and Partitioning of Cadmium in a Tidal Flat Sediment Profile in the Heuckenlock Area Sampled in 1984 and 1988. Sedimentation Rates were Determined by the ^{137}Cs-Method. Cd Pore Water Profile was Determined at Low Tide (Kersten, 1989)

5.2.5 Long-Term Prognosis of Metal-Release from Solid Waste

Because of the reasons mentioned above, long-term behaviour prognosis of metals in the environment based on models alone are of limited value. In such cases direct measuring of changes of chemical forms of typical soil and waste constituents at in-situ conditions with respect to interstitial water composition, e.g. by inserting dialysis bags or PVDF sheets containing typical substrates with varying metal concentrations into boreholes.

Because these approaches neither involve reaction-mechanistic nor kinetic considerations, they are of limited value with regard to prediction of long-term effects in waste-deposits. This lack can be avoided by an experimental approach, originally been used by Patrick et al. (1973) and Herms and Brümmer (1978). Metal mobility can be estimated by comparing sequential extraction results before and after treatment of waste material by controlled significant intensivation of relevant release parameters such as pH-value, redox-potential and temperature using an ion-exchanger system for extracting and analysing the released metals at a adequate frequence (Figure 6).

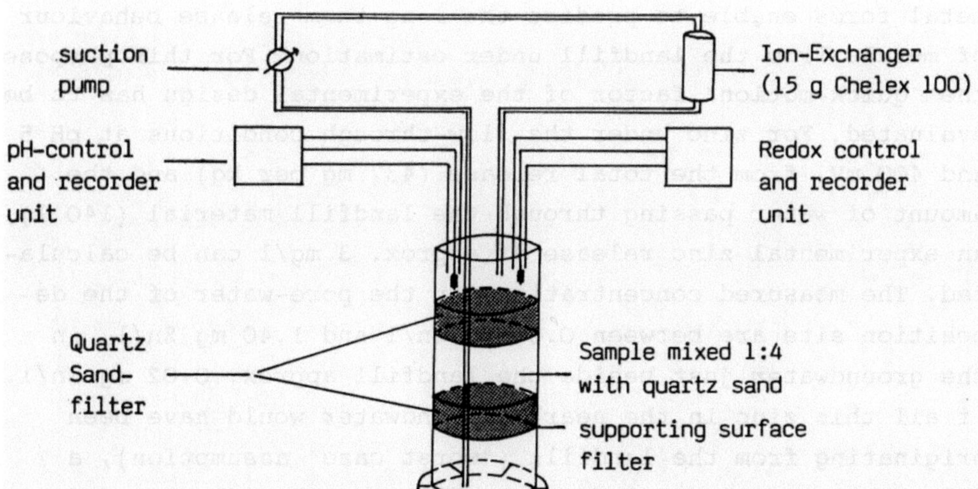

Figure 6 Experimental Design for Long-Term Prognosis of Metal Release (Schoer and Förstner, 1987)

This system was used for the evaluation of the long-term behaviour of metal contaminants (including arsenic, cadmium, thallium and vanadium) in metal oxide residues from heat processing, which were deposited into borrow pits from sand removal along the river Rhine. The way of interaction between the water and the deposit - flow-through at high and flow-by at low permeability of the deposit compared to the surrounding groundwater aquifers; intensive interaction between water and critical components under extreme conditions - which may result from erosion of waste materials - can also be reproduced by the experimental design.

For most elements, the acidity is more effective concerning the release than changes in redox potential. Remobilized element percentages in a typical sample were aproximately 20 to 40 % of total concentration for Zn, Co and Cd at pH 5. Temporal release patterns were also different for the individual elements. The experimental data clearly indicated the much reduced mobilisation of metals in the flow-by experiments (sample in a dialysis bag) compared to flow-through conditions. For cadmium, lead and zinc, maximum release during the flow-by experiment was less than 1 % of the flow-through experiment at the same chemical conditions.

The combination of mobilization kinetics and the chemical metal forms enable to predict the long-term release behaviour of metals from the landfill under estimation. For this purpose the "quick-motion" factor of the experimental design has to be evaluated. For zinc under the flow-through conditions at pH 5 and 400 mV, from the total release (437 mg per kg) and the amount of water passing through the landfill material (140 l), an experimental zinc release of approx. 3 mg/l can be calculated. The measured concentrations in the pore-water of the deposition site are between 0.04 mg Zn/l and 1.40 mg Zn/l, in the groundwater just beside the landfill approx. 0.02 mg Zn/l. If all this zinc in the nearby groundwater would have been originating from the landfill ("worst case" assumption), a "black-box-experimental-factor" of 150 can be calculated from the ratio between predicted and measured zinc concentrations. If this factor is multiplied with the duration of the experiment (10 weeks), a time period of nearly 30 years is represented by the experiment. From solid speciation data it can be concluded, that the release of the rest of "oxalate-bound" zinc would take another 12 experimental weeks; this would be equivalent to nearly 40 years in nature. In total, a time period of at least 50 to 100 years would be covered by the experiment, and a prognosis within this time scale can be drawn from the present results.

6. Demobilization of Metals in Solid Waste Materials

"Geochemical and biological engineering" emphasize the increasing efforts of using natural resources available at the disposal site for reducing negative environmental effects of all types of waste material, in particular of acid mine wastes. Practical examples for improvement of storage quality of metal-containing waste, including measures for recultivation of old and recent mining waste disposal sites and physico-chemical methods to water processing, have been reviewed (Salomons & Förstner, 1988a,b).

Most stabilization techniques aimed for the immobilization of metal-containing wastes are based on additions of cement, water glass (alkali silicate), coal fly ash, lime or gypsum (Calmano, 1988). Generally, maintainance of neutral pH-values beyond favors adsorption or precipitation of soluble metals (Gambrell et al., 1983).

Incorporation in naturally formed minerals, which remain stable over geological times, constitutes favourable conditions for the immobilization of potentially toxic metals in large-volume waste materials both under environmental safety and economic considerations. There is a particular low solubility of metal sulfides, compared to the respective carbonate, phosphate, and oxide compounds (Kersten, 1988). Marine conditions, which are favourable due to the higher production of sulfide ions, in addition, seem to repressing the formation of *mono-methyl mercury* (Craig & Moreton, 1984). This type of waste deposition under stable anoxic conditions, where large masses of polluted materials are covered with inert sediment became known as *"subsediment-deposit"*; the first example was planned for highly contaminated sludges from Stamford Harbour in the Central Long Island Sound following intensive discussions in the U.S. Congress (Morton, 1980). Other advantages of near-shore capped mound deposits (Kester et al., 1983) include the protection of groundwater resources, since the underlying water is saline, and enhanced degradation of organic priority pollutants (Kersten, 1988).

With regard to the immobilization of contaminants in municipal and industrial waste materials the term "*final storage quality*" has been brought into discussion (Baccini, 1989a). Solid residues with final storage quality should have properties very similar to the earth crust (natural sediments, rocks, ores, soil; Table 7). Treatment includes several possible steps, e.g. assortment, thermal, chemical and biological treatments. In most cases this standard is not attained by simply incinerating municipal waste, i.e. by reduction of

Table 7 Comparison of Inventories of Chemical Components in the Two Landfill Alternatives and in the Earth Crust

Reactor Landfill	Final Storage	Earth Crust
Major Solid Constituents		
Solid "Inert" Waste	Silicates, Oxides	Quarz, Fe-Oxide Clay, Carbonates
Putrescible Waste	[Gypsum, NaCl][1]	(Gypsum, NaCl)
Grease Trap Waste	(Char)[2]	Kerogenic Comp.
Minor Solid Constituents		
Organic Micropollutants	Organic Micropoll.	-
Metals in Reactive Chemical Forms	Metal-Rich Minerals Mainly Oxides	Metals Mainly in Inert Forms
Dissolved Constituents		
Protons, Electrons	(Protons)	(pH: Acid Rain)
Organic Compounds	(Organic Residues)	(Humic Acids)
Dissolved Salts	[Dissolved Salts][1]	(Dissolved Salts)

[1] [Partial Extraction by Pretreatment] [2] (Minor Constituent)

organic fractions only. There is, in particular, the problem of easily soluble minerals, such as sodium chloride. *Measures before incineration* include the separate collection of (organic) kitchen and garden wastes (containing some chlorine and sulfur), which can be transferred into compost; a major decrease of chlorine content, however, would require a significant reduction of PVC in municipal solid waste. *After incineration* washing of the residues can be performed either with neutral or acidified water. Another possibility is to put the electrostatic precipitator dusts into a thermal process to remove metal like zinc, cadmium and lead as chlorides at high temperatures (>1200°C). These salts should then be used as educts for metal production (Baccini, 1989b).

7. Conclusions

1. Conceptually, "*mobility*" reflects the flux of metal species in a system of solution/solid interactions, which is influenced by both accelerating and inhibiting factors and processes. Data for description of metal mobility include informations on dominant aqueous species, typical mobilizing/inhibiting system parameters such as pH and pH_2S, and surface properties of the solid phases.

2. Thermodynamic *models* may give suggestions as to the possible species to expect, but due to the important role of kinetically controlled processes, the actual speciation is often different. Particular difficulties with modeling are found for systems rich in organic matter, due to the lack of information on specific ligands and the reduced reversibility of metal sorption processes.

3. Typical "*dynamic*" *macro-environments* are characterized by high-energy mechanical processes, such as bioturbation and dredging activities, by strong chemical gradients, e.g. of redox conditions, and by distinct temporal variations, such as seasonal changes of biological activity in aquatic systems and characteristic successions in metabolic transformations, for example, in waste deposits.

4. In *anoxic* sediment pore water and landfill leachates precipitation as sulfides is considered as the dominant mechanism limiting the solubility of metals exhibiting so-called "B-character", such as copper, lead, zinc and cadmium. While complexation by natural organic ligands does not seem relevant for most metals except for copper, formation of thio-complexes at excess sulfide concentrations could significantly influence metal mobility.

5. Due to the instability of polluted solid materials, *sample handling* and storage prior to analysis is problematic. Many of the analytical techniques for solid speciation are handicapped by disruptive preparation techniques which may alter the chemical speciation of inorganic components or lead to loss of analyte before analysis.

6. Simple "standard" *leaching tests* can be used for easily soluble components such as halides or sulfates, but in most cases are not adequate for assessing mobility of trace metals. With sequential extraction procedures rearrangements of specific solid "phases" can be evaluated prior to its actual remobilisation.

7. Estimation of *long-term release* of metals from solids can be made using a test procedure, which combines data from column experiments applying controlled intensivation of parameters such as pH-value, redox potential and temperature with results from sequential extraction of the contaminated solids before and after exposure.

8. Partitioning studies on materials from *core profiles* are particularly useful, since they provide information on relative variations of elemental phases irrespective of the method applied, and thereby an insight into diagenetic processes taking place after deposition of the sediment or solid waste components.

9. Changes from reducing to oxidizing conditions in *sediments*, which involve transformations of sulfides and a shift to more acid conditions, increase the mobility of critical metals. These results demonstrate the problematic effect of dispersing anoxic waste materials in ecologically productive, high-energy nearshore, estuarine, and inlet zones.

10. Increased concentrations of metals have been observed for in *leachates* during the acidogenic phase of landfill evolution at low pH-values. Recent data from experiments on organic-rich wastes suggest that also subsequent to the methanogenic phase transition from anoxic to oxic conditions could affect a lowering of pH, which in turn could influence mobility of metals like zinc and cadmium. Such effects, together with the long-term emissions of nitrogen compounds, will pose additional questions as to the reliability of "reactor landfills" in the management of municipal wastes.

References

Andreae MO, Asami T, Bertine KK, Buat-Menard PE, Duce RA, Filip Z, Förstner U, Goldberg ED, Heinrichs H, Jernelöv AB, Pacyna JM, Thornton I, Tobschall HJ, Zoller WH (1984) Changing biogeochemical cycles - group report. In: JO Nriagu (Ed) Changing Metal Cycles and Human Health. Dahlem-Konferenzen, Life Sciences Res Rept 28, pp 359-374. Springer, Berlin

Anderson MA (1981) Kinetics and equilibrium control of interfacial reactions involving inorganic ionic solutes and hydrous oxide solids. In: FE Brinckman, RH Fish (eds) Environmental speciation and monitoring needs for trace metal-containing substances from energy-related processes. US Dept Commerce Washington DC, pp 146-162

Aragno M (1989) The landfill ecosystem: a microbiologist's look inside a "black box". In: P Baccini (ed) The landfill - reactor and final storage. Lecture Notes in Earth Sciences 20. Springer Berlin, pp 15-36

Arneth, J-D, Milde G, Kerndorff H, Schleyer, R (1989) Waste deposit influences on ground water quality as a tool for waste type and site selection for final storage quality. In: The Landfill - Reactor and Final Storage, P Baccini (ed). Lecture Notes in Earth Sciences 20, pp 399-415 Springer, Berlin

Baccini P (ed)(1989a) The landfill - reactor and final storage. Lecture Notes in Earth Sciences 20. Springer Berlin

Baccini P (1989b) The control of heavy metal fluxes from the anthroposphere to the environment. In: JP Vernet (Ed) Proc Intern Conf Heavy Metals in the Environment, Geneva 1989, Vol 1, pp 13-23. CEP Consultants Edinburgh

Baccini P, Brunner PH (1985): Behandlung und Endlagerung von Reststoffen aus Kehrichtverbrennungsanlagen. Gas Wasser Abwasser 65: 403-409

Baes CF, Sharp RD (1983) A proposal for the estimation of soil leaching and leaching constants for use in assessment models. J Environ Qual 12: 17-28

Baham J (1984) Prediction of ion activities in soil solutions: Computer equilibrium modeling. Soil Sci Soc Am J 48: 525-531

Benjamin MM, Leckie JO (1981) Multiple-site adsorption of Cd, Cu, Zn, and Pb on amorphous iron oxyhydroxide. J Colloid Interface Sci 79: 209-211

Benjamin MM, Hayes KL, Leckie JO (1982) Removal of toxic metals from power-generated waste streams by adsorption and coprecipitation. J Water Pollut Control Fed 54: 1472-1481

Blakey NC (1984) Behavior of arsenical wastes co-disposed with domestic solid wastes. J Water Pollut Control Fed 56: 69-75

Boulegue J, Lord CJ III, Church TM (1982) Sulfur speciation and associated trace metals (Fe, Cu) in the pore waters of Great Marsh, Delaware. Geochim Cosmochim Acta 46: 453-464

Bradley, SB (1988) Sediment-water interactions. The physical transport of heavy metals in the fluvial system. In: Metals and Metalloid in the Hydrosphere; Impact Through Mining and Industry, and Prevention Technology. Proc. IHP Workshop Bochum, 21-25 Sept 1987. Unesco Techn Doc Hydrol. Unesco Paris

Bradley SB, Cox JJ (1986) Heavy metals in the Hamps and Manifold valleys, North Staffordshire, U.K. - Distribution in floodplain soils. Sci Total Environ 50: 103-128.

Bradley SB, Cox JJ (1987) Heavy metals in the Hamps and Manifold valleys, North Staffordshire, U.K. - Partitioning of metals in floodplain soils. Sci Total Environ 65: 135-153

Brümmer G, Gerth J, Herms U (1986) Heavy metal species, mobility and availability in soils. Z Pflanzenernähr Bodenkde 149: 382-398

Brunner PH, Baccini P (1987) The generation of hazardous waste by MSW-incineration calls for new concepts in thermal waste treatment. In: Second International Conference on New Frontiers for Hazardous Waste Management, Pittsburgh, Pa., Sept. 27-30, 1987.

Brunner PH, Zobrist J (1983): Die Müllverbrennung als Quelle von Metallen in der Umwelt. Müll und Abfall 9: 221-227

Calmano W (1979) Untersuchungen über das Verhalten von Spurenelementen an Rhein- und Mainschwebstoffen mit Hilfe radioanalytischer Methoden. Doctoral Dissertation TH Darmstadt.

Calmano W, Ahlf W, Förstner U (1988) Study of metal sorption/-desorption processes on competing sediment components with a multi-chamber device. Environ Geol Water Sci 11: 77-84

Chester R, Murphy KJT, Towner J, Thomas A (1986) The partitioning of elements in crust-dominated marine aerosols. Chem Geol 54: 1-15

Chian ESK, De Walle FB (1977) Evaluation of leachate treatment. Vol 1: Characterization of leachate. EPA Report 600/2-77-186a

Craig PJ, Moreton PA (1984) The role of sulphide in the formation of dimethyl mercury in river and estuary sediments. Mar Pollut Bull 15: 406-408

Darby DA, Adams DD, Nivens WT (1986) Early sediment changes and element mobilization in a man-made estuarine marsh. In: Sly PG (ed) Sediment and water interactions. Springer-Verlag New York, pp 343-351

De Groot AJ, Zschuppe, KH, Salomons W (1982) Standardization of methods of analysis for heavy metals in sediments. In: Sly PG (ed) Sediment/freshwater interaction. Hydrobiologia 92: 689-692

De Groot GA, Wijkstra J, Hoede D, Van der Sloot HA (1987) Leaching characteristics of hazardous elements from coal fly ash as a function of the acidity of the contact solution and the liquid/solid ratio. Presentation at 4th Int Hazardous Waste Symposium on Environmental Aspects of Stabilization/Solidification of Hazardous and Radioactive Wastes, May 3-6, 1987, Atlanta/Georgia.

Davies-Colley RJ, Nelson PO, Williamson KJ (1985) Sulfide control of cadmium and copper concentrations in anaerobic estuarine sediments. Mar Chem 16: 173-186

Dreesen DR, Gladney ES, Owens JW, Perkins BL, Wienke CL, Wangen LE (1977) Comparison of levels of trace elements extracted from fly ash and levels found in effluent waters from a coal-fired power plant. Environ Sci Technol 11: 1017-1019

Duursma EK (1984) Problems of sediment sampling and conservation for radionuclide accumulation studies. In: Sediments and pollution in waterways. IAEA-TecDoc-302, pp 127-135. International Atomic Energy Agency, Vienna

Ehrig H-J (1983) Quality and quantity of sanitary landfill leachate. Waste Management & Research 1: 53-68.

Ehrig H-J (1989). Water and element balances of landfills. In: P Baccini (Ed) The Landfill - Reactor and Final Storage. Lecture Notes in Earth Sciences 20, pp 83-115, Springer, Berlin

Emerson S, Jahnke R, Heggie D (1984) Sediment-water exchange in shallow water estuarine sediments. J Mar Res 42: 709-730

Engler RM, Brannon JM, Rose J, Bingham G (1977) A practical selective extraction procedure for sediment characterization. In: Yen TF (ed) Chemistry of marine sediments. Ann Arbor Sci Publ, pp 163-171

Fletcher P, Beckett PHT (1987) The chemistry of heavy metals in digested sewage sludge. II Heavy metal complexation with soluble organic matter. Water REs 21: 1163-1172

Förstner U (1985) Chemical forms and reactivities of metals in sediments. In: Leschber R, Davis RD, L'Hermite P (eds) Chemical methods for assessing bio-available metals in sludges and soils. Elsevier Applied Sci, London, pp 1-30

Förstner U (1986) Chemical forms and environmental effects of critical elements in solid-waste materials - combustion residues. In: M Bernhard, FE Brinckman, PJ Sadler (eds) The importance of chemical "speciation" in environmental processes. Dahlem-Konferenzen. Springer-Verlag Berlin. pp 465-491

Förstner U (1987) Metal speciation in solid wastes - factors affecting mobility. In: L Landner (ed) Speciation of metals in water, sediment and soil systems. Lecture Notes in Earth Sciences 11, pp 13-41. Springer Berlin

Förstner U (1989) Soil pollution phenomena - mobility of heavy metals in contaminated soil. In: H Bolt et al (eds) Interactions at the soil colloid - soil solution interface. Chap 17.

Förstner U, Carstens A (1988) In-situ-Versuche zur Veränderung von festen Schwermetallphasen in aeroben und anaeroben Grundwasserleitern. Vom Wasser 71: 113-123

Förstner U, Kersten M (1988) Assessment of metal mobility in dredged material and mine waste by pore water chemistry and solid speciation. In: W Salomons, U Förstner (eds) Chemistry and biology of solid waste - dredged material and mine tailings. Springer Berlin, pp 214-237

Förstner U, Salomons W (1983) Trace element speciation in surface waters: interactions with particulate matter. In: GG Leppard (ed) Trace element speciation in surface waters and its ecological implications. Nato Conf Ser I, Vol 6, pp 245-273. Plenum Press New York

Förstner U, Schoer J (1984) Diagenesis of chemical associations of Cs-137 and other artificial radionuclides in river sediments. Environ Technol Letts 5: 295-306

Förstner U, Kersten M, Wienberg R (1989a) Geochemical processes in landfills. In: P Baccini (ed) The landfill - reactor and final storage. Lecture Notes in Earth Sciences 20. Springer Berlin, pp 39-81

Förstner U, Calmano W, Ahlf W, Kersten M (1989b) Ansätze zur Beurteilung der "Sedimentqualität" in Gewässern. Z Vom Wasser 73: 25-42

Förstner U, Ahlf W, Calmano W, Kersten M (1990) Sediment criteria development - contributions of environmental geochemistry to water quality management. In: D Heling et al (eds) Sediments and environmental geochemistry. Springer Berlin, pp 311-338

Fraser JL, Lum KR (1982) Availability of elements of environmental importance in incinerated sludge ash. Environ Sci Technol 17, 52-54

Frevert T, Galster H (1978) Schnelle und einfache Methode zur In-situ Bestimmung von Schwefelwasserstoff in Gewässern und Sedimenten. Schweiz Z Hydrol 40: 199-208

Fuller WH, Warrick AW (1985) Soils in waste treatment and utilization. Vol II, Chap 4, screening protocol for predicting waste treatability. CRC Press, Boca Raton, pp 77-122

Gambrell RP, Reddy CN, Khalid RA (1983) Characterization of trace and toxic materials in sediments of a lake being restored. J Water Pollut Control Fed 55: 1201-1213

Gendron A, Silverberg N, Sundby B, Lebel J (1986) Early diagenesis of cadmium and cobalt in sediments of the Laurentian Trough. Geochim Cosmochim Acta 50: 741-747

Ham RK, Anderson MA, Stanforth R, Stegmann R (1979) Background study on the development of a standard leaching test. US EPA-600/2-79-109

Harris WR, Silberman D (1988) Leaching of metal ions from fly ash with canine serum. Environ Sci Technol 22: 109-112

Harrison RM, Laxen DPH, Wilson SJ (1981) Chemical associations of lead, cadmium, copper, and zinc in street dusts and roadside soils. Environ Sci Technol 15: 1378-1383

Herms U, Brümmer G (1978) Löslichkeit von Schwermetallen in Siedlungsabfällen und Böden in Abhängigkeit von pH-Wert, Redoxbedingungen und Stoffbestand. Mitt Dt Bodenk Ges 27: 23-43

Herms U, Brümmer G (1980) Einfluß der Bodenreaktion auf Löslichkeit und tolerierbare Gesamtgehalte an Nickel, Kupfer, Zink, Cadmium und Blei in Böden und kompostierten Siedlungsabfällen. Landwirtschaftl Forsch 33: 408-423

Hines ME, Lyons BWM, Armstrong PB, Orem WH, Spencer MJ, Gaudette HE (1984) Seasonal metal remobilization in the sediments of Great Bay, New Hampshire. Mar Chem 15: 173-187

Holmes CW, Slade EA, McLerran CJ (1974) Migration and redistribution of zinc and cadmium in marine estuarine systems. Environ Sci Technol 8: 255-259

Honeyman BD, Santschi PH (1988) Metals in aquatic systems. Environ Sci Technol 22: 862-871

Hunt CD, Smith DL (1983) Remobilization of metals from polluted marine sediments. Can J Fish Aquat Sci 40: 132-142

Jackson DR, Garrett BC, Bishop TA (1984) Comparison of batch and column methods for assessing leachability of hazardous waste. Environ Sci Technol 18: 668-673

Jenne EA, DiToro DM, Allen HE, Zarba CS (1986): An activity-based model for developing sediment criteria for metals. In: JN Lester, R Perry, RM Sterritt (eds) Chemicals in the environment. Selper Ltd London, pp 560-568

Kester DR, Ketchum BH, Duedall IW, Park PK (eds)(1983) Wastes in the ocean. Vol 2: Dredged-material disposal in the ocean. Wiley New York, 299 p

Kersten M (1988) Geochemistry of priority pollutants in anoxic sludges: cadmium, arsenic, methyl mercury, and chlorinated organics. In: W Salomons, U Förstner (eds) Chemistry and biology of solid waste - dredged material and mine tailings. Springer Berlin, pp 170-213

Kersten M (1989) Mechanismus und Bilanz der Schwermetallfreisetzung aus einem Süßwasserwatt der Elbe. Dissertation Technische Universität Hamburg-Harburg

Kersten M, Förstner U (1986) Chemical fractionation of heavy metals in anoxic estuarine and coastal sediments. Water Sci Technol 18: 121-130

Kersten M, Förstner U (1987a) Cadmium associations in freshwater and marine sediment. In: JO Nriagu, JB Sprague (eds) Cadmium in the Aquatic Environment, pp 51-88. Wiley, New York

Kersten M, Förstner U (1987b) Effect of sample pretreatment on the reliability of solid speciation data of heavy metals - implications for the study of early diagenetic processes. Mar Chem 22: 299-312

Kersten M, Förstner U (1989a) Speciation of trace elements in sediments. In: GE Batley (ed) Trace element speciation - analytical methods and problems. CRC Press Boca Raton, p

Kersten M, Förstner U (1989b) Trace element composition of suspended particulate matter in surface waters of the entire North and Irish Sea. In: JP Vernet (Ed) Proc Intern Conf Heavy Metals in the Environment, Geneva 1989, Vol 2, pp 371-374. CEP Consultants Edinburgh

Kersten M, Förstner U, Calmano W, Ahlf W (1985) Freisetzung von Metallen bei der Oxidation von Schlämmen. Vom Wasser 65: 21-35

Kersten M, Dicke M, Kriews M, Naumann K, Schmidt D, Schulz M, Schwikowski M, Steiger M (1988) Distribution and fate of heavy metals in the North Sea. In: W Salomons, BL Bayne, EK Duursma, U Förstner (eds) Pollution of the North Sea: an assessment. Springer-Verlag Berlin, pp 300-347

Khalid RA (1980) Chemical mobility of cadmium in sediment-water systems. In: Nriagu JO (ed) Cadmium in the environment, part I: ecological cycling. Wiley New York, pp 257-304

Kiekens L, Cottenie A (1985) Principles of investigations on the mobility and plant uptake of heavy metals. In Leschber R, Davis RD, L'Hermite P (eds) Chemical methods for assessing bio-available metals in sludges and soils. Elsevier Applied Sci, London, pp 32-47

Krebs J, Belevi H, Baccini P (1988) Long-term behavior of bottom ash landfills. Proc 5th Intern Solid Wastes Exhibition and Conf, ISWA 1988, Copenhagen

Lee FY, Kittrick JA (1984) Elements associated with the cadmium phase in a harbor sediment as determined with the electron beam microprobe. J Environ Qual 13: 337-340

Lewis FM, Voss CI, Rubin J (1987) Solute transport with equilibrium aqueous complexation and either sorption or ion exchange: simulation methodology and applications. J Hydrol 90: 81-115

Lion LW, Altman RS, Leckie JO (1982) Trace-metal adsorption characteristics of estuarine particulate matter: evaluation of contribution of Fe/Mn oxide and organic surface coatings. Environ Sci Technol 16: 660-666

Lu CSJ, Chen KY (1977) Migration of trace metals in interfaces of seawater and polluted surficial sediments. Environ Sci Technol 11:174-182

Luoma SN, Davis JA (1983) Requirements for modeling trace metal partioning in oxidized estuarine sediments. Mar Chem 12: 159-181

Lum KR, Betteridge JS, Macdonald, RR (1982) The potential availability of P, Al, Ca, Co, Cr, Cu, Fe, Mn, Ni, Pb, and Zn in urban particulate matter. Environ Technol Letts 3: 57-62 (1982)

Luther GW, Meyerson AL, Krajewski JJ, Hires R (1980) Metal sulfides in estuarine sediments. J Sediment Petrol 50: 1117-1120

Nriagu JO (1980) Cadmium in the atmosphere and in precipitation. In: JO Nriagu (ed) Cadmium in the environment, part 1, p 71. John Wiley & Sons, New York

Nriagu JO (ed) (1984) Changing metal cycles and human health. Dahlem Konferenzen, Life Sci Res Rep 28. Springer-Verlag, Berlin

Nürnberg HW, Valenta P, Nguyen VD (1983) The wet deposition of heavy metals from the atmosphere in the Federal Republic of Germany. In: G Müller (ed) Proc Int Conf Heavy Metals in the Environment, Heidelberg 1983, pp 115-123. CEP Consultants Edinburgh

Patrick WH, Williams BG, Moraghan JT (1973) A simple system for controlling redox potential and pH in soil suspensions. Soil Sci Soc Amer Proc 37: 331-332

Peiffer S (1989) Biogeochemische Regulation der Spurenmetalllöslichkeit während der anaeroben Zersetzung fester kommunaler Abfälle. Dissertation Universität Bayreuth, 197 p

Peiffer S, Frevert T (1987) Potentiometric determination of heavy metal sulphide solubilization using a pH_2S (glass/-Ag^o,Ag_2S) electrode cell. Analyst 112: 951-954

Plant JA, Raiswell R (1983) Principles of environmental geochemistry. In: I Thornton (ed) Applied environmental geochemistry. Academic Press London, pp 1-39

Prause B, Rehm E, Schulz-Baldes M (1985) The mobilisation of Pb and Cd from contaminated dredge spoil after dumping in the marine environment. Environ Technol Letts 6: 261-266

Rapin F et al (1986) Potential artifacts in the determination of metal partitioning in sediments by a sequential extraction procedure. Environ Sci Technol 20: 836-840

Ronen D, Magaritz H, Levy I (1986) A multi-layer sampler for the study of detailied hydrochemical profiles in groundwater. Water Res 20: 311-315

Rudd T, Campbell JA, Lester JN (1986) Characterization of metal forms in sewage sludges by chemical extraction. In: JN Lester, R Perry, RM Sterritt (Eds) Chemicals in the Environment. Selper Ltd London, pp 756-771

Salomons W (1980) Adsorption processes and hydrodynamic conditions in estuaries. Environ Technol Letts 1: 356-365.

Salomons W (1985) Sediments and water quality. Environ Technol Letts 6: 315-368

Salomons W, Förstner U (1980) Trace metal analysis on polluted sediments. II. Evaluation of environmental impact. Environ Technol Lett 1, 506-517

Salomons W, Förstner U (1984) Metals in the hydrocycle. Springer Berlin, 353 p

Salomons W, Förstner U (Eds)(1988a) Chemistry and Biology of Solid Waste: Dredged Materials and Mine Tailings. Springer Berlin, 305 p

Salomons W, Förstner U (Eds)(1988b) Environmental Management of Solid Waste: Dredged Materials and Mine Tailings. Springer Berlin, 396 p

Salomons W, De Rooij NM, Kerdijk H, Bril J (1987) Sediments as a source for contaminants? In: RL Thomas et al (eds) Ecological effects of in-situ sediment contaminants. Hydrobiologia 149: 13-30

Santschi PH, Honeyman BD (1989) Radioisotopes as tracers for the interactions between trace metals, colloids and particles in natural waters. Proc Intern Conf Heavy Metals in the Environment, Geneva. Vol 1, pp 243-252. CEP Consultants Edinburgh

Sauerbeck D, Styperek P (1985) Evaluation of chemical methods for assessing the Cd and Zn availability from different soils and sources. In: Leschber R, Davis RD, L'Hermite P (eds) Chemical methods for assessing bio-available metals in sludges and soils. Elsevier Appl Sci, London, pp 49-66

Schäfer W, Peiffer, S (1987) Einsatz von Dialysebeuteln zur Charakterisierung der Grundwasserbeschaffenheit im Abstrom einer Mülldeponie. Z Wasser Abwasser Forsch 20: 197-202

Schoer J, Förstner U (1985) Chemical forms of artificial radionuclides in fluviatile, estuarine and marine sediments compared with their stable counterparts. In: EK Duursma (ed) Proc Seminar on the Behaviour of Radionuclides in Estuaries. Commission of the European Communities Luxemburg, pp 27-54

Schoer J, Förstner U. (1987) Abschätzung der Langzeitbelastung von Grundwasser durch die Ablagerung metallhaltiger Feststoffe. Vom Wasser 69: 23-32

Shea D (1988) Developing national sediment quality criteria - equilibrium partitioning of contaminants as a means of evaluating sediment quality criteria. Environ Sci Technol 22: 1256-1261.

Stumm W, Bilinski H (1973) Trace metals in natural waters: difficulties of interpretation arising from our ignorance on their speciation. In: SH Jenkins (ed) Proc 8th Int Conf on Water Pollution Research, Jerusalem 1972. Pergamon Press Oxford, pp 39-49

Stumm W, Morgan JJ (1981) Aquatic Chemistry. Wiley, New York

Takeuchi T et al (1959) Pathological observations of the Minamata disease. Acta Pathol Jpn Suppl 9: 769-783.

Tauchnitz J, Knobloch G, Wiesener G, Schumann H, Kunze V, Mahrla M, Hanrieder M, Kiesel G, Hennig H (1983) Zur Ablagerung der industriellen Abprodukte. 27. Mitteilung: Zum Verhalten von Schwermetallionen in Deponiestandorten. Z Angew Geol 29: 311-317

Tessier A, Campbell PGC, Bisson M (1979) Sequential extraction procedure for the speciation of particulate trace metals. Anal Chem 51: 844-851

Theis TL, Padgett LE (1983) Factors affecting the release of trace metals from municipal sludge ashes. J Water Pollut Control Fed 55: 1271-1279

Thomson EA, Luoma SN, Cain DJ, Johansson C (1980) The effect of sample storage on the extraction of Cu, Zn, Fe, Mn, and organic material from oxidized estuarine sediments. Water Air Soil Pollut 74: 215-233

Turner RR, Lowry P, Levin M, Lindberg SE, Tamura T (1982) Leachability and aqueous speciation of selected trace constituents of coal fly ash. Final Report, Research Project 1061-1/EA-2588. Electric Power Research Institute

Van Den Berg CMG, Dharmvanij S (1984) Organic complexation of zinc in estuarine interstitial and surface water samples. Limnol Oceanogr 29: 1025-1036

Van der Sloot HA, Piepers O, Kok A (1984) A standard leaching test for combustion residues. Studiegroep Ontwikkeling Standaard Uitloogtesten Verbrandingsresiduen (SOSUV). Netherlands Energy Research Centre, Petten

Wadge A, Hutton M (1987) The leachability and chemical speciation of selected trace elements in fly ash from coal combustion and refuse incineration. Environ Pollut 48: 85-99

Wood JM, Chakrabarty AM, Craig PJ, Förstner U, Fowler BA, Herms U, Krull IS, Mackay D, Olson GJ, Russell DH, Salomons W, Silver S (1986) Speciation in systems under stress, Group Report. In: Bernhard M, Brinckman FE, Sadler PS (eds) The importance of chemical speciation in environmental processes. Dahlem Konferenzen, Life Sciences Res Rep 33. Springer Berlin, pp 425-441

THE COMPLEXATION OF METAL IONS BY HUMIC SUBSTANCES IN NATURAL WATERS

Walter Lund
Department of Chemistry
University of Oslo
P.O.Box 1033, N- 0315 Oslo, Norway

Introduction

Metal ions interact in various ways with inorganic and organic dissolved components and particles in natural waters. The relative importance of each of these interactions will change from one sample and locality to another. Many books and review articles have delt with different aspects of these reactions (Aiken et al.,1985; Buffle,1988; Christman and Gjessing, 1983; Kramer and Duinker,1984; Leppard,1983). Special attention has been focussed on dissolved organic compounds, in particular those termed humic substances, because of their potential effect on the bioavailability of toxic metals. It is generally assumed that free metal ions are more toxic to aquatic biota than metal ions bound to large organic molecules like the humic substances. However, the organic matter responsible for an eventual decreased toxicity is seldom well characterized, partly because of the difficulties of such a characterization. Attention has been focussed mostly on the interaction of copper with humic substances, but other trace metals like cadmium and lead have also been studied to some extent. Recently, the complexation of aluminum by humic substances has attracted much attention, because of the possible effects of aluminum on fish mortality, as a result of increased mobilization of this metal due to acid rain.

Different aspects of importance for the study and interpretation of metal-humic interactions in natural waters will be discussed below.

Humic substances

Aquatic humic substances are large organic molecules, which are formed by degradation of biopolymers or polymerization of smaller organic molecules in the environment (Hedges,1988). These substances often contribute more than 50% of the total organic matter in natural waters. The substances may originate from rain-water leaching of soil organic matter, or they can be formed in the water mass itself, as a result of excretion or decomposition of living organisms. The molecular weight may vary greatly, but values in the range 1000 - 3000 are

NATO ASI Series, Vol. G 23
Metal Speciation in the Environment
Edited by J. A. C. Broekaert, Ş. Güçer, and F. Adams

frequently obtained by various experimental techniques (Buffle,1988). The size of the molecules depends on sample parameters such as concentration, pH and ionic strength. Aggregation occurs at high concentrations of humic substances (> 0.1 g/l). A high ionic strength, low pH (Ghosh and Schnitzer,1980) and a high metal ion concentration (Underdown et al.,1981) also promote aggregation. The aggregates presumably consist of units with a moderate molecular weight, which are held together by relatively weak bonds (Wershaw et al.,1986). In natural waters with total organic carbon below 0.1 g/l, the aggregation is probably of minor importance.

Because humic substances are not well defined chemical compounds, the researcher who wants to study the chemical reactivity of these compounds, e.g. the reactions with metal ions, is faced with the difficult task of obtaining a sample which is representative of this class of compounds. There are various approaches to this problem. The most simple solution, which has been chosen by many workers in the field, is to buy a commercial product; humic acids are available from various suppliers. However, this has been heavily criticized by Malcolm and MacCarthy (1986), who claim, on the basis of spectroscopic evidence, that these products are not representative of natural soil or water humic substances. Another objection to the use of commercial humic acids is that they are not furnished with information about their origin, elemental composition or method of extraction.

Unfortunately, the alternative to the purchase of a commercial product is not evident. The most commonly used approach has been to extract the humic substances from soil with dilute aqueous sodium hydroxide; the humic acid fraction is precipitated upon acidification, while the fulvic fraction remains in solution. However, Malcolm and MacCarthy (1986) advocate that humic substances should be extracted from real water samples according to a multistep separation and purification procedure. The International Humic Substances Society (IHSS) now provides water humic and fulvic acids which have been carried through 12-steps procedures, involving various pH adjustments, adsorption on XAD-8 and cation exchange resins and freeze drying (Thurman and Malcolm,1983). However, the chemical treatments involved in such separation procedures may alter the chemical composition of the original compounds, so that the chemical reactivity of the final solid products may not be representative of the substances originally present in the water sample. In addition, the IHSS samples are rather expensive.

For these reasons, some workers prefer to carry out their studies of the water samples directly, eventually after preconcentration by evaporation under reduced pressure at 35°C (John et al.,1988; Lund et al.,1990; Riise and Salbu,1989), without isolating the humic substances. In particular, the determination of the socalled complexation capacity of humic material is usually based on the direct titration of the water sample with a metal ion. Which of the approaches discussed above that is to be preferred may depend on the purpose of the investigation. Recently, we studied the metal complexation of both commercial humic acids, fulvic acid obtained from IHSS, water samples that had been concentrated in a rotavapor at 35°, and wholly untreated fresh water samples (Dundas and Lund; Lund et al.,1990). The metals studied were copper, cadmium, aluminum and iron. Although some differences were noticable, the chemical reactivity of the different humic samples appeared to be quite similar.

Complexation of metal ions

Humic substances are both polyelectrolytes, with a large number of groups that are ionized in solution, and polyfunctional, with a number of different binding sites. The sites may be phenol or carboxyl groups, or a combination of these groups such as phthalic or salicylic groups, where each site may act as a bidentate ligand. The most straightforward approach for studying the metal complexation reactions of humic matter is probably to consider the humic substances as a highly complex mixture of independent organic acids, each of which can bind a given metal ion with 1:1 stoichiometry (Perdue,1988). According to this model, the complexation capacity of each ligand (CC_i) is one mole of metal per mole of ligand, and the CC_i should therefore be independent of pH, ionic strength, nature of metal and concentration of the ligand. At first glance this statement is at variance with much of the published litterature, which suggests that the average metal complexation capacity (CC) varies considerably with experimental parameters - increasing at higher pH, decreasing at higher ionic strength, increasing at higher humic concentrations and varying with the nature of the added metal ion (Perdue,1988).

There can be at least two reasons for the variation of CC with experimental parameters. Either it is a result of inappropriate experimental measurements, or alternatively the original hypothesis of humic substances as a mixture of independent ligands does not provide an adequate description of the system. Perdue (1988) has shown that the extent to which the CC can be realized in experimental measurements is strongly a function of pH, ionic strength, concentration of humic substances and the nature of the metal being complexed. This fact points in the direction of experimental limitations being responsible for the variations in CC. However, if the structure of the humic substances changes with the experimental parameters, owing to aggregation and other conformational changes, there is no reason why the CC should be insensible to changes in pH, degree of metal saturation etc. In accordance with this, Buffle (1988) states that CC should not be considered as a purely stoichiometric parameter. Even so, aggregation is probably of minor importance at the low concentrations frequently encountered in natural waters (< 100 mg TOC/l).

Owing to the uncertainty in the interpretation of the experimental results, CC-values are not very useful for determining to what extent changes in experimental conditions will bring about structural rearrangements in aquatic humic material. Unfortunately, direct evidence for conformational changes is not easily obtained by spectroscopic techniques for the complicated metal-humic interactions. However, there are other experimental evidence that conformational changes may occur as a result of changing chemical conditions. Thus, it has been shown by size-exclusion chromatography that the size of water fulvic acid increases with pH, probably because of an increasing dissociation of the carboxylic and phenolic groups, which results in an increased unfolding of the molecule in solution because of electrostatic repulsion (Varney et al.,1983). Also, light scattering studies have shown that aggregation increases with the degree of site occupation by copper (Langford et al.,1983).

Complexation parameters from titration curves

The capacity of humic acids to bind metal ions is, at most, equal to the number of titratable hydrogen ions. Because of the complicated structure of humic substances, the number of available sites can only be determined by experiment. Many different analytical methods have been used for determining the complexation capacity (CC) of humic substances in natural waters (Neubecker and Allen,1983). The CC is often determined by titration with a metal ion, and copper(II) is usually the metal of choice. The titration curve is obtained by measuring the concentration of free metal as a function of total metal added; an amperometric titration is often the method used (see below). The CC and the complexation constant (K) can be calculated from the titration data using various mathematical procedures. The calculations usually assume 1:1 stoichiometry between the metal ion and each binding site. The mathematical treatment is based on the three equations

$$C_M = [M] + [ML] \quad (1)$$

$$C_L = [L] + [ML] \quad (2)$$

$$[ML] = K\,[M]\,[L] \quad (3)$$

where C_M and C_L are the total concentrations of metal ion and ligand binding site, [M], [L] and [ML] are the concentrations of free metal ion, free ligand binding site and complex, respectively, and K is the stability constant. C_L is equal to the complexation capacity, except that the CC is usually given in terms of mmol metal per gram of carbon, or mmol metal per gram of humic matter. The carbon content of humic matter is ca. 50%. The three equations can be combined to a single equation in different ways, to facilitate the calculation of K and C_L, assuming that C_M is known and [M] is measured. Usually there is no well-defined "end point" of the titration curve, in which case a treatment of all the titration data by non-linear regression analysis is probably better than various extrapolations or other types of linearizations of the original titration curve (Lund et al.,1990).

Published values for the CC of water fulvic acid measured with copper(II) have been collected by Buffle et al.(1984)(Buffle,1988). A majority of the values, which referred to nine different experimental techniques, were in the range 0.5-5 mmol/g C, but values down to 0.01 mmol/g C were obtained by some techniques. If the CC-value is, for example, 1 mmol/g C, this means that one metal atom would be bound per 80 carbon atoms in the saturated complex. Most experimentally determined CC-values appear to be somewhat lower than the total number of binding sites, as defined by the number of acidic functional groups in the fulvic acid. The latter value generally falls within the range 3-10 meqv/g C (Buffle,1988; Stevenson,1985). Of course, the acidity value must be divided by two to give the CC, if each binding site acts as a bidentate ligand with two titratable hydrogen ions.

There may be various reasons for the CC being lower than the total number of binding sites. First of all, the instrumental technique chosen for the estimation of CC may be inappropriate for the purpose. Furthermore, the assumption that there is a 1:1 relationship between each binding site and metal ion will be incorrect, if two or more sites are engaged in the coordination of each metal ion. The primary objection to multi-site binding of each metal ion arises from the relatively large size of the humic molecule, which makes chelate formation with more than one coordination site less likely for sterical reasons.

Another possible reason for the low CC-values sometimes found experimentally could be that some sites are occupied by other metals, e.g. iron(III) or aluminum(III). These metal ions, which form strong complexes with humic substances, are present in natural waters in relatively high concentrations. However, Baccini and Suter (1979) and Dundas and Lund found that the CC for copper is not much influenced by the presence of iron or aluminum. This is rather surprising, and has been explained by the presence of copper-specific binding sites (Baccini and Suter,1979; Hart,1981).

Finally, the reason for low CCs may be that most techniques underestimate this parameter, as already discussed. The procedures should in principle give the maximum amount of metal that can be bound by the humic compounds, but the relatively weak binding sites of these compounds are not easily saturated with metal ions in dilute aqueous solutions. Thus, direct measurements of CCs of unconcentrated humic substances in a water sample may underestimate the total CC of the sample. As pointed out by Perdue (1988), a 100-fold excess of metal ion is required to saturate 90% of the binding sites, if the ligand concentration is 10^{-5}M, and the complexation constant is 10^4.

The interpretation of the CC-measurements is not made more easy when the CC for copper is compared with the values obtained for cadmium. Thus, although the CC for copper may be low compared to the total acidity, it is normally large compared to the CC-value obtained when cadmium is used as titrant (Alberts et al.,1984; Giesy and Briese,1980; Lund et al.,1990). Probably, this indicates that only few of the sites bind cadmium to any appreciable extent. The reported conditional constants for cadmium (John et al.,1988; Mantoura et al.,1978) seem to be of the same order of magnitude as for copper.

Humic substances usually contain both strong and weak binding sites, and a Gaussian distribution of binding sites may therefore be assumed (Perdue and Lytle,1983). A variation in binding strength is also often inferred from the mathematical treatment of the titration data, when the linearizations no longer give straight lines. If the curve has two nearly linear portions, K-values for two types of binding sites may be determined from the respective slopes (Mantoura and Riley,1975; Scatchard et al.,1957). However, a good mathematical fit obtained with a two-site model does not prove the existence of binding sites with different strength. Thus, even if there is only one type of binding site, the K-value for this site may vary during the titration, as a result of conformational changes of the molecule. Generally, a non-linear plot will be obtained, if K depends on the C_M/C_L ratio (Scheinberg,1982).

When humic substances are titrated with a metal ion, the strong binding sites should be saturated first, and then the weaker sites. However, if inverse titrations are carried out, i.e. a metal solution is titrated with the humic substances, the free metal ion concentration should eventuelly approach zero, irrespective of the relative number of strong and weak binding sites. This seems not always to be the case (Lund et al.,1990; Turner et al.,1987), assuming that the electrochemical methods used in these titrations respond only to the free metal ion. However, the fact that the electrochemical peak height approaches a non-zero value at high concentrations of humic matter may also indicate that the metal-humic complex is not completely electrochemically inert.

Measurement of free metal during titration

When a complexation reaction is studied by titration with a metal ion, the concentration of the free metal ion must be monitored. This is frequently done by an electrochemical method, such as potentiometry with an ion selective electrode (ISE), differential pulse polarography (DPP) or differential pulse anodic stripping voltammetry (DPASV). These methods are particularly useful for the study of complexation reactions when the metal is copper, lead or cadmium. However, there are various limitations to the use of the electrochemical techniques for complexometric titrations of natural waters. First, the ISE detection limits of ca.1 µmol/l are too high. In addition, various interference effects, caused by adsorption and dissolution processes at the electrode surface, hampers the use of ISEs for real samples. Also DPP has insufficient detection limits for most natural samples, whereas DPASV can determine free metal concentrations down to the pmol/l level. On the other hand, the latter method may overestimate the free metal ion concentration, owing to dissociation of the metal-humic complex at the electrode surface, and the method is also troubled by adsorption of surface-active substances. Even so, amperometric titration is probably the most frequently used method for studying metal-humic complexation reactions (Turner et al.,1987), and the various pitfalls have been thoroughly dealt with in the litterature (Buffle,1988). Weber (Saar and Weber,1982; Weber,1983; Weber,1988) and Tuschall and Brezonik (1983) have compared ISE and DPASV with other techniques which have be used for the study of metal-humic interactions.

Spectroscopic methods are not easily used for monitoring the concentration of free metal ions in solution, because methods like atomic absorption spectrometry (AAS) and inductively coupled plasma atomic emission spectrometry (ICP-AES) measure only the total concentration of metals in solution. In order to determine free metal ions by these methods, a separation step must usually precede the spectroscopic measurements. Separation methods based on ion exchange resins or solvent extraction are often used for this purpose, as discussed below. Fluorescence spectroscopy has been utilized to study the reaction of humic material with paramagnetic metal ions, because these metals quench the fluorescence of humic matter (Frimmel,1990; Tuschall and Brezonik,1983; Weber,1983; Weber,1988).

Speciation schemes

Because there is no instrumental method available that is truly selective towards individual metal species at the concentration levels normally encountered in natural waters, various schemes have been developed, which allow different operationally defined groups of metal species to be identified. In the most simple schemes, a distinction is primarily made between free metal ions and metal ions bound in inert complexes. In the case of copper, lead and cadmium, the concentration of free metal ions is usually determined directly by anodic stripping voltammetry (ASV), while the total metal concentration is determined after digestion of the sample. The amount of bound metal is then calculated by difference. For aluminum, the concentration of free metal ions is not so easily measured directly. Instead, the free metal concentration is determined by difference, in the following manner. The sample is passed through a column containing a cation exchange resin (e.g. Amberlite IR-120). The resin binds the free metal ions, while the inert metal-humic complexes pass through the column. The sample is analyzed before and after passage through the column, to obtain the total metal and bound metal concentrations, respectively, using an instrumental technique which is not species selective, such as atomic spectrometry. The amount of free metal is finally calculated by difference.

The more elaborate speciation schemes may include a combination of separation methods, which fractionate according to size, charge, complexation or extractability. Separations based on chelating and cation exchange resins, and dialysis and ultrafiltration membranes are among the more frequently used methods. The total metal concentration in each fraction is preferably measured with an instrumental technique which has low detection limits, such as ASV or AAS. The Batley and Florence (1976) scheme for the determination of copper, lead and cadmium species involves four sample treatments: filtration, passage through a Chelex-100 column, irradiation with UV light and UV irradiation followed by a Chelex separation. For each treatment, labile metal is measured by direct ASV, and total metal by ASV after digestion. From the eight measurements, seven groups of metal species are quantified. Later, Florence (1982) also included in his speciation scheme other types of resins, and extraction with hexane-butanol to account for lipid-soluble metal complexes. The results obtained with these schemes will depend on the experimental conditions chosen. For example, the contact time of the water sample with the Chelex resin will normally influence the results; more "free" metal will be found if the separation is done in a batch procedure instead of in a column. This is utilized in the scheme developed by Figura and MacDuffie (1979), where the sample is treated with Chelex-100 in both the column and batch modes. In this way they distinguished between very labile, moderately and slowly labile species. Laxen and Harrison (1981) used ultrafiltration in combination with a Chelex resin, and also Hoffman et al.(1981) used ultrafiltration, whereas Hart and Davies (1977) combined Chelex and dialysis separations. In these schemes, the final measurements of copper, lead and cadmium were done with ASV, while other trace metals were determined by AAS.

All these speciation schemes, which are based on fractionation according to lability and size, will determine operationally defined groups of metal species. To obtain useful results, intercomparison tests should be run for speciation methods that are used by different workers. This has recently been done for aluminum in natural waters. Sullivan et al.(1986) compared two methods currently used for aluminum speciation; the Barnes/Driscoll extraction method (Driscoll,1984), and the pyrocatechol violet method (Seip et al.,1984). The former is based on extraction with 8-hydroxyquinoline in methyl isobutyl ketone, followed by atomic absorption spectrometry or spectrophotometry, while the latter method measures directly by spectrophotometry the colour developed by pyrocatechol violet. Both methods use a cation exchange resin to separate labile from non-labile aluminum. In addition, acid soluble aluminum is normally determined with these methods. Sullivan et al.(1986) found some systematic differences, particularly for the organic aluminum fraction, whereas the results for inorganic aluminum were in good agreement. Similarly, LaZerte et al.(1988) found a good correlation for inorganic aluminum between a dialysis technique and an ion-exchange technique. Unfortunately, most other speciation schemes, like those developed for copper, lead and cadmium, are too different for an intercomparison to be meaningful, although most of these schemes rely on the use of ASV for the final measurements.

Competition of metals

Most studies of metal-humic interactions in natural waters have been concerned with the trace metals copper, lead and cadmium. However, natural waters contain a series of metal ions, and particularly high concentrations of iron and aluminum. The metal binding strength decreases in the order Fe(III) > Al(III) > Cu(II) > Pb(II) > Cd(II) (Schnitzer,1978). It is therefore surprising that high concentrations of iron do not seem to influence the binding of copper to humic substances (Baccini and Suter,1979; Dundas and Lund). However, Rashid (1985) found that copper, zinc, nickel, cobalt and manganese were desorbed from humic acid when iron(III) was used as extractant. Also, iron was recently found to influence the binding of aluminum, and vice versa (Dundas and Lund). Therefore, the concentration of both of these metals should be measured in a natural water sample, even if the investigation is primarily concerned with the binding of only one of the metals.

Quite generally, all metals present in a water sample should be of interest, when the possible effects of humic substances are to be assessed. In a study of the complexation capacity of various fresh waters, Dyrstad and Lund found that the concentration of copper was sufficiently high in some of the samples to saturate all binding sites of the organic material. From the difference between the concentrations of total and free copper, measured by DPASV, the CC was estimated to be below 0.1 μmol/mgTOC in these samples. Also when model studies are carried out with higher concentrations of humics, the original metal content of the humic material should be taken into account. Unfortunately, the metal content of commercially available humic acids is rarely known, although there are some recent results for aquatic humic materials (Riise and Salbu,1989).

Bioavailability

The study of metal-humic interactions is often aimed at predicting the effect of humic substances on the bioavailability of metal ions. In particular, this is the purpose of many of the speciation schemes developed. Therefore, experiments should be carried out to assess the biological toxicity of the different metal fractions determined by these schemes. Considerable evidence now exists that the free ionic forms of metals like copper, lead, cadmium and aluminum are usually the most toxic forms to aquatic biota, and that the complexation by humic substances reduces the toxicity of these metal ions. For example, Sunda and coworkers (Sunda and Lewis,1978; Sunda et al.,1984) were able to correlate copper toxicity with the free concentration of copper ions, using a bacterial bioassay technique, and measurements of copper ion activity with an ISE. The free copper concentration was controlled by adding a well-defined complexing agent (NTA). Later, a number of workers have provided evidence that reduced copper toxicity is observed in the presence of natural organic ligands. It is less well known that synthetic ligands, which form lipid soluble copper complexes, may form species that are more toxic than the free copper ion. These effects have been demonstrated by Florence et al. (1984), who used the depression of growth rate of the marine diatom *Nitzschia closterium* as a measure of toxicity. For lead, it has been shown that the free ionic form is more easily taken up by the mussel *Elliptio complanata* than when the metal is complexed with humus (Campbell and Evans,1987).

Some interesting results have recently been reported for the toxicity of cadmium species. The uptake of cadmium in the green alga *Selenastrum capricornutum* Printz was found to depend on the molecular weight of the humic substances present in the fresh water sample (Sedlacek et al.,1989). The lowest molecular weight fraction reduced cadmium uptake less effectively than the higher fractions. Also Giesy et al. (1977) have reported a similar dependence of cadmium toxicity (to *Sinocephalus serrulatus*) on the molecular size of the organic fraction. However, they found that the smallest molecular fractions actually increased the cadmium toxicity slightly. Further, Winner (1984) found that the addition of humic acid increased the toxicity of cadmium to daphnides, but at the same time a decreased toxicity was observed for copper. In addition, he found that the humic acid had no significant effect on the bioaccumulation of either metal. A complicated relationship between species and toxicity is also indicated by Lægreid et al. (1983), who found that the toxic effect of cadmium towards the alga *Selenastrum capricornutum* Printz showed considerable seasonal variations in an eutrophic lake, with a toxicity during summer which far exceeded what would be expected from the estimated free ion activity. It is hypothesized that qualitative changes in the composition of the dissolved organic matter during the production period are responsible for this effect.

The recent interest in aluminum speciation is related to the research which has shown that acid precipitation can give rise to increased mobilization and transport of aluminum from soil to surface waters. The environmental consequences of increased concentrations of aqueous aluminum species include toxicity to fish and other aquatic organisms, and the inorganic aluminum forms appear to be the primary toxic species (Baker and Schofield,1982).

So far, no speciation scheme has been developed which reliably predicts bioavailability. However, some separation techniques seem to be more relevant for such studies than others. Florence et al. (1984) point out that resins like Chelex-100 and those with thiol groups may grossly overestimate the toxic fraction of copper. Lipid-soluble metal complexes, which may be highly toxic, are presumeably best accounted for by solvent extraction. As indicated above, also size fractionation techniques appear to be relevant for bioavailability studies. Some authors have found a correlation between ASV-labile metal and biotoxicity (Florence et al.,1983; Srna et al.,1980; Young et al.,1979).

Acknowledgement. The author thanks Dr. Egil T.Gjessing for valuable discussions.

References

Aiken GR, McKnight DM, Wershaw RL, MacCarthy P (eds) (1985) Humic substances in soil, sediment and water. Wiley, New York Chichester Brisbane Toronto Singapore

Alberts JJ, Giesy JP, Evans DW (1984) Distribution of dissolved organic carbon and metal-binding capacity among ultrafilterable fractions isolated from selected surface waters of the southeastern United States. Environ Geol Water Sci 6:91-101

Baccini P, Suter U (1979) MELIMEX, an experimental heavy metal pollution study: Chemical speciation and biological availability of copper in lake water. Schweiz Z Hydrol 41:291-314

Baker JP, Schofield CL (1982) Aluminum toxicity to fish in acidic waters. Water Air Soil Poll 18:289-309

Batley GE, Florence TM (1976) A novel scheme for the classification of heavy metal species in natural waters. Anal Lett 9:379-388

Buffle J (1988) Complexation reactions in aquatic systems: An analytical approach: Ellis Horwood, Chichester

Buffle J, Tessier A, Haerdi W (1984) Interpretation of trace metal complexation by aquatic organic matter. In: Kramer CJM, Duinker JC (eds) Complexation of trace metals in natural waters. Nijhoff/Junk Pub, The Hague Boston Lancaster, pp 301-316

Campbell JH, Evans RD (1987) Inorganic and organic ligand binding of lead and cadmium and resultant implications for bioavailability. Sci Total Envir 62:219-227

Christman RF, Gjessing ET (eds) (1983) Aquatic and terrestrial humic materials. Ann Arbor Science, Ann Arbor

Driscoll CT (1984) A procedure for the fractionation of aqueous aluminum in dilute acidic waters. Intern J Environ Anal Chem 16:267-283

Dundas SH, Lund W To be published

Dyrstad O, Lund W Unpublished results

Figura P, McDuffie B (1979) Use of Chelex resin for determination of labile trace metal fractions in aqueous ligand media and comparison of the method with anodic stripping voltammetry. Anal Chem 51:120-125

Florence TM (1982) Development of physico-chemical speciation procedures to investigate the toxicity of copper, lead, cadmium and zinc towards aquatic biota. Anal Chim Acta 141:73-94

Florence TM, Lumsden BG, Fardy JJ (1983) Evaluation of some physico-chemical techniques for the determination of the fraction of dissolved copper toxic to the marine diatom *Nitzschia closterium*. Anal Chim Acta 151:281-295

Florence TM, Lumsden BG, Fardy JJ (1984) Algae as indicators of copper speciation. In: Kramer CJM, Duinker JC (eds) Complexation of trace metals in natural waters. Nijhoff/Junk Pub, The Hague Boston Lancaster, pp 411-418

Frimmel FH (1990) Complexation of paramagnetic metal ions by aquatic fulvic acids. This volume

Ghosh K, Schnitzer M (1980) Macromolecular structures of humic substances. Soil Sci 129:266-276

Giesy JP, Briese LA (1980) Metal binding capacity of northern European surface waters for Cd, Cu and Pb. Org Geochem 2:57-67

Giesy JP, Leversee GJ, Williams DR (1977) Effects of naturally occuring aquatic fractions on cadmium toxicity to *Simocephalus serrulatus* (Daphnidae) and *Gambusia affinis* (Poeciliidae). Water Res 11:1013-1020

Hart BT (1981) Trace metal complexing capacity of natural waters: A review. Environ Tech Lett 2:95-110

Hart BT, Davies SH (1977) A new dialysis-ion exchange technique for determining the forms of trace metals in water. Aust J Mar Freshwater Res 28:105-112

Hedges JI (1988) Polymerization of humic substances in natural environments. In: Frimmel FH, Christman RF (eds) Humic substances and their role in the environment. Wiley, Chichester New York Brisbane Totonto Singapore, pp 45-58

Hoffman MR, Yost EC, Eisenreich SJ, Maier WJ (1981) Characterization of soluble and colloidal-phase metal complexes in river water by ultrafiltration. A mass-balance approach. Envir Sci Tech 15:655-661

John J, Salbu B, Gjessing ET, Bjørnstad HE (1988) Effect of pH, humus concentration and molecular weight on conditional stability constants of cadmium. Wat Res 22:1381-1388

Kramer CJM, Duinker JC (eds) (1984) Complexation of trace metals in natural waters. Nijhoff/Junk Pub, The Hague Boston Lancaster

Laegreid M, Alstad J, Klaveness D, Seip HM (1983) Seasonal variations of cadmium toxicity towards the alga *Selenastrum capricornutum* Printz in two lakes with different humus content. Environ Sci Technol 17:357-361

Langford CH, Gamble DS, Underdown AW, Lee S (1983) Interaction of metal ions with a well characterized fulvic acid. In: Christman RF, Gjessing ET (eds) Aquatic and terrestrial humic materials. Ann Arbor Science, Ann Arbor, pp 219-237

Laxen DP, Harrison RM (1981) A scheme for the physico-chemical speciation of trace metals in freshwater samples. Sci Total Environ 19:59-82

LaZerte BD, Chun C, Evans D (1988) Measurement of aqueous aluminum species: Comparison of dialysis and ion-exchange techniques. Environ Sci Technol 22:1106-1108

Leppard GG (ed) (1983) Trace element speciation in surface waters and its ecological implications. Plenum Press, New York London

Lund W, Helbak IA, Seip HM (1990) Studies of the complexation properties of aquatic humic material by differential pulse polarography. Sci Tot Environ 92:269-281

Malcolm RL, MacCarthy P (1986) Limitations in the use of commercial humic acids in water and soil research. Environ Sci Technol 20:904-911

Mantoura RFC, Dickson A, Riley JP (1978) The complexation of metals with humic materials in natural waters. Est Coast Mar Sci 6:387-408

Mantoura RFC, Riley JP (1975) The use of gel filtration in the study of metal binding by humic acids and related compounds. Anal Chim Acta 78:193-200

Neubecker TA, Allen HE (1983) The measurement of complexation capacity and conditional stability constants for ligands in natural waters. Water Res 17:1-14

Perdue EM (1988) Measurements of binding site concentrations in humic substances. In: Kramer JR, Allen HE (eds) Metal speciation. Theory, analysis and applications. Lewis Pub, Chelsea (Michigan), pp 135-154

Perdue EM, Lytle CR (1983) Distribution model for binding of protons and metal ions by humic substances. Environ Sci Technol 17:654-660

Rashid MA (1985) Geochemistry of marine humic substances. Springer, New York, chap 4 and 7

Riise G, Salbu B (1989) Major and trace elements in standard and reference samples of aquatic humic substances determined by instrumental neutron activation analysis (INAA). Sci Tot Environ 81/82:137-142

Saar RA, Weber JH (1982) Fulvic acid: modifier of metal-ion chemistry. Environ Sci Technol 16:510A-517A

Scatchard G, Coleman JS, Shen AL (1957) Physical chemistry of protein solutions. VII: The binding of some small anions to serum albumin. J Am Chem Soc 79:12-20

Scheinberg HI (1982) Scatchard plots. Science 215:312-313

Schnitzer M (1978) Humic substances: Chemistry and reactions. In: Schnitzer M, Khan SU (eds) Soil organic matter. Elsevier, Amsterdam Oxford New York, pp 1-64

Sedlacek J, Gjessing ET, Kallqvist T (1989) Influence of different aquatic humus fractions on uptake of cadmium to alga *Selenastrum capricornutum* Printz. Sci Total Environ, in print

Seip HM, Muller L, Naas A (1984) Aluminium speciation: Comparison of two spectrophotometric analytical methods and observed concentrations in some acidic aquatic systems in southern Norway. Water Air Soil Poll 23:81-95

Srna RF, Garrett KS, Miller SM, Thum AB (1980) Copper complexation capacity of marine water samples from southern California. Environ Sci Technol 14:1482-1486

Stevenson FJ (1985) Geochemistry of soil humic substances. In: Aiken GR, McKnight DM, Wershaw RL, MacCarthy P (eds) Humic substances in soil, sediment and water. Wiley, New Your Chichester Brisbane Toronto Singapore, pp 13-52

Sullivan TJ, Seip HM, Muniz IP (1986) A comparison of frequently used methods for the determination of aqueous aluminum. Intern J Environ Anal Chem 26:61-75

Sunda WG, Klaveness D, Palumbo V (1984) Bioassays of cupric ion activity and copper complexation. In: Kramer CJM, Duinker JC (eds) Complexation of trace metals in natural waters. Nijhoff/Junk Pub, The Hague Boston Lancaster, pp 393-409

Sunda WG, Lewis JAM (1978) Effect of complexation by natural organic ligands on the toxicity of copper to a unicellular alga, *Monochrysis lutheri*. Limnol Oceanogr 23:870-876

Thurman EM, Malcolm RL (1983) Structural study of humic substances: New approaches and methods. In: Christman RF, Gjessing ET (eds) Aquatic and terrestrial humic materials. Ann Arbor Science, Ann Arbor, pp 1-23

Turner DR, Varney MS, Whitfield M, Mantoura RFC, Riley JP (1987) Electrochemical studies of copper and lead complexation by fulvic acid. II A critical comparison of potentiometric and polarographic measurements. Sci Total Environ 60:17-34

Tuschall JR, Brezonik PL (1983) Complexation of heavy metals by aquatic humus: A comparative study of five analytical methods. In: Christman RF, Gjessing ET (eds) Aquatic and terrestrial humic materials. Ann Arbor Science, Ann Arbor, pp 275-294

Underdown AW, Langford CH, Gamble DS (1981) Light scattering of a polydisperse fulvic acid. Anal Chem 53:2139-2140

Varney MS, Mantoura RFC, Whitfield M, Turner DM, Riley JP (1983) Potentiometric and conformational studies of the acid-base properties of fulvic acid from natural waters. In: Wong CS, Boyle E, Bruland KW, Burton JD, Goldberg ED (eds) Trace metals in sea water. Plenum Press, New York, pp 751-772

Weber JH (1983) Metal ion speciation studies in the presence of humic materials. In: Christman RF, Gjessing ET (eds) Aquatic and terrestrial humic materials. Ann Arbor Science, Ann Arbor, pp 315-331

Weber JH (1988) Binding and transport of metals by humic materials. In: Frimmel FH, Christman RF (eds) Humic substances and their role in the environment. Wiley, Chichester New York Brisbane Totonto Singapore, pp 165-178

Wershaw RL, Thorn KA, Pinckney DJ, MacCarthy P, Rice JA, Hemond HF (1986) Application of a membrane model to the secondary structure of humic materials in peat. In: Fuchsman CH (ed) Peat and water. Elsevier, Barking New York, p 133

Winner RW (1984) The toxicity and bioaccumulation of cadmium and copper as affected by humic acid. Aquatic Toxicol 5:267-274

Young JS, Gurtisen JM, Apts CW, Crecelius EA (1979) The relationship between the copper complexing capacity of sea water and copper toxicity in shrimp zoeae. Mar Environ Res 2:265-273

COMPLEXATION OF PARAMAGNETIC METAL IONS BY AQUATIC FULVIC ACIDS

F. H. Frimmel
Engler-Bunte-Institut der
Universität Karlsruhe
Richard-Willstätter-Allee 5
D-7500 Karlsruhe 1
Federal Republic of Germany

Introduction

Fulvic acids (FA) are the acid and base soluble fraction of the biogenic and refractory organic acids commonly called humic substances (HS) (Stevenson, 1982). Most of the dissolved organic carbon (DOC) in aquatic systems consists of FAs. Due to their macromolecular polydispersity it is impossible to draw a chemical structure in the classical sense. However, there are many results which show that specific functionalities can be quantified. This is most important to understand the diversity of involvement of HS in environmental processes (Frimmel and Christman, 1988). The interaction with metal ions is of importance for the metal cycles in the aquatic systems and has to be considered in most transport processes. In addition the bioavailability of metal ions is strongly influenced by humic matter. The binding and transport of metals by humic materials was reviewed recently (Weber, 1988). Even though many methods have been used for speciation of metals bound to HS there is a severe lack of comparable results. The main reasons for that are the often poorly characterized humic substances and the operationally defined values from the procedures applied. Separation methods and nonseparation techniques often include uncontrolled adsorption effects and changed surface tension. The kinetics of the complexation reaction can be superimposed by kinetic effects caused by the analytical procedure. Therefore many attempts have been made to measure samples from the same origin with independent methods to compare the results and discuss them critically (Donat et al., 1986; Staub et al., 1984; Sterrit and Lester, 1984: Cominoli et al., 1980). Several mathematical models have been elaborated to describe the metal complexes of HS (Fitch and Stevenson, 1984; Perdue, 1985; Sposito, 1986). Even though some models fit the experimental data well, they only reflect average properties and give no information on the metal-HS bonds. Also the predictive capability

NATO ASI Series, Vol. G 23
Metal Speciation in the Environment
Edited by J. A. C. Broekaert, Ş. Güçer, and F. Adams

for effects of ionic strength and temperature are fairly poor. Therefore additional experimental data are needed. Little is known about the nature of the bonding of the metal ion to HS. Electron paramagnetic resonance (EPR) spectroscopic results show that there are at least two different types of chemical bonds in which iron is fixed in HS (Riffaldi and Schnitzer, 1972; Senesi et al., 1977) and that the copper-FA molar ratio has an important influence on the shape of the spectra (Senesi et al., 1988). There is evidence that the bonding of vanadyl ions (VO^{2+}) to FA involves oxygen as donor atoms (Templeton and Chasteen, 1980).

The objective of this work was:

a) to characterize some isolated aquatic FAs
b) to quantify the interaction of the FAs with paramagnetic metal ions by means of polarography and fluorescence
c) to gain some insight into the bond formation by laser flash experiments.

Experimental

Fulvic acids preparation. FAs were isolated from two bog lakes (Brunnensee, BM, Bansee, BAN) in South Bavaria and from aquatic extracts of a soil sample taken at the Institute for Waterchemistry in München-Großhadern (MUC). Details of the sample origin are given in Table 1.

Table 1 Origin of the isolated aquatic FAs

Code	system	date of sampling	ref.
BM4	bog lake	IV 1983	Frimmel et al. (1987)
BM6	bog lake	II 1983	Frimmel et al. (1987)
BM8	bog lake	IV 1984	Frimmel et al. (1987)
BM9	bog lake	II 1986	Frimmel et al. (1987)
BM10	bog lake	IX 1986	Frimmel et al. (1987)
BAN 13	bog lake	I 1985	Frimmel et al. (1987)
MUC1	NaOH extract	X 1987	Frimmel (in press)
MUC2	water extract	X 1987	Frimmel (in press)
MUC3	pyrophosphate extract	X 1987	Frimmel (in press)

From the lake water and the aqueous soil extracts the HS were isolated according to the XAD-2 method (Mantoura and Riley, 1975; Frimmel and Niedermann, 1980). The FA was gained by acidifying (pH = 2.0; HCl) the dissolved HS and thus precipitating the humic acid (HA) fraction. The protocol for the final cleaning step on a small XAD-2 column was done as in Frimmel et al. (1987).

Elemental analysis of the sample was done by LABOR PASCHER, FRG and checked by our own determinations with an elemental analyzer CHN-O Rapid (HAEREUS)

Dissolved organic carbon (DOC) was measured with a UV-DOC analyzer, TOCOR 3 (MAIHAK) together with a MAI 3 integrator.

UV- and VIS-spectra were run with a UV/VIS-spectrometer PYE UNICAM SP 8-100 (PHILIPS) at pH = 11.0 using 1 cm and 5 cm quarz cells.

pH-titrations were done with a digital pH meter pH 535 multical (WTW) in connection with a combined pH glass electrode type E50 (3 M KCl/AgCl; WTW) at 293 K under N_2 atmosphere using an autotitrator DL 25 (METTLER) combined with a pH glass electrode (WTW) (titer: 0.02 M NaOH) (Frimmel et al., 1985)

Polarographic determinations were done with a polarographic analyzer 174 A in connection with the automated electroanalysis controller 315 A, a mercury dropping electrode system SMDE 303 and a X-J-recorder RE 0074 (all from PRINCETON APPLIED RESEARCH) (Frimmel and Geywitz, 1983).

Fluorescence spectra were recorded with a fluorescence spectrometer LS-5 (PERKIN-ELMER) with data system (Frimmel and Hopp, 1986).

Laser flash induced spectra were made with a frequency doubled Nd-YAG laser amplifier and digital storage oscilloscope as described in (Frimmel et al., 1987).

All chemicals were of analytical grade (p. a., MERCK). To exclude oxygen from the reactions all solutions were Ar saturated and the experiments were performed under Ar or N_2 atmosphere.

Results and Discussion

General characterization of isolated FAs

The elemental analysis of the dry samples of the isolated FAs are given in Table 2.

Table 2 Elemental composition (% of dry weight) of the FA samples

Sample	C	H	O	N	S	Ash
BM 4	48.4	4.41	44.7	1.27	0.25	1
BM 6	47.5	4.26	44.7	1.5	~0.1	1.93
BM 8	48.5	4.33	43.6	1.72	<0.2	<1.5
BM 9	48.5	4.18	44.8	1.78	<0.2	<0.48
BM 10	53.4	3.48	38.8	2.05	1.04	1.27
MUC 1	42.0	5.45	43.9	4.35	0.88	3.38
MUC 2	47.9	3.50	38.5	2.85	0.95	6.33
MUC 3	49.4	3.79	40.5	3.91	0.88	1.51

The results are in general agreement with the data from other work (Huffman et al., 1985; Steelink, 1985). The relatively high oxygen content in FAs reflects the donor groups for metal complexation. The relatively low ash content especially in the samples from the bog lake is typical for a successful isolation procedure, and is a necessity for applying fluorescence methods.

The optical properties of the samples are given in Table 3. All data are matched for a DOC of 1 mg/L.

Table 3 Specific spectral absorbances at 254 nm (A 254) and at 436 nm (A 436) of isolated FAs

Sample	pH	$\frac{A(254\ nm)^{1)}}{DOC}$	$\frac{A(436\ nm)^{1)}}{DOC}$	$\frac{A(254\ nm)}{A(436\ nm)}$
BM 4	11.0	5.7	0.5	11.4
BM 6	11.0	5.8	0.5	11.6
BM 8	11.0	5.1	0.5	10.2
BM 9	11.0	6.4	0.6	10.7
BM 10	11.0	6.1	0.6	10.2
MUC 1	11.0	2.3	0.2	11.5
MUC 2	11.0	5.3	0.4	13.2
MUC 3	11.0	5.3	0.6	8.6

1) $\frac{L}{mg \cdot m}$

There is an obvious uniformity of the values for the bog lake samples. The samples from the soil are quite different from the bog lake samples and vary according to the procedure of extraction. There is less optical density in the UV and visible range, especially in the water extracted sample. The acid-base properties of the FAs are reflected in their proton capacity (H^+-CAP). The ranges below and above pH 7.0 and the ratio of the values give useful information on the availability of acidic functional groups. Table 4 shows the H^+-CAP for the isolated FAs. The bog lake samples are fairly similar whereas the soil derived substances vary according to the extraction procedure. It is interesting to note that the alkaline extraction obviously includes a decrease of the acidic groups which can be titrated below pH 7.0.

Quantification of metal complexation

Polarography belongs to the common methods for the determination of complexation capacities (CC) of dissolved organic matter. Copper(II) forms most stable complexes with HS and therefore it can be used to estimate the total CC for divalent metals. The use of isolated FAs allows the application of standardized reaction conditions. They were chosen to be close to natural conditions and lead to comparable results for samples from different origin. At a pH of 6.8 and an adjusted ionic strength of 10^{-2} molar acetate the initial Cu(II) concentration of 10 µmol/L is titrated with a neutral FA solution of known DOC concentration. The difference between the initial Cu(II) signal and the signal at 1 mg/L DOC reflects the formation of polarographically stable complexes and is called CC. Table 4 shows the CC for the FA under investigation.

Table 4 Cu(II) complexation capacities (CC, µmol/mg DOC) of isolated fulvic acids determined by polarography (DPP) and fluorescence quenching (FL.QU)

Sample	DPP	FL.QU
BM 4	1.85	2.3
BM 6	2.0	2.6
BM 8	2.5	n. d.
BM 9	2.3	n. d.
BM 10	2.2	2.8
MUC 1	1.6	0.9
MUC 2	2.8	1.6
MUC 3	3.3	2.6

n. d. not determined

It is obvious that the polarographic data result in smaller CCs for the surface water samples and higher values for the soil samples compared with the data derived from fluorescence quenching. For the interpretation of the operationally defined values it is essential to understand the basic reactions of the methods. Therefore the question whether there is a general difference between the fluorescence of the FAs from surface water and from soil is under further investigation.

Paramagnetic metal ions can effectively quench the fluorescence of FAs (original intensity I_0). As shown in Figure 1 fluorescence quenching by Cu(II) is most effective (Hopp, 1985). Quench experiments with Fe(III) are severely disturbed by precipitation of amorphous $Fe(OH)_3$.

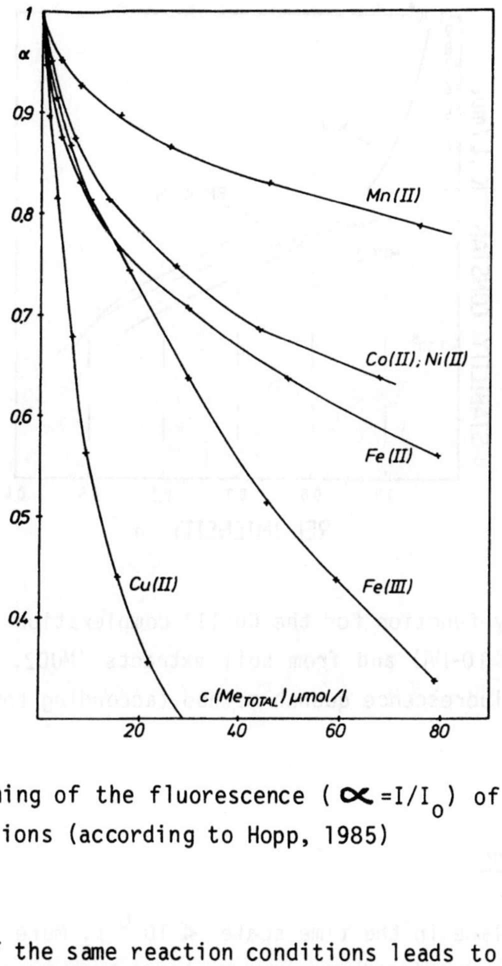

Fig. 1 Quenching of the fluorescence ($\alpha = I/I_0$) of FAs by paramagnetic metal ions (according to Hopp, 1985)

The application of the same reaction conditions leads to values which can be compared with those determined by polarography (Table 4). In all cases the fluorescence quenching produces CCs somewhat different from those of the polarographic procedure, which underlines the operational definition of the values.

In addition, according to the theoretical basis, it is possible to develop stability functions from the plots of conditional stability constants (K') and the metal load on the FAs (Frimmel and Hopp, 1986). Shape and position of the curve are typical for the different samples and reflect the decreasing stability of the metal complexes as the ratio of metal to ligand increases. K'-values cover more than a decade for most samples. A typical example for the FAs from a bog lake and from soil (Table 1) is given in Figure 2.

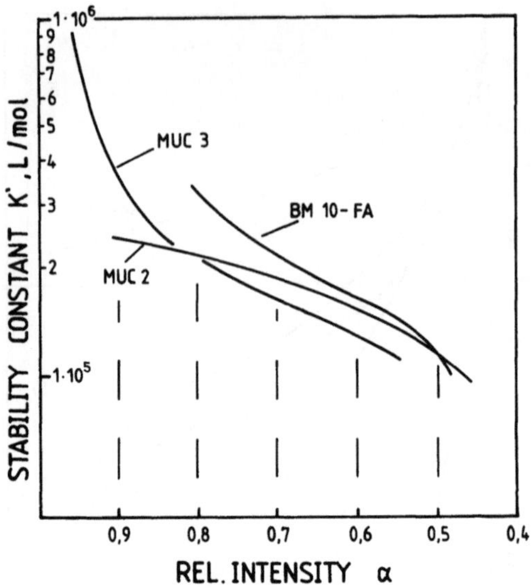

Fig. 2 Stability function for the Cu(II) complexation of FAs from a bog lake (BM 10-FA) and from soil extracts (MUC2, MUC3), determined by the fluorescence quench method (according to Hopp, 1985)

Kinetic considerations

Fluorescence takes place in the time scale $< 10^{-6}$ s. More long-lived excited states are refered to as phosphorescence and can be measured by laser flash experiments. On the basis of first-order kinetics two main transients with different lifetimes (order of 10^{-5} and 10^{-4} s) could be deduced (Frimmel et al., 1987). A typical decay of the transient absorption of BAN 13 is given in Figure 3. The sample was isolated from a boglake (Bansee) near by the Brunnensee according to the same protocol as in Frimmel et al., 1983.

Paramagnetic metal ions can quench the short-lived transients of FA. The intensity of the emission decreased from 20×10^{-5} to 11×20^{-5} in case of addition of Ni(II) (10^{-4} mol/g DOC). The decrease of the emission at different metal concentrations and pH-values is given in Figure 4.

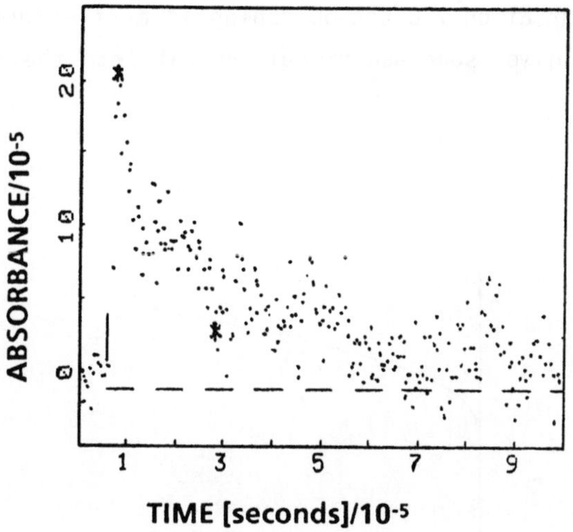

Fig. 3 Decay of the transient adsorption of BAN 13 (FA) (Ar atmosphere); DOC 125 mg/L; filter 590 nm; λ_{exc} 532 nm; λ_{em} 900 nm; I_o 385 mV

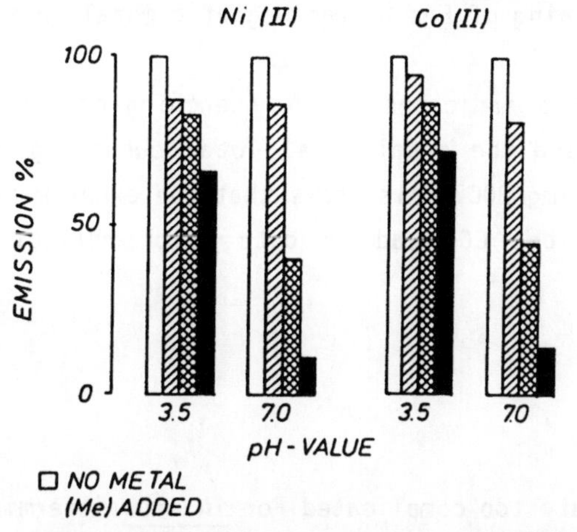

Fig. 4 Quenching effect of Ni(II) and Co(II) on the short lived transients of a FA sample (according to Frimmel et al., 1987)

It is obvious that at pH 7.0 the quenching is most effective. A Stern-Volmer plot of the data gives some additional insight into the bond formation (Fig. 5).

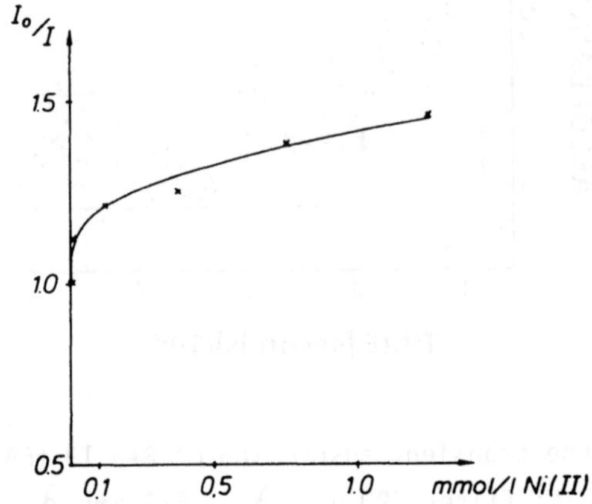

Fig. 5 Typical shape of the Stern-Volmer function for the emission quenching of FAs by paramagnetic metal ions.

The nonlinear curve indicates static quenching caused by coordinative bonds between the FAs and the metal ions. Total quenching occurs for Co(II) and Ni(II) at 10 µmol/mg DOC. This shows that the experimental conditions for the determination of the CC lead to data which reflect the maximum of complexation.

Conclusions

Aquatic systems are too complicated for in situ determination of complexation reactions between metal ions and FAs. The investigation of isolated humic samples under standardized conditions lead to comparable results.

Due to the lack of authentic material with a known chemical structure only operationally defined values can be obtained. The application of independent methods for characterization is therefore advisable.

Complexation capacities for metal ions determined by polarography and fluorescence quenching are in the range of a few µmol/mg DOC. The values for the complexation capacities, determined polarographically, for copper(II) with surface water FAs are lower than the values determined by fluorescence quenching; for the soil FAs the order is reversed. Fluorescence quenching can be done for most of the divalent paramagnetic metal ions with aquatic relevance. The method can be used to determine experimental equilibria functions which show the influence of the metal to ligand ratio on the stability of complexes formed. Laser flash experiments lead to the conclusion that the formation of coordinative bonds between humic ligands and paramagnetic metal ions quench the relatively short-lived excited states ($<10^{-4}$ s) (Frimmel et al., 1987).

Acknowledgements

I thank H. Bauer and W. Hopp for their experimental help and A. M. Braun for fruitful discussions. This work was financially supported by Deutsche Forschungsgemeinschaft, Bonn-Bad Godesberg (Grants Fr 536-5 and Fr 536-8).

References

Cominoli A, Buffle J, Haerdi W (1980) Voltammetric studies of humic and fulvic substances. Part III: comparison of the capabilities of the various polarogrpahic techniques for the analysis of humic and fulvic substances. J. Electroanal. Chem. 110:259-275

Donat JR, Statham PJ, Bruland KW (1986) An evaluation of a C-18 solid phase extraction technique for isolating metal-organic complexes from Central North Pacific waters. Marine Chem. 18:85-99

Fitch A, Stevenson FJ (1984) Comparison of models for determining stability constants of metal complexes. Soil Sci. Soc. Am. J. 48:1044-1050

Frimmel FH, Niedermann H (1980) Komplexierung von Metallionen durch Gewässerhuminstoffe. I: Ein Braunwassersee als Huminstofflieferant. Z. Wasser Abwasser Forsch. 13:119-124

Frimmel FH, Immerz A, Niedermann H (1983) Heavy metal interaction with aquatic humus. J. Environ. Anal. Chem. 14:105-115

Frimmel FH, Geywitz J (1983) Zur koordinativen Bindung von Metallionen an Gewässerhuminstoffe. Fresenius Z. Anal. Chem. 316:582-588,

Frimmel FH, Hopp W, Quentin KE (1985) Titration isolierter aquatischer Huminstoffe und ihrer Calcium-Komplexe mit starken Basen und Säuren. Z. Wasser Abwasser Forsch. 18:259-262

Frimmel FH, Hopp W (1986) Stability spectra for the description of copper-humic complexes - A fluorescence quench study. Fresenius Z. Anal. Chem. 325:68-72

Frimmel FH, Bauer H, Putzien J, Murasecco P, Braun AM (1987) Laser flash photolysis of dissolved aquatic humic material and the sensitized production of singlet oxygen. Environ. Sci. Technol. 21: 541-545

Frimmel FH, Christman RF (eds) (1988) Humic Substances and Their Role in the Environment. John Wiley & Sons, Chichester

Frimmel FH (in press) Polar organic substances and their role in the water saturated and unsaturated zone. In Matthess G et al. (eds) Progress in Hydrogeochemistry - Organics, Carbonates, Silicates, Microbiology, Models. Springer Verlag Heidelberg

Hopp W (1985) Fluoreszenzspektroskopische und potentiometrische Untersuchungen aquatischer Huminstoffe und ihrer Komplexierung mit Metallionen. Dissertation, Technische Universität München

Huffman Jr EWD, Stuber HA (1985) Analytical methodology for elemental analysis of humic substances. In: Aiken GR et al. (eds) Humic Substances in Soil, Sediment and Water. John Wiley & Sons, New York

Mantoura RFC, Riley JP (1975) The use of gel filtration in the study of metal binding by humic acids and related compounds. Anal. Chim. Acta 78:193-200

Perdue EM (1985) Acidic functional groups in humic substances. In: Aiken, G R et al. (eds): Humic Substances in Soil, Sediment and Water. John Wiley & Sons, New York

Riffaldi R, Schnitzer M (1972) Electron spins resonance spectrometry of humic substances. Soil Sci. Soc. Am. Proc. 36:301-305

Senesi N, Chen Y, Schnitzer M (1977) Hyperfine splitting in electron spin resonance spectra of fulvic acid. Soil, Biol. Biochem. 9:371-372

Senesi N, Bocian DF, Sposito G (1988) Electron spin resonance investigation of copper(II) complexation by soil fulvic acid. Soil Sci. Soc. Am. J. 49:114-119

Sposito G (1986) Sorption of trace metals by humic materials in soils and natural waters. In: CRC Critical Reviews in Environmental Control. vol. 16:193-229, CRC Press, Cleveland

Staub C, Buffle J, Haerdi W (1984) Measurements of complexation properties of metal ions in natural conditions by ultrafiltration: influence of various factors on the retention of metals and ligands by neutral and negatively charged membranes. Analyt. Chem. 56:2843-2849

Steelink C (1985) Implications of elemental characteristics of humic substances. In: Aiken GR et al. (eds) Humic substances in Soil, Sediment and Water. John Wiley & Sons, New York

Sterrit RM, Lester JN (1984) Comparison of methods for the determination of conditional stability constants of heavy metal-fulvic acid complexes. Water Research 18:1149-1153

Stevenson FJ (1982) Humus Chemistry; Genesis, Composition, Reactions. Wiley Interscience, New York

Templeton GD III, Chasteen ND (1980) Vanadium-fulvic acid chemistry: Conformational and binding studies by electron spin probe techniques. Geochim. Cosmochim. Acta 44:741-752

Weber JH (1988) Binding and transport of metals by humic materials. In: Frimmel FH, Christman RF (eds): Humic Substances and Their Role in the Environment. John Wiley & Sons, Chichester

STUDY OF LEAD FULVIC ACID INTERACTIONS BY VOLTAMMETRIC METHODS

K. Wallmann, W. Petersen and Pinglin Li[1]
Department of Chemistry
GKSS Research Center
P.O. Box 1160
D-2054 Geesthacht, FRG

INTRODUCTION

Voltammetric methods are frequently used to study complexation reactions in aqueous solutions. Peak currents i_p and peak potentials U_p recorded as functions of metal and ligand concentrations may be used to calculate complex stoichiometries, stability constants and complexation capacities of natural waters. In order to enable a correct interpretation of the measured values it is necessary to investigate the dissociation of complex species at the electrode surface, their diffusion coefficients and adsorption reactions. We studied the interactions between fulvic acid (FA) and Pb^{2+}-ions in aqueous solutions and at the mercury electrode surface using cyclic voltammetry (CV). Solutions containing FA at natural concentrations and the chelating ligands NTA and EDTA were titrated with Pb. Titration curves were recorded by differential pulse anodic stripping voltammetry (DPASV) and differential pulse cathodic stripping voltammetry (DPCSV). The problems arising in the interpretation of conventional DPASV-titration curves obtained in natural fresh waters are discussed. A new procedure based on DPCSV-measurements is presented and recommended for the detection of those ligands forming inert complexes.

EXPERIMENTAL METHODS

FA was extracted by 0.1 M NaOH from a sediment of the river Elbe. The alkaline extract was acidified to pH 2.3 by addition of concentrated HCl and filtered using a 0.2 μm membrane filter. The filtrate was passed through a column packed with adsorption resin (Serva XAD 8). The FA accumulated on the resin was eluted by 0.1 N NaOH and purified by dia-ultrafiltration over a membrane with a molecular weight cutoff limit of 1000 D (Amicon YM 2). The resulting FA concentrate was evaporated to dryness and stored as a solid.

CV was performed on a multi mode electrode (polarographic cell Metrohm VA663) used as a hanging mercury drop electrode (HMDE) with a Metrohm E506 polarograph and Metrohm E612 scanner. Data were recorded using a personal computer (Compaq 386s) and evaluated with a specially designed computer program [*Petersen* 1989]. Measurements were made in a solution containing 500 mg FA/l, 24 μmol Pb/l and 0.02 mol KNO_3/l as supporting electrolyte. A high FA concentration was used to ensure sufficient ligand excess. The time passing between drop formation and the start of the sweep (t_w) was varied to investigate adsorption processes on the electrode surface. The curves were recorded from -0.3 V to -0.8 V and vice versa at sweep rates of between 0.5 V/s and 10 V/s. A FA-free solution containing 24 μmol Pb/l and 0.02 mol KNO_3/l was used as a control.

[1] on leave from Fujian Institute for Science of Environmental Protection, Fuzhou, Fujian, P.R.China

NATO ASI Series, Vol. G 23
Metal Speciation in the Environment
Edited by J. A. C. Broekaert, Ş. Güçer, and F. Adams

DPASV and DPCSV were performed on a HMDE using the polarographic cell Metrohm VA647 and a microprocessor controlled polarograph (Metrohm VA646). An accumulation time and delay time of 60 sec and 20 sec respectively were chosen for both methods. Sweep rates were 10 mV/sec and pulse amplitudes + 50 mV for DPASV and -50 mV for DPCSV. The anodic sweep was recorded after accumulation at - 0.8 V and the cathodic sweep after accumulation at -0.2 V. A supporting electrolyte containing 2 mmol $CaCl_2$/l and 5 mmol NaCl/l was used in the experiments to simulate natural fresh waters. Solutions containing 3 mg FA/l and the chelating ligands NTA or EDTA at concentrations of 0.5 μmol/l were titrated with a Pb^{2+} standard solution. Polarograms were recorded 5 min after each metal addition. The FA concentration was close to the natural concentrations of many fresh waters [*Buffle* 1988, *Thurman* 1985].

Whereas CV measurements were carried out in a teflon vessel at room temperature, DPASV and DPCSV measurements were performed at 25 ±0.2°C in thermostated solutions using a double-walled vessel made of quartz glass. The pH was held at 7.00 ±0.05 in all experiments by the addition of non complexing buffer HEPES (2-[4-(2-hydroxyethyl)-1-piperazino]-ethansulfonic acid (Aldrich)). With the exception of FA all the chemicals used were of analytical grade.

RESULTS AND DISCUSSION

Curves recorded by CV at different adsorption times t_w are shown in fig.1. i_p increased with t_w, and U_p was shifted to negative potentials for the cathodic sweep. These effects may be explained by the adsorption and reduction of Pb-FA complexes at the electrode surface: due to adsorptive complex and ligand accumulation during t_w the surface concentration of Pb and the complexation degree α increased over time resulting in higher i_p- and more negative U_p-values. α is defined as the ratio of total metal concentration to free metal-ion concentration [*Buffle* 1988]. U_p is shifted to negative potentials for labile complexes when α increases [*Crow* 1969]. Oxidation peak currents increased also, but were significantly smaller than i_p-values of reduction for all t_w. An adsorption layer of FA on the electrode surface may hinder the diffusion of re-oxidized Pb into the solution and may thus suppress oxidation i_p.

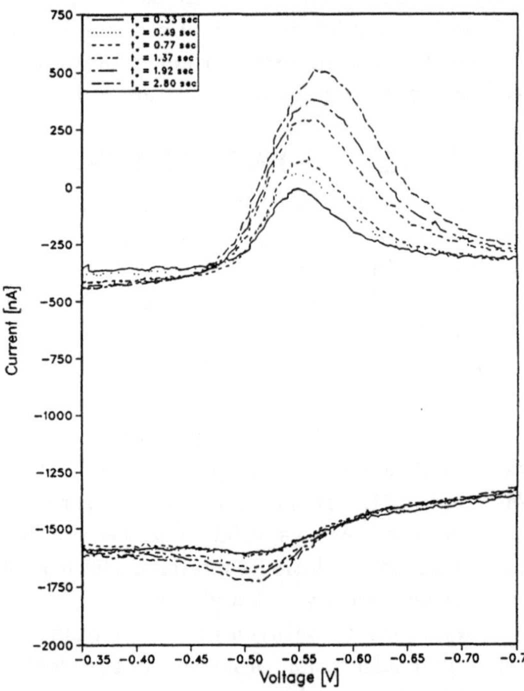

Fig. 1. Current potential curves measured by CV at different adsorption times t_w (sweep rate 5 V/s)

In fig.2 the difference between oxidation and reduction U_p is plotted as function of t_w. For a reversible electrode reaction of a bivalent metal ion bound in a labile complex a potential difference of 29 mV is predicted [*Bond* 1980]. When $t_w = 0$ sec the values approach the theoretical value. We may therefore conclude that Pb-FA complexes are predominantly labile and an irreversible electrode reaction is induced by FA and complex adsorption. Reduction i_p values measured in solutions containing FA were extrapolated to t_w=0. The ratio of the extrapolated value to the corresponding value measured in FA-free solution was 0.5. This i_p- suppression is induced by the complexation reaction: Pb-FA complexes are larger and diffuse slower than Pb^{2+}- aquo ions. Our results confirm those reported by *Buffle et al.* [1976] for a similar investigation.

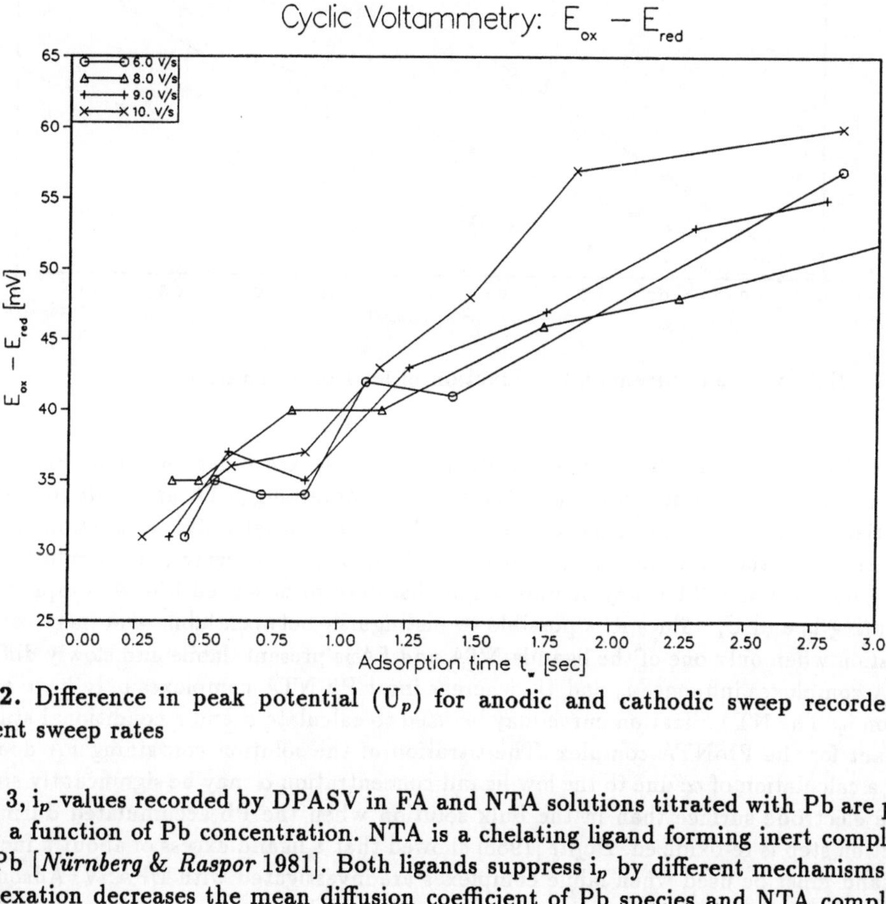

Fig. 2. Difference in peak potential (U_p) for anodic and cathodic sweep recorded at different sweep rates

In fig. 3, i_p-values recorded by DPASV in FA and NTA solutions titrated with Pb are plotted as a function of Pb concentration. NTA is a chelating ligand forming inert complexes with Pb [*Nürnberg* & *Raspor* 1981]. Both ligands suppress i_p by different mechanisms: FA complexation decreases the mean diffusion coefficient of Pb species and NTA complexation introduces Pb species that are not reduced at the electrode under the experimental conditions.

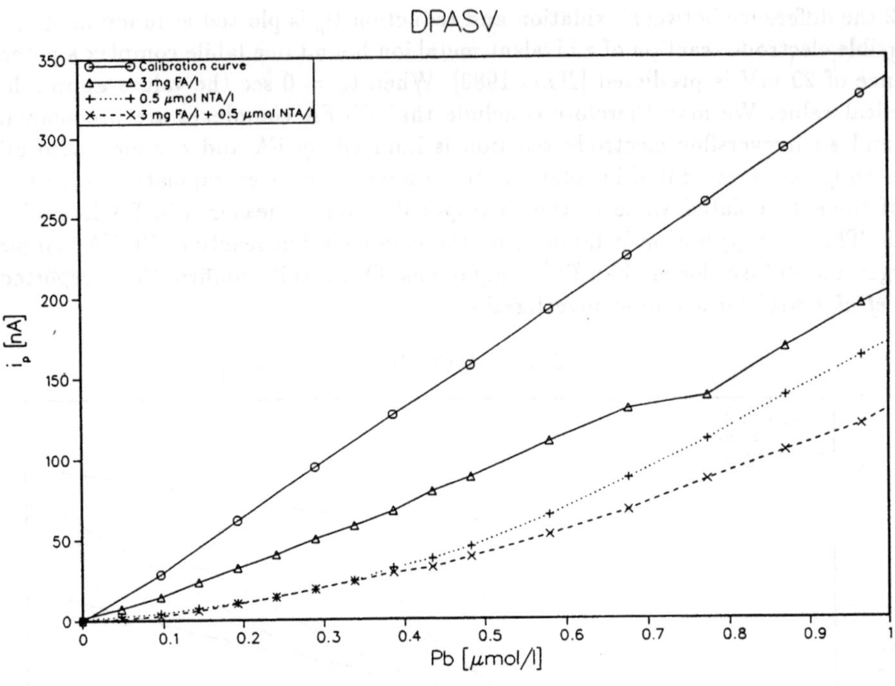

Fig. 3. DPASV: peak current (i_p) as function of lead concentration

In Fig. 4 accompanying U_p-values are depicted. These are independent of Pb concentrations for the calibration curve and the solution containing NTA and shift to negative potentials when labile FA complexes are formed. The potential shift is a function of α and decreases when the Pb concentration increases. The U_p-shift observed in the solution containing both FA and NTA may be induced by dissolved or adsorbed Pb-FA complexes. By comparing i_p and U_p values it is possible to distinguish between labile and inert complex formation when only one of the ligands NTA and FA is present: labile and slowly diffusing Pb-FA complexes influence i_p and U_p whereas inert Pb-NTA complexes only have an impact on i_p. The NTA titration curve may be used to calculate α and a conditional stability constant for the Pb-NTA complex. The titration of the solution containing FA does not allow a calculation of α: due to the low ligand concentration α may be significantly smaller at the electrode surface than in the bulk solution when the Pb accumulated during the reduction step is re-oxidized. *Buffle* [1988] showed that a ligand excess of about a factor of thousand must be used when labile complexes are investigated with DPASV. Adsorption of FA and Pb-FA complexes has an impact on i_p and U_p also, so that interpretation of the titration curve is further complicated. The titration curves of the solution containing both FA and NTA give no information about complex formation. It is not possible to calculate the portions of Pb bound as labile Pb-FA or as inert Pb-NTA complex.

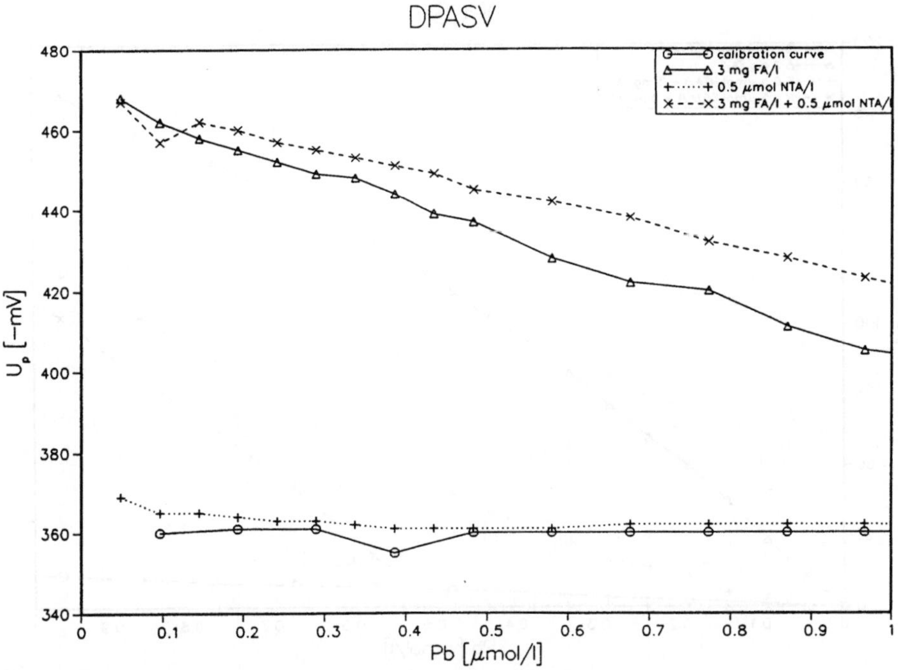

Fig. 4. DPASV: peak potential (U_p) as function of lead concentration

In figure 5 i_p-curves measured with DPCSV are depicted. In DPCSV FA and Pb-FA complexes are accumulated at the electrode surface before a cathodic sweep is applied. The adsorptive accumulation increases i_p, in the presence of Pb-FA complexes, to values higher than in the calibration curve. In solutions containing only NTA or EDTA, i_p is strongly suppressed due to inert complex formation (data not shown). The titration curve for the solution containing FA and NTA shows that FA is able to compete with NTA for Pb^{2+}- ions. In the presence of the strong chelating agent EDTA no FA-Pb complexes are formed.

The slope of all i_p-curves recorded by DPASV increased with increasing Pb-concentration. With DPCSV the shape of the titration curves depends on the solution composition. In solutions containing only FA the slope decreased with increasing Pb-concentrations whereas in the presence of synthetic chelating ligands forming inert complexes an increase in slope was recorded. Thus it is possible to detect ligands that form inert complexes in the presence of surface active ligands that form labile complexes by DPCSV and Pb-titration. In many fresh waters FA and ligands forming strong and inert complexes occur so that DPCSV may be a useful method to characterize the complexation reactions in natural waters.

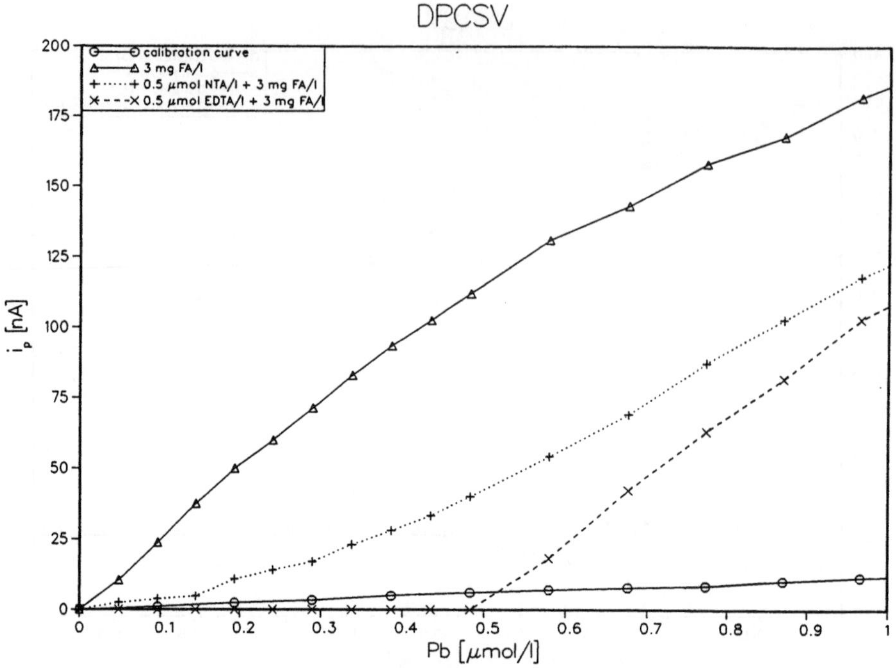

Fig. 5. DPCSV: peak current (i_p) as function of lead concentration

CONCLUSIONS

Pb-FA complexes are surface active and adsorb strongly at the mercury electrode used in voltammetric measurements. They are predominately labile in the time scale of CV-measurements. Adsorption increases reduction-i_p and hinders the re-oxidation of Pb-amalgam. Using DPCSV-procedures high i_p values are obtained in the presence of Pb-FA complexes. Pb-FA complexes are formed in the presence of NTA but not in the presence of EDTA. DPCSV- measurements may be used to detect ligands forming inert complexes in natural waters containing surface active substances.

REFERENCES

Bond AM (1980) Modern polarographic methods in analytical chemistry. Marcel Dekker, New York, Basel

Buffle J, Greter FL, Nembrini G, Paul J, Haerdi W (1976) Capabilities of voltammetric techniques for water quality control problems. Fres Z Anal Chem 282:339-350

Buffle J (1988) Complexation reactions in aquatic systems: an analytical approach. Ellis Howard Ltd., Chichester (GB)

Crow DR (1969) Polarography of metal complexes. Academic Press, London New York

Nürnberg HW, Raspor B (1981) Application of voltammetry in studies of the speciation of heavy metals by organic chelators in the sea. Environ Technol Lett 2:457-483

Petersen W (1989) Computerized electroanalysis: studies of trace metal complexation. Fres Z Anal Chem 334:718

Thurman EM (1985) Organic geochemistry of natural waters. Nijhoff Junk Publ, The Hague Boston Lancaster

ISOTACHOPHORETIC INVESTIGATIONS INTO THE SPECIATION OF NIOBIUM(V) AND PLUTONIUM(IV) IN THE PRESENCE OF CITRATE

S.Lesley Prosser and Robert A.Bulman
National Radiological Protection Board
Chilton
Didcot OX11 0RQ
ENGLAND

1. Introduction

An understanding of the speciation of radionuclides is required so that their movement through the environment can be modelled. Unfortunately, the speciation in the environment of many elements is poorly understood and this is particularly the case for minor elements such as niobium. Examination of the standard chemical texts - eg Cotton and Wilkinson, 1972 - demonstrates that the chemistry of niobium as water-soluble complexes is poorly described. Obviously, the nature of the speciation of such elements in complex environments such as foodstuffs is practically non-existent. An understanding of the speciation of niobium would be of value in modelling the radiation dose which might be received by a critical group eating foodstuffs contaminated with ^{95}Nb. This radionuclide, a daughter of ^{95}Zr which arises in fairly high yield during nuclear fission, has a half-life of 35.15 d which is sufficiently long for it to survive a typical crop growing period. As the gastrointestinal uptake factor for niobium might be as high as 60% for intrinsically incorporated niobium (Schroeder and Balassa, 1965), an investigation of its speciation is justified.

NATO ASI Series, Vol. G 23
Metal Speciation in the Environment
Edited by J. A. C. Broekaert, Ş. Güçer, and F. Adams

The continuing interest in the speciation of the elements at ever lower concentrations calls for new developments in analytical chemistry. In this account we evaluate capillary isotachophoresis, in both its analytical and preparative modes, for identification of the complexed forms of two polyvalent species, Nb(V) and Pu(IV), in citrate which is a relatively simple complexing medium which is widely used when a complexing environment is needed to suppress hydrolysis of polyvalent cations. Isotachophoresis is one of several electrophoretic techniques and it is based upon the migration of charged particles in an electric field. Its distinguishing feature is the presence of sharp boundaries between the separated zones of the individual components of the sample and the migration of zones, in close contact with the adjoining zones, at the same velocity - hence the term *isotachophoresis* (*iso* - same, *tacho* - velocity). The migration sequence of the ions is determined by their electrophoretic mobility.

The use of isotachophoresis as a separation method has the advantages of: a) short analysis time (maximum 40 minutes); b) opportunity to function at low levels ($<10^{-8}$ g is quite possible); c) availability of qualitative and quantitative information; d) variable pH range; e) collection of separated species.

2. Materials and Methods

2.1 Chemicals

Niobium pentachloride (99% purity, Johnson Matthey, Royston, Hertfordshire) caproic acid and β-alanine (Sigma Chemical Co. Ltd.) were of the highest purity available and were used without further purification. All other chemicals were used as supplied by BDH Ltd.

2.2 Radiochemical analysis

^{95}Nb was determined by gamma-ray spectrometry by counting the photons between the energy ranges 600-800 keV using a Canberra-Packard Minaxi Auto-Gamma Series counter.

^{14}C-citrate was determined by liquid scintillation counting using a commercial scintillant (Packard Permaflour^{R}-V) in a Beckman liquid scintillation counter. The samples were later ashed at 500 °C so that ^{239}Pu could be determined.

^{239}Pu was determined in ashed samples by a previously described procedure (Bulman et al, 1984). ^{236}Pu was used as the internal standard. Isotopic abundances were determined by α-spectrometry on silicon-barrier detectors (ORTEC^{R}). In some cases ^{239}Pu was also determined by liquid scintillation counting after the ash had been dissolved in 2M HNO_3 and dispersed in a commercial liquid scintillant (Beckman Readyvalue^{TM}).

2.3 Preparation of citrate complexes

^{95}Nb(V) citrate. ^{95}Nb(V) was isolated from an aqueous medium into which ^{95}Zr(IV) had been introduced 70 d previously. As the radionuclide ^{95}Zr is no longer commercially available, ^{95}Zr was obtained by extraction from natural uranium which had been irradiated at Harwell Laboratory in reactor beam at a flux of 10^{13} neutrons sec^{-1} cm^{-2} for 1000 min. ^{95}Zr was isolated from the mixed fission products on an anion exchange resin column (CG400 (Biorad) 6 x 0.6 cm) and further purified by co-precipitation with stable zirconium hydroxide. The purified extract was taken up in 5% oxalic acid and allowed to stand for 10 weeks to allow ^{95}Nb to grow in before it was extracted by the method of Jurriaanse and Moore (1966). The extract was taken to dryness and the charred residue destroyed by oxidation in a muffle furnace at 500 °C. The residue was dissolved in conc. HNO_3 and evaporated to dryness. Finally, $^{95}Nb(V)$ citrate was prepared by taking up the radionuclide in 0.01**M** HNO_3 (0.25 ml) to which was added an equal volume of 0.02**M** trisodium citrate. The resulting solution was ultrafiltered through a membrane of porosity 0.025 μm to minimize the presence of polymeric materials. This procedure has been used by others to produce a solution containing niobium-95 in a citrate-containing medium which was then used to determine the gastrointestinal uptake factor for $^{95}Nb(V)$ (Harrison et al, 1990). As both ^{14}C and ^{95}Nb emit β^- particles with a decay energy of 0.2 MeV it was not posible to use ^{14}C-citrate for rapid location of $^{95}Nb(V)$ citrate on the cellulose acetate strip.

<u>^{239}Pu(IV) citrate</u>. An established procedure was used to prepare this complex (Smith et al, 1976). Basically, ^{239}Pu (IV) was dried down from stock solution in 4**M** HNO_3 and the citrate complex prepared in the manner described above. A small quantity of ^{14}C-citrate was added to aid detection of citrate eluted on to the cellulose acetate strip during preparative isotachophoresis.

<u>Stable niobium(V) citrate</u>. The procedure of Grigoreva and Golubeva (1979) was used. Essentially niobium pentachloride (2 g, 7.4 mmoles) was dissolved in conc. HCl (6 ml) and trisodium citrate (6.53 g, 22.2 mmoles) added. The resulting viscous solution was diluted with a small volume of water and the solution adjusted to pH 3.2 by careful addition of NaOH solution.

<u>2.4 Isotachophoresis</u>

<u>Analytical isotachophoresis</u>. Isotachophoresis was performed using an LKB Tachophor (2127) with an LKB potentiometer

(2210) recorder. Isotachophoretic separation was achieved in a poly(tetrafluoroethylene) (PTFE) column (25 cm x 0.1 cm, internal diameter) in the standard manner described in the manual. The leading electrolyte, prepared from 10 **mM** HCl and containing 0.3% hydroxymethylcellulose, was buffered at pH 3.4 with β-alanine. The terminating electrolyte was prepared by diluting caproic acid with distilled water (0.01**M**, pH 4.2). Both buffers were filtered through a membrane of average porosity 0.25 μm. Samples - typically 1-3 μl containing either 5 to 10 Bq ^{239}Pu or <u>ca</u>. 250 Bq ^{95}Nb - were injected into the column using a gas-tight microsyringe (10 μl, Scientific Glass Engineering, Australia). Isotachophoretic separation was achieved with a current of 100 μA. The migrating zones, except those of ^{95}Nb(V) citrate and ^{239}Pu(IV) citrate were characterized by conductivity and UV detectors.

<u>Preparative isotachophoresis</u>. Similar instrumental procedures were used for preparative isotachophoretic separations but with the addition of the *tachofrac*, the fraction collecting device (LKB tachofrac 2127) which was used for the collection of the species eluted from a PTFE column (43 x 0.1 cm internal diameter). The counter flow of leading electrolyte was delivered from a gas tight syringe (1701lt, Hamilton) mounted on a 50 ml microsyringe pump (Braun Perfusor IV). The citrate point-of-elution during isotachophoresis was determined from the recorded UV signals which had been accompanied by simultaneously event-marking of the cellulose acetate strip. When the position of the eluted radionuclides had been determined on the event-marked cellulose acetate strips they were carefully dissected into discrete regions and analyzed for radionuclide deposition.

3. Results

The preparative isotachophoretic distribution of ^{95}Nb and ^{239}Pu between the eluate, the column contents and the electrode wells is summarized in Tables 1 and 2. The elution profile of the radionuclides on to the cellulose acetate strip is depicted in Figs. 1 and 2.

Table 1. Percentage distribution of ^{95}Nb between eluate, column contents and electrode compartments

	% Injected ^{95}Nb			
	Run 1	Run 2	Run 3	Mean
Cathode compartment	0.0	0.0	0.1	<0.1
Anode compartment	0.0	0.0	0.1	<0.1
Column	47.8	51.1	48.0	49.0
Eluate	48.0	46.6	50.2	48.0
Sum	95.8	97.7	98.2	97.0

Table 2. Percentage distribution of ^{239}Pu between eluate, column contents and electrode compartments

Terminating reservoir	Leading reservoir	Capillary
0.39	0.0	11.8

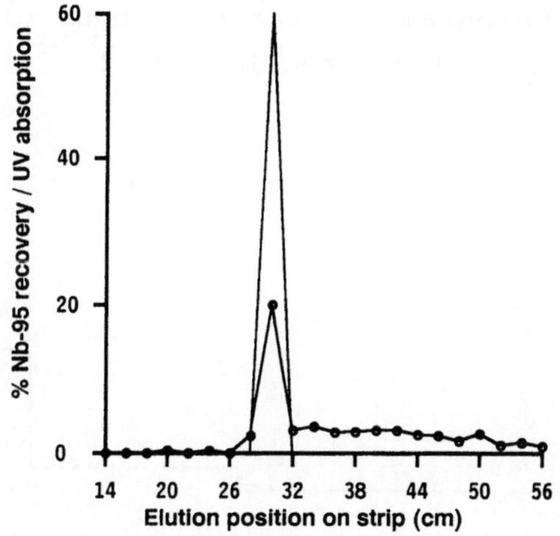

Figure 1. Elution profile of ^{95}Nb (•—•) and 0.02M citrate (—) on to the cellulose acetate strip

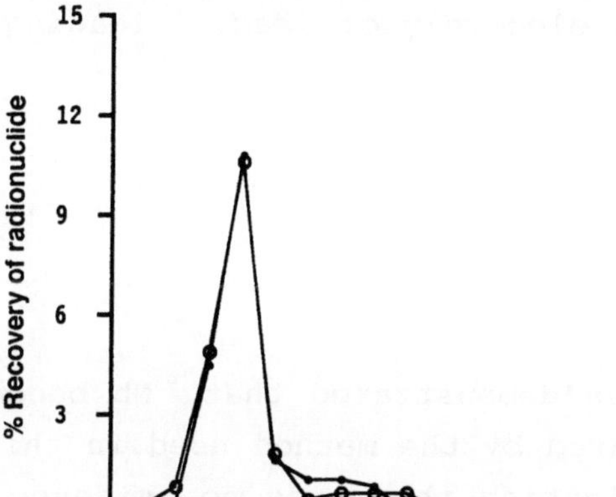

Figure 2. Elution profile of ^{239}Pu (•—•) and ^{14}C-citrate (o-o) on to the cellulose acetate strip

An isotachopherogram of stable niobium citrate showing a two plateaux plot is illustrated in Fig. 3.

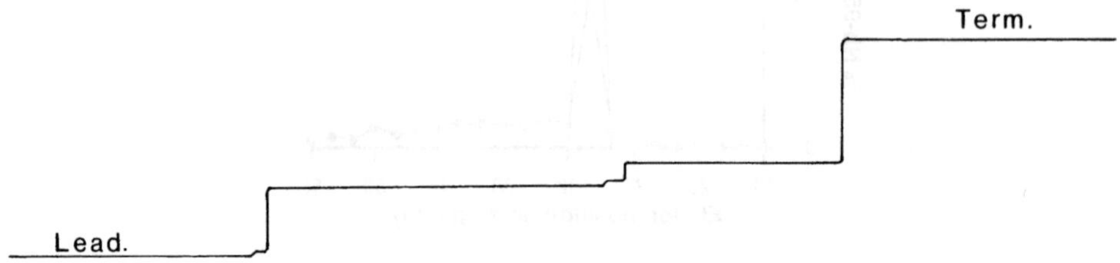

Figure 3. Isotachopherogram of 0.02**M** niobium (V) citrate: *term.*, terminating electrolyte; *lead.*, leading electrolyte

4. Discussion

This study has demonstrated that ^{95}Nb becomes complexed by citrate when prepared by the method used in this investigation. Until this investigation there was no evidence of the chemical form which would be produced; it was, for instance, not even possible to exclude the formation of niobate. The co-migration of ^{95}Nb and the citrate anion through the column, as depicted in Fig. 1, can only come about if ^{95}Nb(V) and citrate form a complex anion. At first consideration it might be expected that the Nb (V) citrate anion and citrate should have quite distinguishable electrophoretic mobilities. The phenomenon we have observed has been reported by others for a series of polyvalent cations complexed by chelating agents such as oxalate and EDTA

(Yoshida et al, 1979; Frederickson, 1980; Gebauer et al, 1980) and explained by Gebauer et al (1980) who have proposed that kinetically labile complexes decompose and reform as the zone migrates. A detailed analysis of this phenomenon as it applies to niobium (V) citrate will not be presented here as this communication is intended solely for presentation of our investigations of the speciation of niobium in a citrate medium.

By using isotachophoresis it has been possible to demonstrate around 50% of ^{95}Nb migrates as an anionic citrate complex and that 50% carries no charge at pH 3.4 (Table 1). From an investigation of the ultrafilterability of the column retentate (Prosser and Bulman, to be published) almost 70% of the retentate is retained on a membrane with a nominal molecular weight limit cut off of 500 Da. From their investigations of stable niobium(V) citrate, Grigoreva and Golubeva (1979) have proposed the existence of polymeric citratoniobate complex ions at acidic solutions. It is possible that the oligomeric material we have observed is the same chemical form as their complex ions.

Our investigations of the speciation of stable niobium(V) citrate at 0.02M indicate the formation of two forms of niobium (V) citrate as depicted in Fig. 3. Our demonstration of two ionic forms might be at variance with those Grigoreva and Golubeva (1979) who proposed that only $[Nb(OH)_3citrate]^-$ was formed at pH 5, and $[Nb(OH)_2citrate]^-$ at pH 6.5, for we have two observed two plateaux at pH 3.4.

This study with stable Nb (V) citrate indicates that $^{95}Nb(V)$ citrate might be expected to migrate as two citrate complexes. Although ^{95}Nb (V) citrate is collected as one band on the cellulose acetate strip, isotachophoretic separation into two bands might be possible if an anionic complex of intermediate electrophoretic mobility was available.

The isotachophoretic investigation of the speciation of ^{239}Pu (IV) in citrate confirms the migration of the radionuclide as a citrate complex. However, unlike the above study with ^{95}Nb there is much lower overall recovery of ^{239}Pu on the cellulose acetate strip. In an endeavour to trace the missing ^{239}Pu, preparative isotachophoretic separations were run for 20 min at 50 μamp, 10 min at 100 μamp and 27.5 μamp and the distribution of ^{239}Pu between the electrode reservoirs and the capillary determined. The recovery of ^{239}Pu declined from 86%, through 53% to 47%. It must be concluded that the missing ^{239}Pu had become associated with either the PTFE column or some compoment in the tachofrac.

5. Conclusions

This study has demonstrated that isotachophoresis will contribute much to our investigations of the speciation of the elements. It has demonstrated that Nb (V) can exist in a variety of anionic complexes. This study has also demonstrated that at least 50% of ^{95}Nb (V) in a citrate medium is present in an oligomeric form, a form which might have a low uptake across the gut wall. ^{239}Pu has been shown to migrate as a citrate complex but the recovery of the radionuclide from the apparatus falls as the time of the isotachophoretic run increases.

Acknowledgements

We thank Dr D S Popplewell for providing the mixed fission products, for his advice on the preparation of ^{95}Zr-containing solutions and also the subsequent isolation of ^{95}Nb. This work was funded in part by the research contract BI6-B-048-UK from the Commission of the European Communities.

References

Cotton F A, Wilkinson G (1972) Advanced Inorganic Chemistry, 2nd edn. Interscience Publishers, New York.

Fredriksson S (1980) Complexation with ferric ions - a potential source of error in the isotachophoretic determination of oxalate. J Chromatogr 188: 266-269.

Gebauer P, Bocek P, Deml M, Janak J (1980) Isotachophoresis of kinetically labile complexes J Chromatogr 199: 81-94

Grigoreva V V, Golubeva I V (1979) Determination of the composition and dissosciation constants of citrate complexes of niobium(V) by the photometric method. Sov Progr Chem 45:39-43.

Harrison J D, Haines J W, Popplewell D S (1990) The gastrointestinal absorption and retention of niobium in adult and newborn guinea pigs. Int J Radiat. Biol, *submitted*.

Jurriaanse A, Moore F L (1966) Fast separation of niobium-95 from zirconium-95 in commercial radioisotope solutions. Anal Chem 38: 964-965.

Schroeder H A, Balassa J J (1965) Abnormal trace metals in man: niobium. J Chron Dis 18: 229-241.

Smith H, Stradling G N, Bulman R A, Ham G J (1976) Experimental studies on the use of citrate to enhance the urinary excretion of plutonium in the rat. Health Phys 30: 318-320

Yoshida H, Nukatsuka I, Hikime S (1979) Fast separation of some EDTA complexes on capillary tube isotachophoresis. Bunseki Kagaku (Jap Anal) 28: 382-385

THE EFFECT OF CAFFEIC ACID ON THE SPECIATION OF METAL IONS IN PLANT NUTRIENT SOLUTIONS

Peter W Linder, Alexander Voyé and Sarah Cocks
Department of Chemistry
University of Cape Town
Rondebosch
7700
South Africa

1. Introduction

An understanding of factors which determine the speciation of metals in soils is important for a variety of reasons. In particular, the bioavailability of essential metal micronutrients and the biotoxicity of metals, in general, are governed by the speciation of the metals concerned. (Florence, 1982) In considering the chemical equilibria involving the metals occurring in soils, account should be taken of interactions between the soil itself and the aqueous phase with which it is in contact as well as the atmosphere. The dynamic nature and complexity of the soil/air/water system is depicted in Figure 1.

Thus, the speciation of metals in the vicinity of the roots of plants is affected by multifarious homogeneous and heterogeneous processes. Further, there is a great deal of variation from one location to another.

In order to add, effectively, to the present knowledge of the role of speciation in determining the bioavailability or biotoxicity of metals in soils, it is necessary to investigate model systems which are considerably simpler than that suggested in Figure 1.

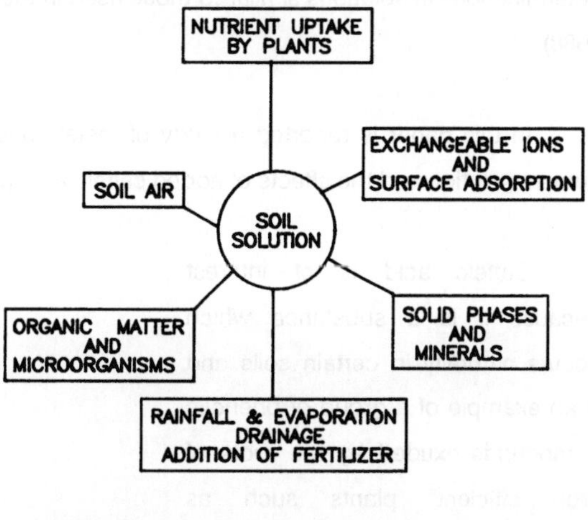

Figure 1. *Schematic diagram illustrating the dynamic equilibria that occur in soils.(From Lindsay WL (1979) Chemical Equilibria in Soils, 1st edn. John Wiley and Sons, New York*

NATO ASI Series, Vol. G 23
Metal Speciation in the Environment
Edited by J. A. C. Broekaert, Ş. Güçer, and F. Adams

The approach adopted in the work reported here is to focus attention on the aqueous phase, to consider interactions with atmospheric gases and with solids which may precipitate from the solution but to exclude interactions with refractory particulate solids, humic materials, detritus and other chemically indefinable constituents of soil. In particular, the focus is on models consisting of hydroponic plant nutrient solutions. The feasibility of cultivating plants in hydroponic solutions is well established. (Benton Jones, 1982) Further, such systems lend themselves, in principle, to the tractable investigation of factors affecting metal uptake by plants from the solution.

Investigations of this nature would, ideally, include first biological experiments in which the yield of a crop, the health of the plants as well as the uptake of metallic micronutrients are evaluated and second chemical estimations of the concentrations and changes in concentration of the metals in the solution. Preferably the latter should involve determining the concentrations of the individual physico-chemical forms in which each metal exists in the solution, that is, the speciation of each metal. One such study of this nature involved finding biological results in the literature (Wallace, 1980) and comparing these with the results of chemical speciation determinations in solutions similar to those used in the biological experiments. (Linder and Voyé, 1988)

In this paper is reported a study of metal speciation in a very simple hypothetical plant nutrient solution and the effects of added caffeic acid (trans 3,4 dihydroxy-cinnamic acid, I).

Caffeic acid is of interest because it is a substance which occurs naturally in certain soils and is an example of a group of phenolic compounds exuded by the roots of "iron efficient" plants such as soybean, tomato and maize. (Brown and Ambler, 1973) These plants are believed to exude the phenolic compounds when deprived of iron.

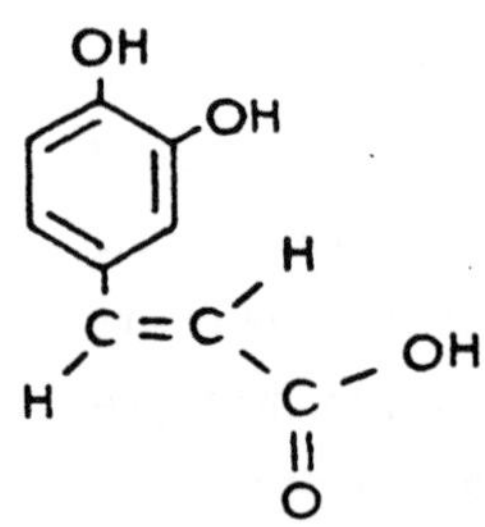

(I)

The biological function of the phenolic exudates is not clear, however. Brown and Ambler, (1973) and Olsen *et al* (1982) suggest a teleological explanation whereby the phenolic compounds act by reducing iron(III) in the rhizosphere, to iron(II). In this regard, the iron(II) aqua ion is considered to be the bioavailable form of the element. Olsen *et al* (1982) conclude, further, that the iron reduction process is inhibited by manganese(II), copper(II), zinc(II), hydroxyl, orthophosphate, pyrophosphate and molybdate ions. In contrast with the aforementioned authors, Chaney *et al* (1972) and Bienfait *et al* (1982) assert that the essential iron reduction process is effected through enzyme action at the root surface; the implication is that caffeic acid is not involved.

The purpose of the present study is to use metal speciation, determined by computer simulation, to test the hypothesis of Brown and Ambler (1973) and Olsen *et al* (1982) that caffeic acid functions as a reducing agent for iron in aqueous solution. Also investigated is the influence of manganese(II), copper(II) and zinc(II) upon the iron speciation, including the reduction process.

2. Method

In multicomponent systems such as soil solutions, ground waters, sea water or plant nutrient solutions, as form the subject of the present paper, the total concentrations of trace metals and ligands tend to be rather small and it is only in the past decade or so that analytical techniques have been perfected which allow total concentrations of the trace metals to be determined reliably. Individual complexes of the trace metals tend to occur in even substantially smaller concentrations. Further, the complexes are generally numerous so that the task of separating and determining, reliably, the concentrations of the individual complexes usually represents a formidable problem. Even in circumstances in which determination of the complex concentrations can be made, there is sometimes the risk of perturbing the equilibria one is attempting to investigate. Nevertheless, other papers in the present volume indicate that advances are being made towards determining individual metal species in multicomponent systems. Much promise, in this regard, is developing through the approach of interfacing a sensitive technique such as atomic absorption spectroscopy or plasma emission spectroscopy with a separation technique for example one of the forms of chromatography. Other types of hybrid technique such as inductively coupled plasma-mass spectrometry are also showing promise.

A very useful complementary approach is that of equilibrium modelling carried out with a computer (Jenne, 1979). Computer models which simulate real multicomponent systems provide the advantage of facilitating the understanding of processes which govern the composition of the system. Examples are dissolution, precipitation, oxidation, reduction, acid-base interactions and metal complexation. Of course, there are disadvantages with the method such as its dependence on the reliability of the thermodynamic data base used and on the specific assumptions made regarding the system being modelled. Nevertheless, the method permits estimates to be made of the concentrations of individual aqua cations, the free ligands and the complexes in a multicomponent system. Computer modelling is the method used in the work described in this paper.

The principles of computer modelling, in the present context, may be summarised as follows. Suppose there are n components which react to form m complexes. Given that either the total concentration or the free (i.e. uncomplexed) concentration of every component is known, the problem is simply one of calculating the concentration of every conceivable complex that can be formed. In general, the free concentration of only a few of the components is known, for example the concentration of free hydrogen ions as determined from glass electrode measurements and perhaps the concentrations of the aqua ions of one or two of the metals or free simple ligands such as fluoride or ammonia, as obtained from ion selective electrode measurements or anodic stripping voltammetry. For most of the components, only the total concentrations will be known, however.

Also required for the determination of speciation is the value of the protonation constants for every ligand and the formation constant for every conceivable complex that could potentially be formed. These constants, denoted by β_{pqr}, are defined for a general complex-forming reaction, (1), by equation (2).

$$pM + qL + rH \rightleftharpoons M_pL_qH_r \qquad (1)$$

M = metal ion; L = ligand; H = proton; M_pLqH_r = complex;

charges are omitted, for clarity; for protonation of the ligand, the stoichiometric coefficient, p = o; the stoichiometric coefficient, r, may be positive, zero or negative, when negative the complex is of the hydroxy variety.

$$\beta_{pqr} = [M_pL_qH_r]/[M]^p[L]^q[H]^r \qquad (2)$$

β_{pqr} is expressed in terms of concentrations of the reacting species and is termed a stoichiometric formation constant; the value of β_{pqr} is strongly dependent on the ionic strength of the solution.

Yet another requirement for the speciation calculations is a knowledge of the solubility product for every partially soluble salt that could potentially precipitate.

To illustrate the calculation of speciation, a system consisting of a solution phase only is considered initially. The concentration, C_i, of complex, i, is given by equation (3), which is simply an expression of the Law of Mass Action.

$$C_i = \beta_i \prod_{j=1}^{n} X_j^{a(i,j)} \tag{3}$$

$\beta_i = \beta_{pqr}$ for complex, i; X_j = the concentration of free component, j; a(i,j) = the stoichiometric coefficient of component, j, in complex, i.

For calculating the speciation in the system, m equations of form (3) are needed. Since for the majority of the components, only the total concentrations, not X_j values, are known, estimates of the latter have to be found. To do this, the Law of Mass Balance (equation 4) is invoked.

$$^{c}T_j = X_j + \sum_{i=1}^{m} a(i,j)C_i \tag{4}$$

^{c}Tj = the calculated total concentration of component, j, and is to be distinguished from ^{r}Tj, the real or analytical total concentration of component, j. The combination of the Mass Action, (3) and Mass Balance, (4), equations yields a set of n simultaneous non-linear equations with most of the X_j's as unknowns. In principle, these can be solved to yield values of the unknown X_j's and hence the speciation for the system can be determined by calculating each of the m C_i's. In practice, analytical solutions to the simultaneous equations do not exist and therefore resort has to be made to numerical methods. The approach, generally, is one of adjusting the unknown X_j values in a series of successive approximations until an appropriate objective function, for example

$$\sum_{j=1}^{n} (^{c}T_j - {}^{r}T_j)^2,$$

becomes minimised. The procedure is to use a suitable least squares computer program of which several are currently available, for example, ECCLES, (May *et al*, 1977) HALTAFALL, (Ingri *et al*, 1967) REDEQL, (Morel and Morgan, 1972) MINEQL. (Westall *et al*, 1975) Each individual program utilises a particular numerical procedure for determining the X_j's. Once convergence has been obtained, the next stage is to determine which of the solids may precipitate. This is accomplished by comparing the calculated free concentration product with the solubility product value, provided by the thermodynamic data base, for each solid which potentially may precipitate. A solid which precipitates, of course, effectively removes some of the components involved from the solution. Thus a new equilibrium speciation is computed using the component concentrations which have been modified by solid precipitation, as the starting values in the iteration procedure. Hence, a new set of precipitating solids is determined and the whole procedure repeated re-iteratively until there is no change in the set of precipitated solids. It must be noted that during the nested re-iteration procedure, a solid which has precipitated could re-dissolve.

In order to carry out speciation modelling of any given system, the program which the present authors would choose would be JESS (Joint Expert Speciation System). (Murray and May) Indeed JESS comprises a suite of programs and utilities for carrying out equilibrium calculations easily and efficiently. Unfortunately, the development of JESS is not yet complete, so for the work described in this paper, the second choice, namely, MINEQL was used.

3. Feasibility of Reduction of Iron in Soil by Caffeic Acid

In order to provide a raison d'être for the present study consideration can be given as follows to the feasibility of iron(III) being reduced to iron(II) by caffeic acid under typical soil conditions. Equation 5 shows the redox half reaction for the reduction of 3(O-benzoquinon-4-yl)prop-2-enoic acid (benzoquinone acrylic acid, BQA) to caffeic acid for which the standard potential is 0.793 V at 25°C. (Horner and Geyer, 1965)

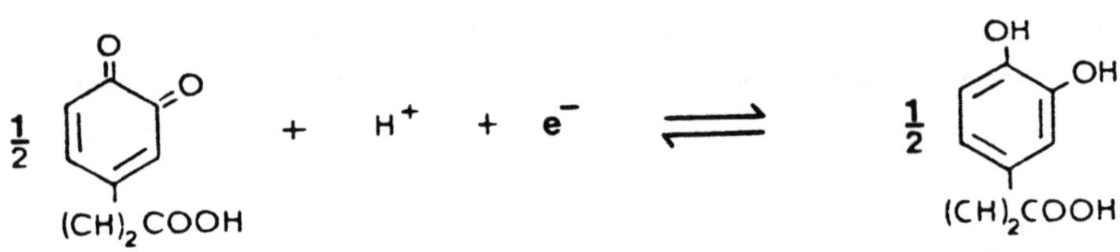

(5)

BQA H$_3$CAFF

The standard potential, at 25°C, for the reduction of Fe^{3+} to Fe^{2+} is 0.771 V. It follows that the emf, E_{rxn}, at 25°C, for reaction (6) is given by equation (7).

$$\tfrac{1}{2}\,H_3CAFF + Fe^{3+} \rightarrow \tfrac{1}{2}\,BQA + Fe^{2+} \qquad (6)$$

$$E_{rxn} = -0.022 - 0.026\ \ln\{BQA\}^{\frac{1}{2}}\{Fe^{2+}\}/\{H_3CAFF\}^{\frac{1}{2}}\{Fe^{3+}\} + 0.06pH \qquad (7)$$

The concentration of BQA in soil is likely to be smaller than that of caffeic acid. The concentration of iron(III) ions will certainly exceed that of iron(II) ions. It follows that the second term on the right hand side of equation (7) is likely to be negative. Therefore, provided the pH is greater than 0.4 (which must be the case in soil), E_{rxn} is positive, demonstrating that the reduction of iron(III) by caffeic acid in soil is thermodynamically feasible.

4. The Model Nutrient Solution and Thermodynamic Data Base

To investigate the effects of caffeic acid and of manganese(II), copper(II) and zinc(II) ions, the relatively simple composition of Table 1 was used as a model nutrient solution. The intention behind using a model of this simplicity was to ensure that the effects being sought would be emphasised through omission of possible obscuring features.

TABLE 1

Composition of the hypothetical plant nutrient solution used in the present study

COMPONENT	TOTAL CONCENTRATION/M	COMPONENT	TOTAL CONCENTRATION/M
K^+	2.06 E-3	Mn^{3+}	0.00
Fe^{3+}	5.00 E-5	H^+	range
Fe^{2+}	0.00	electrons	0.00
Mn^{2+}	1.00 E-5 or 0.00	CO_3^{2-}	0.00
Cu^{2+}	1.00 E-5 or 0.00	$EDTA^{4-}$	5.00 E-5
Zn^{2+}	1.00 E-5 or 0.00	$HCAFF^{2-}$	1.00 E-4 or 0.00
Cu^+	0.00	BQA^-	0.00

Although Table 1 reflects an excess of positive ionic charge, electrical neutrality is considered to be established by the implied presence of an inert counter anion such as chloride. The component, $HCAFF^{2-}$, represents the dianion of caffeic acid with one proton of very high pK_a (> 13) being considered strongly bound by sharing between the two oxygen atoms of the catechol moiety. BQA^- denotes the anion of BQA. Fe^{2+}, Cu^+, Mn^{3+}, electrons and BQA^- are included as components to allow for the respective redox half reactions involving Fe^{3+}, Cu^{2+}, Mn^{2+} and $HCAFF^{2-}$. The initial concentration of the members of the former set is zero because the model nutrient solution is not prepared with these as constituents. Carbonate CO_3^{2-}, is included, with an initial concentration of zero, to allow for interactions with atmospheric carbon dioxide and the consequent formation of dissolved carbonate, bicarbonate and carbonic acid.

The authors have compiled a general thermodynamic data base for application to speciation computations in plant nutrient solutions and similar systems. Currently this consists of the equilibrium constants for 299 protonation and complexation reactions, four redox half reactions, 39 equilibria between the solution phase and precipitated solid phases that potentially could form and interactions between the solution phase and the atmospheric gases, carbon dioxide and oxygen. Most of the constants were taken from the literature; (Sillén and Martell, 1964, 1971; Högfeldt, 1982; Perrin, 1979; Martell and Smith, 1976; Smith and Martell, 1976; Baes and Mesmer, 1976) those pertaining to caffeate complexation were determined by the authors but only some of these have, hitherto, been published. (Linder and Voyé, 1987) The constants were adjusted to a temperature of 25°C, when necessary, and to zero ionic strength. The temperature corrections were obtained by means of the van't Hoff equation and ionic strength corrections were carried out by the procedure of Linder and Murray (1982). The reason for selecting zero ionic strength for the data base is because the program MINEQL includes ionic strength adjusted via the Davies equation (Davies, 1962) thus providing the facility of permitting the user to choose any desired value of the ionic strength. The data base is in the format suitable as input to MINEQL and is available on request from P W Linder.

5. Results and Discussion

In Figure 2 is shown part of the iron speciation computed for the solution of Table 1 with all the components present except caffeate, in the pH range 4 to 8. It can be seen that although the initial total concentration of iron(II) species was zero, upon reaching equilibrium, the concentrations of the iron(II) species become significant.

This is attributed principally to the reduction of iron(III) species by manganese(II) ions. The concentration of iron(II) aqua ion at equilibrium ranges from $10^{-10.6}$ M at pH 4 to $10^{-16.9}$ M at pH 8. Towards the lower pH values, the total iron(II) concentration is made up virtually entirely from the iron(II) aqua ion concentration. On the other hand, towards higher pH, the iron(II)-EDTA complex makes the major contribution to the total iron(II) concentration.

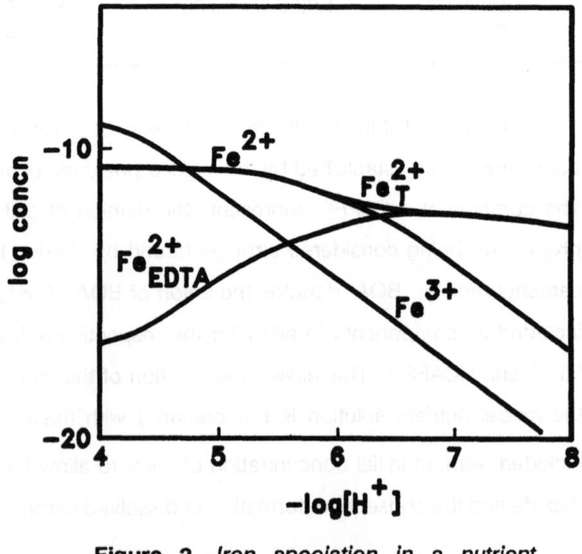

Figure 2. *Iron speciation in a nutrient solution with the composition of Table 1, but with caffeate excluded*

Upon adding caffeate at a concentration of 10^{-4} M to the solution of Figure 2, the iron speciation shown in Figure 3 is obtained. It is seen that the addition of caffeate leads to a dramatic increase, over several orders of magnitude, in the concentrations of the iron(II) species. By contrast with the solution of Figure 2, the presence of caffeate brings the iron(II) aqua ion concentration up to $10^{-5.6}$ M at pH 4 and $10^{-13.9}$ M at pH 8.

In a third speciation computation, caffeate was retained as a component of the solution of Table 1 but BQA was removed thereby disabling the caffeate redox half reaction. Thus, the speciation computed incorporates the complexing characteristics but not the reducing property of caffeate. The results are displayed in Figure 4.

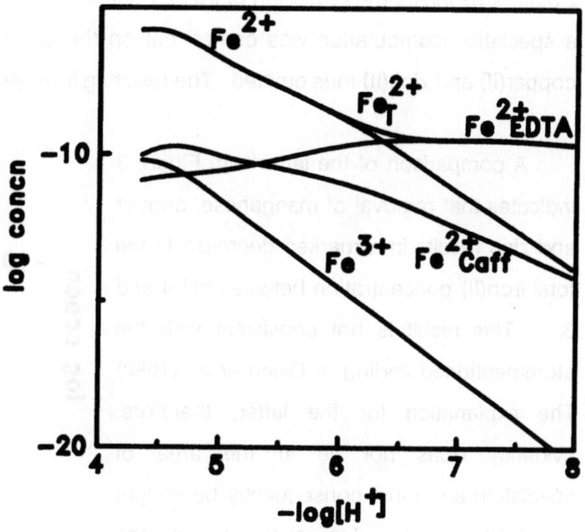

Figure 3. *Iron speciation in a nutrient solution having the composition of Table 1*

Figures 2 and 4 show that caffeate complexes iron(II) in the solution, significantly, but above pH 5.3 the complexation is markedly weaker than with EDTA. Further, the caffeate complexation does not influence the concentrations of the iron(II) aqua ion, the iron(II)-EDTA complex or the total iron(II) concentration, perceptibly.

It is noteworthy that the iron speciation in Figure 4 is identical to that in Figure 2 except, of course, for the appearance of the iron(II)-caffeate complex.

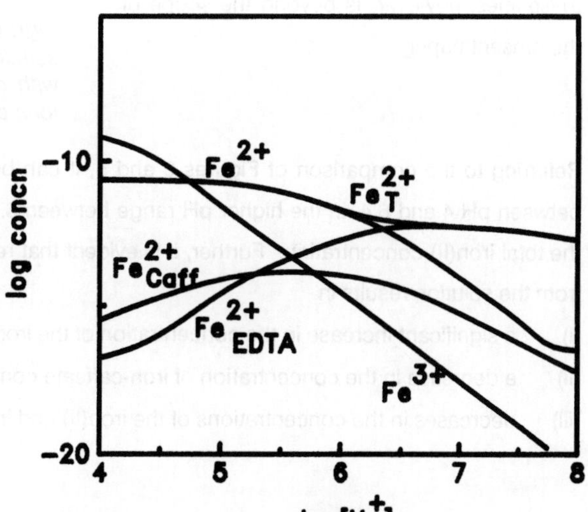

Figure 4. *Iron speciation in a nutrient solution with the composition of Table 1 but with the caffeic acid-BQA redox half reaction disabled*

Thus, it would appear that the dramatic increase in the iron(II) species concentrations in Figure 3 compared with Figure 2 is due solely to the reducing property of caffeate. It follows that the results in Figures 2, 3 and 4 are consistent with the hypothesis of Brown and Ambler (1973) and of Olsen *et al* (1982) that the function of root-exuded caffeic acid is to reduce iron(III) to iron(II).

To investigate whether the finding of Olsen *et al* (1982) that manganese(II), copper(II) and zinc(II) ions inhibit the reduction of iron by caffeic acid, can be explained in terms of speciation, a speciation computation was carried out on the solution of Table 1 but with manganese(II), copper(II) and zinc(II) ions omitted. The resulting iron speciation is shown in Figure 5.

A comparison of the latter with Figure 3 indicates that removal of manganese, copper and zinc results in a marked decrease in the total iron(II) concentration between pH 4 and 6.4. This result is not consistent with the aforementioned finding of Olsen *et al.* (1982) The explanation for the latter, therefore, evidently does not lie in the area of speciation and must consequently be sought in another area, for example, in the mechanism of processes occurring at root membrane surfaces. Any speculation along these lines, however, is beyond the scope of the present paper.

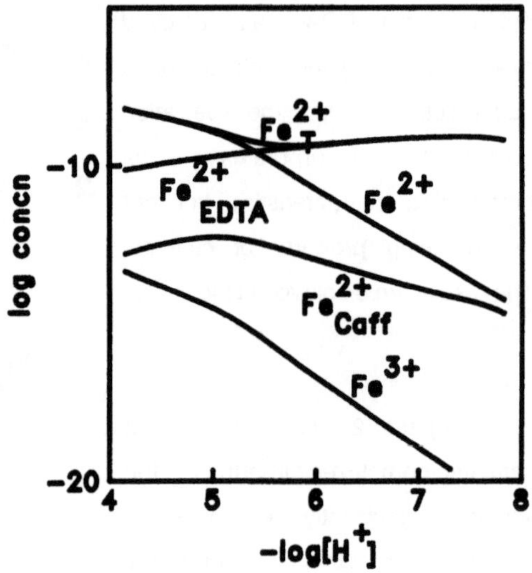

Figure 5. *Iron speciation in a nutrient solution with the composition of Table 1 but with manganese(II), copper(II) and zinc(II) ions excluded*

Referring to the comparison of Figures 3 and 5, it can be seen that, in contrast with the effect between pH 4 and 6.4, in the higher pH range between 6.4 and 8 there is a marginal increase in the total iron(II) concentration. Further, it is evident that removal of manganese, copper and zinc from the solution results in

(i) a significant increase in the concentration of the iron(II)-EDTA complexes;

(ii) a decrease in the concentration of iron-caffeate complexes;

(iii) decreases in the concentrations of the iron(II) and iron(III) aqua ions over the whole pH range investigated. These observations are readily explained by assuming that removal of manganese(II) ions, copper(II) ions and zinc(II) ions from the solution of Figure 3 releases a portion of the total EDTA which had hitherto been complexed by these metallic micronutrients. The released EDTA binds additional iron(II) and iron(III) including a portion of the iron(II) which had previously been bound in the iron(II)-caffeate complexes. In the latter regard, EDTA binds iron(II) considerably more strongly than caffeate does. Thus, the iron(II) aqua ion, iron(III) aqua ion and iron(II)-caffeate concentrations drop upon removing manganese(II), copper(II) and zinc(II) ions. The aforementioned effect of pH on the change in total iron(II) concentration is clarified upon realising that the total iron(II) concentration is made up predominantly from the iron(II) aqua ion towards the lower pH values and from the iron(II)-EDTA complexes towards the higher pH values.

A further speciation computation was carried out on the solution of Table 1 but with the EDTA omitted. The purpose here was to seek further clarification of the competition between EDTA and caffeate as complexing agents of the metals present. The results indicate barely perceptible increases in the concentrations of iron(II) aqua ion, iron(III) aqua ion and iron(II)-caffeate complexes. Removal of EDTA, however, yields marked increases in the manganese(II)-caffeate, copper(II)-caffeate and zinc(II)-caffeate complexes above pH 6.2, 4.3 and 4.7, respectively, as shown in Figure 6. These results are consistent with the conclusions drawn in respect of Figure 5.

With EDTA removed, caffeate remains as the only complexing agent and Figure 6 shows copper(II) and zinc(II) to have a much stronger affinity for this ligand than iron(II) or manganese(II), although the affinity of manganese(II) increases substantially when the pH exceeds 6.2. The copper(II) present and to a lesser extent, the zinc(II), restrict the availability of caffeate to the iron(II). This occurs to such a degree that the iron(II)-caffeate concentration remains at about as low a level as when EDTA is present.

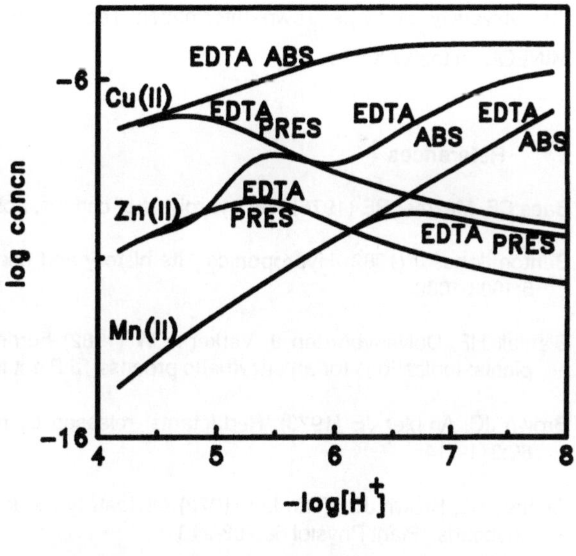

Figure 6. *Concentrations of the manganese(II)-caffeate, copper(II)-caffeate and zinc(II)-caffeate complexes in a solution having the composition of Table 1, with EDTA present and absent*

When the caffeate concentration was varied over the range, 0.8×10^{-4} M to 10×10^{-4} M, in the absence of EDTA, no significant changes in the iron(II) speciation became apparent. Indeed, throughout this range, the total iron(II) concentration comprised essentially the iron(II)aqua ion concentration with the iron(II)-caffeate complexes making a negligible contribution.

Finally, it is necessary to give an explanation of the decline in the concentrations of the iron(II) and iron(III) aqua ions with increases in pH, as is evident from Figures 2 to 6. This is attributed to precipitation of $Fe(OH)_3$. In contrast, Schwab and Lindsay (1983) suggest that the solid phase controlling iron concentrations in the solution is ferrosic hydroxide, $Fe_3(OH)_8$. This discrepancy arises from the fact that $Fe_3(OH)_8$ is metastable and, in time, tends to become converted to the thermodynamically stable $Fe(OH)_3$. Schwab and Lindsay (1983) point out that soils constitute dynamic systems in which not every process is at equilibrium. In particular the $Fe_3(OH)_8/Fe(OH)_3$ system would not be at equilibrium, thus accounting for the presence of $Fe_3(OH)_8$ in significant quantities. By contrast, the speciation computations of the present work apply to systems at equilibrium; hence $Fe(OH)_3$ is detected as a precipitated solid but $Fe_3(OH)_8$ is not.

Acknowledgements

The authors thank the Foundation for Research Development for financial assistance and the University of Cape Town Information Technology Services for the smooth running of MINEQL on the VAX 6330.

References

Baes CF, Mesmer RE (1976) The hydrolysis of cations, John Wiley and Sons, New York

Benton Jones J (1982) Hydroponics: Its history and use in plant nutrient studies. J Plant Nutr 5:1003-1030

Bienfait HF, Duivenvoorden J, Verkerke W (1982) Ferric reduction by roots of chlorotic bean plants: Indications for an enzymatic process. J Plant Nutr 5:451-456

Brown JC, Ambler JE (1973) "Reductants" released by roots of Fe-deficient soybeans. Agro J 65:311-314

Chaney RL, Brown JC, Tiffin LO (1972) Obligatory reduction of ferric chelates in iron uptake by soybeans. Plant Physiol 50:208-213

Davies CW (1962) Ion Association, Butterworths, Washington

Florence TM (1982) The speciation of trace elements in waters. Talanta 29:345-364

Högfeldt E (1982) Stability constants of metal-ion complexes : Part A, Inorganic Ligands, IUPAC Chemical Data Series No 21. Pergamon Press, Oxford New York Toronto Sydney Paris Frankfurt

Horner L, Geyer E (1965) Zur Kenntnis der o-Chinone, XXVII. Redoxpotentiale von Brenzcatechin-Derivaten. Chem Ber 98:2016-2027

Ingri N, Kakolowicz W, Sillén LG, Warnqvist B (1967) High-speed computers as a supplement to graphical methods V. Haltafall, a general program for calculating the composition of equilibrium mixtures. Talanta 14:1261-1286

Jenne EA (ed) (1979) Chemical modeling in aqueous systems. American Chemical Society, Washington DC

JESS (Joint Expert Speciation System) Murray K, CSIR South Africa, May PM, Murdoch University Western Australia

Linder PW, Murray K (1982) Correction of formation constants for ionic strength, from only one or two data points : An examination of the use of the extended Debye-Hückel equation. Talanta 29:377-382

Linder PW, Voyé A (1987) Potentiometric investigations of the equilibria between caffeic acid and copper(II), zinc(II), iron(II) and hydrogen ions in aqueous solution. Polyhedron 6:53-60

Linder PW, Voyé A (1988) Influence of synthetic chelating agents on the speciation of copper, zinc and manganese in plant nutrient solutions. J Coord Chem 17:381-387

May PM, Linder PW, Williams DR (1977) Computer simulation of metal-ion equilibria in biofluids : Models for the low-molecular-weight complex distribution of calcium(II), magnesium(II), manganese(II), iron(II), copper(II), zinc(II) and lead(II) ions in human blood plasma. J Chem Soc Dalton Trans 588-595

Morel F, Morgan J (1972) A numerical method for computing equilibria in aqueous chemical systems. Environ Sci Tech 6:58-67

Martell AE, Smith RM (1976) Critical stability constants. Volume 3: Other organic ligands. Plenum Press, New York London

Olsen RA, Brown JC, Bennett JH, Blume D (1982) Reduction of Fe^{3+} as it relates to Fe chlorosis. J Plant Nutr 5:433-445

Perrin DG (1979) Stability constants of metal-ion complexes : Part B, Organic Ligands, IUPAC Chemical Data Series No 22. Pergamon Press, Oxford New York Toronto Sydney Paris Frankfurt

Schwab AP, Lindsay WL (1983) Effect of redox on the solubility and availability of iron. Soil Sci Soc Am J 47:201-205

Sillén LG, Martell AE (1964) Stability constants of metal-ion complexes, Special Publication No 17. The Chemical Society, London

Sillén LG, Martell AE (1971) Stability constants of metal-ion complexes, Special Publication No 25. The Chemical Society, London

Smith RM, Martell AE (1976) Critical stability constants. Volume 4: Inorganic Complexes. Plenum Press, New York London

Wallace A (1980) Effect of excess chelating agent on micronutrient concentrations in bush beans grown in solution culture. J Plant Nutr 2:163-170

Wallace A, Wallace GA, Alexander GV (1983) Effect of excess chelating agent in nutrient solution at low levels of iron, zinc, copper and manganese. J Plant Nutr 6:507-511

West RC (ed) (1982-1983) CRC handbook of chemistry and physics, 63rd edn. CRC Press, Florida

Westall JC, Zachary JL, Morel FMM (1975) MINEQL : A computer program for the calculation of chemical equilibrium composition of aqueous systems, Technical Note No 18. MIT, Massachusetts

Complexes of Monosaccharides with Metal Ions: Some Bioinorganic and Environmental Inorganic Chemistry Aspects

Carlos F.G.C.Geraldes and M.Margarida C.A.Castro
Chemistry Department
University of Coimbra
3000 Coimbra, Portugal

1. Introduction

The formation of complexes by interaction of carbohydrates with metal ions has long been used as an analytical tool in separation methods such as chromatography (Weigel,1963; Ferrier, 1978) and electrophoresis (Foster,1957). Recent increased interest in this field (Angyal, 1980; Tsubomura *et al.*, 1983; Blunden *et al.*, 1983; Cervilla *et al.*, 1983; Bunel and Ibarra, 1985) results from developments in various areas involving their coordination chemistry as ligands, such as in bioinorganic chemistry in connection with metal-containing enzymes, or in environmental inorganic chemistry.

The complexation reactions of metal ions in natural waters are very complex (Buffle, 1988). However, small organic molecules such as aminoacids, sugars or small hydroxyacids, do not play a significant role in complexing trace elements because their concentrations are too small. The vast majority of the organic aquatic components are macromolecules, including polysaccharides, PROM, AROM and fulvic-humic compounds (Buffle, 1988). All the chemical models of these, sometimes ill defined, compounds contain sugars as part of their basic units (Buffle, 1988), and therefore it is important to characterize the complexation ability of those units using model compounds.

Another important recent development in environmental inorganic chemistry is the possible role played by root polysaccharides in the uptake of metal ions by plants (Farago, 1981; Irgolic and Martell, 1985). In particular, it has been found that an extracellular apparatus, rich in polyuronates, allows plant roots to remove nutrient cations from soils (Leppard and Ramamoorthy, 1975; Ramamoorthy and Leppard, 1977). Specific interactions such as those

NATO ASI Series, Vol. G 23
Metal Speciation in the Environment
Edited by J. A. C. Broekaert, Ş. Güçer, and F. Adams

found between polygalacturonic acid and metal ions, could have relevant functional and physiological implications (Deiana *et al.*, 1980, 1983; How *et al.*, 1969).

In spite of all these facts, the field of sugars or their related compounds- metal complexes is still to be fully explored. In this short review we describe, in a non-systematic way, recent studies on the synthesis, characterization and stereochemistry of complexes containing monosaccharides and related compounds, with special emphasis on our own studies with oxoions of transition metals. These studies may be of relevance to help characterize the complexation ability of large organic compounds containing sugar units.

2. Importance of Metal Ion - Carbohydrate Interactions in Bioinorganic Chemistry and Related Fields

Metal ions are particularly important for healthy plant life, as excesses or deficiences of metal ions have marked effects on plant growth and morphology (Sanchelli, 1969; Rorison, 1969; Epstein,1972; Hewett and Smith, 1975). The physiology of metal toxicity and tolerance in plants has been reviewed (Farago, 1981; Foy *et al.*, 1978).Excessive concentrations of some metals in soils, producing toxic effects, may have a variety of causes, such as natural outbodies near the surface (geochemical anomalies), exploitation of mineral resources or agricultural and waste disposal practices.

Chelating agents in soil can have two opposing effects on toxic metal ions (Dykeman and Sousa, 1966; Lindsay and Carson, 1974): they can promote solubilization and movement of metal ions, thus increasing their toxicity; or they can sequester the toxic metal ion and make it less available to the plant. For instance, it wasfound (Bloomfield, 1969; Bloomfield *et al.*, 1976) that some insoluble metal oxides were rendered soluble by decomposing organic plant material, partly through formation of soluble metal complexes by humic acids. Although the exact composition of these complexes is unknown, humic and fulvic acids and their complexes have been studied by a variety of physical techniques, including IR (Stevenson and Goh, 1971), ESR (Gamble *et. al.*, 1971; Goodman and Cheshire, 1975) and NMR (Stuerman and Paym, 1976; Gamble *et. al.*, 1976) spectroscopies and electron microscopy (Chen and Schnitzer, 1976).

The uptake of ions by plant roots and transport to the aerial parts of the plant has been reviewed (Epstein, 1973; Bowling, 1976). Plant cells are surrounded by a wall consisting of cellulose and other polysaccharides, which, as the plant cell ages, becomes rich in lignin. The walls are cemented together by pectates, in which Ca^{2+} acts as a structure former (pectic acid, poly(α-galacturonic acid), is made up from ß(1-4) linked α-galacturonic acid monomers). Metal ions and water are taken into the plant by the root, easily crossing its outer portion, the cortex, where the cellulose walls and intercellular spaces are freely available to aqueous solutions. However, they cannot pass freely to the internal vascular system (xylem) (Bowling, 1976). They must cross at least one membrane, possibly by active transport. They cross the membrane in various chemical forms, often as complexes, coordinated to small molecules such as malic acid (for Al^{3+}) (Jayman and Sivasubramaniam, 1975), biomolecules such as carbohydrates (for Pb^{2+}, Cu^{2+}, Ni^{2+}) (Hofner, 1970), aminoacids, carbohydrates or protein fractions (for Mn^{2+}) (Hofner, 1970; Dieckert and Rozacky, 1969) or simply in the form of aquoions, like for Zn^{2+} and Mn^{2+} (Dekock and Mitchell, 1957). The uptake of iron involves the reduction of Fe^{3+} to Fe^{2+} (Chaney *et al.*, 1972), but there is reoxidation in the cytoplasm (Tiffin and Brown, 1962) and translocation from the roots to the shoots in the form of ferric citrate (Tiffin, 1970).

Plant roots are rich in cellulose and other polysaccharides like acid poly-α-glacturonic and its methyl esters, and these are naturally the preferred biochelating agents of most of the absorbed ions. In fact, in root surfaces, the polygalacturonic components of cell walls are responsable for ion uptake, providing about 90% of cation exchange capacity (Stevenson and Ardakani, 1972). The study of such specific interactions, such as the ESR study of the absorption and reduction of chromate(VI) to Cr(III) by plant roots, is of functional and physiological relevance (Micera *et al.*, 1981). Recent developments (Deiana *et al.*, 1983) have been concerned with redox processes occurring upon interaction of polygalacturonic acid with soil mineral species. Reduction of Fe(III) to Fe(II), V(V) to V(IV) and Mo(VI) to Mo(V) occurs, followed by complexation of the reduced ions. The reaction is due to the reducing properties of the polysaccharide end units. This reduction process is particularly significant in relation to iron uptake, as ferric species are not available for plants unless reduction to the divalent state occurs.

The biological importance of polygalacturonic acid is related to their presence in the bacterial polysaccharides (Goebel and Barber, 1983) and in the fact that its formation can have an important role in detoxification and absorption of various substances (Furberg *et al.*, 1963). There are differences in properties of uronic acids and the corresponding aldohexoses, such as their rates of glycosidic hydrolysis and epimerization reactions (Siddiques and Punoes, 1963; Whisler and Richards, 1958), possibly due to the inductive effect of the carboxyl group, which stabilizes the glycosidic bond.

The interaction of carbohydrates with metal ions are of high biological importance. Exposure to carcinogenic metals, such as arsenic, chromium and cadmium (Sunderman, 1979) is fought by renal excretion. In the case of nickel, it was found that 70% of the low molecular weight nickel in the renal soluble fraction of rat, cow and man is bound to an oligosaccharide fraction, containing sulphate and uronic acids as the probable metal-binding components (Templeton and Sarkar, 1985).

The interaction of carbohydrates with alkali and alkaline-earth metal ions has also been investigated. These interactions are of non-ionic nature, as shown by X-ray crystal diffraction and vibrational spectroscopy studies (Beevers and Cochran, 1946; Einspaln and Bugg, 1980). Complexes of sugars with Ca^{2+} seem to participate in biological processes of adhesion and agglutination, in particular those occurring at cell surfaces (Weiss, 1973). The gelation process of low methoxyl pectins in low sugar jams is also induced by Ca^{2+} ions, which stabilize chain association of pectin molecules through multiple bonding to their oxygen-containing groups (Rees, 1969).

Since the discovery of the anti-tumour activity of cisplatin (Rosenberg *et al.*, 1969), a great number of platinum complexes have been examined in order to decrease their toxicity and increase their solubility. Recently, many anti-cancer reagents that have sugar residues, e.g. bleomycin, adriamycin and many antibiotics, have been widely used clinically. Therefore, the action of platinum compounds having sugar residues has been examined. A series of complexes of cisplatin type complexes of amino-sugars, [$PtCl_2$(amino- sugar)] was synthesized and fully characterized (Tsubomura *et al.*, 1986). Some of those complexes are reasonably soluble in water and show promisingly good anti-tumour activity in vivo.

The interaction of metal ions with carbohydrates is also of chemical interest. Since long ago, borate esters of polyhydroxy compounds have been used for analytical purposes in separation techniques using electrophoresis (Foster, 1957) or chromatography (Ferrier, 1978). This interaction is also a process of configurational analysis of carbohydrates (Böeseken, 1949). Metal complexes of N-glycosides (glycosylamines), derivatives of cyclic sugars in which the glycosidic hydroxyl group has been replaced by an amino group, have been studied (Yano, 1988). In particular, Ni(II) and Co(III) complexes of various ligands, including N-glycosides derived from diamines and aldoses or amino sugars, Schiff bases derived from amino-sugars and aldonic acids, have been prepared and structurally characterized (Yano, 1988). It is of interest to develop methods whereby metals catalyse transformation of sugars in bioinorganic chemistry and in industry. It has been shown that the presence of borate increases the selectivity of the hydrogenation reaction of fructose to yield mannitol (Makkee *et al.*, 1985), and inhibits the alkaline epimerization of various aldoses, acting as an efficient stereoselective catalytic agent in the synthesis of monosaccharides (How *et al.*, 1969). The stereoselective catalytical action of molybdenum trioxide in the formation of aldoses through hidroxylation of glycals is due, in part, to the formation of complexes between the hydroxyl group in position 3 and peroxomolybdenic acid (Bilik and Kucan, 1970). The oxoion molybdate in mildly acid solutions is an efficient catalyst of the selective epimerization (at C-2) of aldoses with the formation of a thermodynamic equilibrium mixture of the two epimers (Bilik *et al.*, 1978), without the production of ketoses. The mechanism of this reaction was investigated by using ^{13}C and ^{2}H-enriched aldoses and ^{13}C NMR spectroscopy, demonstrating that the reaction involves a 1,2-shift of the carbon skeleton, resulting in inversion of configuration at C-2

(Haynes *et al.*, 1982). The epimerization and stereoselective uptake of aldoses by Ni(II) complexes and by combination of various metal ions (Ca^{2+}, Co^{2+}, Sr^{2+}, Pr^{3+}, Ce^{3+}) and amines has also been studied ((Yano, 1988).

The glycol cleavage oxidation of starch and related polysaccharides, such as amylose, cellulose and ß-cyclodextrin, to their 2,3-dicarboxy derivatives was studied and patented (Floor *et al.*, 1988a, 1988b, 1989, in press; Schwegler *et al.*, 1988). Such dicarboxy polysaccharides were shown to be potentially environmentally attractive phosphate substitutes as detergent builders owing to their excellent calcium complexation properties, stability under alkaline conditions of the washing process and biodegradability in the acidic waste water (Floor *et al.*, 1988a, 1988b, 1989, in press; Schwegler *et al.*, 1988).

3. Studies of the Structure of Complexes of Monosaccharides with Metal Ions

The type of coordination interaction which takes place between monosaccharides and metal ions depends, among other factors, on the configuration and conformation of the ligand, which determine the spatial orientation of its hydroxyl groups, and on the stereochemical preferences of the metal center.

Monosaccharides, in the free form or as simple carbohydrate units, generally occur as 5 or 6 membered ring structures, known respectively as furanose or pyranose forms. These forms have various conformational possibilities, some of which with lower internal energy than others, as described by their pseudorotational cycles (Stoddart, 1971; Altona and Sundaralingam, 1973). In the case of pyranose rings, the chair forms are generally more stable than boat or skew conformations, but the orientation of bulky substituents may change this. Furanose rings also have non-planar conformation, known as envelope and twist forms. In general, polysaccharides containing furanose rings are more flexible than those containing chains of pyranose units in the chair conformation. Another important structure of sugars is the open chain form, which contains a free aldehyde group and is quite flexible.

All monosaccharides may isomerize through interconversions of forms with different configuration at carbon atoms. A difference of configuration at C-1 originates α- and ß - anomers, and their interconversion (mutarotation) occurs spontaneously in solution through the open form. Inversion of configuration at any of the other ring carbons (epimerization) requires, however, specific chemical or enzymatic reactions.

We now describe the main results of some recent structural studies of monosaccharide complexes in the crystal form and in solutions. X-ray diffraction and NMR spectroscopy have proved to be the most informative techniques relative to the geometry of the metal center and the coordination sites of the ligand.

A particularly favorable arrangement for complexing cations is a cis-axial, equatorial, axial sequence of three hydroxyl oxygens on consecutive carbon atoms in a six-membered ring (Angyal and Davies, 1971; Angyal, 1974). Using the values of the observed proton NMR complexation shifts, this type of coordination has been found to occur in the interaction of sugar derivatives wih Na^+ in acetone solution (Haines *et al.*, 1975), in the complexation of aldoses and cyclitols with Ca^{2+} and lanthanide(III) cations in water (Lenkinski and Reuben, 1976; Angyal and Greeves, 1976) and in the formation of tridentate complexes of monosaccharides with periodate (Perlin and Rudloff, 1965) and molybdate (Alfoldi *et al.*, 1978, 1980) in aqueous solution.

The interaction of alkali and alkaline earth metal ions with carbohydrates is generally presumed to be of the simple ionic type, with metal ion binding non-specifically to such anionic residues as ionized sugar hydroxyl groups. However, X-ray structural analysis of systems like Na(sucrose)Br.$2H_2O$ (Beevers and Cochran, 1946) and several other alkaline earth metal-sugar complexes (Einspahr and Bugg, 1980) demonstrated the non-ionic nature of such metal-sugar interactions. These conclusions were supported by various vibrational spectroscopic studies, where the I.R. spectra of structurally known sugar complexes were compared with those of the corresponding free sugars and of other salts of the same metal ion.

Infrared spectra of the isomorphous Na^+, K^+ and Rb^+ salts of ß-D glucuronic acid (Tajmir-Riahi, 1984a), of formula M(D-glucoronate).$2H_2O$ (M=Na^+, K^+, Rb^+), were interpreted in terms of six -fold coordination of each metal cation, surrounded by oxygens O-1, O-3, O-6 and O-7 of the sugar residue, as well as by two oxygen atoms of the water molecules. The crystal structures have an extensive hydrogen bonding network, and there are direct binding forces between the positive ion and the electronegative oxygen atoms. Vibrational spectra of crystalline alkali metal halide adducts of sucrose (Tajmir-Riahi, 1986a) Na(sucrose)X.$2H_2O$ (X=Cl, Br) and Na_3(sucrose)$_2I_3$.$3H_2O$, were interpreted in terms of six coordination of Na^+ by O-4, O-6, O-6' of the sugar, by two water molecules and by a halide anion, forming a distorted octahedron.

X-ray crystal diffraction, vibrational spectroscopy and molar conductivity techniques were used to study the interaction of various sugars with a series of alkali and transition metal ions, showing the formation of complexes of the types M(sugar)X_2.nH_2O and M(sugar)$_2$.$2H_2O$ (M=Mg, Ca, Zn, Cd, Hg; X=Br^-, Cl^-; sugar=D-glucuronic acid, L-arabinose, D-fructose, sucrose and ß-D-glucurono-1,4-lactone; n=0, 2, 4) (Tajmir-Riahi, 1983, 1984b, 1984c, 1985, 1986b, 1986d, 1987a, 1988a, 1988b). As far as the alkaline-earth ions are concerned, while Mg^{2+} has a preferential hexacoordinated geometry, Ca^{2+} has a preference for a coordination number of eight. Therefore, complexes are formed from interactions of those ions with hydroxyl groups of the sugar molecules, with a number of inner-sphere water molecules necessary to complete their coordination spheres (Tajmir-Riahi, 1988a). Due to the observed spectral similarities, it was included that the Zn^{2+} and Cd^{2+} ß-D-fructose complexes are isomorphous to those of Mg^{2+}, with a hexacoordinated geometry composed of two consecutive hydroxyl groups of each of the two ligands and two water molecules (Tajmir-Riahi, 1988a). The Hg^{2+} ions bind to two sugar moieties in the same fashion as do the Zn^{2+} and Cd^{2+} ions, resulting in a 4 form coordination geometry, which means that Hg^{2+} does not bind any water molecule. These studies with ß-D-fructose showed that, while the Ca^{2+} coordination occurs via the ß-D-fructopyranose, Mg^{2+}, Zn^{2+}, Cd^{2+}, Hg^{2+} and $UO_2{}^{2+}$ binding can occur with the ß-D-fructofuranose and ß-D-fructopyranose.

Compounds of the type Ag(D-glucono-∂-lactone)$NO_3.H_2O$ and Ag(D-gluconate), resulting from the interaction of those sugars with the Ag^+ cation, have been isolated and characterized as solids by X-ray diffraction and IR spectroscopy, and in solution by this later technique (Tajmir-Riahi, 1986c, 1986f, 1987c). These spectral observations suggested that, in the first compound the Ag^+ ion has a coordination number of two and binds the sugar molecule through the lactone carbonyl oxygen atom and to a bridging nitrate group, while the second compound is a dimeric complex where Ag^+ binds to two sugar anions via the carboxylate O-6 of the one and O-6' of the other. In both cases, the coordination geometry of the Ag^+ ion is linear. It was also shown that the hydrolysis of the lactone is facilitated in aqueous solution by the presence of silver acetate and silver carbonate, while silver nitrate and silver perchlorate cause lactone coordination.

The interaction of inorganic tin(IV) and tin(II) compounds with sugars and related polyhydroxylated molecules has been studied in the solid state and in coordinating and non-coordinating organic solvents (Dudycz *et al.*, 1981; Alföldi *et al.*, 1982; Blunden *et al.*, 1983). These later studies used ^{13}C and ^{119}Sn NMR. Unstable 1:1 adducts of $SnCl_4$ with sugars can be isolated from non-coordinating solvents, and, in these complexes, the sugar molecule acts as a neutral bidentate oxygen donor ligand through a vicinal diol interaction, where the HO--Sn interactions are symmetrically disposed in he octahedral tin(IV) environment. Monomethylation of vicinal diols by $SnCl_2$ in diazomethane/methanol involves formation of an intermediate tin(II) species in solution which is a neutral donor adduct $LSnX_2$ (X=Cl or OMe) with the organic hydroxyl ligand L, (Blunden *et al.*, 1983), rather than cyclic tin(II) ether LSn (Dudycz *et al.*, 1981). Since tin(II) compounds have a marked tendency to occupy a 3-coordinate pyramidal geometry (Donaldson, 1967), it is probable that only one hydroxyl group of L is directly coordinated to the metal, with a very weak second OH--Sn interaction.

Optically active amino complexes of Co(III), containing monosaccharides such as D-ribose, L-sorbose or D-glucosamine, were prepared and the observed Cotton effects observed in solution were correlated to their structural characteristics (Bunel and Ibarra, 1985). Those observations led to the proposal that, among the various possible cis-diol sugar-metal ion interactions, binding through OH groups of carbon atoms C-1 and C-2, when available, leads to the most stable species because it is best adapted to the distorted octahedral coordination metal site. This type of Co(III) complexes have also been studied using ^{13}C NMR (Reuben, 1984).

The complexation of sugars with uranyl ions has been studied using various methods. The interaction between D-glucuronic acid and hydrated uranyl salts has been studied in aqueous solution and in the solid state, and complexes of the type UO_2(D-glucuronate)X.2H_2O and UO_2(D-glucuronate)$_2$.2H_2O, where X=Cl^-, Br^- or NO_3^-, were isolated and characterized by means of FT-IR, proton NMR and X-ray diffraction (Tajmir-Riahi, 1986e). Six coordination of UO_2^{2+} was observed, with bipyramidal geometry, where equatorial coordination occurs preferentially through the two ionized carboxylate O-6 and O-6' oxygens, although binding to only one of the oxygens of each carboxylate and one hydroxyl or ring oxygen of each sugar ring was also observed (Tajmir-Riahi, 1986e). Similar studies were also undertaken for the UO_2(D-fructose) X_2.2H_2O (X=Cl^-, Br^-, NO_3^- and 0.5 SO_4^{2-}) adducts (Tajmir-Riahi, 1987b), which indicated that UO_2^{2+} binds to he two sugars moieties via two cis diol interactions, resulting in six-coordination.

The interaction of various monosaccharides and related compounds with UO_2^{2+} was also studied in aqueous solution using proton and ^{13}C NMR, and complexation was found to occur above pH 10 (Geraldes *et al.*, 1988b). In the case of the aldoses D-mannose and D-ribose, it was found that the furanose and pyranose forms are complexed by UO_2^{2+} using various cis-diol modes of interaction forming five or six-membered rings (Geraldes *et al.*, 1988b). In the case of sugar derivatives like D-ribose 5-phosphate or mononucleotides (Agarwall and Feldman, 1968, 1969; Feldman and Rich, 1970; Castro and Geraldes, 1987), polynuclear complexes are formed above pH 11 involving chelation of UO_2^{2+} both by the sugar hydroxyl groups and the phosphate groups, whereas below pH 10 the interaction occurs exclusively with the phosphate groups.

Proton and ^{11}B NMR has been used to detect specific interactions of borate with cis-diol groups forming five or six membered rings. For example, D-glucose forms 1,2-furanose and 1,2-pyranose complexes of tetrahedral geometry, with the borate ion $[B(OH)_4]^-$ in alkaline (pH 10-12) aqueous solution (Kennedy and How, 1973)) in accordance with studies involving borate and model hydroxyl and hydroxycarboxylate compounds (Van Duin *et al.*, 1984, 1985). This type of sugar complexation capacity of borate is reflected in catalytic effects of borate on various sugar reactions, such as inhibition of the alkaline epimerization of aldoses or the improved stereoselectiity of aldol synthesis of N-acetylneuraminic acid (How *et al.*, 1969).

Several studies of the complexation of molybdenum and tungsten oxoions by sugars have been reported. Various complexes of Mo(VI) and Mo(V) with monosaccharides have been detected in aqueous solution, and studied by optical rotary dispersion and circular dichroism (Brown and McPherson, 1970; Bayer and Voelter, 1966a, 1966b; Voelter *et al.*, 1968). The Mo(V) complexes formed at pH 6-8 are diamagnetic, which indicates formation of dimeric dioxo bridge metal centers (Brown and McPherson, 1970). Mo(VI) forms complexes at lower pH 4.5-6.5. Both Mo(V) and Mo(VI) yield octahedral complexes, mainly with the pyranose forms of the hexoses and pentoses. Aldoses like D-mannose, D-lyxose and D-ribose act as terdentate ligands by using three favorably oriented consecutive hydroxyl groups of the pyranose form in binding to the Mo(VI) and W(VI) oxocations, forming very strong rigid complexes with very well defined Cotton effects (Bayer and Voelter, 1966a, 1966b; Voelter *et al.*, 1968). These complexes yield very well defined proton and ^{13}C NMR spectra, whose complexation shifts were interpreted in terms of a single rigid skew complexed pyranose conformation (Alföldi *et al.*, 1978, 1980; Geraldes *et al.*, 1988b).

Pyranoses with cis axial hydroxyl groups with a trans intermediate hydroxyl group, like D-arabinose, D-galactose, D-xylose and D-glucose, complex weakly Mo(VI) and W(VI) through a six-membered cis-diol chelate, yielding poorly defined Cotton effects (Bayer and Voelter, 1966a, 1966b; Voelter *et al.*, 1968) and only weak, poorly defined proton NMR signals of complexed forms (Geraldes *et al.*, 1988b). The 1,3-cis diol binding efficiency was tested by studying the interaction of Mo(VI) with cyclic polyols, like all-cis-cyclohexane-1,3,5-triol and myo-inositol (Geraldes *et al.*, 1988b). It was found that 1,3,5-triaxial interactions are not allowed by steric reasons.

The reaction mechanism, proposed by Hayes et al. (Hayes *et al.*, 1982) for the molybdate catalysis of selective C-2 epimerization of aldoses, involves formation of catalytic anionic complexes of the open aldehyde aldose form with dimolybdate through binding to he hydroxyl oxygen atoms at C-2, C-3 and C-4. Independent work was also undertaken (Taylor and Waters, 1981) on the molybdate-catalysed C-2 epimerization of pentoses and stereoselective complexation of one of the epimers. A complex, composed of two molybdate and one sugar moiety, was derived from the reaction between ammonium molybdate and D-xylose. Its crystal structure showed a binuclear metal center with a triple-oxygen bridge, with the sugar D-lyxose in an unusual furanose form. Therefore, Mo(VI) catalyses C-2 epimerization of the starting D-xylose and establishes an equilibrium with its C-2 epimer, D-lyxose. This sugar preferentially complexes the metal center.

The interaction of sugar derivatives, like D-ribose 5 phosphate (Geraldes *et al.*, 1988b) and nucleosides or mononucleotides (Geraldes and Castro, 1986, 1988a), was studied by multinuclear NMR. The interaction of Mo(VI) with the furanose form of the sugar is very weak, and preferential binding occurs to the phosphate groups. Therefore, the sugar binding capacity of the angular MoO_2^{2+} and WO_2^{2+} oxocations and of the linear UO_2^{2+} ion are very different.

The interaction of various monosaccharides with vanadate(V) in aqueous solution had been studied using proton and ^{51}V NMR (Geraldes and Castro, 1989a; Tracey and Gresser, 1988a; Tracey *et al.*, 1988b). Complex formation occurs in the pH range of 5-7, through interaction with the sugar hydroxyl groups. Various types of complexes are formed, including monodentate tetrahedral vanadate esters as well as cyclic bidentate and tridentate species of trigonal bipyramidal geometry, resulting from cis-diol and cis-triol interactions. Strongest complexation occurs through the later mode, in cases like D-ribose and D-mannose. These results are in accordance with studies on vicinal diols (Gresser and Tracey, 1986) and hydroxycarboxylates (Tracey *et al.*, 1987; Caldeira *et al.*, 1987). The interaction of vanadate with sugar derivatives like nucleosides (Van Duin *et al.*, 1984) and nucleotides (Geraldes and Castro, 1989b) occurs preferentially through the ribose cis-diol hydroxyls, rather than the phosphate groups.

The complexation of oxovanadium(IV) by sugar derivatives like D-galacturonic and D-glucuronic acids and their polymers has been studied as a function of pH using potentiometry, ESR and ENDOR spectroscopy (Micera *et al.*, 1989; Branca *et al.*, 1989). The carboxylate group initiates coordination of the metal ion at pH>3 and one or two deprotonated sugar hydroxyl groups are then involved. Parallel studies with Cu^{2+} (Micera *et al.*, 1989) establishes that this ion has similar, but weaker, coordination modes.

4. General Conclusions

The complexation of metal ions by complex carbohydrates, present in biological systems or in organic matter from the environment, is an important research field due to the need of fundamental knowledge of the mechanisms of the impact of metal pollution on the environment. The processes of the transport of elements in soil and ground water, of their absorption by plants and other organisms and of their consequences to the metabolism of living organisms are not totally understood (Micera *et al.*, 1981).

In this context, structural studies of simple monosaccharide complexes have led to the identification of the main factors which control the types of complexes formed and their stability, namely the preferential coordination geometry of the ion and the spacial orientation of the sugar hydroxyl groups. The conclusions on these topics are summarized in Table 1.

In the case of sugar derivatives, where the hydroxyl groups coexist with a more basic group, such as a phosphate or carboxylate, this latter donor usually becomes the primary binding site. In some cases like VO_2^{2+} or VO^{2+}, the hydroxyl groups compete efficiently at high pH for complexation. Similar situations might occur in the case of polyuronates and algal or connective tissue polysaccharides.

Table 1. Binding Characteristics of Metal Ions to Monosaccharides

Metal ion	Coordination number	Coordination geometry	Preferential -OH interaction
Na^+, K^+, Rb^+	6	octahedral(?)	mono and polyol(variable)
Mg^{2+}	6	octahedral	mono and cis-diol
Ca^{2+}	7,8	variable	mono and polyol(variable)
Ln^{3+}	8,9	variable	mono and polyol(variable)
Co^{3+}	6	octahedral	cis-diol, cis-triol
Ni^{2+}, Cu^{2+}	6	octahedral tetragonal	cis-diol, cis-triol
Zn^{2+}, Cd^{2+}	6	octahedral	cis-diol
Hg^{2+}	4	tetrahedral(?)	cis-diol
Ag^+	2	linear	monodentate
Sn^{2+}	3	pyramydal	monodentate(+weak int.)
Sn^{4+}	6	octahedral	cis-diol
$B[OH]_4^-$	4	tetrahedral	cis-diol
IO_4^-	6	octahedral	cis-diol, cis-triol
UO_2^{2+}	6	bipyramidal	cis-diol
MoO_2^{2+}	6	octahedral	cis-diol, cis-triol
WO_2^{2+}	6	octahedal	cis-diol, cis-triol
VO_4^{3-}	4,5	tetrahedral bipyramidal	mono and cis-diol,cis-triol
VO^{2+}	5	pyramidal	cis-diol

Acknowledgement

Support from INIC, Portugal, is gratefully acknowledged.

References

Agarwall RP, Feldman I (1968) J Am Chem Soc 90:6635; (1969) 91:2411.
Alfoldi J, Petrus L, Bilik V (1978) Coll Czech Chem Commun 43:1476.
Alfoldi J, Bilik V, Petrus L (1980) Coll Czech Chem Commun 45:123.
Alfoldi J, Toman R, Peciar C (1982) Carbohydr Res 105:258.
Altona C, Sundaralingam M (1973) J Am Chem Soc 95:2333.
Angyal SJ (1974)Tetrahedron 30:1695.
Angyal SJ (1980) Chem Soc Rev 9:415.
Angyal SJ, Davies KP (1971) Chem Commun 500
Angyal SJ, Greeves D (1976) Aust J Chem 29:1223.
Bayer E, Voelter W (1966a) Justus Liebigs Ann Chem 696:194.
Bayer E, Voelter W (1966b) Biochim Biophys Acta 113:632.
Beevers CA, Cochran W (1946) Proc Roy Soc London 190:257.
Bilik V, Kucan S (1970) Carbohydr Res 13:311.
Bilik V, Petrus L, Farkas V (1978) Collect Czech Chem Commun 43:1163.
Bloomfield C (1969) Chem Ind 1633
Bloomfield C, Kelso WI, Pruden G (1976) J Soil Sci 27:16.
Blunden SJ, Cusack PA, Smith PJ, Barnard PWC (1983) Inorg Chim Acta 72:217.
Böeseken J (1949) Adv Carbohydr Chem 4:189.
Bowling DJF (1976) "The Uptake of Ions by Plant Roots" Chapman and Hall, London
Branca M, Micera G, Kozlowski H (1989) J Chem Soc Dalton Trans 1283
Brown DH, McPherson J (1970) J Inorg Nucl Chem 32:3309.
Buffle J (1988) "Complexation Reactions in Aquatic Systems; an Analytical Approach", Ellis Horwood, Chichester.
Bunel S, Ibarra C (1985) Polyhedron 4:1537.
Caldeira MM, Ramos ML, Oliveira NC, Gil VMS (1987) Can J Chem 65:2434. Castro MMCA, Geraldes CFGC (1987) Inorg Chim Acta 140:377.
Cervilla A, Ramirez JA, Beltran-Porter A (1983) Trans Met Chem 8:21.
Chaney RL, Brown JC, Tiffin LO (1972) Plant Physiol 50:208.
Chen Y, Schnitzer M (1976) Soil Sci Soc Am 40:682.
Deiana S, Ene L, Micera G, Piu P, Gessa C (1980) Inorg Chim Acta 46:249.
Deiana S, Dessi A, Micera G, Gessa C, de Cherchi ML (1983) Inorg Chim Acta 79:231
Dekock PC, Mitchell RL (1957) Soil Sci 84:55.
Dieckert JW, Rozacky E (1969) Arch Biochem Biophys 134:473.
Donaldson JD (1967) Progr Inorg Chem 8:287.
Dudycz L, Kotlicki A, Shugar D (1981) Carbohydr Res 91:31.

Dykeman WR, de Sousa AS (1966) Can J Bot 44:871.

Einspahr H, Bugg CE (1980) Acta Chryst 36A:264.

Epstein EE (1972) "Mineral Nutrition in Plants", Wiley, London.

Epstein E (1973) Sci Am 228:48.

Farago ME (1981) Coord Chem Rev 36:155.

Feldman I, Rich KE (1970) J Am Chem Soc 92:4559.

Ferrier RJ (1978) Adv Carbohydr Chem Biochem 35:3.

Floor M, Kieboom APG, Van Bekkum H (1988a) Recl Trav Chim Pays-Bas 107:362.

Floor M, Kieboom APG, Van Bekkum H (1988b) Dutch Pat Appl NL 88.02907.

Floor M, Kieboom APG, Van Bekkum H (1989) Recl Trav Chim Pays-Bas 108:128.

Floor M, Koek H, Smeets FLM R, Niemantsverdriet E, Peters JA, Van Bekkum H, Kieboom APG. Carbohydr Res *in press*.

Foster AB (1957) Adv Carbohydr Chem 12:81.

Foy CD, Chaney RL, White MC (1978) Ann Rev Plant Physiol 29:511.

Furberg S, Hamener H, Mostad A (1963) Acta Chem Scand 17:2444.

Gamble DS, Schnitzer M, Skinner DS (1971) Can J Soil Sci 35:471.

Gamble DS, Lanford CH,Tong JKP (1976) Can J Chem 54:1239.

Geraldes CFGC, Castro MMCA (1986) J Inorg Biochem 28:319.

Geraldes CFGC, Castro MMCA (1988a) Inorg Biochem 33:47.

Geraldes CFGC, Castro MMCA, Saraiva ME, Aureliano M, Dias BA (1988b) J Coord Chem 17:205.

Geraldes CFGC, Castro MMCA (1989a) J Inorg Biochem 35:79.

Geraldes CFGC, Castro MMCA (1989b) J Inorg Biochem 37:213.

Goebel WF, Barber FH (1983) J Biol Chem 100:573.

Goodman BA, Cheshire MV (1975) Geochim Cosmochim Acta 39:1711.

Gresser MJ, Tracey AS (1986) J Am Chem Soc 108:1935.

Haines AH, Symes KC, Wells AG (1975) Carbohydr Res 41:85.

Hayes ML, Pennings NJ, Serianni AS, Barker R (1982) J Am Chem Soc 104:6764.

Hewett EJ, Smith EA (1975) "Plant Mineral Nutrition", English Universities, London

Hofner W (1970) Physiol Plant 23:673.

How MJ, Halford MDA, Stacey M, Vickers E (1969) Carbohydr Res 11:313.

Irgolic KJ, Martell AE ,Eds.(1985) "Environmental Inorganic Chemistry", VCH Publishers, Deerfield Beach, Fla.

Jayman TCZ, Sivasubramaniam S (1975) J Sci Food Agric 26:1895.

Kennedy CR, How MJ (1973) Carbohydr Res 28:13.

Lenkinski RE, Reuben J (1976) J Am Chem Soc 98:3089.

Leppard GG, Ramamoorthy S (1975) Can J Bot 53:1729.

Lindsay WL, Carson EW , Eds.(1974) "The Plant Root and its Environment", Charlotville University Press, Virginia.
Makkee M, Kieboom APG, Van Bekkum H (1985) Carbohydr Res 138:225.
Micera G, Deiana S, Gessa C, Petrera M (1981) Inorg Chim Acta 56:109.
Micera G, Dessi A, Kozlowski H, Radomska B, Urbanska J, Decock P, Dubois B, Olivier I (1989) Carbohydr Res 188:25.
Perlin AS, von Rudloff E (1965) Can J Chem 43:2071.
Ramamoorthy S, Leppard GG (1977) J Theor Biol 66:527.
Rees DA (1969) Adv Carbohydr Chem Biochem 24:267.
Reuben J (1984) J Am Chem Soc 106:6180.
Rorison IH ,Ed.(1969) "Ecological Aspects of the Mineral Nutrition of Plants", Symp Br Ecol Soc, Sheffield, Blackwell, Oxford.
Rosenberg B, Vancamp L, Trosko JE, Mansom VH (1969) Nature (London) 222:385.
Sanchelli V (1969) "Trace Elements in Agriculture" Van Nostrand, New York.
Schwegler M, Floor M, Van Bekkum H (1988) Tetrahedron Lett 29:823.
Siddiques IR, Punoes CB (1963) Can J Chem 41:382.
Stevenson FJ, Goh KM (1971) Geochim Cosmochim Acta 35:471.
Stevenson FJ, Ardakani MS (1972) "Micronutrients in Agriculture", J.J.Mortvedt, P.Giordano and W.L.Lindsay, Eds., Soil Sci.Am., Madison, WI p.79
Stoddart JF (1971) "Stereochemistry of Carbohydrates", Wiley, New York.
Stuerman HD, Paym JR (1976) Geochim Cosmochim Acta 40:1109.
Sunderman Jr FW (1979) Biol TraceElem Res 1:63.
Tajmir-Riahi HA (1983) Carbohydr Res122:241.
Tajmir-Riahi HA (1984a) Carbohydr Res 125:13.
Tajmir-Riahi HA(1984b) Carbohydr Res127:1.
Tajmir-Riahi HA (1984c) J Inorg Biochem 22:55.
Tajmir-Riahi HA(1985) J Inorg Biochem 24:127.
Tajmir-Riahi HA (1986a) J Coord Chem 15:95.
Tajmir-Riahi HA(1986b) J Inorg Biochem 27:123 .
Tajmir-Riahi HA (1986c) J Inorg Biochem 27:205.
Tajmir-Riahi HA (1986d) Biophys Chem 23:223.
Tajmir-Riahi HA (1986e) Inorg Chim Acta 119:227.
Tajmir-Riahi HA (1986f) Inorg Chim Acta 125:43.
Tajmir-Riahi HA(1987a) J Inorg Biochem 31:255.
Tajmir-Riahi HA (1987b) Inorg Chim Acta 135:67.
Tajmir-Riahi HA(1987c) Inorg Chim Acta136:93.
Tajmir-Riahi HA (1988a) Carbohydr Res 172:1.
Tajmir-Riahi HA (1988b) Carbohydr Res 172:327.

Taylor GE, Waters JM (1981) Tetrahedron Lett 22:1277.

Templeton DM, Sarkar B (1985) Biochem J 230:35.

Tiffin LO, Brown JC (1962) Science 135:311.

Tiffin LO (1970) Plant Physiol 45:280.

Tracey AS, Gresser MJ, Parkinson KM (1987) Inorg Chem 26:629.

Tracey AS, Gresser MJ (1988a)Inorg\Chem 27:2695.

Tracey AS, Gresser MJ, Liu S (1988b) J Am Chem Soc 110:5869.

Tsubomura T, Yano S, Yoshikawa S (1983) Polyhedron 2:123.

Tsubomura T, Yano S, Kobayashi K, Sakurai T, Yoshikawa S (1986) J Chem Soc Chem Commun 459

Van Duin M, Peters JA, Kieboom APG, Van Bekkum H (1984) Tetrahedron 40:2901.

Van Duin M, Peters JA, Kieboom APG, Van Bekkum H (1985) Tetrahedron 41:3411.

Voelter E, Bayer E, Records R, Bunnenberg E, Djerassi C (1968) Justus Liebigs Ann Chem 718:238.

Weigel H (1963) Adv Carbohydr Chem 18:61.

Weiss L (1973) J Natl Cancer Inst 50:3.

Whisler RL, Richards GN (1958) J Am Chem Soc 80:4888.

Yano S (1988) Coord Chem Rev 92:113.

KINETIC STUDIES ON THE INTERACTION OF METALS BETWEEN WATER AND CLAY MINERAL

D. Donnert, S.H. Eberle, J. Horst
Kernforschungszentrum Karlsruhe
P.O. Box 3640
D-7500 Karlsruhe 1
Federal Republic of Germany

1. Summary

Calculations have shown that the trace species Cu, Zn, Pb and Ni, as well as the main constituents Ca and Mg, exist in surface waters as carbonate complexes which are transformed to NTA-complexes at concentrations of more than 20 μg/l NTA.
Measurements of the immobilization kinetics of zinc and copper showed that this process can be mathematically described by a diffusive transport across the interstitial water of the clay material. The immobilization is a slow process lasting some hundred hours and takes place only in a small upper layer of the undisturbed clay material. The calculations lead to diffusion coefficients for zinc and copper which are in the same order of magnitude than those given in the literature for the transport of these two cations in the free solution.

2. Introduction

One of the main proposed solutions for reducing the phosphorus concentrations of the waters in the Federal Republic of Germany is the substitution of phosphate in detergents by chelators like nitrilotriacetic acid (NTA) or ethylenediamine tetraacetic acid (EDTA). But, by doing this, anòther problem may occur: discharge of the complexing agents via the waste water path and remobilization of heavy metals from the sediments in rivers. Therefore until now restrictions regarding the use of NTA have been layed down by the government of the Federal Republic of Germany based upon an experimental study (Bernhardt, 1984).

In order to be able to decide whether this limitation should be continued, research work was started with the aim to develop a kinetic model of the remobilization process of heavy metals from the sediments caused by chelators, which allows mathematical pre-

NATO ASI Series, Vol. G 23
Metal Speciation in the Environment
Edited by J. A. C. Broekaert, Ş. Güçer, and F. Adams

dictions. For this it is necessary to investigate the equilibrium chemical speciation in the water and to perform kinetic experiments on immobilization as well as on remobilization of heavy metals from the sediments by chelators.

3. Equilibrium studies

A computer program MULTI (Hennes, 1982; Eberle et al., 1982) based on a complex-chemical model was developed to get more information about a water system than analytical data can provide. The result obtained is a speciation of the complete water system which for example allows the prediction of water purification measures or of the discharge of chemicals like nitrilotriacetic acid.

Although some commercially available programs (Tuesdell, 1974; Parkhurst, 1980) already exist, the own program has advantages regarding the flexibility - it easily allows the adding or omitting of species as well as a change of the constants - and the velocity of the calculations.

3.1 Physical-chemical fundamentals

MULTI uses the mass balance of each component (i.e. a constituent or compound) which is present in the water. Each compound is considered to be formed from its constituents.

The equilibrium of the formation of a compound is defined by

$$CA(j) = K(j) \prod_{i} CA(i)^{N(j;i)} \tag{1}$$

with

K(j) = formation constant of the species j listed in (Eberle, 1989) which are calculated for temperatures other than 298 K with the Van't Hoff equation,

CA(j) = activity of the species j which are calculated from the Debye-Hückel equation,

i = number of the constituent species,

j = number of the compound species,

N(j;i) = stoichiometric index of the compound,

For practical purposes the constituents are also treated as compounds with a formation constant of 1. The total concentration of a constituent is defined as:

$$tC(k) = \sum_j \frac{N(j;k)}{F(j)} CA(j) \tag{2}$$

with

$tC(k)$ = total concentration of the constituent k

$F(j)$ = activity coefficient of the species j

This equation system (2) is nonlinear and is solved in MULTI by a Newtonian iteration. Thus, the concentrations of all components are obtained.

MULTI iteratively calculates all concentrations of the constituents until the calculated total concentration of a species $tC(k)_0$ is equal to its real total concentration determined analytically.

3.2 *Results of calculations with MULTI*

Table 1 shows the influence of the addition of different amounts of NTA on the calculated speciation of heavy metals in River Rhine water.

Table 1.

Heavy metal speciation in River Rhine water at pH = 7.5

	0 μg/l NTA	100 μg/l NTA	500 μg/l NTA
Zn	Zn^{2+}, $Zn(CO_3)^0$	50 % NTA-Komplex	96 % NTA-Komplex
Cd	Cd^{2+}, $Cd(OH)^+$, $Cd(SO_4)^0$	12 %	58 %
Pb	Pb^{2+}, $Pb(CO_3)^0$	92 %	100 %
Ni	Ni^{2+}, $Ni(CO_3)^0$	90 %	98 %
Cu	Cu^{2+}, $Cu(CO_3)^0$	100 %	100 %

Regarding zinc at a pH of 7.50 about 85 % is free Zn^{2+}, and about 10 % is complexed as $Zn(CO_3)^0$.

After the addition of 100 μg/l H_3NTA about 50 % of the zinc is complexed by NTA, 40 % is still Zn^{2+}, and about 5 % remained as $Zn(CO_3)^0$. If 500 μg/l H_3NTA are added, 96 % of the Zn is complexed by NTA.

Table 1 also shows the changes of the calculated speciation of some other heavy metals in River Rhine water caused by different concentrations of NTA. According to these results the chemical state of Cu, Pb and Ni, and to a less extent of Cd, is particulary influenced by NTA.

Figure 1 shows the NTA speciation in the River Rhine water. With an increase of the NTA-concentration a larger amount is bound as a Ca-(NTA)$^-$-complex, but primarily the heavy metal complexes of NTA are formed. Only about 5 % are "free NTA", which is not bound.

Regarding the Ca-concentration, the amount of the NTA-complex is very small (table 1) because Calcium is present in a high surplus with respect to NTA.

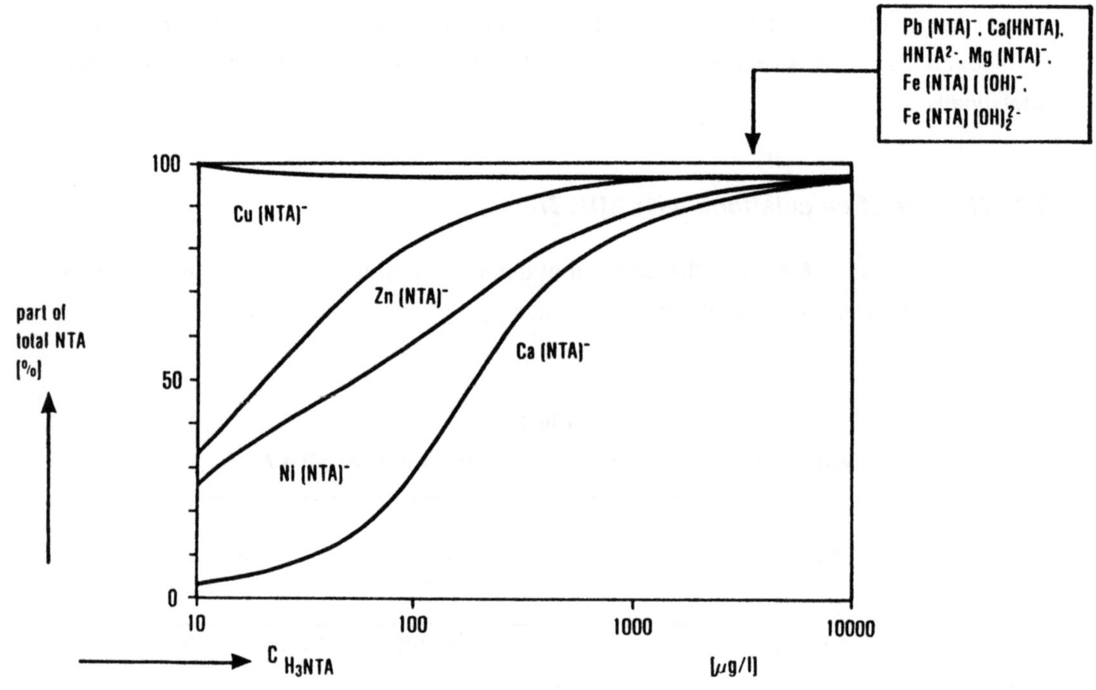

Figure 1. Complexation of NTA in River Rhine water

3.3 Conclusions

The calculations show that a NTA-concentration of 100 μg/l causes a decrease of the Zn^{2+} - concentration of 50 %. This in turn creates a driving force for a reestablishment of the water/sediment equilibrium in the direction of remobilization. The new equilibrium concentration may be calculated on the basis of an adsorption isotherm based on the knowledge of the sediment loading. However, only the possible upper limit of the concentration can be calculated using the adsorption isotherm; the actual value mainly depends on the transport kinetics and the relation of water volume to the sediment affected by the exchange process.

4. Kinetic studies on the water/clay mineral exchange of heavy metals

4.1 Introduction

If a complex forming substance is introduced into natural water, all equilibria in the water as well as in the clay mineral are disturbed and the system tends towards a new equilibrium state. The concentrations in the water phase of a river are time dependent, whereas the concentrations in the clay mineral are time as well as site dependent. This is valid even if the complex forming agent is in the form of a certain complex, e.g. the calcium complex, which may be assumed for a waste water containing NTA.

4.2 Development of a mathematical model of remobilization kinetics

The derivation of the equations is demonstrated in this section through a simple example, the remobilization of copper in a stationary solution phase containing only $CuCl_2$, NaCl and Na_2HNTA. Of course, for a realistic system more species (e.g. Ca^{2+}, HCO_3^-) must be regarded.
The following points must be considered:

1. Mass balance equations describing the clay mineral system

2. Changes of the species concentrations in the water phase

3. Reaction and exchange equilibria

4. Electroneutrality in the solution and in the clay mineral

4.2.1 Assumptions for modelling the system

1. If the solution contains sodium and chloride in addition to copper and NTA, the following species which are listed with abbreviations must be regarded for the calculation:

   $1 = Na^+$, $2 = Cu^{2+}$, $3 = Cl^-$, $4 = NTA^{3-}$, $5 = CuNTA^-$

   Further influences of H^+ - and OH^- -ions or of other CuNTA- complexes may be disregarded as a first assumption in the case of natural surface waters at a pH range of 6.5 to 7.5.

2. The dominating step of the reaction is the diffusion which needs a considerable time compared to the chemical reactions.

3. The adsorption of NTA complexes onto the clay mineral phase is disregarded.

4. The clay mineral layer is considered as a homogeneous phase.

5. The diffusion is only effective perpendicular to the clay mineral surface, i.e. no convective transport is regarded.

6. The mass transport between solution and clay mineral takes place across a film of thickness δ_F with a linear concentration gradient.

7. The diffusion coefficients are constant for all species and are the same in the film and the interstitial water, respectively.

8. For the exchange equilibria of the metal ions between clay mineral and the interstitial water a Langmuir-Isotherm is assumed.

9. No other complex forming substances (e.g. humic acid for copper) are present in the water.

A scheme of the modelling system is shown underneath in figure 2.

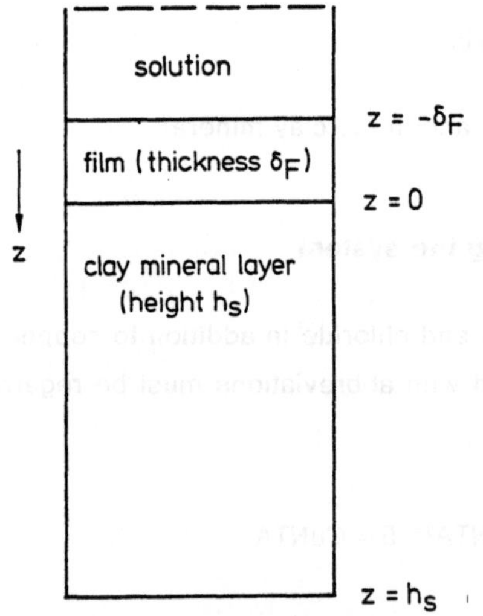

Figure 2. Scheme of the modelling system (z = site variable)

4.2.2 Mass balance equations describing the diffusion across the clay mineral system

For the diffusion within the clay mineral layer the differential equations (3) - (6) are obtained if each constituent is balanced around a differential layer. The following symbols are used:

$c_{i,s}$ = concentration of the component i in the interstitial water, t = time, ε = porosity of the layer, ρ_T = density of the clay mineral, $q_{i,s}$ = loading of the clay mineral with the component i, $J_{i,s}$ = mass flux of the component i in the layer relative to its surface area

$$\text{Na:} \quad \frac{\partial c_{1,S}}{\partial t} + \frac{1-\varepsilon}{\varepsilon} \rho_T \frac{\partial q_{1,S}}{\partial t} + \text{div } J_{1,S} = 0 \tag{3}$$

$$\text{Cu:} \quad \frac{\partial c_{2,S}}{\partial t} + \frac{\partial c_{5,S}}{\partial t} + \frac{1-\varepsilon}{\varepsilon} \rho_T \frac{\partial q_{2,S}}{\partial t} + \text{div } J_{2,S} = 0 \tag{4}$$

$$\text{Cl:} \quad \frac{\partial c_{3,S}}{\partial t} + \text{div } J_{3,S} = 0 \tag{5}$$

$$\text{NTA:} \quad \frac{\partial c_{4,S}}{\partial t} + \frac{\partial c_{5,S}}{\partial t} + \text{div } J_{4,S} + \text{div } J_{5,S} = 0 \tag{6}$$

These are coupled partial differential equations of second order which describe the changes of the species concentration and loadings with respect to time and site within the clay mineral layer. The fluxes $J_{i,s}$ are calculated from the Nernst-Planck-Equation (D_i = individual diffusion coefficients, z_i = ionic charge).

$$J_{i,S} = -D_i \text{ grad } c_{i,S} + D_i \frac{z_i\, c_{i,S}}{\sum_i D_i z_i^2 c_{i,S}} \sum_i D_i z_i \text{ grad } c_{i,S} \tag{7}$$

This equation already contains the condition of electroneutrality demanding identical ionic fluxes:

$$\sum_i z_i J_{i,S} = 0 \tag{8}$$

- Boundary conditions

In order to solve the equation system described above the boundary conditions for z = 0 und z = h_s (depth of the clay mineral layer) must be known.

Boundary conditions at z = 0

The boundary conditions at z=0 are determined from the equality of the mass flux at the boundary line between the film ($J_{i,F}$) and the clay mineral ($J_{i,s}$), i.e.

$$J_{i,F}(z=0) = J_{i,s}(z=0) \tag{9}$$

1. The mass flux in the film results from equation (7) by substituting grad $c_{i,s}$ by the linear concentration gradient in the film $\frac{(c_{i,L} - c^{\star}_{i,s})}{\delta_F}$, and $c_{i,s}$ is replaced by $c^{\star}_{i,s}$ ($c_{i,L}$ = concentration in the solution, $c^{\star}_{i,s}$ = boundary concentration, δ_F = film thickness).

2. Mass flux in the outer clay mineral layer results again from equation (7) by replacing $c_{i,s}$ by $c^{\star}_{i,s}$.

Boundary conditions at z = h_s

No mass transfer takes place across the boundary z = h_s , therefore the mass flux, i.e. the change of the concentrations in the interstitial water, is set to zero:

$$[\,\text{grad } c_{i,s}]_{z=h_s} = 0 \tag{10}$$

4.2.3 Change of the species concentration in the water phase

If the volume of the solution is not infinitely large, the changes of the species concentrations in the solution must be regarded. The corresponding equations may be derived from the mass balance at the interface between solution and film:

$$\dot{N}_{i,L}(z=-\delta_F) = \dot{N}_{i,F}(z=-\delta_F) \tag{11}$$

with

$\dot{N}_{i,L}$, $\dot{N}_{i,F}$ = mass flux in the solution and film,

z = site variable (figure 2).

The mass flux in the solution is defined as:

$$\dot{N}_{i,L} = V_L \frac{dc_{i,L}}{dt} \tag{12}$$

with V_L = volume of the solution.

The mass flux in the film is defined as:

$$\dot{N}_{i,F} = A_S \, \varepsilon \, J_{i,F} \tag{13}$$

with

A_s = surface area of the clay mineral,

ε = porosity of the sediment layer.

The following equation (14) is obtained by substituting with (12),(13) in (11) and allows the calculation of the species concentrations in the solution.

$$\frac{dc_{i,L}}{dt} = \frac{A_S \, \varepsilon}{V_L} \, J_{i,F} \tag{14}$$

4.2.4 Reaction and exchange equilibria

Besides the mass balances described in the previous sections the equilibria concerning the reactions and exchanges must be fulfilled simultaneously as well. The complex formation may be described by the mass action law.

1. The following equilibrium relationships exist for the free solution and the interstitial water:

$$c_{5,L} = K_B \, c_{2,L} \, c_{4,L} \tag{15}$$

$$c_{5,S} = K_B \, c_{2,S} \, c_{4,S} \tag{16}$$

K_B is the equilibrium constant of the complex formation.

2. The exchange equilibrium between the interstitial water and the clay mineral is described by the Langmuir-Isotherm (see 4.2.2 equation (4)):

$$q_{2,S} = \frac{q_{max}^{x}}{z_2} \, \frac{c_{2,S}}{c_{2,S} + K_L} \tag{17}$$

with

K_L = Langmuir-Constant,

q_{max}^{x} = maximum loading of the clay mineral in meq/g

4.2.5 Electroneutrality in the solution and in the mineral phase

Finally the electroneutrality of the solution as well as of the interstitial water must be maintained. Thus as additional conditions the following two equations (z_i = ionic charges) are obtained:

$$\sum_i z_i \, c_{i,L} = 0 \tag{18}$$

$$\sum_i z_i \, c_{i,S} = 0 \tag{19}$$

and a third one for the loading of the clay mineral:

$$q_{max}^{x} - \sum_i z_i \, q_{i,S} = 0 \tag{20}$$

4.2.6 Calculation procedure of the computer program

In order to solve the equations described above initial concentrations and loadings must be assumed, then a finite difference method is applied. The diffusion and the adjustment of the equilibria (both the equilibria of the reaction and the exchange) are calculated separately.

For each time increment the concentration profile in the interstitial water is calculated, which is affected by diffusion. As the next step the local equilibria are calculated using the mass balance equations.

By means of this method, the concentrations of the species in the solution as well as the concentration or loading profiles within the clay mineral phase may be calculated as a function of the reaction time. Five diffusion coefficients D_i and one film thickness δ_F are necessary which can be obtained either from literature or by fitting the calculated concentration vs time curves to the experimental data.

The computer program uses FORTRAN on a PC, e.g. AT80386, with a mathcoprocessor.

4.3 Experiments

In order to describe the whole system, experiments were conducted investigating the immobilization of heavy metals onto kaolinite from aqueous solutions not containing NTA. Thus the calculation is simplified, e.g. only the equations (10) to (12) must be used, equation (11) is simplified because NTA (term 5) is omitted. The equations (15) and (16) are only describing complex formation reactions of NTA and are therefore omitted as well.

This immobilization process may be described in two steps:

- diffusion of the metal ion into the clay mineral
- sorption/desorption of the metal ion by the kaolinite

A concentration gradient in the water phase may be disregarded because the solution was slightly stirred without resuspending the clay mineral layer (figure 3).

4.3.1 Sorption equilibria

Kaolinite was used as clay mineral which was pretreated with a concentrated NaCl-solution in order to eliminate exchangeable foreign ions from the raw material.
Afterwards the kaolinite was equilibrated with a solution of copper or zinc chloride for about one week. The ionic strength was adjusted by adding NaCl to 0.005 moles/l.

4.3.2 Immobilization kinetics

As the first step experiments were carried out regarding immobilization of copper and zinc ions onto kaolinite.

The kaolinite, which had 77 % by weight of interstitial water, was contacted in a device shown in figure 3 with a 0.005 molar NaCl-solution containing copper or zinc. The solution was slightly stirred to affect a mixing, but not to resuspend the kaolinite into the water. The reason for this measure is the fact that in natural aquatic systems like rivers and lakes the clay mineral is only resuspended in storm events. The decrease of the corresponding metal concentration was measured as a function of time. It was necessary to use teflon holders to avoid sorption effects on the walls of the beakers.

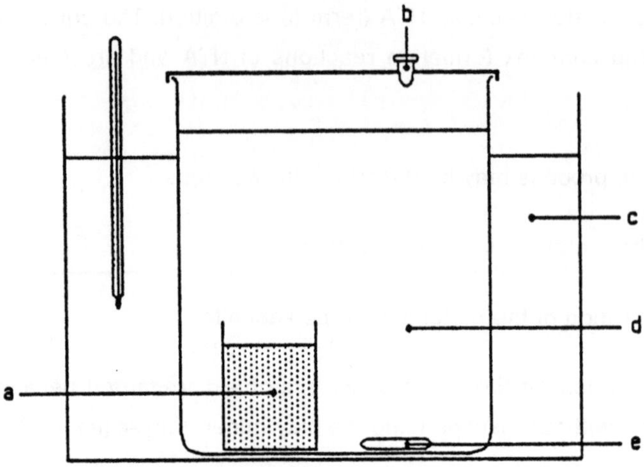

Figure 3. Experimental device for the measurement of the immobilization kinetics: (a) sample holder with kaolinite layer, (b) opening for taking samples, (c) water bath, (d) solution, (e) magnetic stirrer

4.4 Results

4.4.1 Sorption equilibria

The results of the experiments described in the previous chapter were evaluated by the Langmuir-Equation (17). The constants obtained are listed in table 2.

Table 2

Constants of the Langmuir-Equation

metal ion	q_{max} [μg/g]	q_{max} [μmol/g]	K_L [μg/l]	K_L [μmol/l]
Cu^{2+}	556	8.7	263	4.1
Zn^{2+}	814	12.5	728	11.1
Ni^{2+}	789	13.4	705	12.0

4.4.2 Immobilization kinetics

4.4.2.1 Immobilization of copper

The figures 4 and 5 show experimental data of the immobilization of copper ions on kaolinite as a function of the reaction time together with the calculated residual concentration. The fit is as satisfactory as may be expected for this type of experiment and may be regarded as a proof of the appropriateness of the model assumptions. The model is at least able to calculate the diffusive exchange within the experimentally investigated range.

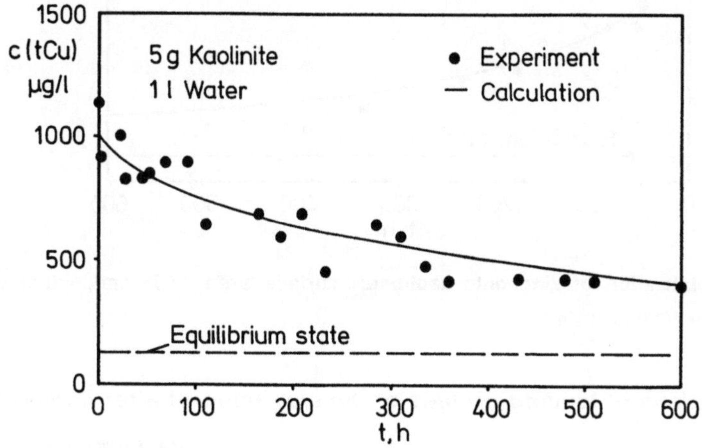

Figure 4. Immobilization of copper onto kaolinite: surface area of the kaolinite 10.75 cm², initial concentration 1000 μg/l Cu

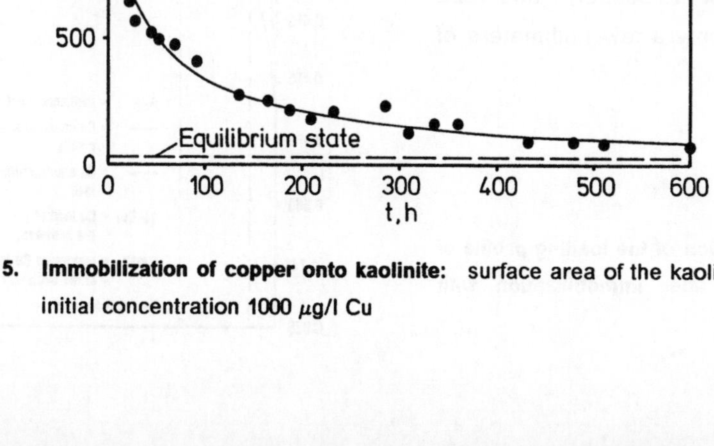

Figure 5. Immobilization of copper onto kaolinite: surface area of the kaolinite 43.00 cm², initial concentration 1000 μg/l Cu

4.4.2.2 *Immobilization of zinc*

The comparison of calculation and experiment for the immobilization of zinc onto kaolinite is shown in figure 6 for an initial concentration of 1000 μg/l Zn. The calculation fits quite well with the experiments.

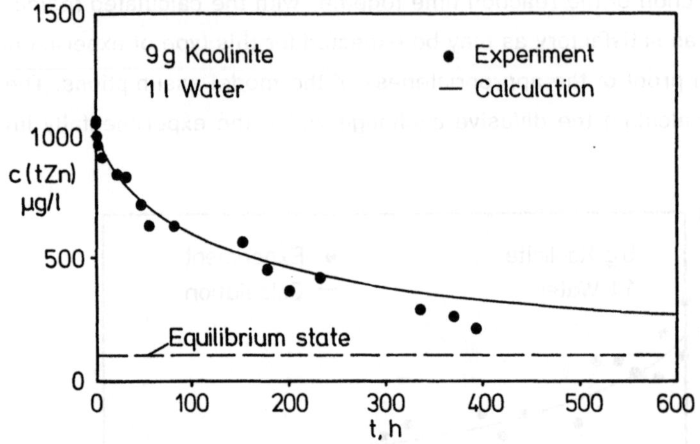

Figure 6. Immobilization of zinc onto kaolinite: surface area 17.35 cm², initial concentration 1000 μg/l Zn

4.4.2.3 *Calculations of loading profiles of the clay mineral after immobilization with copper*

The model calculation describes the development of the concentration in the aqueous phase as well as the loading of the clay column, the latter one not being tested experimentally so far. But it is interesting to look at the results in figure 7 which show that after a reaction time of 600 hours most of the copper is adsorbed into a depth of only a few millimeters of the kaolinite layer.

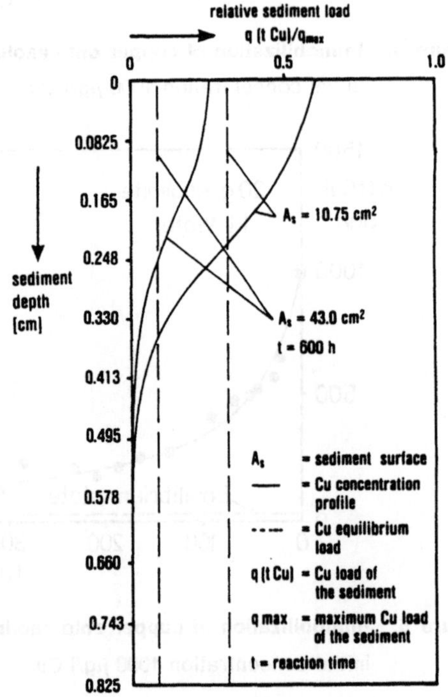

Figure 7. Calculation of the loading profile of a kaolinite layer after immobilization with copper

4.4.2.4 *Comparison of the results of the kinetic experiments with the literature*

Table 3 shows the diffusion coefficients which were calculated from our experimental data together with results from the literature. Both diffusion coefficients, for copper as well as for zinc, are practically equal to the values given in the literature for the transport in a free solution.

Table 3
Comparison of the diffusion coefficients with results from the literature

diffusion coefficients		
	our experiments	literature (Landolt-Börnstein, 1969)
$D(Cu^{2+})$	0.72×10^{-5} cm²/s	0.72×10^{-5} cm²/s in a 0.0056 m $CuSO_4$ solution at 25°C
$D(Zn^{2+})$	1.0×10^{-5} cm²/s	0.71×10^{-5} cm²/s in a 0.0044 m $ZnSO_4$ solution at 25°C
		1.38×10^{-5} cm²/s in a 0.0001 m $Zn(NO_3)_2$ solution at 20°C

4.5 *Conclusions*

Our conclusion is that that the mathematical model leads to realistic results and is therefore suitable for the description of the experiments, because the diffusion coefficients obtained have reasonable values compared with the literature, and, because the calculated values of the immobilization reactions fit well with the measured experimental data.

The suitability of the model for the remobilization predictions will be validated in the course of further experiments.

References

Bernhardt H (1984) Studie über die aquatische Umweltverträglichkeit von Nitrilotriacetat (NTA). Hans Richarz St. Augustin - ISBN 3-88345-376-5

Hennes ECh (1982) Entwicklung und experimentelle Überprüfung eines komplexchemischen Gleichgewichtsmodells für Gewässer am Beispiel des Rheinwassers. Thesis, Technische Universität Karlsruhe, FRG

Eberle SH, Hennes ECh, Dehnad F (1982) Berechnung und experimentelle Prüfung eines komplexchemischen Modells der Hauptkonstituenten des Rheinwassers. Z Wasser Abwasser Forsch 15(5):217-229

Eberle SH (1989) A correctness test of the computation of chemical speciation for the main constituents of natural waters. Wat Res 23(11):1373-1382

Landolt-Börnstein (1969) Zahlenwerte und Funktionen. Springer Verlag Heidelberg New York. In: Bd 5a, S 631-632

Parkhurst, DL (1980) Phreeqe - a computer program for geochemical calculations. In: Water-Resources Investigations p 90-96

Tuesdell, AH (1974) A computer program for calculating chemical equilibria of natural waters. J Res US Geol Surv 2(2):233-248

The occurrence of heavy metals in Antarctic and Greenland ancient ice and recent snow

C.F. Boutron and U. Görlach
Laboratoire de Glaciologie et Géophysique de l'Environnement du CNRS
54, rue Molière, Domaine Universitaire, B.P. 96
38402 St Martin d'Hères Cedex, France

1. Introduction

The investigation of the occurrence of heavy metals such as Pb, Cd, Hg, Cu and Zn in the successive dated snow and ice layers deposited in Antarctica and Greenland provides a unique opportunity to reconstruct the past and present day changes in the large scale tropospheric fluxes of these toxic metals. Data for ancient ice, whose age can extend back to a few hundred thousand years, will indeed allow to determine the past natural pre-man fluxes, hence establishing a firm reference against which to evaluate modern trends. Once these references established, data for snow or ice deposited during the last centuries, especially since the industrial revolution, will allow to directly determine the human impact on these fluxes both in the northern hemisphere (Greenland data) and in the southern hemisphere (Antarctic data).

Such studies have unfortunately been hampered because the concentrations of heavy metals found in polar snow and ice are the lowest detected on Earth. As an illustration, typical concentrations of Pb and Cd in Antarctic ice several thousand years old are as low as about 0.4 pg Pb/g and 0.1 pg Cd/g, respectively, (Boutron and Patterson, 1986 ; Batifol et al., 1989), i.e. concentrations which are much lower than those in ultrapure water produced in most research laboratories. Consequently, many of the published data are erroneous by up to several orders of magnitude because of improper control of contamination during field sampling or/and laboratory analysis (Wolff and Peel, 1985a ; Boutron, 1990a). It is only during the past few years that refined enough procedures have been developed which have allowed to get reliable albeit very incomplete data.

2. Assessing the accuracy of published data

For researchers outside the immediate field, it is often very difficult to evaluate seriously whether a published data set is reliable or has suffered

NATO ASI Series, Vol. G 23
Metal Speciation in the Environment
Edited by J. A. C. Broekaert, Ş. Güçer, and F. Adams

contamination problems. This is mainly due to the fact that the reports often lack details of the sampling and analytical procedures used by the investigators.

2.1. Decontamination of the samples

Whatever precautions are taken to collect the snow or ice samples in the field (Boutron, 1990a), most if not all of them are more or less contaminated on their outside during field collection, transportation and storage. The severity of this outside contamination varies greatly from one kind of sample to another and from one heavy metal to another (Boutron et al., 1989). It is generally very limited, but however significant, for shallow snow samples excavated from the walls of clean pits or drilled using acid cleaned specially designed augers (Boutron and Patterson, 1983). On the other hand, it is extremely high for deep ice cores drilled in holes filled with wall retaining fluids (Ng and Patterson, 1981 ; Boutron and Patterson, 1986 ; Boutron et al., 1987, 1989) : as an illustration, Pb and Zn contamination has been found to be as high as about 20,000 and 300,000 pg/g, respectively, in the outside layer (first 5 mm from the outside or so) of typical sections of the Vostok deep Antarctic ice core, which was drilled in a hole filled with a mixture of kerosene and freon (Boutron et al., 1987, 1989). This is approximately up to five orders of magnitude higher than the real concentration of these metals in the original ice.

If reliable data are to be obtained from the analysis of such samples, it is then of utmost importance to study in full details for each heavy metal how deep the outside contamination has intruded into the samples in order to determine whether the inner parts of the samples are free of contamination or not and are then suitable for heavy metals analysis. It will then be necessary to develop ultraclean laboratory procedures to get these inner parts of the samples without transferring outside contamination to them.

The best available procedure, suitable both for shallow snow samples and for deep ice cores, is to mechanically chisel successive veneers of snow or ice in progression from the outside to the inside of each sample (Ng and Patterson, 1981 ; Boutron and Patterson, 1983, 1986, 1987 ; Boutron et al., 1987, 1989). For low density snow samples, another suitable decontamination procedure consists in pushing ultraclean cylindrical Teflon tubes into a virgin untouched section of the sample obtained by splitting it (Wolff and Peel, 1985b, 1988). Finally, ice cores can also be decontaminated by successive rinsings with ultrapure water (Batifol et al., 1989).

Whatever procedure is used, it is essential to analyse separately for each investigated heavy metal the successive snow or ice layers from the outside surface to the center of each individual sample. If an unambiguous plateau of concentration is observed for at least two consecutive inner layers, see Figure 1A, it clearly indicates that outside contamination has not penetrated to the central part of the sample ; the heavy metals concentrations data obtained from the analysis of these inner layers will then allow to get the genuine heavy metals concentrations in the snow or ice, providing

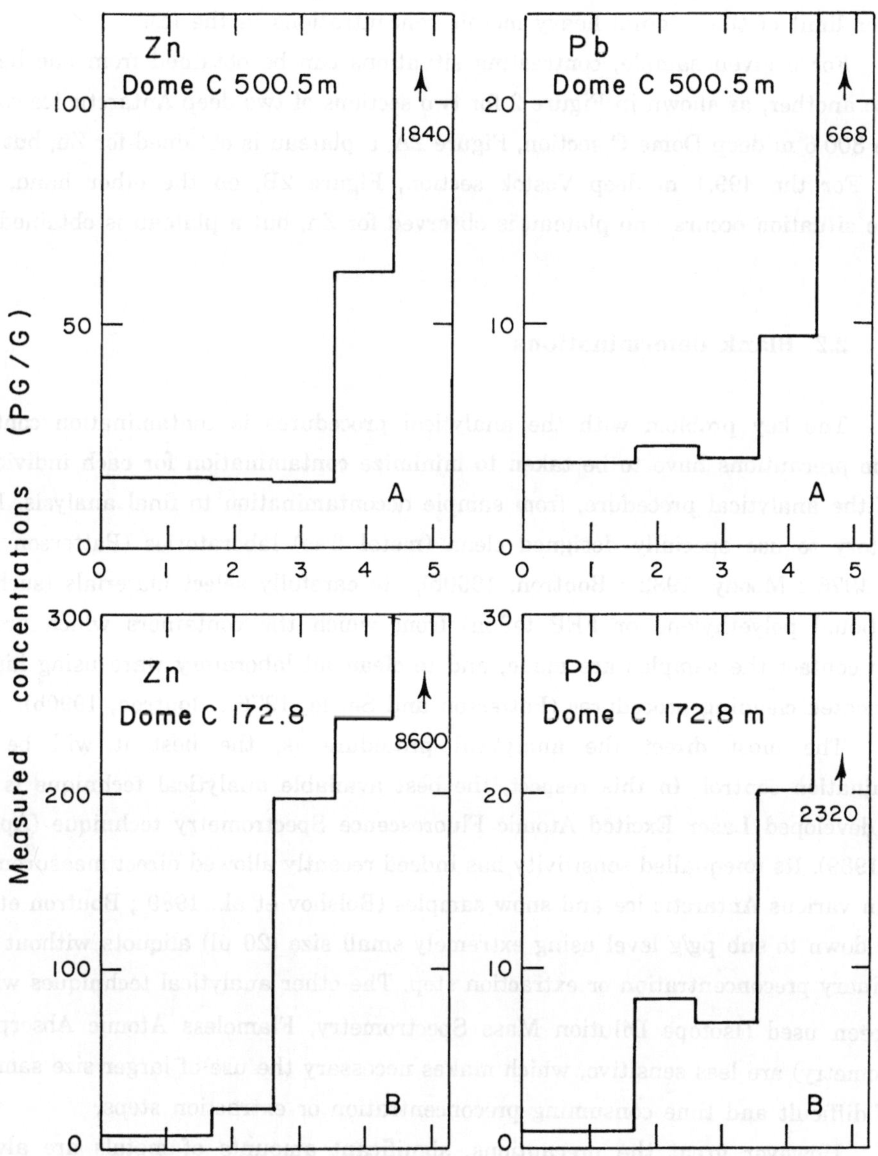

Figure 1 . *Variations in measured concentrations of Zn and Pb as a function of radius in two sections of the 905 m Dome C Antarctic ice core obtained by thermal drilling in a dry hole. (A) : 500.5 m, (B) : 172.8 m*

of course that satisfying blank corrections have been made for the decontamination and analytical procedures. On the other hand, if there is a continuous decrease of concentrations from the outside to the inside, see Figure 1B, this indicates that outside contamination has penetrated to the very center of the samples : the heavy metals data obtained from the analysis of the central part of the sample will then allow to get only an upper limit of the original heavy metals concentrations in the ice.

For a given sample, contrasting situations can be obtained from one heavy metal to another, as shown in Figure 2 for two sections of two deep Antarctic ice cores. For the 300.6 m deep Dome C section, Figure 2A, a plateau is obtained for Zn, but not for Pb. For the 499.1 m deep Vostok section, Figure 2B, on the other hand, the opposite situation occurs : no plateau is observed for Zn, but a plateau is obtained for Pb.

2.2. Blank determinations

The key problem with the analytical procedures is contamination control. Extreme precautions have to be taken to minimize contamination for each individual step of the analytical procedure, from sample decontamination to final analysis. It is mandatory to use specially designed clean (metal free) laboratories (Patterson and Settle, 1976 ; Moody, 1982 ; Boutron, 1990b), to carefully select materials (such as conventional polyethylene or FEP teflon) from which the containers which are to directly contact the samples are made, and to clean all laboratory ware using highly sophisticated cleaning procedures (Patterson and Settle, 1976 ; Boutron, 1990b).

The most direct the analytical procedure is, the best it will be for contamination control. In this respect, the best available analytical technique is the newly developed Laser Excited Atomic Fluorescence Spectrometry technique (Apatin et al., 1989). Its unequalled sensitivity has indeed recently allowed direct measurement of Pb in various Antarctic ice and snow samples (Bolshov et al., 1989 ; Boutron et al., 1990a) down to sub pg/g level using extremely small size (20 µl) aliquots without any preliminary preconcentration or extraction step. The other analytical techniques which have been used (Isotope Dilution Mass Spectrometry, Flameless Atomic Absorption Spectrometry) are less sensitive, which makes necessary the use of larger size samples and of difficult and time consuming preconcentration or extraction steps.

However great the precautions, significant amounts of metals are always added to the samples during analysis. It is therefore essential to be able to accurately determine for each of the investigated heavy metal the exact contribution from the walls of each container, from the air in the clean laboratory, and from each separate

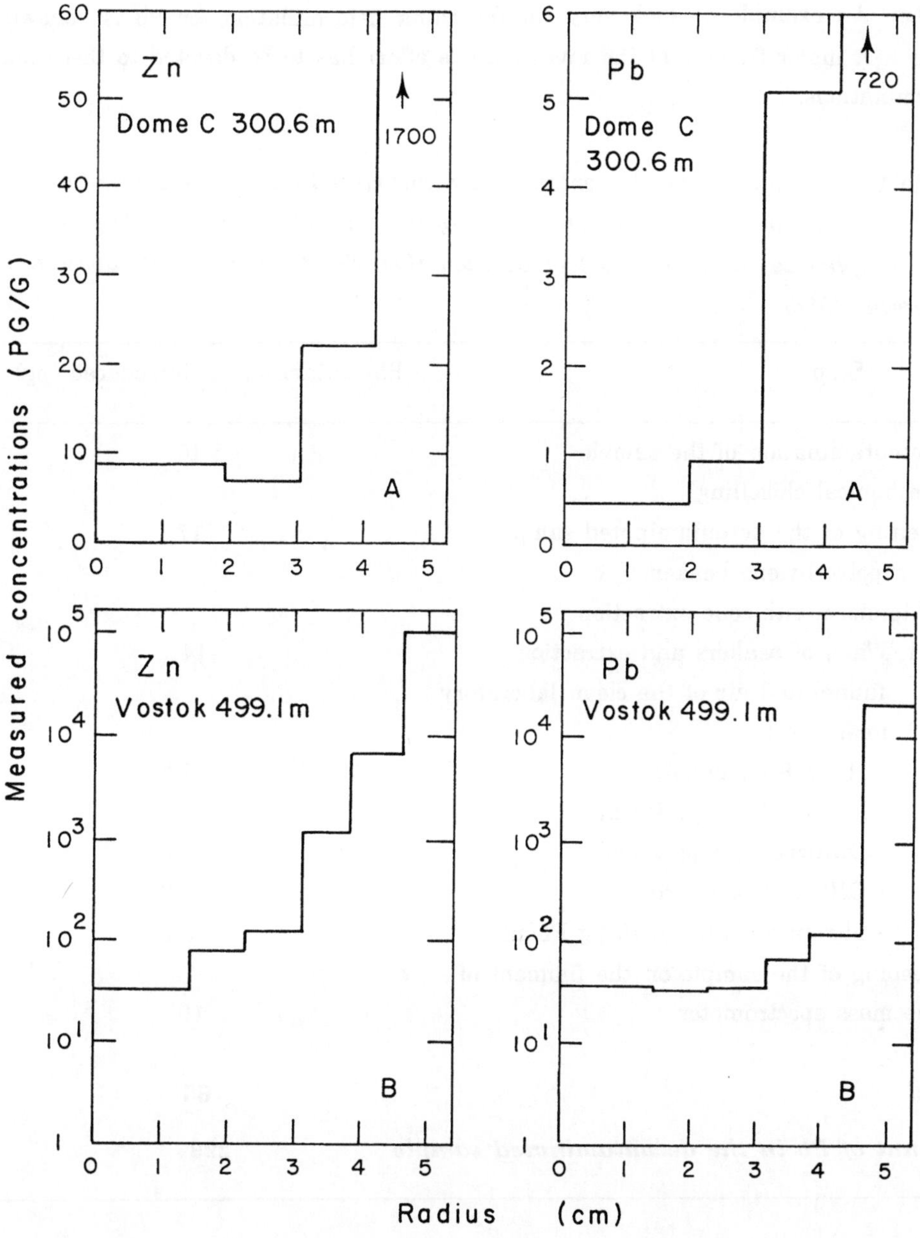

Figure 2 . *Variations in measured concentrations of Zn and Pb as a function of radius in (A) : one section (300.6 m) of the 905 m Dome C Antarctic ice core obtained by thermal drilling in a dry hole, (B) one section (499.1 m) of the 2,083 m Vostok Antarctic ice core obtained by thermal drilling in a fluid-filled hole*

reagent. An example of such very careful blank determination for Pb is shown in Table 1. A major fraction of the investigator's effort has to be devoted to these blank determinations.

Table 1 . *Estimation of Pb contamination introduced during the successive steps of the analysis of an Holocene core section of the Antarctic Dome C deep ice core This sample was analysed by Isotope Dilution Mass Spectrometry. From Boutron and Patterson (1986)*

Step	Pb contamination introduced (pg)
1. Decontamination of the sample (mechanical chiselling)	10
2. Melting of the decontaminated sample in a polyethylene beaker	17
3. Chloroform-dithizone extraction	
3.1. Walls of beakers and extraction funnel and air of the clean laboratory	14
3.2. Reagents :	
- HNO_3(8 pg Pb/ml)	2.3
- NH_4OH (1.28 pg Pb/ml)	0.7
- Dithizone (11 pg Pb/ml)	5.5
- $CHCl_3$ (0.14 pg Pb/ml)	2.7
- ultrapure water (0.16 pg Pb/ml)	3.2
4. Loading of the sample on the filament of the mass spectrometer	10
Total	**65**
Amount of Pb in the decontaminated sample	**223**

3. Reliable data for ancient ice

3.1. Pb and Zn in Antarctic ancient ice

Reliable data are now available on the variations of Pb and Zn concentrations in Antarctic ice for the 155,000 to 4,000 years BP time period, thanks to a series of extremely careful studies on the Dome C and Vostok deep ice cores (Boutron and

Patterson, 1986 ; Boutron et al., 1987 ; Batifol et al., 1989 ; Boutron et al.,1990b). This time interval includes the present Holocene interglacial period (back to about 13,000 years B.P.), the last ice age (from about 13,000 to 115,000 years BP), the next to last interglacial (from about 115,000 to 140,000 years BP) and the last part of the next to last ice age (before about 140,000 years BP), Figure 3A. This period was characterized by major changes in soil and rock dust and sea-salt loading in the global troposphere

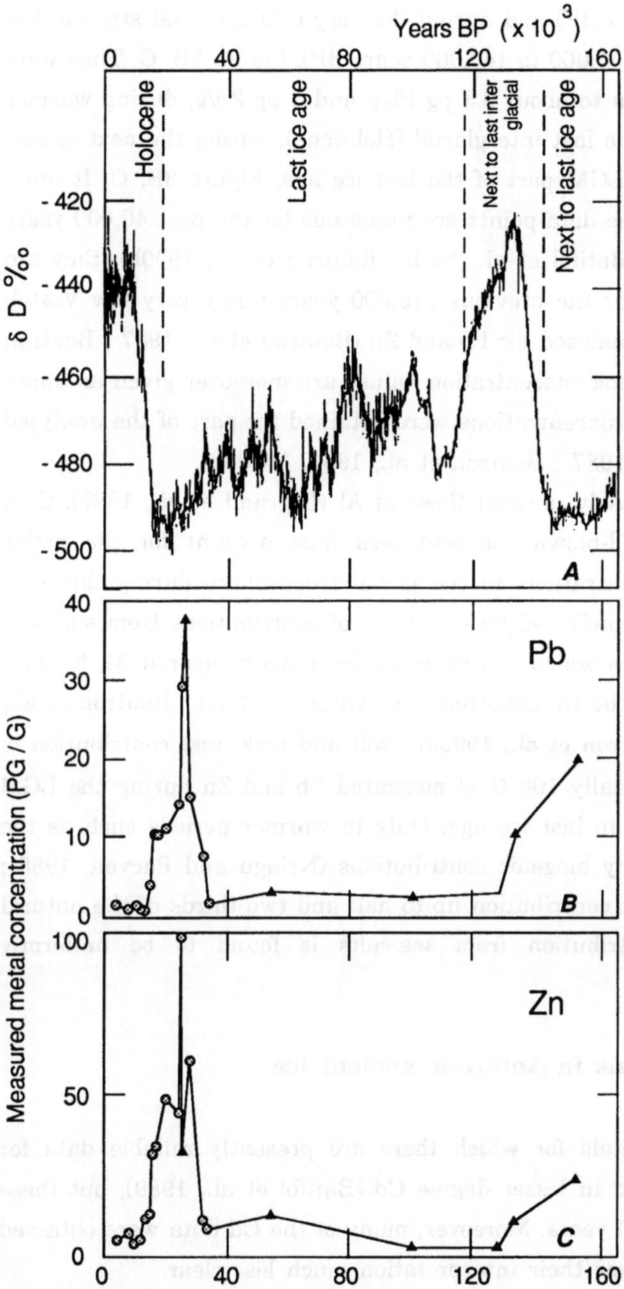

Figure 3 . *(A) : Changes in climate during the past 160,000 years as depicted by measurement of hydrogen isotopes in the Vostok Antarctic ice core (from Jouzel et al., 1987). Approximate limits of the successive climatic stages are also shown. (B) and (C) : Changes in Pb and Zn concentrations during the past 155,000 years in Antarctic ice as seen in the Dome C and Vostok ice cores (from Boutron and Patterson, 1986 ; Boutron et al., 1987 ; Boutron et al., 1990b). Dots are used for data from the Dome C core sections, and triangles for data from the Vostok core sections. Ages assignement for the Dome C core sections is from Jouzel et al. (1989)*

(Legrand et al., 1987), and is therefore an ideal period to determine the relative contribution of the various natural sources of Pb and Zn to the global troposphere.

As shown in Figure 3B, C, Pb and Zn concentrations are found to have strongly varied in Antarctic ice and therefore in the global troposphere during the past 155,000 years. They were rather high, up to about 40 pg Pb/g and 100 pg Zn/g, during the Last Glacial Maximum (LGM, i.e. the very cold terminal stage of the last ice age from about 30,000 to 13,000 years BP) and during the very cold terminal stage of the next to last ice age (from about 155,000 to 140,000 years BP), Figure 3B, C. They were on the other hand very low, down to about 0.3 pg Pb/g and 3 pg Zn/g, during warmer time periods, especially during the last interglacial (Holocene), during the next to last interglacial, and during the pre-LGM part of the last ice age, Figure 3B, C. It must however be emphasized that if the data points are numerous for the past 40,000 years (Boutron and Patterson, 1986 ; Batifol et al., 1989 ; Boutron et al., 1990b), they are on the other hand very sparse for the previous 115,000 years : only very few Vostok core sections have indeed been analysed for Pb and Zn (Boutron et al., 1987 ; Boutron et al., 1990b) ; part of these Vostok concentration values are moreover given as upper limits only, since no plateaus of concentrations were obtained for part of the analysed Vostok sections (Boutron et al., 1987 ; Boutron et al., 1989, 1990b).

These time variations closely parallel those of Al (Legrand et al., 1987), then indicating that changes in wind-blown soil and rock dust account for the major fluctuations of Pb and Zn concentrations in the global troposphere during this pre-man's influence period. This is confirmed by estimates of contributions from soil and rock dust, sea-salts and volcanoes which can be made from the measured Al, Na and excess sulfate concentrations in the ice (Boutron and Patterson, 1986 ; Boutron et al., 1987 ; Batifol et al., 1989 ; Boutron et al., 1990b) : soil and rock dust contribution is indeed found to account for virtually 100 % of measured Pb and Zn during the LGM and during the end of the next to last ice age. Only in warmer periods such as the Holocene do volcanic and possibly biogenic contributions (Nriagu and Pacyna, 1988 ; Nriagu, 1989) make a significant contribution up to half and two thirds of the natural Pb and Zn, respectively. Contribution from sea-salts is found to be uniformly insignificant.

3.2. Other heavy metals in Antarctic ancient ice

The only two other metals for which there are presently reliable data for ancient Antarctic ice are Cu and in lesser degree Cd (Batifol et al., 1989), but these data are only for the past 40,000 years. Moreover, many of the Cd data were obtained as upper limits only, which makes their interpretation much less clear.

During the past 40,000 years, changes in Cu concentrations are found to closely parallel those observed for Pb and Zn during that time period : Cu concentrations are high, up to about 30 pg Cu/g, during the LGM ; they are on the other hand much lower, down to about 1 pg Cu/g, during the Holocene and the pre-LGM period of the last ice age (Batifol et al., 1989). For Cd, the variation profile is not so clear because many of the concentration values are given as upper limits only (< 0.1 pg Cd/g or < 0.3 pg Cd/g). It however suggests that Cd concentrations were possibly higher during the LGM, up to about 0.6 pg Cd/g, than during the Holocene.

As for Pb and Zn, soil and rock dust contribution is found to explain virtually 100 % of measured Cu and possibly Cd during the LGM. During the Holocene and the pre-LGM part of the last ice age, this last contribution explains only part of the measured concentrations : the observed excesses above that contribution are explained by volcanic and biogenic contributions. The sea-salts contribution is again found to be extremely small, Batifol et al. (1989).

3.3. Greenland ancient ice

There are unfortunately presently almost no reliable data on the occurrence of heavy metals in ancient Greenland ice, since most published data have been conclusively shown to be highly erroneous because of major contamination problems, as discussed by Wolff and Peel (1985a) and Boutron (1990a). The only reliable data are for Pb for four Holocene samples whose age ranges from 2,700 to 5,500 years BP. Three of them were sections of the Camp Century deep ice core (Ng and Patterson, 1981), the last one was collected in a coastal ablation area (Murozumi et al., 1969). The corresponding concentration values are moreover given as upper limits only, ranging from < 0.8 pg Pb/g to < 4.1 pg Pb/g. These upper limits are consistent with those obtained for this metal for Holocene Antarctic ice.

4. Reliable data for recent snow

4.1. Pb

For Antarctica, there are now some reliable data for present day (post-1980) snow at a few locations in East Antarctica and in the Antarctic Peninsula area (Wolff and Peel, 1985b ; Boutron and Patterson, 1987). Part of these values are however probably affected by local emissions of Pb by human activities in Antarctica. Detailed inventory of Pb emissions from use of leaded gasoline, diesel fuel and kerosene and from waste burning (Boutron and Wolff, 1989) have indeed conclusively shown that these emissions (1,800 kg Pb yr^{-1}) are probably contaminating very large areas around scientific stations and around flying routes of piston-engined aircrafts. This is illustrated by changes in Pb concentrations in surface snow with increasing distance from Dumont d'Urville and Amundsen Scott scientific stations, Figure 4 (Boutron and Patterson, 1987) : Pb concentrations are found to decrease with increasing distance from the stations.

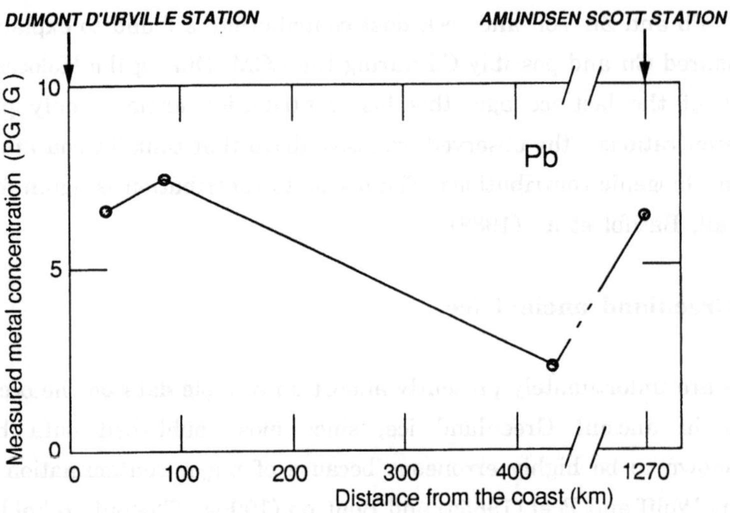

Figure 4 : *Changes in Pb concentrations in East Antarctic surface snow as a function of the distance from Dumont d'Urville and Amundsen Scott scientific stations (from Boutron and Patterson, 1987).*

Only the lowest concentration values (about 2.3 pg Pb/g) are then considered to be representative of large scale pollution. When these last values are compared (Figure 5A) with the few available values for Pb in Holocene Antarctic ice, which average around 0.4 pg Pb/g, it appears that Pb concentrations have probably increased about 6-fold in Antarctic ice and snow and therefore in the remote areas of the southern hemisphere during the past few centuries, Figure 5A. This is further supported by a few additional reliable data points for the 1910-1950 time period, Figure 5A.

For Greenland, the only reliable time series, Figure 5B, is still that reported by Murozumi et al. (1969) which gave unchallenged evidence for a tremendous increase

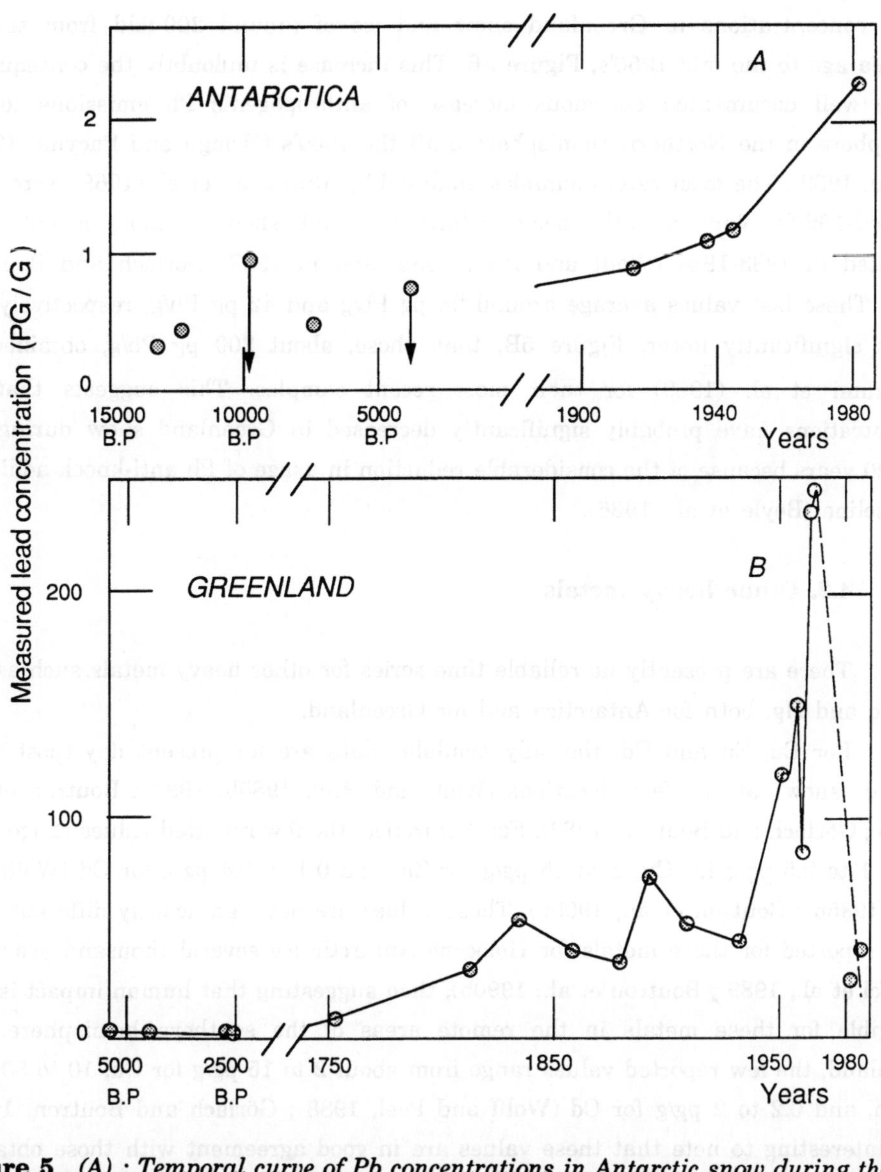

Figure 5 . *(A) : Temporal curve of Pb concentrations in Antarctic snow during the past century (from Boutron and Patterson, 1983, 1987). Concentrations measured in Holocene Antarctic ice several thousand years old (Boutron and Patterson, 1986) are also shown. (B) Temporal curve of Pb concentrations in Greenland snow and ice during the past two centuries (from Murozumi et al., 1969 ; Wolff and Peel, 1988 ; Görlach and Boutron, 1989). Concentrations measured in Holocene Greenland ice several thousand years old (Murozumi et al., 1969 ; Ng and Patterson, 1981) are also shown*

in Pb concentrations in Greenland snow and ice of around 200-fold from several millenia ago to the mid 1960's, Figure 5B. This increase is undoubtly the consequence of the well documented enormous increase of anthropogenic Pb emissions to the atmosphere in the Northern Hemisphere until the 1960's (Nriagu and Pacyna, 1988 ; Nriagu, 1989). The most recent samples analysed by Murozumi et al. (1969) were from the mid-1960's. Very recently, some values were published for more recent snow deposited in 1983-1984 (Wolff and Peel, 1988) and in 1987 (Görlach and Boutron, 1989). These last values average around 28 pg Pb/g and 42 pg Pb/g, respectively, i.e. values significantly lower, Figure 5B, than those, about 200 pg Pb/g, obtained by Murozumi et al. (1969) for their most recent samples. This suggests that Pb concentrations have probably significantly decreased in Greenland snow during the past 20 years because of the considerable reduction in usage of Pb anti-knock additives in gasoline (Boyle et al., 1986).

4.2. Other heavy metals

There are presently no reliable time series for other heavy metals such as Cu, Zn, Cd and Hg, both for Antarctica and for Greenland.

For Cu, Zn and Cd, the only available data are for present day (post-1980) surface snow at a few locations (Wolff and Peel, 1985b, 1988 ; Boutron et al., 1990b ; Görlach and Boutron, 1989). For Antarctica, the few reported values range from about 1 to 2.5 pg/g for Cu, 1 to 15 pg/g for Zn, and 0.1 to 0.4 pg/g for Cd (Wolff and Peel, 1985b ; Boutron et al., 1990b). These values are not significantly different from those reported for these metals for Holocene Antarctic ice several thousand years old (Batifol et al., 1989 ; Boutron et al., 1990b), then suggesting that human impact is still negligible for these metals in the remote areas of the southern hemisphere. For Greenland, the few reported values range from about 2 to 15 pg/g for Cu, 10 to 80 pg/g for Zn, and 0.2 to 2 pg/g for Cd (Wolff and Peel, 1988 ; Görlach and Boutron, 1989). It is interesting to note that these values are in good agreement with those obtained for surface snow from the eastern Arctic Ocean (Mart, 1983). There are unfortunately no values for ancient Greenland ice to which to compare these values, so that it is difficult to determine how important is human impact. Rough estimates of the contribution of soil and rock dust, volcanoes and sea salts to these measured Greenland Cu, Zn and Cd present day concentrations however suggest that human impact is probably limited for Cu, but rather important for Zn and Cd in present day (post-1980) Greenland snow.

For Hg, most published data (see for instance Appelquist et al. (1978) and Murozumi et al. (1978)) are now thought to be highly erroneous, as confirmed by the new preliminary data recently obtained by Sheppard (1989) and Dick et al. (1990) for a few Antarctic surface snow samples from the Horseshoe Mountain area and the Ross Ice Shelf, using improved ultraclean procedures : these new data indicate that the true Hg concentration values for recent Antarctic snow are probably less than 1 pg/g, i.e. values considerably lower than all those previously obtained by other investigators.

5. Conclusions

From this review, it appears that our understanding of the occurrence of heavy metals in Antarctic and Greenland ancient ice and recent snow is still limited, with the exception of Pb for which a high quality but still incomplete data set is now available.

In the near future, the most urgent need will be to obtain reliable and detailed time series tc cover the past few hundred years both in Antarctica and in Greenland at single sites remote from local contamination sources, especially for Cu, Zn, Cd and Hg. The scientists involved will however have to be able to demonstrate at the outset that their data are fully reliable : it will be mandatory to study the variations of the concentrations of the investigated metals from the outside to the inside of each individual sample, and to perform detailed and accurate blank determinations for the entire analytical procedure from sample decontamination to final analysis.

It will also be very interesting to determine the speciation of the investigated heavy metals, both for ancient ice and for recent snow. Such data on the speciation of heavy metals are presently not available for Antarctica and Greenland.

References

Apatin VM, Arkhangelskii BV, Bolshov MA, Ermolov VV, Koloshnikov VG, Kompanetz ON, Kuznetsov NI, Mikhailov EL, Shishkovskii VF, Boutron CF (1989) Automated laser excited atomic fluorescence spectrometer for determination of trace concentrations of elements. Spectrochim Acta 44B : 253-262

Appelquist H, Jensen KO, Sevel T, Hammer C (1978) Mercury in the Greenland ice sheet. Nature 273 : 657-659

Batifol F, Boutron CF, De Angelis M (1989) Changes in copper, zinc and cadmium concentration in Antarctic ice during the past 40,000 years. Nature 377 : 544-546

Bolshov MA, Boutron CF, Zybin AV (1989) Direct determination of lead in Antarctic ice at picogram per gram level by laser atomic fluorescence spectrometry. Anal Chem 61 : 1758-1762

Boutron CF (1990a) Atmospheric lead, cadmium, mercury and arsenic in Antarctic and Greenland recent snow and ancient ice. In Hutchinson TC, Gordon CA, Meema K (eds) National Perspectives on Global Metal Cycling, Scope Rpt 31, Wiley Eastern, New Delhi, India, in press

Boutron CF (1990b) Clean laboratories for ultralow concentration metal analysis. Fresenius Z Anal Chem, in press

Boutron CF, Patterson CC (1983) The occurrence of lead in Antarctic recent snow, firn deposited over the last two centuries and prehistoric ice. Geochim Cosmochim Acta 47 : 1355-1368

Boutron CF, Patterson CC (1986) Lead concentration changes in Antarctic ice during the Wisconsin/Holocene transition. Nature 323 : 222-225

Boutron CF, Patterson CC (1987) Relative levels of natural and anthropogenic lead in recent Antarctic snow. J Geophys Res 92 : 8454-8464

Boutron CF, Wolff EW (1989) Heavy metal and sulphur emissions to the atmosphere from human activities in Antarctica. Atmos Envir 23 : 1669-1675

Boutron CF, Patterson CC, Petrov VN, Barkov NI (1987) Preliminary data on changes of lead concentrations in Antarctic ice from 155,000 to 26,000 years BP. Atmos Envir 21 : 1197-1202

Boutron CF, Patterson CC, Barkov NI (1989) Assessing the quality of thermally drilled deep Antarctic ice cores for trace elements analysis. Proceedings Third Int Workshop on Ice Core Drilling 10-15 Oct 1988 Grenoble, France : 182-197

Boutron CF, Bolshov MA, Koloshnikov VG, Patterson CC, Barkov NI (1990a) Direct determination of lead in Vostok Antarctic ancient ice by laser excited atomic fluorescence spectrometry. Atmos Envir, in press

Boutron CF, Patterson CC, Barkov NI (1990b) The occurrence of zinc in Antarctic ancient ice and recent snow. Earth Planet Sci Lett, in press

Boyle EA, Chapnick SD, Shen GT (1986) Temporal variability of lead in the Western North Atlantic. J Geophys Res 91 : 8573-8593

Dick AL, Sheppard DS, Patterson JE (1990) Mercury content of Antarctic surface snow : initial results. Atmos Envir, in press

Görlach U, Boutron CF (1989) Heavy metal concentration in surface snow from central Greenland. In Vernet JP (ed) Proceedings 7th Int Conf on Heavy Metals in the Environment, 10-15 Sept 1989, Geneva, Switzerland, CEP Consultants Ltd, Edinburgh, UK : 24-27

Jouzel J, Lorius C, Petit JR, Genthon C, Barkov NI, Kotlyakov VM, Petrov VN (1987) Vostok ice core : a continuous isotope temperature record over the last climatic cycle (160,000 years). Nature 329 : 403-408

Jouzel J, Raisbeck G, Benoist JP, Yiou F, Lorius C, Raynaud D, Petit JR, Barkov NI, Korotkevich YS, Kotlyakov VM (1989) A comparison of deep Antarctic ice cores and their implications for climate between 65,000 and 15,000 years ago. Quaternary Res 31 : 135-150

Legrand MR, Lorius C, Barkov NI, Petrov VN (1987) Vostok (Antarctica) ice core : atmospheric chemistry changes over the last climatic cycle (160,000 years). Atmos Envir 22 : 317-331

Mart L (1983) Seasonal variations of Cd, Pb, Cu and Ni levels in snow from the eastern Arctic ocean. Tellus 35B: 131-141

Moody (1982) NBS clean laboratories for trace element analysis. Anal Chem 54 : 1358A-1376A

Murozumi M, Chow TJ, Patterson CC (1969) Chemical concentrations of pollutant lead aerosols, terrestrial dusts and sea salts in Greenland and Antarctic snow strata. Geochim Cosmochim Acta 33 : 1247-1294

Murozumi M, Nakamura S, Yoshida Y (1978) Chemical constituents in the surface snow in Mizuho plateau. Mem Nat Inst Polar Res Tokyo Spec Issue 7 : 255-263

Ng A, Patterson CC (1981) Natural concentrations of lead in ancient Arctic and Antarctic ice. Geochim Cosmochim Acta 45 : 2109-2121

Nriagu JO (1989) A global assessment of natural sources of atmospheric trace metals. Nature 338 : 47-49

Nriagu JO, Pacyna JM (1988) Quantitative assessment of worldwide contamination of air, water and soils by trace metals. Nature 333 : 134-139

Patterson CC, Settle DM (1976) The reduction of orders of magnitude errors in lead analysis of biological materials and natural waters by evaluating and controlling the extent and sources of industrial lead contamination introduced during sample collection and analysis. In : La Fleur P (ed) Accuracy in Trace Analysis, Nat Bur Stand Spec Publ 422 : 321-351

Sheppard DS (1989) Mercury in Antarctic snow 1988-89, New-Zealand Antarctic Record 9 : 26

Wolff EW, Peel DA (1985a) The record of global pollution in polar snow and ice. Nature 313 : 535-540

Wolff EW, Peel DA (1985b) Closer to a true value for heavy metal concentrations in recent Antarctic snow by improved contamination control. Ann Glaciol 7 : 61-69

Wolff EW, Peel DA (1988) Concentrations of cadmium, copper, lead and zinc in snow from near Dye 3 in South Greenland. Ann Glaciol 10 : 193-197

ELEMENTAL SPECIES ANALYSES WITH ISOTOPE DILUTION MASS SPECTROMETRY

K. G. Heumann
Institut für Anorganische Chemie
der Universität Regensburg
Universitätsstraße 31
D-8400 Regensburg
Federal Republic of Germany

1. Introduction

The accuracy of results is a great problem in trace element analyses. This problem becomes even larger when elemental species must be determined instead of the total concentration of a trace element. Even if only a small number of interlaboratory studies for speciation have been carried out up to now, one must assume from present experience that the variation of results between different laboratories will normally be much higher in speciation than in the case of the determination of total element contents. However, in both cases the differences in analytical results will rapidly increase with decreasing concentration. Isotope dilution mass spectrometry (IDMS) is a "definitive" method which produces relatively accurate analysis results (Heumann, 1988; Fassett and Paulsen, 1989). Although most of the ionization methods used in elemental mass spectrometry are not able to differentiate between different species, IDMS can be applied under certain preconditions.

2. Mass spectrometric methods and isotope dilution technique

The most important mass spectrometric ionization methods used for element analyses are summarized in Table 1. For the first group of methods, which are typical mono- and oligoelement methods, the quantification of an element is preferably carried out with IDMS. For the second group of methods, which are multielement methods, quantification is normally achieved by the use of relative sensitivity factors. In a few cases the isotope dilution technique was also applied for spark source mass spectrometry (Jochum et al., 1988) and ICP-MS (Thompson and Houk, 1986; McLaren et al., 1987), respectively. Up to now, thermal ionization mass spectrometry is the most frequently used ionization method in IDMS

NATO ASI Series, Vol. G 23
Metal Speciation in the Environment
Edited by J. A. C. Broekaert, Ş. Güçer, and F. Adams

for element analyses. In this case positive single-charged thermal ions can be produced for most of the metals in the periodic table of the elements on the hot surface of a metal filament (Re, W or Ta), whereas non-metal and non-metal compounds with electron affinities >2 eV can form single-charged negative thermal ions (Heumann, 1988; Heumann et al., 1985).

Table 1. Ionization methods for element analyses in mass spectrometry

Ionization methods	Preferred compounds
1. IDMS preferred (mono- and oligoelement)	
Thermal ionization:	
- positve ions	Metals with low first ion. potential
- negative ions	Nonmetals with high electron affinity
Electron impact	Noble gases
2. Quantification preferred without IDMS (multielement)	
Spark source	All elements
Plasma source (ICP,MIP,glow discharge)	All metals and semimetals
Laser	All elements
SIMS	All elements

The fundamental principles of the isotope dilution technique and details of application are described elsewhere (Fassett and Paulsen, 1989; Heumann, 1986, 1988). The isotope dilution process is schematically shown in Figure 1. The element - in the case of speciation the element of the species to be determined - must have at least two stable or long-lived radioactive isotopes which can be handled in a mass spectrometer. Usually, the sample with the unknown concentration has a known isotopic composition which is identical with the natural isotopic distribution of this element. An exactly known quantity of a spike isotope - in the case of speciation of the labeled species - is added to the sample. The isotopic composition of the spike must be different from the sample. After equilibration between the sample isotopes and the spike isotopes part of the isotope diluted element or species is chemically separated. This isolated portion need not be quantified. From the measured isotope ratio of the isotope diluted sample one can calculate the element or element species concentration in the sample.

The fact that only a non-quantitative isolation of the isotope diluted sample is necessary is one of the main advantages of IDMS. Additionally, the use of a stable spike isotope for element analysis or of a labeled spike compound for species analysis is the most ideal form of an internal calibration. When a thermal ionization technique is applied, selective ionization of most of the elements can be achieved depending on the ionization conditions used. This selective ionization,

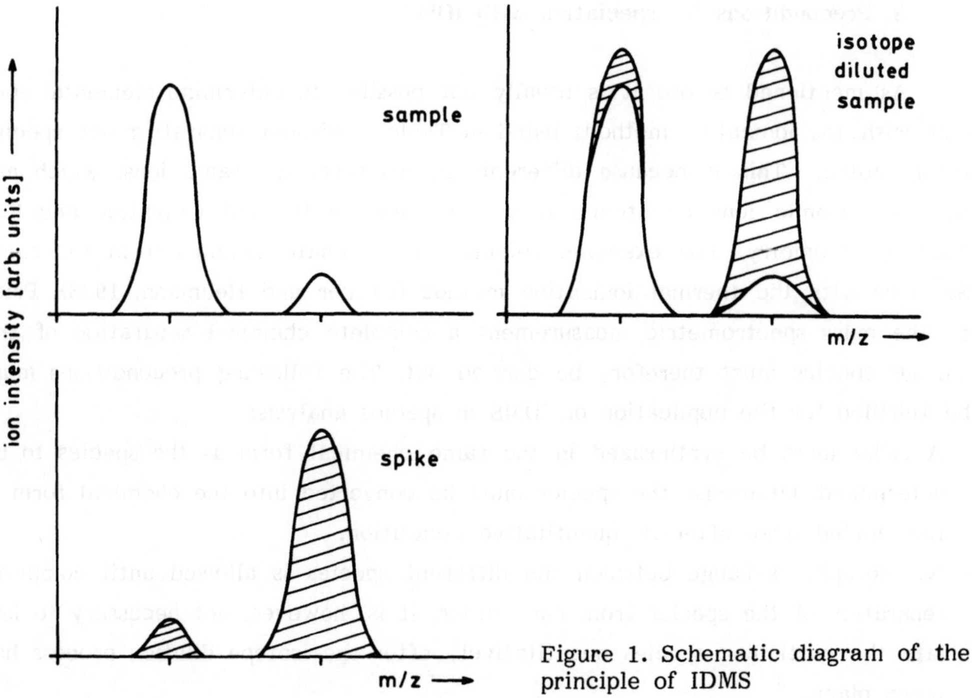

Figure 1. Schematic diagram of the principle of IDMS

which prevents mass spectrometric interferences, is not possible in the case of multielement methods like ICP-MS. The described advantages of IDMS result in relatively precise and accurate analyses and that is why IDMS is internationally accepted as a "definitive" method. However, it must be pointed out that also IDMS can result in wrong analyses, especially at very low concentration levels if contamination is not under control.

There are a number of disadvantages using IDMS. The time-consuming process for sample preparation is - aside from the expense of instruments - one of these disadvantages. However, if one needs sensitive and accurate information about elemental species there is not a great selection of analytical methods. The isotope dilution technique using thermal ionization mass spectrometry in combination with a selective chemical separation of the species to be determined or the use of an ICP-MS system connected with liquid chromatography might therefore become powerful methods in speciation.

3. Preconditions for speciation with IDMS

As mentioned before it is usually not possible to determine elemental species with the ionization methods listed in Table 1 without separating one species from another. This is because different species form the same ions, which are normally atomic ions or atomic oxide ions, e.g. in thermal ionization and ICP mass spectrometry. For example, selenite and selenate samples form the same Se^- ions with the thermal ionization method (Großer and Heumann, 1988). Prior to the mass spectrometric measurement a complete chemical separation of the various species must therefore be carried out. The following preconditions must be fulfilled for the application of IDMS in species analysis:

- A spike must be synthesized in the same chemical form as the species to be determined. Otherwise the species must be converted into the chemical form of the labeled spike after its quantitative separation.
- No isotopic exchange between the different species is allowed until complete separation of the species from each other. It is, however, not necessary to isolate the particular species quantitatively after the isotope dilution process has taken place.

The listed preconditions have to be carefully checked for each species analysis with IDMS. Up to now, it was possible to analyse the following species with this method by using thermal ionization mass spectrometry: nitrite and nitrate in food samples (Unger and Heumann, 1985), selenite, selenate and organoselenium in fresh water samples. (Heumann and Großer, 1989; Tanzer, 1990; Tanzer and Heumann, 1990), iodide, iodate and organoiodine compounds in mineral water and fresh water samples (Heumann and Seewald, 1985; Reifenhäuser and Heumann, 1990). Additionally, Götz and Heumann (1988a) were able to differentiate between chromium complexes with humic substances in aquatic systems which are kinetically inert for isotopic exchange and those where the spike and sample compounds are in equilibrium. Köppe (1989) applied an ultra filtration procedure to determine the association of vanadium and molybdenum in humic substances of different molecular masses.

Using different solvents for the leaching of sediments it was also possible to characterize different forms of the chemical bond of heavy metals in these solid samples (Götz and Heumann, 1988b; Götz, 1989). A special feature of IDMS in liquid/solid systems is the possibility to identify adsorbed compounds on the surface of the solid phase. Beauchemin et al. (1988) used the isotope dilution technique in combination with flow injection ICP-MS to determine methyl mercury in a marine biological standard reference material. Selected examples, which were developed in my laboratory at the University of Regensburg, are presented in the following.

4. Iodine speciation in aquatic systems

Radiotracer experiments have shown that three different iodine species could be chromatographically separated from various river and lake water samples by using a column filled with the strongly basic anion exchanger Dowex AG1-X8 (Reifenhäuser and Heumann, 1990; Heumann and Reifenhäuser, 1990). These are the two inorganic iodine species iodate and iodide and one organic compound, which was called the "anionic organoiodine" because of its behaviour at an anion exchanger resin (see Figure 2). Another group of organoiodine compounds, which could not be eluted from the chromatographic column under the elution conditions used, could also be identified. Due to this chromatographic behaviour this fourth iodine species was called "non-elutable organoiodine". Membrane filtration procedures have shown that a high portion of the non-elutable organoiodine compound consists of molecules with masses >60 000 D. From the fact that high concentrations of humic substances in the aquatic system normally increase the content of the two organoiodine compounds it could be concluded that in this case iodine is preferably bound to the humic substances.

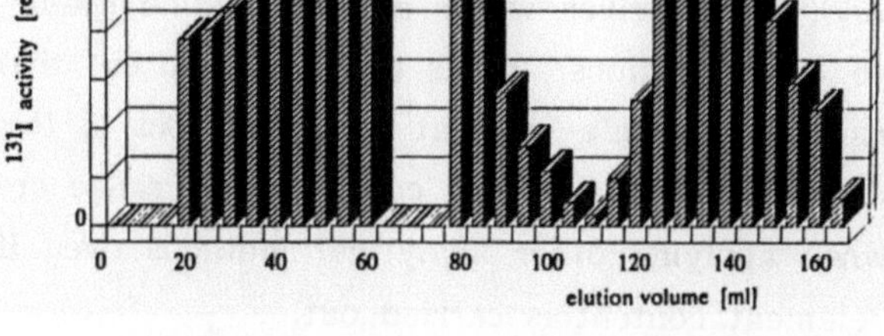

Figure 2. Separation of ^{131}I labeled iodine species from a natural fresh water sample using a strongly basic anion exchanger resin

In Figure 3 the chemical treatment of an aquatic sample for iodine speciation with IDMS is schematically represented as an example. To determine the three iodine species iodide, iodate and anionic organoiodine, an exactly known amount of a $^{129}I^-$ and of a $^{129}IO_3^-$ spike solution are added to the water sample. The iodine species are then separated chromatographically as shown in Figure 2. Up to this step the isolation of the isotope diluted compounds iodide and iodate must not be carried out quantitatively, which is opposite to the situation for the non-isotope diluted anionic organoiodine compound. After reduction of the iodate fraction to iodide, the two inorganic iodine fractions could be precipitated as AgI and measured in the mass spectrometer. The total fraction of the anionic organoiodine compound is then mixed with a well known amount of the $^{129}I^-$ spike solution and after reduction of the organoiodine with sulfite to iodide this species could also be determined.

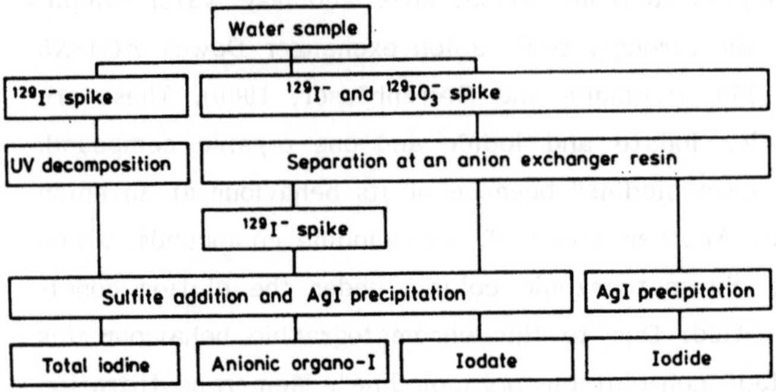

Figure 3. Sample treatment for the determination of inorganic and organic iodine species in aquatic systems with IDMS

The determination of the total iodine concentration is carried out after an UV decomposition of the organic substances in the sample. The following procedure is similar to that described before. The concentration of the non-elutable organoiodine compound could be calculated by the difference between the total iodine and the sum of the three elutable iodine species. A more detailed description of the chemical procedure is given by Reifenhäuser and Heumann (1990).

Figure 4 represents the very different distribution of the iodine species in a moorland lake compared with a river water sample for two sampling dates. Whereas in the moorland lake water the only detectable iodine species were the organoiodine compounds (iodide and iodate are less than the detection limit of 0.5 µg/L and 0.1 µg/L, respectively), in the river water all four iodine species could be analysed with comparable concentrations. It has to be pointed out that these iodine species determinations were possible even at concentrations of less than 1 µg/L. Accurate iodine determinations in this low concentration range are not an easily solvable problem when applying other analytical methods even if only a determination of the total element content is carried out.

The application of the described sensitive and accurate IDMS method was the essential precondition for obtaining positive correlations between iodine species concentrations and other important water parameters. Such a correlation was found between the oxygen and the iodate content, the redox potential and the

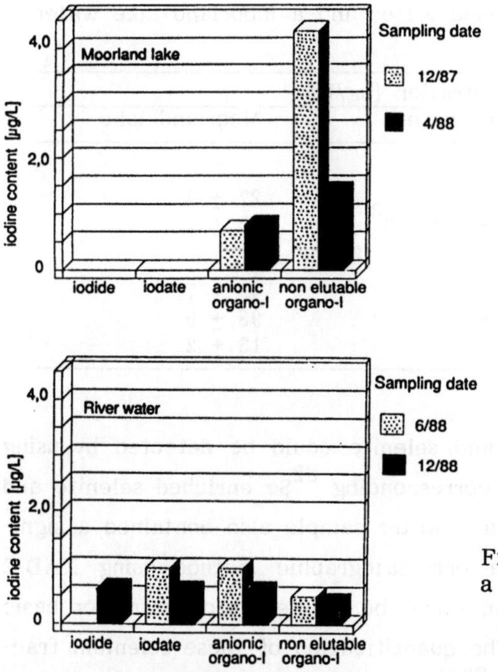

Figure 4. Distribution of iodine species in a moorland lake and a river water sample

anionic organoiodine compound for fresh water samples of different origins as well as between the DOC content and the total iodine concentration in moorland lake water samples (Heumann and Reifenhäuser, 1990; Reifenhäuser and Heumann, 1990).

5. Selenium speciation in aquatic systems

Selenium concentrations in natural aquatic systems are generally very low, normally below 500 ng/L (Robberecht and van Grieken, 1982). The analytical problems for selenium speciation due to sensitivity and accuracy are therefore at least comparable with those pointed out during the discussion of iodine speciation. The selenium species which can be expected in natural water systems are the inorganic compounds selenite and selenate and different organic compounds, especially dimethyl selenide, dimethyl diselenide and the trimethyl selenonium ion. It is postulated that emissions from the biosphere, e.g. from biological activities in the aquatic system, are the most important natural sources for possible organoselenium compounds in the environment (Ross, 1990).

Applying comparable methods for the sample treatment as described for iodine speciation it was possible to determine selenite, selenate and different organoselenium compounds in various fresh water samples. A selection of results is listed in Table 2 (Tanzer, 1990; Tanzer and Heumann, 1990). In a ground water

Table 2. Selenium speciation in a ground water and a moorland lake water sample with IDMS

Se species	Concentration [ng/L]	
	Ground water	Moorland lake
Inorganic species:		
Selenite	590 ± 10	27 ± 4
Selenate	$(15.3 \pm 0.1)\times10^3$	35 ± 7
Organic species:		
Neutral and basic compounds	-	49 ± 7
Acidic compounds	-	93 ± 9
Trimethyl selenonium ion	-	13 ± 2

of a limestone formation only selenate and selenite could be detected by using anion exchange chromatography and the corresponding ^{82}Se enriched selenite and selenate spike solutions. The moorland lake water sample also contained a significant amount of organoselenium. By a chromatographic method using XAD-2 (Suzuki et al., 1980) the organic fraction could be divided into neutral or basic compounds and into acidic compounds. The quantification of these selenium fractions were carried out by addition of a ^{82}Se enriched selenite spike followed by a decomposition of the organic material with $HClO_4/HNO_3$. Trimethyl selenonium (TMSe) was determined by using a cation exchanger column filled with Dowex 50W-X8 and the corresponding ^{82}Se enriched TMSe spike solution. This is one of the first times that TMSe could be determined in the environment. The extremely low concentration of 13 ng/L shows the high sensitivity of IDMS in speciation. Also the sum of all selenium species of 217 ng/L in the moorland lake water agrees very well with the result of 213 ng/L for the total selenium determination in this sample after a $HClO_4/HNO_3$ decomposition of organic material. This additionally shows the consistence of the IDMS speciation.

6. Metal speciation with IDMS

6.1 General aspects

With the isotope dilution technique all species are analysed which are in isotopic exchange with the spike solution. This is the reason why only kinetically stable metal species can be characterized with this method.

If one leaches a solid sample with a solvent under the presence of a spike solution the sum of the following metal species are determined with IDMS in the liquid phase after separation:

a) Free metal ions in the liquid phase.
b) All metal complexes, which are kinetically unstable (isotopic exchange between the metal complex and the spike solution within the experimental time range).
c) Metal ions, which were originally dissolved but then precipitated, e.g. precipitation of barium sulfate after dissolving barium carbonate by sulfuric acid. This is due to the fact that during an interim period barium is dissolved and therefore also isotope diluted. This is a special feature of IDMS.
d) Adsorbed metal ions on the surface of the solid phase because the ion exchange reaction between a solution and the adsorbed ions is generally very fast. This topic is also a special feature of IDMS.

As one can see, special information can be obtained about the chemical form of metal ions by the use of IDMS, which cannot normally be achieved by other analytical methods. However, the most detailed information will become available by the combination of IDMS with other analytical procedures.

6.2. Characterization of the chemical form of heavy metals in sediments by leaching experiments

The total heavy metal concentrations and the aqua regia soluble portion of Pb, Cr, Zn, Cd, and Cu were determined in three different standard reference sediments with IDMS (Götz and Heumann, 1988b). The results are presented in Figure 5. Dependent on the type of sediment, 65%-95% of the total Pb is dissolved in aqua regia. This means that 5%-35% of the total Pb is bound in the undissolved phase. 60%-90% of Cr is extractable with aqua regia but for Zn only very small differences between the total concentration and the aqua regia soluble portion were measured. For Cd and Cu no significant differences were found. These results permit the conclusion that Cu, Cd and - a little less completely - also Zn are preferably bound to carbonates and organics, whereas a significant part of the total Cr and Pb amount in sediments is fixed in the silicates or as insoluble compounds, e.g. $PbSO_4$.

The previously discussed results with IDMS are accurate because a "definitive" method was applied. However, the same results can also be obtained by other reliable analytical methods. As mentioned before a special feature of IDMS is the possibility also to identify adsorbed compounds on the surface of the solid phase of a liquid/solid system. This could be shown by Götz (1989) when sediments were leached with different solvents adding the spike solution to the system before the separation of the liquid phase was carried out in a first expe-

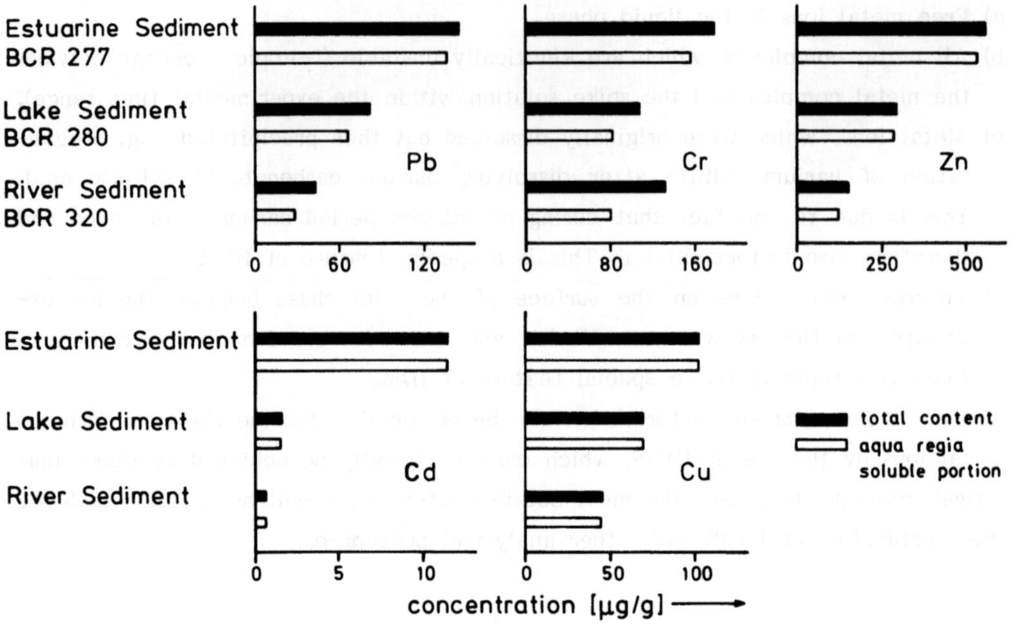

Figure 5. Comparison of the total content and the aqua regia soluble portion of heavy metals in three sedimentary reference materials

riment. In a second parallel experiment the liquid phase was first separated from the undissolved solid residue and only then was the spike solution added to the liquid phase. The difference between both experiments should be the amount of metal adsorbed at the solid phase.

Table 3. Determination of the adsorbed metal portion in the solid phase during leaching experiments with the standard reference sediment BCR 277

Element	Concentration [μg/g]			
	Total	Aqua regia	HNO_3 (Spike before)	HNO_3 (Spike after)
Cu	101.8	101.9	73.5	71.1
Pb	145.0	135.6	122.8	111.7

The results, which were obtained for the corresponding leaching experiments with a HNO_3 solution of pH=1 for Cu and Pb with one of the standard reference sediments are summarized in Table 3. The total concentration and the aqua regia soluble part are also listed for comparison. The results show that 3.4% of the HNO_3 soluble Cu and 9% of the HNO_3 soluble Pb were adsorbed at the undissolved residue. Although other analytical methods cannot normally identify adsorbed metals in a liquid/solid system, the knowledge of this portion is important information, at least because of the possible mobilization of toxic heavy metals in the environment.

6.3. Kinetically stable and unstable chromium complexes

Most of the metal complexes are kinetically unstable, which means that a relatively fast isotopic exchange between the metal ions in the complex and the free ions in a solution will occur. In this case the metal ions of a spike solution will even exchange with the metal ions in the complex and the determination of the total metal concentration (complexed and free metal ions) will be the result with IDMS. Cr(III) is one of the well known exceptions which is able to form kinetically stable complexes (Basolo and Pearson, 1973). If it is possible to separate the stable complexes from the unstable ones, a direct characterization of the latter species can be obtained with the IDMS method. The determination of the total Cr concentration in the system, e.g. by using a decomposition method for the organic compounds, will yield the kinetically stable species as the difference of the results.

Table 4. Determination of kinetically unstable and stable Cr and Cd complexes in a filtered moorland lake water sample with IDMS

Element	Concentration [µg/L]		
	Unstable complex*	Total metal**	Stable complex
Cr	1.64 ± 0.01	2.27 ± 0.02	0.63
Cd	0.023 ± 0.001	0.022 ± 0.001	0

* Determined by electrolytic deposition after isotope dilution
** Determined after a H_2O_2/HNO_3 decomposition procedure

Götz and Heumann (1988) found for the first time kinetically unstable and stable chromium complexes in aquatic systems with high concentrations of humic substances. In Table 4 one example of a moorland lake water sample is listed. The cadmium results for the same sample are also given for comparison. As one can see, only in the case of Cr is there a difference of about 30% between the unstable complex and the total metal concentration. This difference must be due to the portion of kinetically stable chromium complexes with humic substances in this aquatic system. It is interesting to point out that the same results were also obtained when the moorland lake water sample was stored together with the added spike solution for more than three months.

6.4. Association of V and Mo with humic substances

The association of V and Mo traces in a moorland lake water sample with humic substances of different molecular masses was investigated by Köppe (1989) using ultrafiltration processes with filters of different pore sizes. Even if the range of molecular masses is not very exactly determinable with this method, the results given in Table 5 show an individual distribution of both elements in the different fractions of the humic substances.

Table 5. Association of V and Mo with humic substances of different molecular masses in a moorland-lake water sample

Sample preparation	Range of masses [D]	Portion of total amount [%]*	
		V	Mo
Filter 0.015 µm	>60000	6	14
0.005 µm	10000 - 60000	6	10
YM 500	1000 - 10000	11	16
Filtrate	<1000	77	60

*Total concentration in the original sample: V = 2.2 µg/L, Mo = 0.082 µg/L

The total V and Mo concentrations in natural water samples are mostly in the low µg/L level or below. This means that only very sensitive methods like IDMS are able to detect interactions between humic substances and these trace metals of geochemical interest.

6.5. Determination of Tl(I) and Tl(III) in a manganese nodule

It is well known that Tl(I) is the most stable oxidation state of thallium which is the reason why Tl(I) is predominantly found in the environment. Under certain conditions also the oxidation state +3 can be stabilized, e.g. in seawater (Batley and Florence, 1975). It could also be expected that Tl(III) is stabilized in a number of solid samples, especially in those where other elements also exist in higher oxidation states. However, the direct determination of oxidation states for trace elements in a solid sample is a general problem unsolved up to now. This is the reason why only indirect determination methods, e.g. by leaching processes, are used today. Applying these indirect methods one has to make sure that the species to be determined in the solid is not converted into another compound during the chemical process.

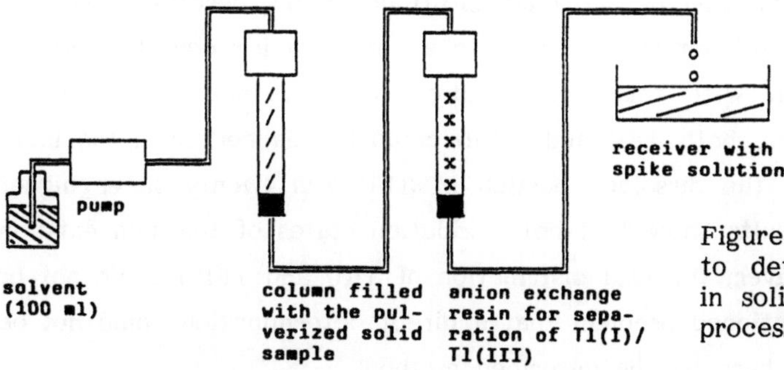

Figure 6. Analytical procedure to determine Tl(I) and Tl(III) in solid samples by a leaching process

A manganese nodule from the Pacific Ocean and, for comparison, a river sediment from the river Neckar were investigated with the system represented schematically in Figure 6 (Arweck, 1989). The pulverized samples were filled in a first column and then leached by 0.4 mol/L chloride solutions (HCl or NaCl). In a second column filled with the strongly basic anion exchanger resin Dowex AG1-X8 a separation of Tl(I) and Tl(III) could be obtained because under the conditions used only Tl(III) forms a stable anionic chlorocomplex which was fixed by the resin. After Tl(I) was collected in a first receiver, where a known amount of spike solution was added, Tl(III) could be reduced by a sulfite solution and then collected in the same way.

Table 6. Tl(I) and Tl(III) determined in solvents after leaching a manganese nodule and a river sediment

Sample	Solvent [mol/L]	Total Tl conc. in sample [µg/g]	Dissolved portion [%]	
			Tl(I)	Tl(III)
River sediment	0.4 HCl	0.49	10	0
Manganese nodule	0.4 HCl	76.3	5.5	61
	0.4 NaCl		0.4	0.02

The results are listed in Table 6. Using HCl as a solvent 10% of the total thallium in the river sediment was leached. In this case only Tl(I) could be detected. Using the same conditions for the manganese nodule 66.5% of the total thallium was dissolved. In this case the predominant oxidation state measured in the solution was +3. It must be assumed, however, that the high portion of Tl(III) did not originally come from the solid sample rather from an oxidation reaction under the conditions used. This oxidation is due to the formation of elemental chlorine by manganese dioxide, which is the major compound of manganese nodules, in the HCl acidic solution. This could be confirmed by a model experiment using a synthetic mixture of pure manganese dioxide and traces of a Tl(I) salt.

Because of the formation of artificial Tl(III) under the described leaching conditions with HCl the solvent was changed to a non-oxidizing condition with a 0.4 mol/L NaCl solution. With this NaCl solution, which is nearly identical in concentration to seawater, both Tl(I) and Tl(III) could be leached from the manganese nodule. Although the dissolved portion of Tl(I) was twenty times higher than for Tl(III) these results show that both oxidation states of thallium exist in manganese nodules. However, the real distribution of Tl(I) and Tl(III) could not be analysed due to the mentioned problem that a direct determination could not be carried out in the solid phase by the described method.

7. Conclusion

It could be shown that IDMS is not only suitable for accurate determinations of total trace element content but also for accurate speciation in very low concentration levels. Essential preconditions for the application of IDMS in speciation are the facts that no isotopic exchange between different species may occur and that a complete separation of all species must be carried out before the mass spectrometric measurement. Taking these aspects into account thermal ionization mass spectrometry in combination with selective species separation methods and ICP-MS connected with liquid chromatography might become powerful methods for speciation.

In comparison with results also obtainable with other suitable speciation methods, IDMS can yield additional information, e.g. about adsorbed ions in a heterogeneous system or about kinetically stable and unstable complexes in a homogeneous system.

References

Arweck F (1989) Untersuchungen zum Verhalten von Thallium im Untergrund bei Infiltration von Oberflächenwasser. Thesis, University of Regensburg

Basolo S, Pearson R (1973) Mechanismen in der anorganischen Chemie. Thieme, Stuttgart, p 119

Batley GE, Florence TM (1975) Determination of thallium in natural waters by anodic stripping voltammetry. Electroanal Chem Interfacial Electrochem 61: 205-211

Beauchemin D, Siu KWM, Berman SS (1988) Determination of organomercury in biological reference materials by inductively coupled plasma mass spectrometry using flow injection analysis. Anal Chem 60:2587-2590

Fassett JD, Paulsen PJ (1989) Isotope dilution mass spectrometry for accurate elemental analysis. Anal Chem 61:643A-649A

Götz A (1989) Entwicklung und Anwendung von Meßtechniken zur Bestimmung von Schwermetallspuren und ihrer Bindungsformen in unterschiedlichen Umweltmaterialien mit Hilfe eines THQ. Thesis, University of Regensburg

Götz A, Heumann KG (1988a) Chromium trace determination in inorganic, organic and aqueous samples with isotope dilution mass spectrometry. Fresenius Z Anal Chem 331:123-128

Götz A, Heumann KG (1988b) Determination of heavy metals (Pb, Cd, Cu, Zn, Cr) in sedimentary reference materials with IDMS: total concentration and aqua regia soluble portion. Fresenius Z Anal Chem 332:640-644

Großer R, Heumann KG (1988) Negative thermal ionization mass spectrometry of selenium. Part 1. Isotope ratio measurements and determinations in aquatic systems with the isotope dilution technique. Fresenius Z Anal Chem 331:268-272

Heumann KG (1986) Isotope dilution mass spectrometry of inorganic and organic substances. Fresenius Z Anal Chem 325:661-666

Heumann KG (1988) Isotope dilution mass spectrometry. In: Adams F, Gijbels R, van Grieken R (eds) Inorganic mass spectrometry. Wiley, New York, pp 301-376

Heumann KG, Großer R (1989) Negative thermal ionization mass spectrometry of selenium. Part 2. Selenite and selenate species determination in ground waters with the isotope dilution technique. Fresenius Z Anal Chem 332: 880-883

Heumann KG, Seewald H (1985) Simultaneous determination of Cl^-, Br^- and I^- in water samples by isotope dilution mass spectrometry. Fresenius Z Anal Chem 320:493-497

Heumann KG, Reifenhäuser C (1990) Interaction of humic substances with iodine. In: Mattheß G (ed) Progress in hydrogeochemistry. Springer, Heidelberg, in press

Heumann KG, Schindlmeier W, Zeininger H, Schmidt M (1985) Application of an economical and small thermal ionization mass spectrometer for accurate anion trace analyses. Fresenius Z Anal Chem 320:457-462

Jochum KP, Seufert HM, Midinet-Best S, Rettmann E, Schönberger K, Zimmer M (1988) Multielement analysis by isotope dilution spark source mass spectrometry. Fresenius Z Anal Chem 331:104-110

Köppe M (1989) Entwicklung einer massenspektrometrischen Isotopenverdünnungsanalyse zur Bestimmung von Vanadium, Molybdän und Wolfram in Wasser-, Boden- und Aerosolproben. Thesis, University of Regensburg

McLaren JW, Beauchemin D, Berman SS (1987) Application of isotope dilution inductively coupled plasma mass spectrometry to the analysis of marine sediments. Anal Chem 59:610-613

Reifenhäuser C, Heumann KG (1990) Development of a definitive method for iodine speciation in aquatic systems. Fresenius Z Anal Chem 336:559-563

Robberecht H, van Grieken R (1982) Selenium in environmental waters: determination, speciation and concentration levels. Talanta 29:823-844

Ross HB (1990) Biogeochemical cycling of atmospheric selenium. In: Adams F, Broekaert JAC, Gücer S, Klockow D (eds) Metal speciation in the environment. NATO ASI Series. Springer, p

Suzuki Y, Sugimura Y, Miyake Y (1980) The content of selenium in sea water in the western North Pacific and its marginal seas. Proc 4th Symp for the cooperative study of the Kuroshio and adjacent regions, pp 396-414

Tanzer D, Heumann KG (1990) GC determination of dimethyl selenide and trimethyl selenonium ions in aquatic systems using element specific detection. Atmos Environ, submitted

Tanzer D (1990) Chromatographische und massenspektrometrische Bestimmung anorganischer und organischer Selenverbindungen in der Umwelt. Thesis, University of Regensburg

Thompson JJ, Houk RS (1986) Inductively coupled mass spectrometric detection for multielement flow injection analysis and elemental speciation by reversed phase liquid chromatography. Anal Chem 58:2541-2548

Unger M, Heumann KG (1985) New calibration method for the nitrate and nitrite trace determination in food samples. Fresenius Z Anal Chem 320:525-529

SPECIATION OF TRACE ELEMENTS IN SERUM OR PLASMA

Rita Cornelis

Laboratory of Analytical Chemistry
Institute for Nuclear Sciences, Rijksuniversiteit Gent
Proeftuinstraat 86
B-9000 Gent, Belgium

1. Introduction

Blood plasma or serum is a complex mixture of more than 100 compounds (proteins, fats, low molecular weight compounds, ions,...). It functions primarily as a transport medium that conveys nutrients to cells, removes their metabolites, and carries hormones to their specific site of action. In addition plasma is involved in the regulation of physiological processes. Its composition and volume influences the activities of other organ-systems which, in turn, maintain the homeostasis of the entire organism (Versieck and Cornelis, 1989). Serum differs from plasma chiefly in that it contains no fibrinogen, which, due to polymerization, forms the blood clot. To obtain plasma an anticoagualant is added to the blood.

Since serum or plasma reflect most of the metabolic activities in the body, they are frequently analysed for diagnostical purposes. Trace element determinations are one of the most recent trends. The essentiality of the elements Co, Cu, Se and Zn for human beings has been well documented. Other elements such as Cr, Mn, Mo, Ni, V, etc. are known to be essential but have not yet been identified as an essential biocomponent in the human body. Nevertheless these elements are investigated for diagnostical or epidemiological purposes.

Determination of the total amount of a trace element in serum or plasma does, however, not provide a complete picture of the dynamic status of the important biochemical processes

NATO ASI Series, Vol. G 23
Metal Speciation in the Environment
Edited by J. A. C. Broekaert, Ş. Güçer, and F. Adams

involving the element. So, besides determining how much of an element is present, it is necessary to find "the molecular nature of the trace element complexes". Only then we can make statements about mobility, storage, accumulation and toxicity of the element.

2. Guidelines for trace element speciation work

Speciation of a trace element in serum or plasma implies the identification of the biologically active compounds to which the trace element is bound and the quantification of the element in relation to those particular molecules. These investigations entail first of all separation of the biocomponents, followed by their identification and quantitation. After the separation comes the determination of the trace element in the different fractions. This kind of work presents many challenges (Gardiner, 1987).

Requirements for fractionation work can be summarized as follows:

- Realistic assumptions have to be made about the system and appropriate experimental conditions must be chosen to fractionate the biomolecules,
- It is fundamental that the methods used do not disturb the specific metal-protein associations. This can be partly assured by choosing appropriate experimental conditions, such as the buffer, pH, ionic strength,
- Besides the separation of the biomolecules it is also necessary to identify them unequivocally and quantify them in the different fractions and in the original sample. Ideally the recovery should be 100%,
- Any changes that occur in the sample either after collection and/or during storage should be known.

Requirements for trace element determination in speciation work are very demanding:

- The analytical method must not only be precise and accurate, but there is need for low detection limits. Levels over 100 times lower than the original total content in serum can be anticipated in the fractions,

- The fractionation work should involve no contamination with the trace element under study. This implies a stringent control of extraneous additions,
- Finally the sum of the species (minus the blank) should equal the total content of the trace element,
- Ideally the method should be on-line, compatible with e.g. the flow rate of liquid chromatography, with ease of operation and automation. The precision and accuracy of the measurements have to be guaranteed.

The range of techniques needed (biochemical and analytical) is quite extensive and neither the equipment nor the overall expertise is usually present in a single laboratory. An interdisciplinary approach in planning the experiments is required to solve the aspects of the separations with concurrent considerations about the trace element investigations.

The separation methodology (e.g.chromatography) requires careful decontamination procedures of the buffer, the gel or other packing material, the column as such and as a matter of fact any material coming into contact with the sample. Adequate strategies are described for different trace elements. What should be avoided at all cost is that the serum serves as the ultimate scavenger for trace element impurities. This was clearly illustrated with the aid of radiotracers by Wallaeys et al. (1987). They investigated the behaviour of Na, Zn, Se, Rb, Sb and Cs ions on a Sephacryl S200 column. The gel permeation chromatography (GPC) showed a virtually complete retention of Zn on the column, clearly indicating its interaction with the medium. (Se, Rb, Sb and Cs all eluted together at the end of the GPC-spectrum). A subsequent run with human serum on this contaminated column led to the binding of Zn with the serum proteins in the albumin-transferrin region. Overlooking such a phenomenon may lead to artifactual conclusions in Zn speciation studies.

3. Methods and techniques

3.1. Biocomponents (Alberts et al., 1983)

3.1.1. Fractionation techniques

The techniques most commonly used for fractionation of biomolecules in the field of speciation of trace elements in serum are affinity chromatography, gel permeation chromatography, ion exchange chromatography, precipitation and ultrafiltration.

Affinity chromatography

Affinity chromatography is a very efficient technique as it takes advantage of the biologically important binding interactions that occur on protein surfaces. For example in order to separate albumin, an inert matrix is first loaded with specific albumin antibodies. These are complexed to the column packing. When serum is eluted, only albumin (the biomolecule of interest) is specifically adsorbed to the matrix, whereas the other substances are washed through the column. Once the biomolecule of interest is adsorbed free from contaminants, it can be eluted. The most common way to do this is to lower the pH. There is a high degree of specificity involved in affinity chromatography.

Gel permeation chromatography

Gel permeation chromatography (GPC) or size exclusion chromatography is achieved by exploiting differences in molecular size. Large molecules having a diameter greater than that of the largest pores are excluded from the pore structure of the gel and pass straight through the column. Very small molecules (e.g. solvent molecules) can penetrate into even the smaller pores and will therefore take the longest time to be eluted from the column. Molecules of intermediate size diffuse into some of the pores and will be eluted at an intermediate time.

In GPC there are theoretically neither specific nor non-specific interactions between the sample molecules and the stationary phase. This is of major concern for speciation work, because it gives some guarantee that the integrity of the species may be maintained.

Ion exchange chromatography

Proteins usually have a net positive or negative charge due to the negatively and positively charged groups of amino acids on their surface. If a protein has a net positive charge at pH 7, it will bind to a column of beads containing negatively charged carboxylate groups whereas negatively charged proteins will not. A positively charged protein bound to such a column can then be eluted or released by increasing the concentration of NaCl in the eluting buffer or the pH. The Na^+ ions compete with the positively charged groups for binding on the column. Proteins that have a low density of net positive charge will tend to emerge first, followed by those having a higher charge density.

Positively charged proteins (cationic species) can be separated on a negatively charged carboxymethyl cellulose (CM-cellulose) column.

Negatively charged proteins (anionic proteins) can be separated by chromatography on positively charged diethylamino ethyl-cellulose (DEAE-cellulose) columns.

Precipitation

Proteins are fractionated through precipitation after addition of certain chemicals, e.g. with 5% trichloroacetic acid. Such a separation will generally not be very specific.

Ultrafiltration

Ultrafiltration is based on the concept of molecular retention or separation effected by permselective membranes (cellulose acetate). These have a particular filtration threshold, e.g 25 - 30 000 D (pore size about 10 nm). Small molecules, such as amino acids are 10 times smaller, a few nm, and pass through the membrane. Usually centrifugation is applied.

3.1.2. Detection of the biomolecules

The detector for the chromatographical separations measures the absorption of light in the ultraviolet (UV) or visible (VIS) region. It is based on the fact that large molecules, such as proteins, contain chromophores (peptide bonds, amino acid chains, any prosthetic group {helper group}) which absorb part of the light (peptide absorption 210 - 220 nm; aromatic amino acids: (phenylalanine, tyrosine, tryptophan, hystidine) 230 - 300 nm (near UV)). This type of detection offers very little specificity and is semi-quantitative.

3.1.3. Identification of the biomolecules

The identification and purity of a fraction can be checked with electrophoretic measurements: e.g. iso-electrofocusing, SDS-PAGE (sodium dodecyl sulphate - polyacrylamide gel electrophoresis).

3.1.4. Quantification of the biocomponents

The quantification of the proteins occurs with techniques such as nephelometry, radial immunodiffusion, kinetic immunoturbidity, etc.

3.2. Trace elements

Several optical methods (atomic absorption spectrometry (Welz, 1985), atomic emission spectrometry (the inductively coupled plasma ICP - AES) (Schramel, 1983), fluorimetry (Kirkbright et al., 1974)) and neutron activation analysis (Heydorn, 1984) have been used for trace element determinations in speciation work. Atomic absorption spectrometry (especially the graphite furnace type, GFAAS) is the leading technique.

4. Speciation studies

Until now only a limited number of trace elements in serum or plasma have been investigated for speciation purposes. The results published for Zn, Cu, Se, Ni and Cr will be reviewed.

4.1. Zinc

Reliable reference values for Zn lie at 1.0 μg/ml (Versieck and Cornelis, 1989). Whereas the determination of the total Zn concentration now occurs on a routine basis in clinical laboratories, profiling remains still limited to research work. It is, however, the best documented trace element as far as knowledge about the distribution among the different biocomponents is concerned.

As cited in the article by Giroux et al. (1976), the earlier work published by Himmelhich et al. in 1966 and by Parisi and Vallee in 1970, could be confirmed by their own findings in 1976. It showed that Zn in human serum is partitioned between 2 major Zn binding proteins: albumin (M.W. 70,000 D, conc. 30 - 55 mg/ml serum) and α_2-macroglobulin. The latter was fractionated through addition of ethylglycol 6,000 up to 10%; next all other proteins (including albumin) were precipitated with 5% trichloroacetic acid. Zn was measured by atomic absorption spectrometry. Giroux et al. (1976) using atomic absorption spectrometry for the Zn determinations and appropriate protein determination techniques, concluded albumin- bound Zn accounted for 80% of total serum Zn of healthy individuals. The remaining 20% was found in the α_2-macroglobulin fraction.

Nowadays GPC and ion exchange chromatography are the methods of choice for exploring proteins in serum or plasma.

In 1981 Gardiner et al. (1981) published very interesting work on the speciation of protein-bound Zn in human blood serum. The graph in Fig. 1 shows the UV profile of the proteins after a GPC separation of serum. As expected the resolution of the proteins is rather poor. The distribution of Zn is indicated in the figure as well.

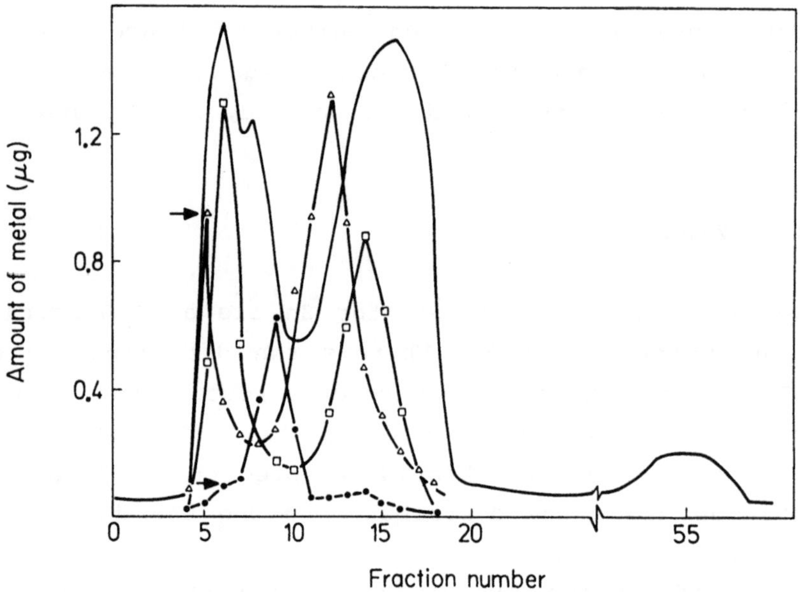

Fig. 1. Distribution of (□) zinc, (●) copper and (△) iron after gel filtration of a serum sample from a healthy donor on a Sephadex G-150 column (2.6 x 40 cm) using a gel superfine grade beads (1 ml of serum eluted with 0.05 M Tris-HCl buffer, pH 7.4, 40 °C at 10 ml h^{-1}). The continuous line is the serum protein profile at 280 nm. (Gardiner et al., 1981)

Gardiner et al. (1981) assumed that in these relatively complex fractions obtained after GPC, the association between a metal and protein ligand could be said to have been established if their Gaussian distributions coincide. This is more likely to be true if the elution volumes of the constituents are in the fractionation range rather than in the void volume. This reasoning holds true as far as the identification of the Zn binding proteins is concerned. The exact distribution of Zn in relation to these proteins is much more demanding. It entails: - the establishment of an unequivocal relation to the protein (with the exclusion of any other component present) in the separated fraction, - the quantification of the protein and - the recovery.

Because of the ubiquity of trace elements, it is a rule of thumb that if no precautions are taken, the recovery will be abnormally high. Zn is one of the more hazardous elements to work with in this respect. An anti-coagulant added to the sample may not only be susceptible to contaminate the serum but it can also act as complexing and binding agent for the trace elements. Some comments will be given in the subsequent text, when the data for the ultra filtrable amount of Zn in serum and plasma are compared. It is also evident that speciation work should start as soon as possible after collection.

Whereas Gardiner et al. in the development stage encountered a recovery of > 200%, this came down to 100% with the precautions described in paragraph 2.

The distribution of Zn was not clear cut. Four areas of interest could be identified. As can be seen in Fig. 2, the two most outspoken Zn areas coincided with the distribution of α_2-macroglobulin and albumin. The region in between is subject to many assumptions such as IgG (immunoglobulin), pre-albumin.... It is also conceivable that the integrity of the original Zn-protein complex may not have been maintained. In this study it was found that two thirds of the Zn is associated with albumin. The other third looks more complex to pin down.

It is evident that a single fractionation technique, such as GPC, is not adequate for a refined separation of the many molecules present. A better resolution may be obtained when properties other than differences in size are exploited. These include differences in ionic characteristics, in affinity for other molecules and hydrophobicity.

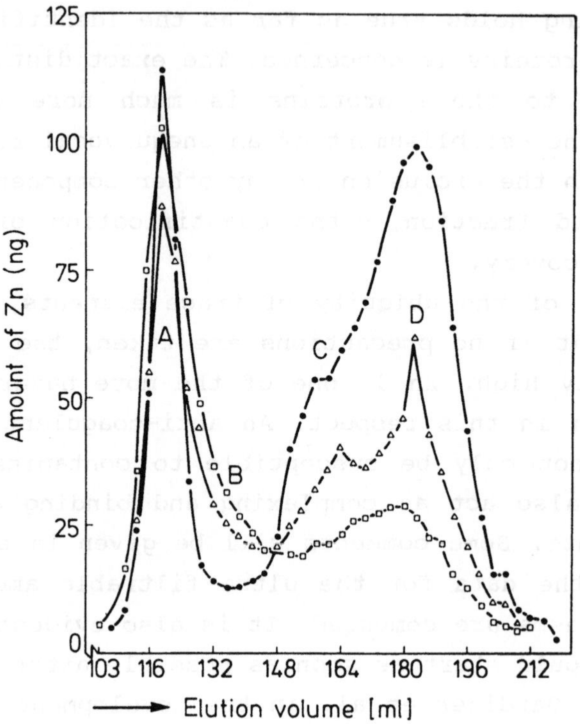

Fig. 2. Elution profile of zinc after gel filtration on a Sephadex G-100 column from (●) a normal subject and (△,□) two patients on total parenteral nutrition; letters A, B, C and D denote areas of suspected zinc binding to protein. (Gardiner et al., 1981)

Chilvers and coworkers (1984) published very interesting work on the speciation of Zn in plasma proteins after a simple chromatographic separation on a DEAE-Sepharose column. They used a Tris.HCl buffer of constant ionic strenght (0.2 M NaCl) at pH 8.6. All necessary precautions were taken to avoid Zn contamination. Fig. 3 shows the elution profile of the proteins at 280 nm. The identity of the proteins in the individual fractions was determined by radial immunodiffusion and by use of purified "markers". Zn was determined in the different fractions with GFAAS. In order to match the matrices of the standards and those of the samples, the Zn standard solution was prepared in the Tris.HCl buffer with 0.2 M NaCl.

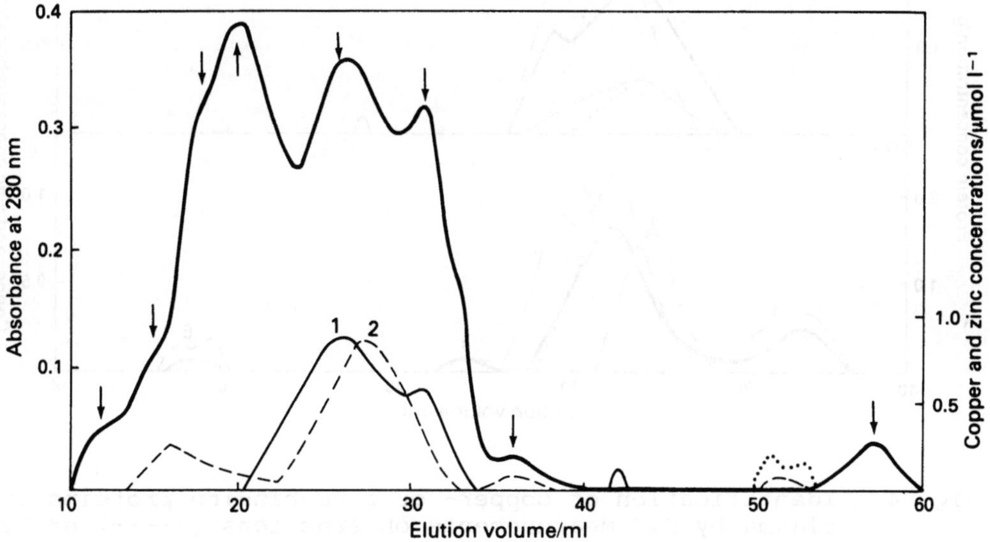

Fig. 3. Elution profiles of proteins, copper and zinc from normal plasma. Bold curve, absorbance at 280 nm and dotted curve at 206 nm. Curves 1 and 2 are copper and zinc concentrations, respectively, as measured by atomic-absorption spectroscopy. Vertical arrows show the location of the major peaks. (Chilvers et al., 1984)

Through deconvolution techniques Chilvers and coworkers (1984) could associate 6 major and 7 minor UV absorbing proteins with Zn. Fig. 4 shows the elution profiles of 5 proteins that were proposed as Zn binding proteins: α_2-macroglobulin, α_2-HS-glycoprotein (x10), albumin A (x0.025), retinol binding protein, low relative molecular mass compound. The bold curve shows the Zn concentration in the fractions. The recovery from the column (all amounts added) was 100.7 ± 4.5% for 72 chromatographic runs. Equally good recoveries were obtained for the proteins, except for retinol binding protein. The distribution of Zn in these various fractions was examined in 36 healthy normal men and women. The results are summarized in Table 1.

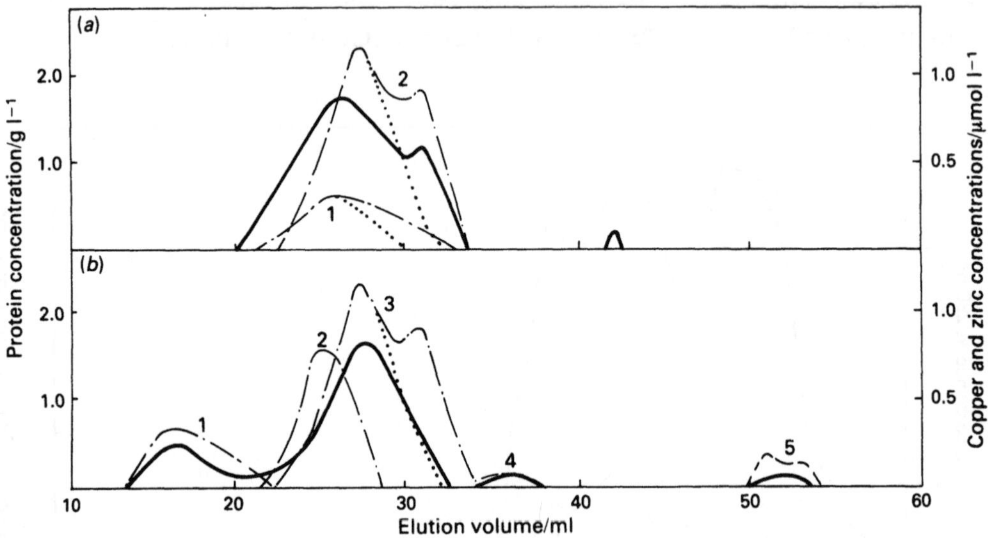

Fig. 4. Identification of copper- or zinc-binding proteins in plasma by RID measurements on fractions (-·-·-) or by the use of purified materials (....). (a) Bold curve, copper concentration in fraction. Peaks: 1, caeruloplasmin (x5); and 2, albumin (x0.25). (b) Bold curve, zinc concentration in fraction. Peaks: 1, α_2-macroglobulin; α_2-HS-glycoprotein (x10); 3, albumin A (x0.025); 4, retinol-binding protein; and 5, low relative molecular mass compound absorbing at 206 nm (- - -). Figures in parentheses (here) are the scaling factors used to produce comparable peak amplitudes in the diagram. (Chilvers et al., 1984)

The mean total Zn concentration is significantly lower in women than in men, as was the concentration of Zn bound to albumin. In contrast, the mean concentration of Zn associated with the retinol complex was significantly higher in women than in men. The mean concentration of Zn bound to α_2-macroglobulin was also higher in women than in men though the difference did not reach the 5% level of significance.

Table 1. The distribution of Zn over different plasma components according to Chilvers et al. (1984)

Zn-fraction	Zn μg/ml, mean ± standard error men (n=36)	women (n=36)
total	0.987 ± 0.017	0.934* ± 0.019
albumin	0.864 ± 0.015	0.789+ ± 0.018
α_2-macroglobulin	0.109 ± 0.006	0.123 ± 0.005
retinol complex	0.017 ± 0.001	0.026+ ± 0.003

* $p < 0.05$ + $p < 0.01$

Zn bound to albumin accounted for 87.6% of the total Zn in men and for 84.5% in women, while Zn bound to α_2-macroglobulin accounted for 11.1 and 13.2% respectively. Zn bound to retinol binding protein accounted for less than 3% of the total plasma Zn.

The authors also compared the Zn distribution in paired plasma and serum samples. They expressed the opinion that the higher concentration of Zn in plasma is not due to contamination by heparin. On the contrary the lower value in serum is due to Zn loosely bound to albumin, which is lost during the clotting of the blood. Chilvers et al. (1984) claim that the material of choice for Zn distribution when using DEAE Sepharose CL-6B is blood plasma.

Another method which was successfully applied for Zn speciation work was affinity chromatography. Foote and Delves (1983, 1984-a, 1984-b) used Blue Sepharose CL-6B to separate albumin. They determined the albumin and α_2-macroglobulin amounts by kinetic immunoturbidity and the Zn concentration by GFAAS. The sum of the concentrations of Zn found in association with albumin and α_2-macroglobulin was generally in good agreement with the total Zn concentration with a total recovery of 100.0 ± 6.4%.

The distribution of Zn between the different components in serum is given in Table 2. None of these parameters showed a significant difference when data relating to the male and female populations were analysed separately.

Table 2. Distribution of Zn between different serum components according to Foote and Delves (1984-b)
n=123 86 men 37 women

	Zn μg / ml mean ± standard deviation	% distr.
total Zn	0.968 ± 0.157	
α_2-macroglobulin bound Zn	0.157 ± 0.039	16.2%
albumin bound Zn	0.811 ± 0.144	83.8%

When considering the molar ratios of albumin to albumin bound Zn, there appeared to be only 1.1 - 2.6% albumin present in the serum sample engaged in the Zn transport. On the contrary nearly all binding sites of the α_2-macroglobulin appeared to be occupied by Zn.

More recently, Foote and Delves (1988) persevered in the determination of non-protein bound Zn in human serum and plasma using ultrafiltration and GFAAS. No albumin, α_2-macroglobulin nor retinol binding protein was detected in samples of ultra filtrate. The Zn was quantitatively determined in the 2 fractions.

The total Zn and ultrafilterable Zn concentrations in serum and plasma samples of 17 healthy volunteers read as follows:

total Zn :		1.014 ± 0.163	μg/ml
non protein bound Zn:	in serum	0.0053 ± 0.0010	μg/ml
	in plasma	0.0022 ± 0.0008	μg/ml

The ultrafilterable Zn in serum amounted to no more than 0.5 ± 0.1% of the total Zn present. They concluded that heparin exhibits a specific affinity for Zn and disrupts the interaction of Zn with albumin, its major carrier protein. Zn-heparin complexes have a relative molecular mass of 5,000 - 20,000 D and may therefore be expected to be retained. The Zn concentrations in the plasma ultrafiltrates were approximately 50% of those measured in paired ultrafiltrates of serum. Measurement of ultrafilterable serum Zn is useful because it is considered to be the carrier-mediated mechanism which controls the influx of the metal into cells. This small non-protein bound fraction is almost certainly the crucial and final component of the pathway which maintains cellular Zn nutrition.

4.2. Copper

The Cu value in human serum and plasma is well established since more than 3 decades. It amounts to approximately 1 μg Cu/ml (Versieck and Cornelis, 1989). In 1948 Holmberg and Laurell with the aid of precipitation techniques and 5 years later Gubler et al. (1953) by measuring the amount of Cu available to the chelating agent DDC, concluded that about 90 - 95% of Cu is associated with ceruloplasmin, an α_2-glycoprotein of MW 132,000 D (concentration 300 μg/ml). The remaining part was assumed to be bound to albumin, mainly on the basis that, at the time, no other Cu proteins were known. The earlier experiments using gel permeation chromatography may have been biased by those findings of Holmberg and Gubler. Gardiner et al. (1981) using GPC (Fig. 1 & 2) on human serum found in their elution pattern for Cu versus the protein profile a small Cu peak in the void volume associated with a first batch of heavy proteins. It could not be explained and was therefore dismissed. It was attributed to proteins scavenging traces of Cu deposited on the gel. The percentage Cu found in the fractions containing ceruloplasmin was 92.7 ± 7.3% (mean ± 1 s.d.) and with albumin 7.3 ± 2.7% (n=7). The Cu recovery was slightly higher than 100%.

Chilvers and coworkers (1984), using ion exchange chromatography, (Fig. 3 & 4) published the results given in Table 3. There are age related changes in these distributions for women but not for men.

The proportions of Cu attributable to these various components took a fundamental turn when Wirth and Linder (1985) came up with an additional Cu transport protein in blood plasma, named transcuprein (MW 270,000 D, eluted in the void volume of the Sephadex G-150. They concluded that Cu is not distributed among serum components in the proportions originally thought. Their results (n=6) showed that Cu is bound to transcuprein for 14 ± 2%, to ceruloplasmin (65 ± 11%), to albumin (14 ± 4%) and to low molecular weight (13,000 - 30,000 D) molecules for 8 ± 5%.

Table 3. Distribution of Cu between plasma components according to Chilvers et al. (1984)

	Cu μg/ml, mean ± standard error	
Cu-fraction	men (n=36)	women (n=36)
total Cu	0.866 ± 0.015	1.040 ± 0.027
ceruloplasmin Cu	0.753 ± 0.023	0.911 ± 0.038
albumin Cu	0.096 ± 0.007	0.119 ± 0.010
UCBC Cu	0.030 ± 0.002	0.028 ± 0.002

(UCBC : unidentified Cu-binding compounds)

It can be expected that research with more refined biochemical separation and detection techniques will confirm these latest findings, maybe even bring new developments in Cu speciation of human serum.

4.3. Selenium

The element is present in human serum at a concentration of about 0.1μg/ml, apart from recognised disparities attributable to geographical and age dependent factors (Versieck and Cornelis, 1989). In contrast to most other trace elements, the hazard to contaminate the serum samples or the separated fractions with measurable amounts of Se appears rather remote.

Stiefel et al. (1980) using preparative isotachophoresis found that Se was present for 6% in prealbumin, 38% in albumin, 32% in α_2-glycoprotein (?), 24% in τ-globulin.

Beilstein and Whanger (1983) investigated with GPC and fluorimetry the distribution of Se versus that of glutathione peroxidase (GPx) in human plasma. GPx chromatographed in 2 peaks, the Se in 3, only 1 coinciding somewhat with GPx proteins. The distribution of the Se content was: 60% of total Se in the first peak, then 35% in the second and finally 5% of the total Se associated with GPx.

Brätter and coworkers (1988) reported some interesting findings on the speciation of Se in serum. Their method consisted of an on-line combination of GPC (HPLC) and ICP-AES detection. In order to increase the sensitivity of the Se-determination, the eluted fractions were first submitted to an acid digestion procedure and formation of Se(VI) with the subsequent hydride generation in the connecting flow. The Se in serum was found to be distributed among 3 different molecular fractions: 90,000 D, 200,000 D, and > 600,000 D. The peak at 90,000 D is mainly albumin. The authors express the opinion that it may also hold the Se-containig enzym glutathion peroxidase (GPx) (MW 88,000 D) which elutes in the same region. The Se distribution among the different protein fractions was investigated in serum samples of 7 subjects. The results read as follows: the 90,000 D fraction carried 77 ± 4% of the total Se content, the 200,000 D 10 ± 3% and the > 600,000 D the remaining 13 ± 5%. Plasma transport of Se appears to be facilitated by albumin. In this study there is no mention of any attempt to identify and quantify the biocomponents.

To my knowledge no large scale studies on the speciation of Se in human serum have been undertaken to date.

4.4. Nickel

Nickel is present in human serum at a concentration of 0.5 ng/ml (Versieck and Cornelis, 1989). No physiological function for Ni has yet been established in man, there is neither any specific enzyme nor any other Ni-compound considered essential. Interest in Ni is initiated by its hazard to people who are occupationally exposed to this metal and recently also by the use of Cr-Co-Ni alloy prostheses and the possible harmful effect they may cause in the long run, especially in patients who have a decreased renal function.

The research group of Nomoto and Sunderman (1988) recently renewed their efforts to define the Ni-containing components in human serum. They first ran the serum on a GPC column with a fast protein liquid chromatography system (FPLC). In the protein UV-profile 6 regions could be discerned (see Fig. 5 & 6). The Ni concentration profile exhibited 3 Ni-containing regions (A, B, and C). The 6 peaks in the protein profile and the original serum were submitted to iso-electrofocusing and revealed the different proteins which were present. Their association with the Ni-content could be interpreted as follows:

A coinciding with 2 and 3, could be associated with α-globulins

B with albumin and τ-globulins

C eluted beyond the bed volume, contained ultrafilterable constituents, such as amino acids, binding Ni.

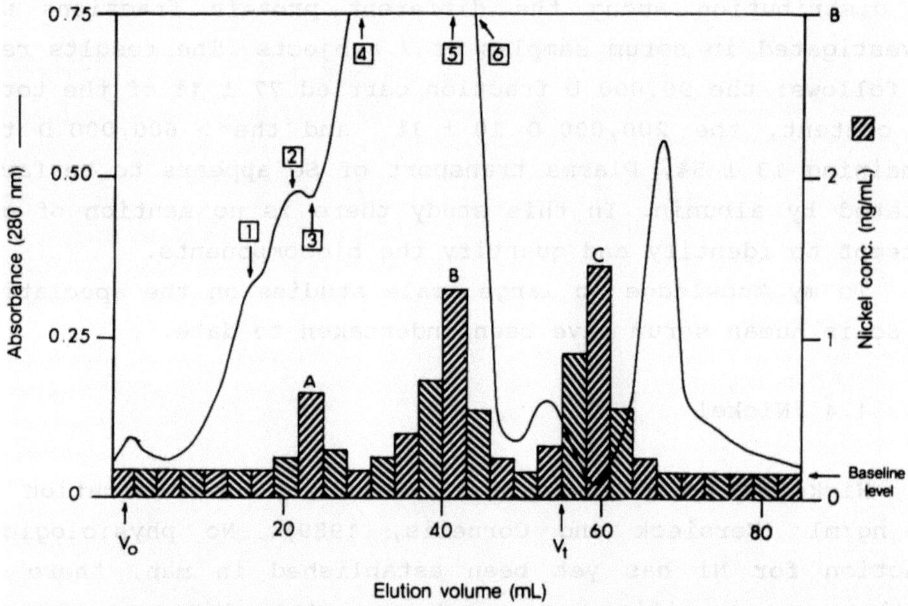

Fig. 5. Chromatographic profiles of Ni and A_{280} nm in effluent from a gel-filtration column of "Superose 6" (void volume V_0; bed volume V_t). The starting material was human serum (2 mL). Three Ni-containing peaks ("A", "B", and "C") were detected. The fractions labelled 1 , 2 , 3 , 4 , 5 , and 6 correspond to samples analyzed by agarose gel electrophoresis. (Nomoto and Sunderman, 1988)

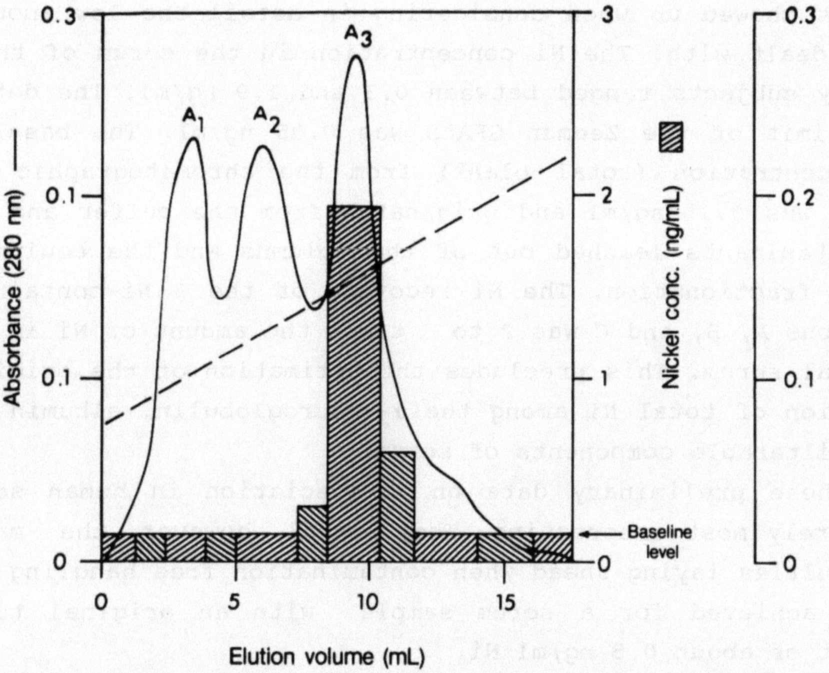

Fig. 6. Chromatographic profiles of Ni and A_{280} nm in effluent from an ion-exchange column of "Mono-Q", with elution in tris buffer containing a linear gradient of sodium nitrate (0.075 to 0.27 mol per L). The starting material was peak "A" obtained by gel-filtration chromatography of human serum. The A_{280nm} plot has been corrected for background absorbance of the sodium nitrate gradient. The chromatographic profile shows a single nickel-containing peak ("A_3"). (Nomoto and Sunderman, 1988)

The A fraction was then submitted to Mono Q ion exchange chromatography and eluted with a linear gradient of $NaNO_3$. (This to avoid the use of NaCl which would produce matrix interferences later on during GFAAS). This separation resulted in 3 components A_1, A_2 and A_3. A_3 was the single Ni-containing peak above the base line concentration. The A_3 peak was further identified by submitting it to SDS-PAGE (sodium dodecyl sulfate polyacrylamide gel electrophoresis) which yielded an approximate MW of 700,000 D. Immunoelectrophoretic analysis with specific horse antibody to human serum α_2-macroglobulin yielded the final proof for its unequivocal identity. Difficulties showed up when considering in detail the low amounts of Ni dealt with. The Ni concentration in the serum of the 4 healthy subjects ranged between 0.3 and 1.9 ng/ml. The detection limit of the Zeeman GFAAS was 0.05 ng/ml. The baseline Ni-concentration (total blank) from the chromatographic effluent was 0.15 ng/ml and originates from the buffer and the Ni contaminants leached out of the columns and the equipment during fractionation. The Ni recovery of the 3 Ni-containing fractions A, B, and C was 2 to 3 times the amount of Ni in the original serum. This precludes the estimation of the original partition of total Ni among the α_2-macroglobulin, albumin and ultrafilterable components of serum.

These preliminary data on Ni-speciation in human serum are surely most interesting. They reveal, however, the major difficulties laying ahead when contamination free handling has to be achieved for a serum sample with an original total content of about 0.5 ng/ml Ni.

4.5. Chromium

The concentration of Cr in serum of healthy individuals is about 0.2 ng/ml i.e. five thousand times lower than that of Zn and Cu (Versieck and Cornelis, 1989). Speciation of Cr in serum was initiated by the finding of extremely high Cr levels (50 times) in the serum of dialysis patients with terminal renal failure (Wallaeys et al., 1986). The underlying reason for this increased concentration is that the small amount of Cr, 1 ng/ml, present as an impurity in the dialysate liquid, is irreversibly taken up by certain substances in serum.

Previous studies by Hopkins and Schwarz (1964) and Wallaeys et al., (1988) have already shown that Cr is mainly bound to transferrin and, to a lesser degree, to albumin.

Because of the tedious workload to determine the very low concentrations in the various fractions and of the immense hazard of contaminating with exogenous Cr, Wallaeys et al. (1987, 1988) and Borguet et al. (1989) started with in vitro and in vivo labelled ^{51}Cr plasma in CAPD (continuous ambulatory peritoneal dialysis) patients and in rats. This allows an

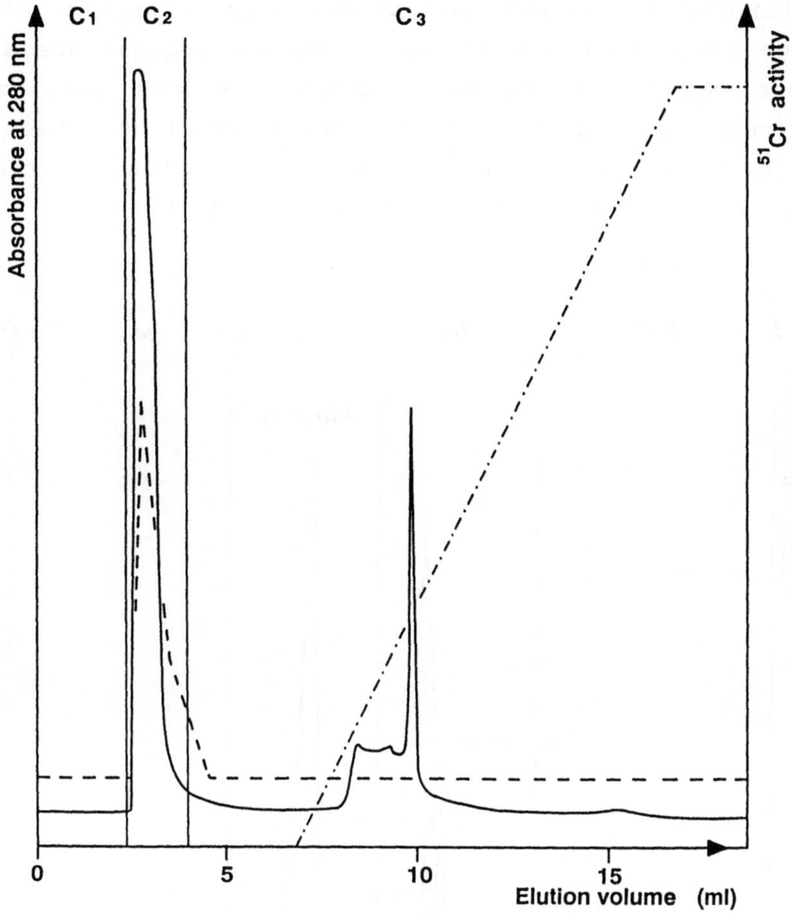

Fig. 7. UV-chromatogram (——), ^{51}Cr elution pattern (- -), and NaCl gradient (-·-) of ^{51}Cr labelled human plasma. Sample: 0.2 ml ^{51}Cr incubated healthy human plasma; Gel: Mono S; eluent: 0.025 M Tris.HCl pH 8; bed volume: 1 ml; elution time: 20 min

easy tracking of the element on a molecular level. This approach reduced the detection of the element Cr to a mere radioactivity measurement. These researchers were aware that the use of labelled ^{51}Cr-compounds whether in vivo or in vitro, to study the distribution of added Cr to plasma or serum binding components, may create many artefacts. Care was taken to avoid addition of large, unphysiological doses of Cr as these would bind to those components with many binding sites (such as albumin) and only these would be detected. In order to allow the Cr to form part of a molecule time was given to the system to reach a certain equilibrium.

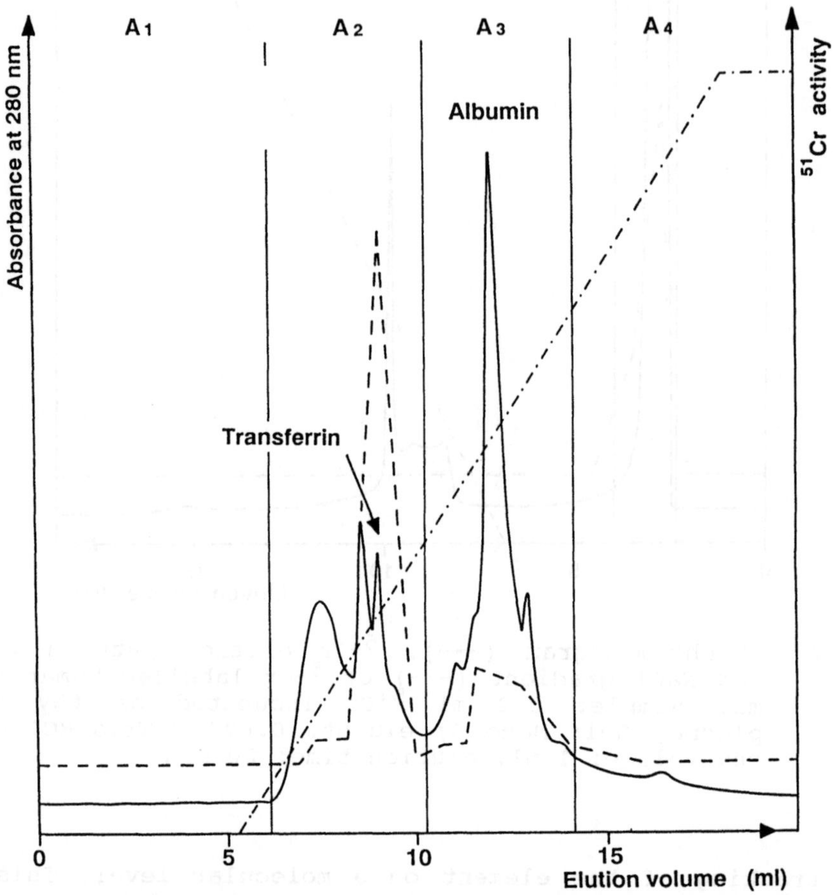

Fig. 8. UV-chromatogram (——), ^{51}Cr elution pattern (- -), and NaCl gradient (-··-) of 0.2 ml of C_2 fraction of Mono S elution. Gel: Mono Q; eluent 0-05 M NaCl gradient in 0.025 M Tris.HCl pH 8; bed volume: 1 ml; elution time: 20 min

Preliminary experiments with GPC ended in consecutive failures, as it appeared impossible to keep the original Cr-protein binding intact. This phenomenon was related to the long residence time (10 h) of the plasma compounds on the resin. A much shorter separation scheme was designed consisting of ion exchange chromatography (Fig. 7 & 8). When searching anionic proteins, superior separations can be achieved by passing the plasma first over a cation exchanger. The negative, or least positive, fractions are immediately washed out by the buffer and are then put on an anionic exchanger. These are then eluted with a NaCl gradient. Fig. 6 shows the separation on Mono Q with FPLC system of the anionic species present in ^{51}Cr incubated human serum, together with the ^{51}Cr-distribution. The latter is mainly linked to tranferrin and to a minor degree to albumin. The purity of the biomolecule in the different fractions was checked with iso-electrofocusing. The proteins were quantified with the aid of nephelometry and Cr was followed by measuring the ^{51}Cr activity. The results of in vivo labelling differed substantially from those obtained from in vitro labelled rat serum. In the first case only binding to transferrin could be discerned.

5. Final considerations

This short account of research about speciation of trace elements in human serum and plasma reveals how indispensable is the interdisciplinary approach.

In order to design a separation scheme, one needs to make some realistic assumptions regarding to which serum components the trace element may possibly be linked and appropriate experimental conditions must be chosen. The applied biochemical fractionation techniques need more specificity and selectivity than what shown up to now, but the basic knowledge exists. The more elaborate the separation scheme for the biocompounds, however, the greater the hazard that changes may occur in the original metal-ligand binding. A biochemist's interest lies mainly in the molecule as such, with little interest for the conservation and the authenticity of the metal-ligand binding. He is also not really worried about the complete recovery of the biomolecule, but mainly about its purity. This is essentially different thinking from that of an analytical chemist, who pursues a 100% recovery of the metal-ligand complex with neither losses nor contamination with extraneous amounts of the trace element. The more complex the separation procedure and the lower the trace element concentration, the more tedious and endless the task becomes.

6. References

Alberts B, Bray D, Lewis J, Raff M, Roberts K, Watson JD (1983) Molecular biology of the cell. Garland Publ., Inc., New York, London

Beilstein MA, Whanger PD (1983) Distribution of selenium and glutathione peroxidase in blood fractions from humans, rhesus and squirrel monkeys, rats and sheep. American Institute of Nutrition, 2183-2146

Borguet F, Cornelis R (1990) Speciation of chromium in plasma of endstage renal failure patients on continuous ambulatory peritoneal dialysis. Biol Trace Element Res, in press

Brätter P, Gercken B, Tomiak A, Rösick U (1988) Combination of HPLC and ICP-AES for the speciation of selenium and other trace elements in body fluids.In Brätter P, Schramel P (eds) Trace element analytical chemistry in medicine and biology. 5. De Gruyter, Berlin, New York, p 119-135

Chilvers DC, Dawson JB, Bahreyni-Toosi MH, Hodgkinson A (1984) Identification and determination of copper- and zinc-protein complexes in blood plasma after chromatographic separation on DEAE-Sepharose CL-6B. Analyst 109: 871-876

Foote JW, Delves HT (1983) Distribution of zinc amongst human serum proteins determined by affinity chromatography and atomic-absorption spectrophotometry. Analyst 108: 492-504

Foote JW, Delves HT (1984-a) Distribution of zinc amongst human serum globulins determined by gel filtration - affinity chromatography and atomic-absorption spectrophotometry. Analyst 109: 709-711

Foote JW, Delves HT (1984-b) Albumin bound and α_2-macroglobulin bound zinc concentrations in the sera of healthy adults. J Clin Pathol 37: 1050-1054

Foote JW, Delves HT (1988) Determination of non-protein bound zinc in human serum using ultrafiltration and atomic absorption spectrometry with electrothermal atomisation. Analyst 113: 911-915

Gardiner PE, Ottaway JM, Fell GS, Burns RR (1981) The application of gel filtration and electrothermal atomic absorption spectrometry to the speciation of protein-bound zinc and copper in human blood serum. Anal Chim Acta 124: 281-294

Gardiner PE (1987) Species identification for trace inorganic elements in biological materials. Topics in Current Chemistry 141: 145-174

Giroux EL, Durieux M, Schechter PJ (1976) A study of zinc distribution in human serum. Bioinorganic Chem 5: 211-218

Gubler CJ, Lahey ME, Cartwright GE (1953) Studies on copper metabolism. IX. The transportation of copper in the blood. J Clin Invest 32: 405-414

Heydorn K (1984) Neutron activation analysis for clinical trace element research, Vol I & II. CRC Bota Raton, Florida

Hopkins LL, Schwarz K (1964) Chromium(III) binding to serum proteins, specifically siderophilin. Biochim Biophys Acta 90: 484-491

Kirkbright GF, Sargent M (1974) Atomic absorption and fluorescence spectroscopy, Academic Pres, New York

Nomoto S, Sunderman FW (1988) Presence of nickel in alpha-2 macroglobulin isolated from human serum by high performance liquid chromatography. Ann Clin Lab Sc 18: 78-84

Schramel P (1983) Consideration of inductively coupled plasma spectroscopy for trace element analysis in the bio-medical and environmental fields. Spectrochim Acta Part B 38: 199-206

Stiefel T, Schulze K, Tölg G (1980) Separation of localized trace elements (Be, Cd, Se) in serum protein fractions under physiological pH conditions by preparative isotachophoresis. In: Brätter P, Schramel P (eds) Trace element analytical chemistry in medicine and biology, De Gruyter, Berlin, New York, p 451-462

Versieck J, Cornelis R (1989) Trace elements in human plasma or serum, CRC Bota Raton, Florida

Wallaeys B, Cornelis R, Lameire N (1986) Trace elements in serum, packed cells, and dialysate of CAPD patients. Kidney Int 30: 599-604

Wallaeys B, Cornelis R, Belpaire F, Lameire N (1987) Speciation study of chromium during peritoneal dialysis. In: Brätter P, Schramel P (eds) Trace element analytical chemistry in medicine and biology, 4, De Gruyter, Berlin, New York, p 451-461

Wallaeys B, Cornelis R, Sabbioni E (1988) Kinetics of chromium during peritoneal dialysis. Sci Tot Environm 71: 401-410

Welz B (1985) Atomic absorption spectrometry, 2nd edn., Verlag Chemie, Deerfield Beach, Florida

Wirth PL, Linder MC (1985) Distribution of copper among components of human serum. JNCI 75: 277-284

Acknowledgements

R. Cornelis is a Research Director of the National Fund for Scientific Research.

Chemical Sensors in Environmental Analysis: Ammonium and Cadmium Sensors

J. Reichert, R. Czolk, W. Sellien and H.J. Ache

Kernforschungszentrum Karlsruhe
Institut für Radiochemie
D - 7500 Karlsruhe, F.R.G.

Introduction

In order to recognize the great potential of the application of suitable chemical sensors in the area of environmental analysis it seems appropriate to recall the various tasks which we face in environmental analysis (Fig. 1).

The first stage would be the analytical detection of materials or compounds which are suspected to have an adverse effect on the environment, the determination of the magnitude or the extent of their distribution in the environment and their contribution to the overall burden caused by anthropogenic materials to the environment. The analytical work needed in this area requires highly sensitive methods and high performance equipment of the more classical type, such as MS, AAS, AES, GC, HPLC, IR, etc. Especially the second part which is concerned with the distribution of the materials in the environment will require many test series. These investigations will have to be accompanied by toxicological studies which should reveal the tolerance levels of these compounds in the environment. These investigations will then result in the definition and fixation of maximum tolerance levels. On this basis in the next step regulations and laws will be set up to be instrumental in the removal of these materials from the environment or even better to prevent their emission to the environment.

The monitoring of the compliance with these regulations will in turn require numerous test stations where environmental samples can be analyzed. This would be only one admittedly important and expensive aspect; in addition, however, from the great number of samples to be tested and the great variety of classes of compounds it will become quite clear that the classical off-line analysis techniques namely taking samples in the environment and to transfer them to

NATO ASI Series, Vol. G 23
Metal Speciation in the Environment
Edited by J. A. C. Broekaert, Ş. Güçer, and F. Adams

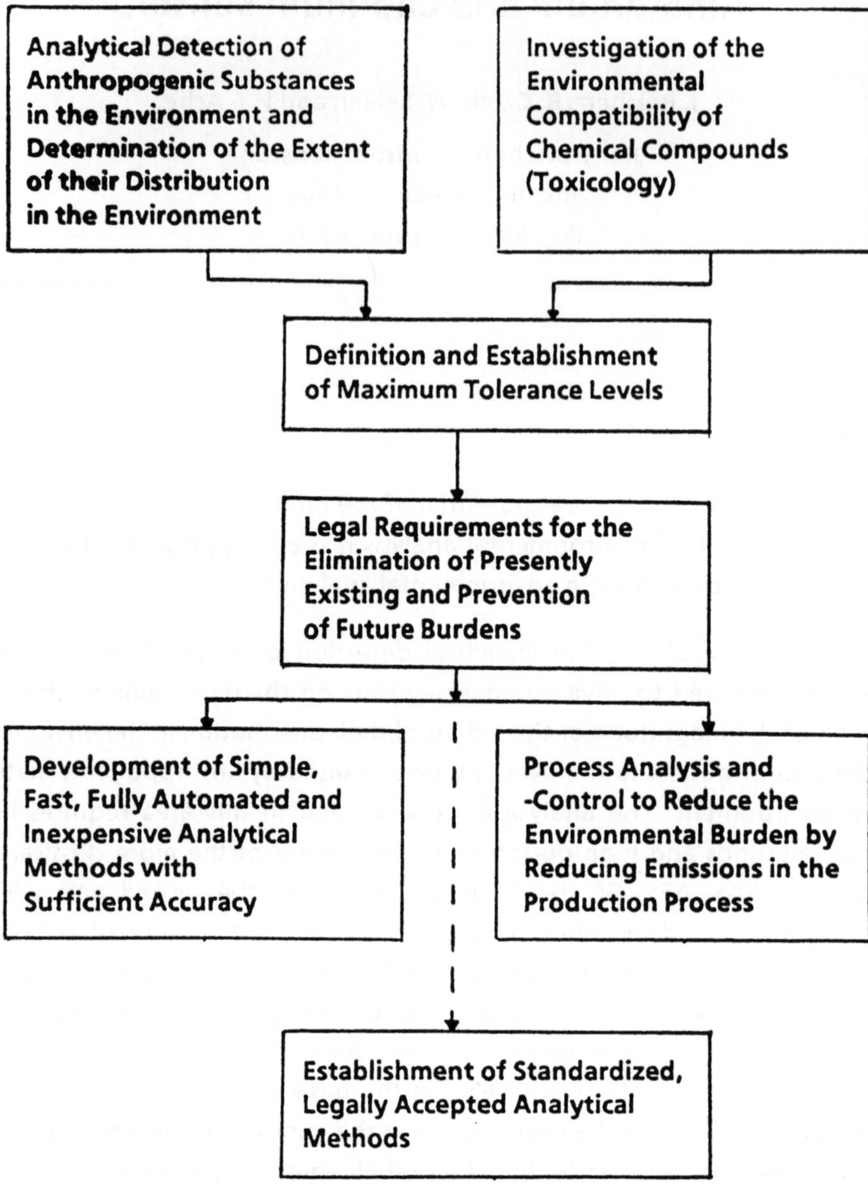

Fig. 1 Tasks in environmental analysis

central laboratories will not be practicable anymore. In view of the fact that more and more requests for analyses are made this procedure is too costly and labor intensive to be continued, in addition, the composition and concentration of the compounds to be analyzed may change during the time necessary to transport the samples to the analytical laboratory. Consequently, there is an urgent need to design in-situ analytical techniques, fully automated and inexpensive, with sufficient accuracy for this monitoring task. In a similar way it would be necessary to improve process analysis again by fast responsive in-line analysis which allows the reduction of emission of relevant compounds to the environment by a better control of the production process; perhaps in this way expensive retention systems can become redundant.

While the environmental monitoring can also be considered as a kind of process analysis with test points in the environment, usually there will be a much higher sensitivity required, in the ppb range, while in the second case in a processing plant, this will be more the exception than the rule. But there is a number of features which these two analytical applications have in common and developments in one area may also be of interest to the work in the other one.

A different issue is the development of standard analytical methods or reference methods, whose results will hold up and be accepted in legal procedures.

There are basically two approaches to solve this problem. The first one can be considered a centralized approach in which case signals from numerous test points (up to one hundred or so) are relayed e.g. via fiber optic systems to a sophisticated central instrument and processed (1). Such a system could be e.g. the *guided wave spectral photometer*, used in our own laboratory (2). It is equipped with a Xenon lamp from which via fiber optics white light is carried to the test points where the presence of analytes will lead to changes in the absorption or fluorescence. This information is brought back to the central instrument via fiber optics and analyzed. The data processing of the spectra usually requires chemometric methods (3). In this way the large number of testpoints which can be covered will justify the relatively high costs of the central instrument.

The ultimate goal, and that would be the second approach, will be the development and application of miniaturized chemical sensors which can be fabricated by means of microstructure techniques similar to the production of microchips.

A chemical sensor is a device to be fabricated with microtechniques which can transform chemical parameters into electrical signals in this way permitting an identification and quantification of chemical compounds.

If sensors of this type become available a number of problems in environmental analysis could be solved in a cost-efficient manner. Some of these tasks would involve e.g. ground water monitoring, survey of an economic application of fertilizer in agricultures, control of household machine e.g. tensides distribution in washing machines, exhaust gas controls in automobiles, monitoring leaks in storage tanks and, of course, the monitoring of the level of toxic gases in plants, laboratories, machine shops etc.

All sensors presently in use or in development are based on a two-step detection mechanism. In the first step chemical selectivity is achieved by a chemical reaction or chemi- or physisorption on a chemically selective surface. In the second step a physical change, which is the result of the first selective chemical step, is transformed by a suitable transducer into an electrical signal. This physical change can be a variation of the chemical potential, caused by the reaction or sorption process, a change in optical properties, a change of mass, conductivity, surface resistance or conductance.

The simplest type of an optical sensor is a quartz or glass fiber, to which a reagent has been attached (Fig. 2). If this reagent forms reversible complexes with an analyte its optical properties, light absorption or fluorescence, will change, e.g. light, produced via a Xenon lamp, laser or photodiode of a given wavelength will then be more strongly absorbed, an effect which can be monitored in a suitable way. Light of a different wavelength, not absorbed by the complex can be used as reference. Problems occurring with that type of detector are again deficiencies in the long term stability caused by photochemical decomposition of the reagent, and, of course, the unsatisfactory permanent fixation of the reagent to the fiber.

Some of these developments shall be demonstrated by reporting on sensor studies carried out in our own laboratory with respect to environmental problems (4).

Ammonium sensor

The first example is the development of an ammonium sensor for ground water monitoring. Optical ammonia sensors have been described in the literature (5,6). They are based on the fact that ammonia as a base can bind protons which in

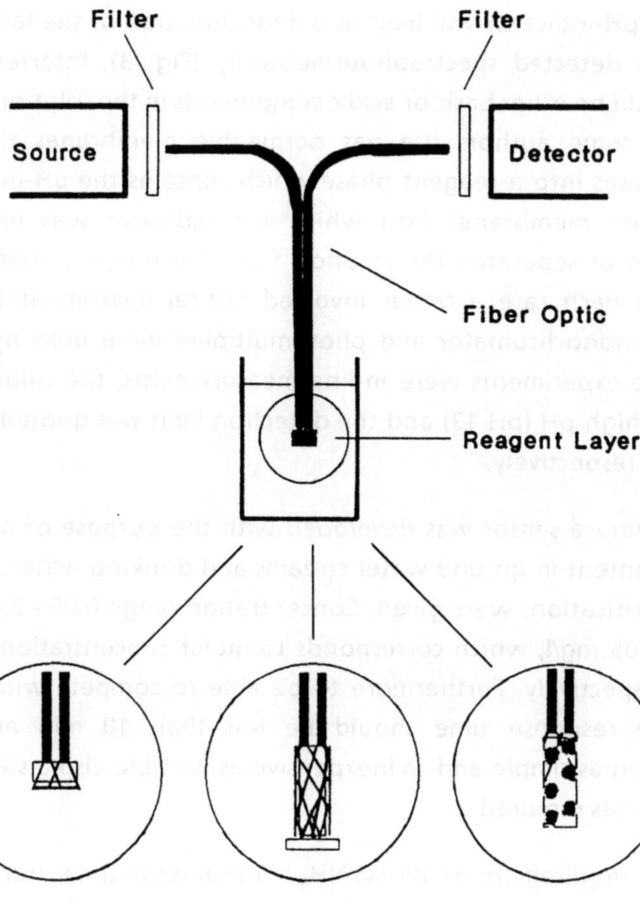

Fig. 2 Principal components of an optrode: Possible arrangements for immobilized reagent layers are indicated

presence of a pH-indicator will lead to a deprotonation of the latter, a reaction which can be detected spectrophotometrically (Fig. 3). Interfering with this detection would be other basic or acidic components in the solution. To eliminate these effects some authors use gas permeable membranes through which ammonia diffuses into a reagent phase which contains the pH-indicator. Other authors applied membranes into which the indicator was immobilized by polymerization or separated the solution from the reagent phase via a teflon membrane. In each case a rather involved optical instrumentation including Xenon lamp, monochromator and photomultiplier were used by the authors. None of these experiments were in-line measurements, the solution has to be adjusted to a high pH (pH 13) and the detection limit was quoted as 5 µM or 10 µM ammonia, respectively.

In our laboratory a sensor was developed with the purpose of monitoring the ammonium content in ground water streams and drinking water. Therefore the following specifications were given. Concentration range 0.05 - 0.5 mg/l with an accuracy of 0.05 mg/l, which corresponds to molar concentrations of 3 - 28 µM and 3 µM, respectively. Furthermore to be able to compete with ion selective electrode the response time should be less than 10 min and its optical instrumentation as simple and as inexpensive as possible also a standing time of several weeks was required.

The detection requirement of 10 µM NH_4^+ translates in a neutral solution to an available ammonia concentration of ca. 100 nM. If an acidic environment prevails in the reagent phase the ammonia molecules diffusing into this phase will pick up a proton. Consequently the pH will be shifted toward the neutral region. The ammonia flow, of course, will stop when in both phases the same concentration is reached. Calculations will show that with indicators, whose pK is in the range from 3 - 4, the above mentioned detection limit should be obtained. The protonation of the ammonia is followed by a deprotonation of the indicator. Considering the small concentration of ammonia and the fact that for optical detection a certain amount of indicator has to react, it seems sensible to keep the indicator concentration small so that short response times can be obtained. Indicator pK and indicator concentration have to be optimized for a given sensor design.

In the following several approaches were tried. In the first experiment bromocresolgreen (pK = 4.6) was used as indicator. It was emulgated into a silicon caoutchouc membrane; its concentration was about 30 mM. Light sources were two light emitting diodes (LED's), 595 and 660 nm (Fig. 4). The detector was a

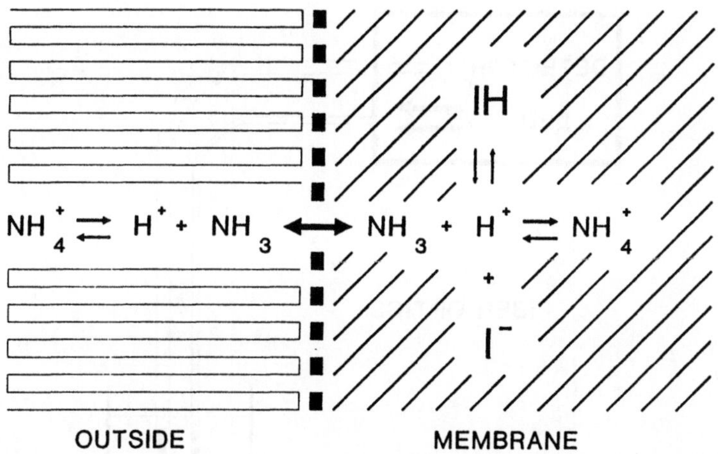

Fig. 3 Reaction scheme for the ammonium sensor: Ammonia diffuses into the membrane phase and deprotonates the pH-indicator which can be observed spectroscopically

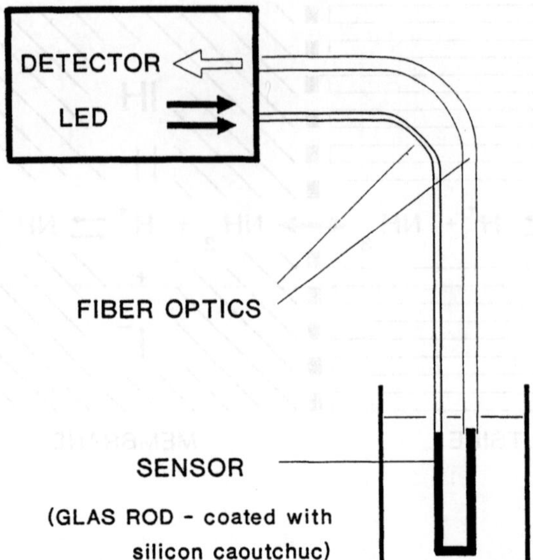

Fig. 4 Schematic arrangement of the ammonium-optrode: The pulsed light of two LED's (595 and 660 nm, resp.) is sent via plastic fibers to the sensor (coated glass rod) and the received light is detected by a photodiode

highly sensitive Si-photodiode. The observed absorption was evaluated by sample and hold processing of the pulsed light. The absorption A at 595 or 660 nm could be either separately measured or the difference A_{595} - A_{660} could be determined (Fig. 5). The measurements were carried out in the reflection mode by applying plastic fiber optics. The detection limit was found to be 3 µM ammonia, i.e. within the specifications. Problems were encountered with the response time, e.g. it took the sensor 1 - 2 hours to reach the maximum signal output and the recovery time of 10 hours was completely unsatisfactory. An attempt to improve the situation by using a PVC-ionophor membrane for better selectivity in combination with a smaller reagent phase volume did not succeed. A faster response time was observed by using thinner e.g. 20 µm polyvinylacetate membranes. A similar improvement was obtained by dipcoating the optical fiber with a thin (20 µm) silicon caoutchouc layer which contained bromophenolblue (pK = 3.8, 5 mM). With this arrangement (see figure 4) utilizing multireflections an increase in the optical thickness was obtained without any loss of response time which was 10 and 30 min for on- and off-time (Fig. 6). The detection limit was determined to be 10 µM ammonium or 160 nM ammonia which was within the required concentration range (Fig. 7). Possible interferences caused by HCO_3 and other ions are presently under investigation and long term experiments have been initiated in order to evaluate the standing time of this device.

Cadmium sensor

A second example which illustrates another facet of the problems involved in the developing of chemical sensors is the work on a cadmium optrode in our laboratory. Here first attempts were made utilizing the shift in the optical properties of certain complexes such as morin (7) (2',3,4',5,7-pentahydroxy-flavon), 8-hydroxychinolin-5-sulfonic acid and 3,5'-bis(dicarbocymethylamino-methyl)-4,4'-dihydroxy-trans-stilben especially on their fluorescence behaviour when they form a complex with cadmium. Since they are not very selective for Cd it was tried to provide the necessary selectivity by using membranes with a Cd^{++} specific ionophor (8).

Unlike in the case of ionselective electrodes where basically the potential difference on both sides of such a membrane is measured the detection of the complexation of Cd^{++} by the complexing reaction mentioned above requires substantial quantities of Cd-ions to move through the membrane.

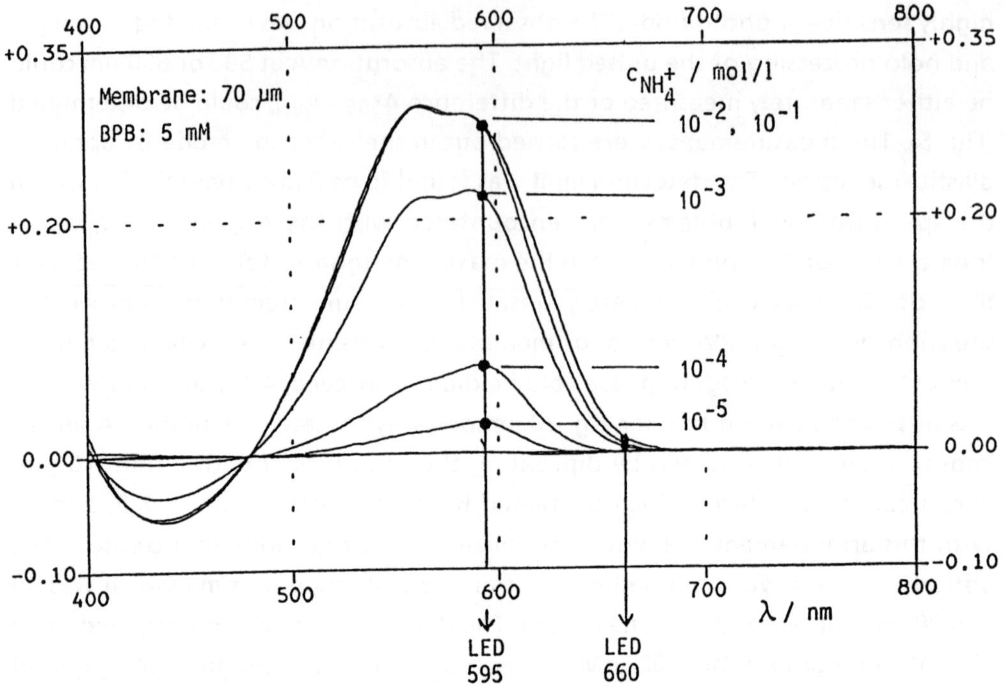

Fig. 5 Difference spectra of a caoutchuc membrane in dependence of various concentrations (10^{-5} M - 10^{-1} M) of ammonium (50 mM phosphate buffer, pH 7.5)

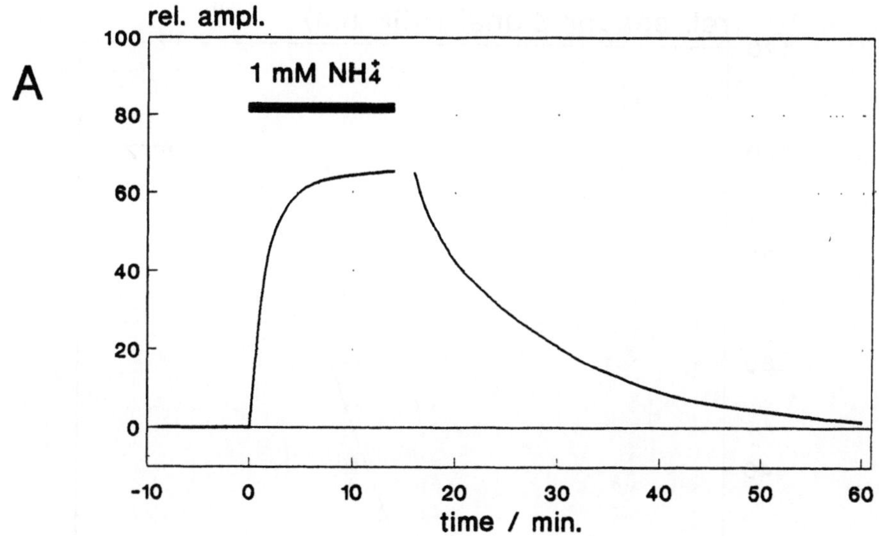

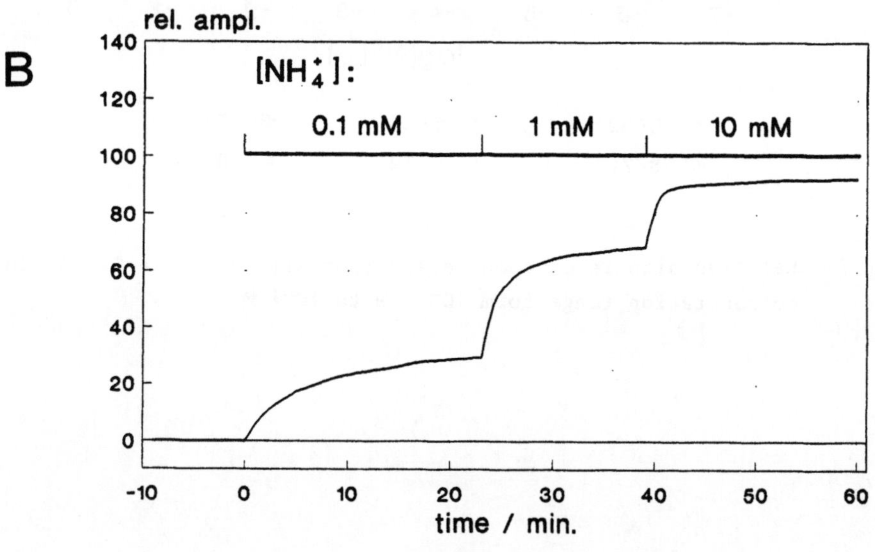

Fig. 6 Time course of the signal (595 nm) of the ammonium sensor: (A) 1 mM NH_4^+, on- and off response, and (B) response to 0.1 mM - 10 mM NH_4^+ in 50 mM phosphate buffer (pH 7.5)

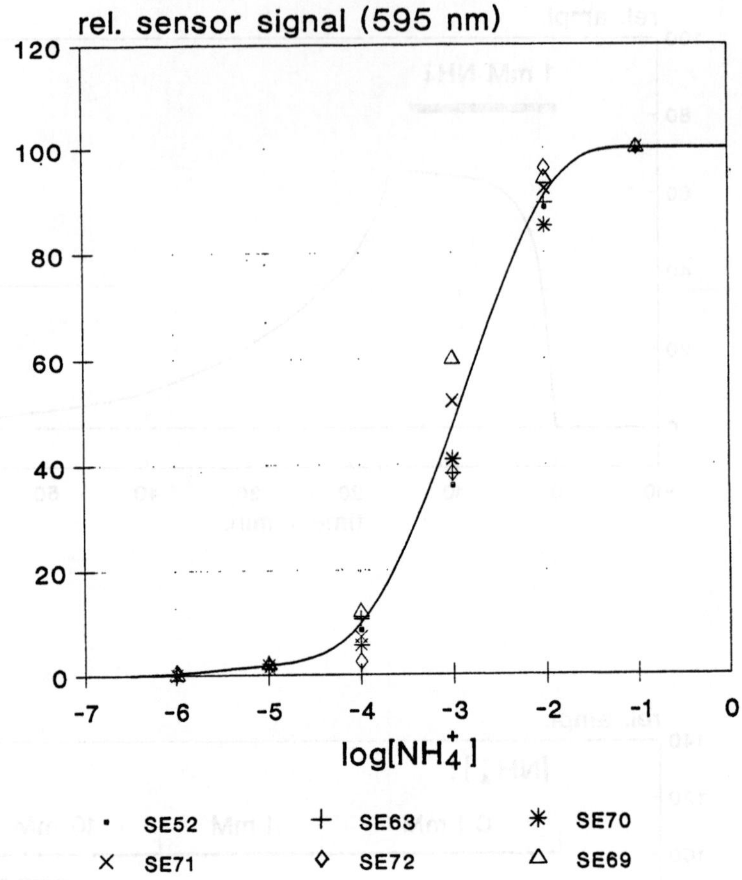

Fig. 7 Relative signals of some sensor (normalized at c = 0.1 M) in the concentration range form 10^{-6} M to 10^{-1} M

1.) COUNTER-TRANSPORT:

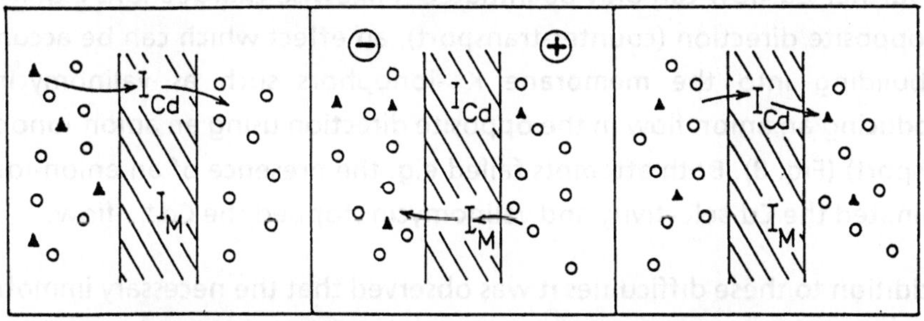

I_{Cd} = cadmium-ionophor ▲ = cadmium

I_M = cation-ionophor o = cation

2.) CO-TRANSPORT:

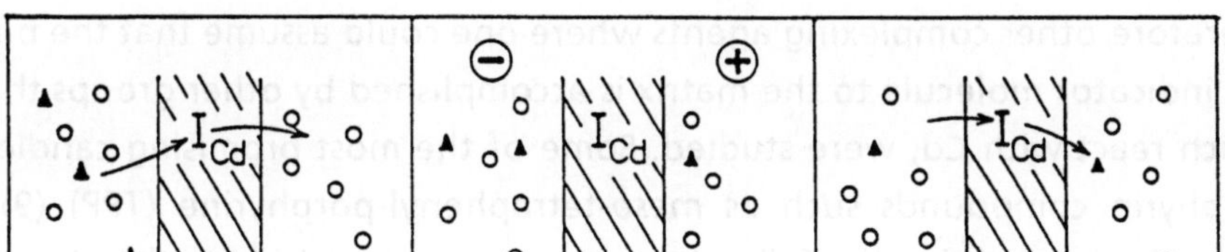

I_{Cd} = cadmium - ionophor ▲ = cadmium

I_A = anion - ionophor o = anion

Fig. 8 PVC - membrane for selecting Cd^{++} with Cd^{++} ionophors

Such a flow, however, will stop if the potential has reached a characteristic value and further Cd-flow can only be initiated if likewise charged ions e.g. K^+ flow in the opposite direction (counter-transport), an effect which can be accomplished by building into the membrane K^+-ionophors such as valinomycin or by introducing an anion flow in the opposite direction using an anion-ionophor (co-transport) (Fig. 8). Both attempts failed e.g. the presence of an anion-ionophore eliminated the Cd-selectivity and valinomycin stopped the Cd^{++} flow.

In addition to these difficulties it was observed that the necessary immobilization of the complexing agent in a matrix such as polyvinylchloride (PVC), polyvinyl-alcohol (PVA), polyallylamin (PAA), cellulose, VA-Epoxy (PVA with oxiran groups), Lichrosorb-NH_2 (modified silicagel with amino groups), and Nylon, lead to a reduction of the complexing capability of the substances. This is probably due to the fact that the group which are instrumental in complexing the Cd-ions are also those which are responsible for the immobilization.

Therefore other complexing agents where one could assume that the binding of the indicator molecule to the matrix is accomplished by other groups than those which react with Cd, were studied. Some of the most promising candidates are porphyrin compounds such as *meso*-tetraphenyl-porphyrine (TPP) (9) and its derivatives containing the following groups in para position to the phenyl group: -COOH (TPPC), -SO_3Na (TPPS), -NH_2 (TPPN), -CH_3 (TPPMe), and -NO_2 (TPPNO) (Fig. 9). The large molar extinction coefficients ($\varepsilon > 10^5$ $l \cdot mol^{-1} \cdot cm^{-1}$) of these indicators allow the photometric detection by absorption measurements and the amount of indicator can be kept small, which is important with regard to the reversibility of the system.

With this indicator different absorption spectra caused by the presence of other ions in the sample offer the possibility also to eliminate interferences using statistical treatment of the data. These indicators were immobilized on VA-Epoxy, PVA and PVC.

A sensor with TPPC immobilized on VA-Epoxy showed the required reversibility, however, the response time was too long with ca. 3 hours (Fig. 10). To reduce the response time an attempt was made to regain the required mobility of the complexing molecule for the reaction with Cd in the matrix by placing "spacers" (organic molecules with hydrocarbon chains C_3 to C_{12}) between the anchoring groups of the TPPN and the matrix (PVA). A chain length of C_5 proved to be the

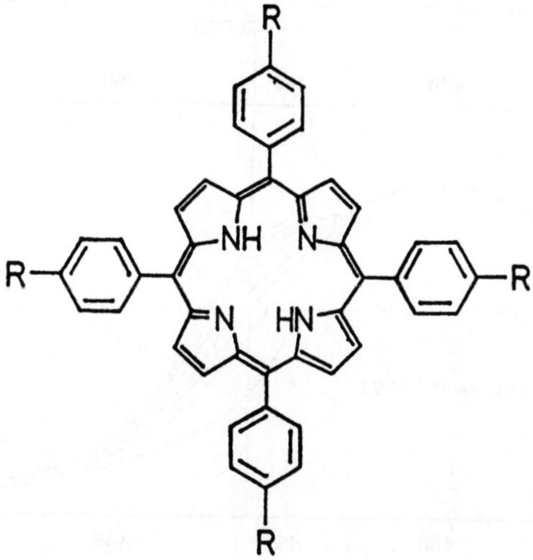

R =	COOH	(TPPC)
	SO_3Na	(TPPS)
	NH_2	(TPPN)
	NO_2	(TPPNO)
	CH_3	(TPPMe)

Fig. 9 Structure formula of tetraphenylporphyrine(TPP)-derivatives

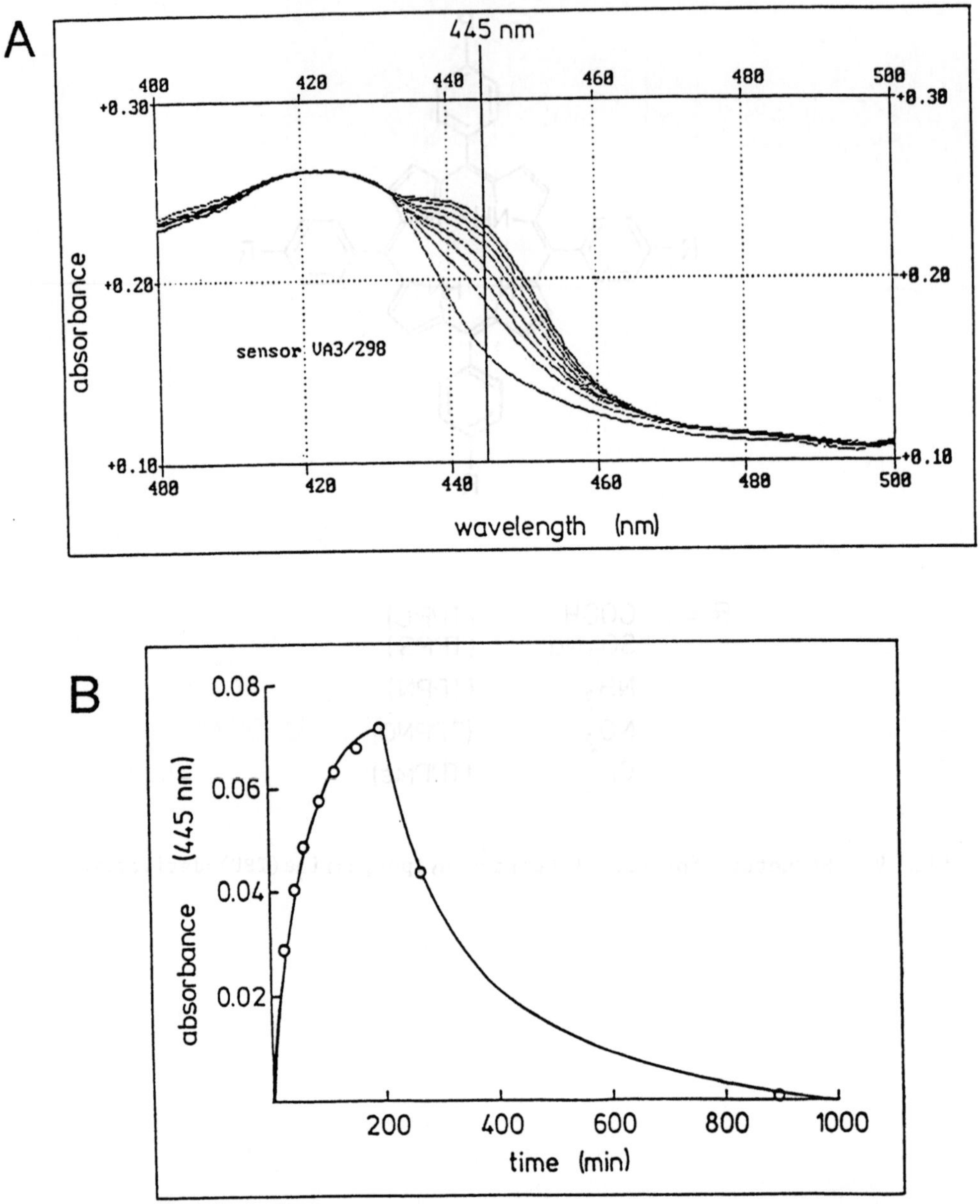

Fig 10 Effect of 10^{-4} mol/l Cd^{++} to the absorbance of TPPC immobilized on VA-Epoxy (pH 8.8, 0.2 mol/l Tris-HCl): (A) Spectra (on-response, 0 - 180 min). (B) Time course of the absorption at 445 nm

optimum size. In this way a reduction of the response time was expected. This investigation is still in progress.

These two examples should demonstrate the complexity of the problems to produce a chemical sensor with the required specifications. More systematic work will be needed to accomplish this goal.

References

1. Hirschfeld T, Haugen G, Milanovich F (1983) Instrumentation for remote sensing over fiber optics. In: Lyon WS (ed) Analytical Spectroscopy. Elsevier, Amsterdam, pp 13 - 18
2. Fith P, Gargus AG (1985) Remote UV-VIS-NIR spectroscopy using fiber optic chemical sensing. American Laboratory 17:64-74
3. Sharaf AS, Illman DL, Kowalski BR (1986) Chemometrics. In: Elving PJ, Winfordner JD (eds) Chemcial Analysis. Vol 83, Wiley & Sons, New York
4. Ache HJ, Reichert J, unpublished results
5. Wolfbeis OS, Posch HE (1986) Fiber-optic fluorescing sensor for ammonia. Anal Chim Acta 185:321-327
6. Rhines TD, Arnold MA (1988) Simplex optimation of a fiber-optic ammonia sensor based on multiple indicators. Anal Chem 60:75-81
7. Zhujun Z, Seitz WR (1985) A fluorescent sensor for aluminium(II), magnesium(II), zinc(II) and cadmium(II) based on electrostatically immobilized quinolin-8-ol sulfonate. Anal Chim Acta 171:251-258
8. Schneider JK, Hofstetter P, Ammann D, Simon W (1980) N,N,N',N'-Tetrabutyl-3,6-dioxaoctan-dithioamid, Ionophor mit Selektivität für Cd^{++}. Helv Chim Acta 63:217-224
9. Komata M, Itoh J (1988) A highly sensitve spectrophotoemtric determination of cadmium with α, β, γ, δ-Tetrakis(4-N-trimethylaminophenyl)porphine. Talanta 35:723-724

Use of ICP-Spectrometry for Environmental Analysis

J.A.C. Broekaert
Institut für Spektrochemie und angewandte Spektroskopie (ISAS)
Postfach 10 13 52
D-4600 Dortmund 1, Federal Republic of Germany

1. Introduction

For the solution of problems in the environmental sciences elemental determinations are of crucial importance. They are indispensible for studies of the mobility of elements and their compounds in complex systems. Because trace elements interact and collectively influence biological activities, knowledge about the concentrations of as many elements as possible is very desirable. Therefore, trace analytical methodologies with multielement capability are needed. Among the modern analytical methods for the determination of trace elements, plasma atomic spectrometry is very powerful (Broekaert and Tölg, 1987). The power of detection and the reliability are high and it may be used for the determination of trace element compounds.

The inductively coupled plasma (ICP), the most important plasma atomic spectrometric source, was first used as a radiation source for optical emission spectrometry (OES) in 1964 (Greenfield et al., 1964; Wendt and Fassel, 1965). This method rapidly developed to maturity. Sequential as well as simultaneous ICP emission spectrometers are now available worldwide from more than 15 manufacturers (Broekaert et al., 1986). In 1980 the use of the ICP as ion source for elemental mass spectrometry (MS) was described (Houk et al., 1980). ICP mass spectrometers are now commercially available from several manufacturers. The power of detection is higher for ICP-MS than for ICP-OES. The possibility of performing dilutions with stable isotopes makes the method attractive for tracer studies as well. The significance of ICP-OES and ICP-MS for environmental investigations will be discussed with analyses of water, airborne dust, and soils as examples. Trends in the development and the

NATO ASI Series, Vol. G 23
Metal Speciation in the Environment
Edited by J. A. C. Broekaert, Ş. Güçer, and F. Adams

outlook of less conventional techniques such as fluorescence spectrometry will be included.

2. Inductively coupled plasma optical emission spectrometry

2.1 Instrumentation (Boumans, 1987; Montaser and Golightly, 1987)

In the inductively coupled plasma the energy from a generator operating at a frequency between 5 and 100 MHz is transferred via a working coil to a working gas flowing in a quartz tube assembly. A stable plasma is formed with an argon flow of 6 - 15 L/min and a power of 0.6 - 3 kW. When the dimensions of the quartz torch are optimized, a toroidal discharge is obtained into the center of which wet aerosols can be introduced. The ICP finds its main use in the analysis of liquids and solids subsequent to sample dissolution. Nowadays, low-gas-consumption plasmas are also available. For environmental matrices, they have the same performance at a gas consumption of 6 L argon/min and a power of 600 W as the conventional systems that are operated with much higher power and gas consumption.

For the production of the aerosol pneumatic nebulizers are most frequently used. The nebulizer should produce an aerosol with a low nebulization gas flow, to avoid cooling of the analytical zone of the plasma. Moreover, the nebulizer should be as efficient as possible, because the detection limits are directly related to the amount of analyte introduced into the plasma per unit of time. The performance of the nebulizer should be as independent as possible from the physical properties of the analyte liquids. Concentric glass nebulizers (Meinhard Assoc.), cross-flow nebulizers, but also Babington nebulizers, that operate trouble-free with solutions of high salt concentrations and even work with suspensions are now available for ICP work. Apart from pneumatic nebulization several further techniques such as ultrasonic nebulization, hydride generation, electrothermal evaporation, and direct sample insertion were investigated for sample introduction. These techniques may be particularly useful for environmental samples.

Because all the elements brought into the plasma are excited, ICP-OES can be used for simultaneous multielement determinations when the radiation source is combined with a polychromator. Because of the stability of the excitation elemental line intensities can also be measured one after the other with a monochromator. Thus, flexible multielement determinations can be performed. Nowadays, a strong trend toward combined instruments can be observed, with which a number of elements are determined simultaneously with preselected, fixed lines and other elements sequentially, when required for a particular problem. In addition, alternate lines for an element can be selected, when spectral interferences occur for the fixed line. Every element emits a large number of lines, making the emission spectra complex. Accordingly, the spectrometers must have a sufficiently high optical resolving power to isolate the analytical line, the intensity of which is proportional to the element number density in the analyte. Moreover, the lines are superimposed on a structured spectral background that cannot be ignored when trace analyses are performed. Therefore, an emission spectrometer must have a procedure for background measurement. Such a procedure may be based on digitized slew scans, slewing of the entrance or exit slit, or on the use of a refractor plate in the optical path inside the spectrometer. With these techniques the spectral environment of an elemental line can be measured and corrections for spectral background or spectral interferences can be made. For these purposes diode-array detectors, charge coupled devices (CCD) and charge injection devices that allow parts of the spectrum at all wavelengths to be recorded simultaneously are very promising. However, their spectral characteristics, resolution, dynamic range, and wavelength coverage still need further development. In the instrumentation now available the plasma ignition, plasma operation, the line selection, and the data acquisition are computer-controlled.

2.2 Analytical figures of merit

In ICP-OES with pneumatic nebulization the detection limits are in the 1 - 10 ng/mL range (Table 1). Especially for elements that form refractory oxides and elements with rather low ionization energies for which the most sensitive lines are ion lines (magnesium, calcium, beryllium, zirconium, hafnium, titanium, rare earths, ..), the ICP-OES detection limits are considerably lower than in flame atomic absorption spectrometry (AAS) and in some cases also lower than in graphite furnace AAS.

Table 1. Detection limits of ICP-OES (Winge et al., 1979) and ICP-MS after single element optimization (Gray, 1985) and under compromise conditions (Gray and Date, 1983)

Element	Detection limit (ng/ml)		
	ICP-OES	ICP-MS single-element-optimization	Multielement-optimization
Ag	7	0.03	0.2
Al	23	-	0.6
As	35	0.04	7
Au	17	0.06	0.2
Ba	0.2	-	0.3^{++}
B	5	0.4	1
Cd	2	0.05	0.5
Ce	50	-	0.2^{++}
Co	6	0.05	0.5
Cr	6	0.06	0.2
Cs	-	-	0.1
Ge	40	0.02	1
Hg	25	0.02	0.1
In	6	0.06	0.2
La	10	0.05	0.2^{++}
Li	2	0.1	3
Mg	0.1	0.7	0.5
Mn	1	0.1	0.8
Mo	8	0.04	-
Ni	10	0.1	-
Rb	-	-	0.2
Se	75	0.7	15
Sn	25	0.06	0.06
Te	40	0.08	0.5
Pb	40	0.05	0.3
Th	65	0.02	0.2^{++}
Ti	4	-	0.3
U	250	0.03	0.4
V	30	-	0.4
W	30	0.05	0.5
Zn	2	0.2	3

++ doubly charged ions

To achieve high power of detection, the nebulizer gas flow (around 1 L/min), the operation power, and the observation height must be carefully optimized. Because the optimal conditions differ from one element to another, compromise conditions have to be selected for multielement determinations. The detection limits under compromise conditions do not differ by more than a factor of 5 relative to an optimal single element detection limit.

The analytical precision of ICP-OES is high. Relative standard deviations of measured intensities are 0.2 - 1 % and are constant within a large range of concentration. Considerable improvements can be realized by the use of a matrix line or of a line of an element that was added to all samples to be measured as a reference signal.

Systematic errors in ICP-OES may result from different steps in the analytical procedure.

The method is influenced by blank contributions from all steps during sampling and sample preparation. The blank might be considerable and must be controlled in work at trace concentrations (Tschöpel and Tölg, 1982). During the measurement procedure errors may come from nebulization effects. Differences in viscosity, density or surface tension lead to differences in the droplet size distribution and to differences in nebulization efficiency. Such effects can be minimized by optimization of the aerosol gas flow and by a liquid flow forced with a peristaltic pump. Further matrix effects may result from differences in the concentrations of easily ionized elements in the solutions. The geometry of the aerosol channel, the ionization equilibrium for analyte species, and for instance the plasma temperatures may change (De Galan, 1984) or the metastable argon which contributes to the excitation may be quenched. However, easily ionized elements such as Na, K, Ca,.. at concentrations of up to 500 μg/ml did not cause any matrix effect. At higher concentrations matching between analysis and calibration samples is required (Broekaert et al., 1979).

Considerable systematic errors in ICP-OES may arise from spectral interferences. Many elements such as Fe, Zr, Ti and W have very line-rich spectra increasing the potential for line coincidences. Therefore, emission spectrometers should have a high practical resolution (Laqua, 1980). It is a compromise between the speed as well as the stability of the spectrometer and the demand for high resolution. To minimize spectral interferences the analytical lines must be carefully selected with respect to the power of detection required. For this task in the development of ICP working procedures, compilations of spectral lines with their ICP sensitivities (Winge et al., 1985; Boumans, 1984) are very helpful. Not only matrix lines but also contributions from structured spectral background (resulting from molecular bands), from stray radiation caused by intense matrix lines (for instance from Ca in water analysis) and from the background continuum may cause spectral interferences and must be measured by suitable spectrometric procedures and corrected for with appropriate algorithms.

2.3 Applications to environmental analysis

2.3.1 Water analysis

Because of the low detection limits, the high multielement capacity, and the ease of calibration, ICP-OES has become a most important technique for water analysis. The capabilities and limitations of ICP-OES for the analysis of biological and environmental samples were discussed (Que Hee et al., 1988) and compared with those of DCP-OES (Schramel, 1988). In one of the first reports on water analysis by ICP-OES (Winge et al., 1977), interferences from elements such as calcium and magnesium were pointed out.

2.3.1.1 Waste waters

ICP-OES allows a large number of elements to be directly determined in waste waters (Dietz, 1986). B, Ba, Cd, Cu, Fe, Mn, Mo, Ni and Zn concentrations as low as 0.05 to 0.4 mg/L could be determined in waste waters of different origin after their decomposition with HNO_3/H_2O_2 (Broekaert and Leis, 1979). Because of varying amount of calcium in the samples, background correction had to be performed. Instrumental as well as software-based correction techniques (Kempster et al., 1983) are available. Also nebulization effects caused by variations in the salt concentrations have to be considered. Babington nebulizers (Garbarino and Taylor, 1980; Garbarino et al., 1985) were particularly useful because they work without clogging with high-salt samples.

Based on EPA prescriptions, a DIN Norm for the determination of 24 elements in waste water and sludge was workied out (DIN 38 406 Teil 22, 1987). Also determinations of Cd, Cr, Cu, Pb, Zn, As, and Hg in sludge by ICP-OES after sample decomposition in a closed PTFE vessel at 110 °C were described (Hawke and Lloyd, 1988). Elements such as B and P that cannot be determined easily by atomic absorption spectrometry (Ishizuka and Nakajima, 1980) can be directly determined with ICP-OES at the concentrations encountered in waste waters. For the case of P, extraction of the molybdoantimonylphosphoric acid into diisobutylketone was applied (Miyazaki et al., 1981). For heavy metals, ICP-OES and AAS can both be used (Moselhy and Vijan, 1981). For preconcentration of heavy metals, adsorption on activated charcoal (Berndt et al., 1985), might be very useful. To eliminate systematic errors arising from high concentrations of alkali and alkaline earth metals, the heavy metals can be extracted as dithiocarbamates into methylisobutylketone and the extracts analyzed by ICP-OES (Broekaert et al., 1981).

For elements such as arsenic, selenium, and antimony, hydride techniques improve the detection limits considerably to levels of $\mu g/L$. For waste water analysis, mineralisation of the samples, for instance by treatment with H_2SO_4/H_2O_2, and standard addition techniques must be used to avoid interferences (Broekaert and Leis, 1980). In the case of As corrections for spectral interferences may also be required (Davies and Kempster, 1986).

Thermal-spray techniques, developed for the coupling of liquid chromatographs to mass spectrometers can also be used in ICP-OES (Elgersma et al., 1986). These techniques significantly improve the sampling efficiency and the power of detection. However, concomittants cause high matrix effects. Therefore, even for the analysis of natural water the technique at present can be used only after the alkali and alkaline earth elements are separated from the analyte, for instance by ion exchange on chelating resins. Such a procedure was used for the determination of Fe, Cu, Mn, Pb, and Cd in river water (Vermeiren, 1989).

Flow-injection analysis in conjunction with ICP-OES increases the sample throughput and makes work with small-volume samples possible (Christian and Ruzicka, 1987). Under these conditions the risks of clogging are much reduced, when solutions with high salt concentrations - often encountered in highly loaded waste waters, for instance, from the galvanic industry - are aspirated. Combinations of thermal-spray techniques and flow injection are very attractive as they limit solvent and concomitant effects (Koropchak and Winn, 1986). Flow injection analysis is also useful for on-line pre-enrichment (Hartenstein et al., 1985) and for the determination of anions (Cox et al., 1987).

Techniques for the determination of species employ chromatography with ICP-OES (Fuwa, 1986). Several papers describe the determination of Cr(III) and Cr(VI) species. Preconcentration of Cr(III) on an Al_20_3 column (Cox et al., 1985; Cox and McLeod, 1986) and subsequent flow injection, high-performance liquid chromatography (HPLC) (Krull et al., 1982), separation on poly(dithiocarbamate) chelating resin (Miyazaki and Barnes, 1981) and ion chromatography (Nakata et al., 1985) were used. Phosphorus (Morita and Uehiro, 1981) and arsenic (Low et al., 1986) compounds were determined by HPLC/ICP-OES.

2.3.1.2 Natural water

In fresh water samples only a very limited number of elements (Ca, Mg, (Fe),..) can be determined by ICP-OES without preconcentration. As shown by a multi-method comparative study of hydrogeological samples from granitc regions (Sansoni et al., 1988), ICP-MS is more appropriate. ICP-OES was used to determine sulfide in ground water (Lewin et al., 1987) and sulfate in seawater (Miles and Cook, 1982).
Many types of preconcentration techniques can be used. Coprecipitation with $In(OH)_2$ (Hiraide et al., 1980), reductive precipitation with borohydride (Skogerboe et al., 1985), and on-line ion exchange on various resins (Horvath et al., 1985; Horvath and Barnes, 1986; Hirata et al., 1986) were described. Sorption on cellulose derivates followed by ICP-OES was used to determine uranium in mineral, river, ground and sea waters (Burba et al., 1984). Extraction of complexes with different ligands into organic solvents (Sugiyama et al., 1986; Miyazaki et al., 1982b) is a powerful tool for the determination of low concentrations of many metals in natural waters, as shown for Cd (Kumamura et al., 1986), for Ag in tap water (Baucells et al., 1986), and for phosphorus (Miyazaki et al., 1982a).

Special sample introduction techniques are helpful for achieving the power of detection required. Ultrasonic nebulization (Fassel and Bear, 1986) improves the power of detection by a factor of at least ten, provided desolvation is applied. Flow injection with discrete samples and ultrasonic nebulization still leads to lower detection limits than direct injection nebulization (Lafreniere et al., 1985). Hydride generation is very useful for the determination of As, Se, Te, and Sb (Goulden et al., 1981; Huang et al., 1987; Hutton and Preston, 1983) and even of lead (Ikeda et al., 1981) in natural waters. Electrothermal evaporation coupled to ICP-OES also enabes it to improve the power of detection as shown by the determination of mercury in drinking water after fixation with $(NH_4)_2S$ (Matusiewicz et al., 1985).

The determination of the various species often becomes possible by the use of chemical reactions, such as in the known methods for NH_4^+ (Alder et al., 1977) and total nitrogen (Butler-Sobel, 1982) or by the coupling of ICP-OES with chromatography. This is shown at the example of a separate determination of inorganic and organomercury by HPLC coupled to ICP-OES and cold vapour generation (Krull et al., 1986) and by the distinction of magnesium and calcium present in inorganic salts and in organic matter by size exclusion chromatography on a Sephadex G-10 column (Gardner et al., 1982).

Direct analyses of seawater by ICP-OES are hampered by the high salt concentrations of the samples. The high salt concentrations may not only hamper the nebulization but may cause deposition of solids at the burner top (Cobbold, 1986). Babington-type nebulizers (Cecconie et al., 1988) are very useful for high-salt samples. Corrections for long-term drifts may be performed with the aid of a reference line, as shown for the determination of calcium and phosphates in waters and brines (Brenner et al., 1984). Recently, also a grid nebulizer has been used for the analysis of waters with high salt concentrations (Brotherton and Caruso, 1987).

Some preconcentration procedures used for seawater are based on coprecipitation with the hydroxides of zirconium, which leads to a preenrichment factor of 20 for Al, As, Cd, Co, Cr, Cu, Fe, La, Mn, Mo, Ni, Pb, Sb, Ti, V, Y, and Zn)(Akagi et al., 1985a), or of gallium (Akagi et al., 1985b). The extraction of dithiocarbamates (McLeod et al., 1981) and other complexes was also used. When subsequently applying electrothermal evaporation coupled to ICP-OES a very high power of detection is obtained. Further, chelating resins (Cheng et al., 1987) and ion exchange with subsequent ICP-OES using ultrasonic nebulization (Berman et al., 1980) were used for the analysis of seawater. Hydride generation ICP-OES was used for the determination of As (10 μg/g level), Sb (0.1 μg/g level) and Se (0.1 - 1 μg/g level) in marine biological samples (tissues) and sediments (De Oliveira et al. 1983).

2.3.2 Analysis of airborne dust

In airborne dust samples a large number of environmentally relevant elements can be determined by ICP-OES after digestion. In airborne dust sampled with a cascade impactor, Al, Ca, Cr, Cu, Fe, Mg, Mn, Pb, Sr, V, and Zn could be determined in 0.1 - 1 mg amounts of samples collected from the different stages. The samples were mineralized with $HN0_3$/HF, stabilized with H_3B0_3 and analyzed by ICP-OES (Broekaert et al., 1982). Dynamic background correction and careful optimization of the aerosol carrier-gas flow is very important in these analyses for avoiding systematic errors.

The sampling procedure applied and the metereological factors during the sampling are of paramount importance. The decomposition procedure must be optimized to avoid systematic errors (Dannecker et al., 1986; Xiao-Quan et al., 1987). The samples can be treated with a mixture of $HN0_3$, $HCl0_4$, and HF in a sealed PTFE vessel as described for the determination of Sc in coal fly ash (Bettinelli et al., 1987), for instance, or in an open vessel below 150°C as shown for the case of Cr, Ni, and V (Schlieckmann und Umland, 1984). Direct nebulization of a slurry may be possible. Sytematic errors due to the loss of the larger particles are likely to occur. Slurry techniques are confined to the determination of the major elements such as Fe, Si, and Ca (Sugimae and Mizoguchi, 1982). Other samples, such as city waste incinerator ash may be mineralized with $HN0_3$/HF (for Cd, Cr, Cu, Ni and Zn) or by fusion with $Li_2B_40_7$ (for Cr) (Taylor et al., 1985). In coal products components producing gaseous pollution (S)(Caroli et al., 1988) can be determined by ICP-OES subsequent to combustion in an oxygen flow.

2.3.3 Analysis of soils and plant materials

A series of environmentally and nutritionally relevant trace elements can be determined in plants and soils by ICP-OES (Dahlquist and Knoll, 1978; Liese, 1985). The dissolution procedures applied in the analysis of soils by ICP-OES are of prime importance, especially when volatile elements such as Pb, Tl, As, and Cd need to be determined. A decomposition with $HN0_3$ and subsequently with $HCl0_4$ and HF has been proposed (McQuaker et al., 1979b). The residual $HCl0_4$ contents needed to be matched in the standard solutions (McQuaker et al., 1979a). Slurry nebulization of finely ground soils as known from DCP-OES (Sparkes and Ebdon, 1986) might be very useful. For the determination of Cd and Mo in soils, extraction using diethylenetriaminepentaacetic acid in the presence of triethanolamine and $CaCl_2$ was proposed (Baucells et al., 1985). For the determination of boron in soils, the samples can be decomposed by treatment with a mixture of $HN0_3$, $HCl0_4$, HF, and H_3P0_4 in a sealed PTFE vessel. When the B 249.678 nm line is used, a careful background correction is required (Li-Qiang and Zhu, 1986). For the determination of Mg, Zn, and Fe (Aten et al., 1983) and B, Ca, Cu, Fe, Mg, Mn, P, K, and Zn (Jones Jr, 1982) in fertilizers, ICP-OES is an alternative to atomic absorption spectrometry.

For trace element determinations in plant materials sampling and sample pretreatment may be complicated. For raw agricultural crops treatment of a composite of sample with 6 M HCl at 80°C in 60-mL polyethylene vessels was proposed. Calibration was performed with synthetic standards and with appropriate standard reference materials (Kuennen et al., 1982). ICP-OES has sufficiently low detection limits for the determination of molybdenum in plants (0.2 - 1 μg/g)(Lyons and Rooyafel, 1982). Also special sample introduction techniques were proposed. Direct sample insertion allowed the volatile trace elements P, Cd, Pb, As, and Zn to be determined (Abdullah and Haraguchi, 1985). A multi-element technique using ICP-OES and hydride evolution atomic absorption spectrometry for the analysis of plant and animal tissues was developed (Jones et al., 1982). The trace elements Cd, Cu, Mo, Ni, V, and Zn were concentrated on a Chelex-100 column and separated from the alkali and alkaline earth elements. As, Se, and Sb were not retained. ICP-OES was compared with energy-dispersive XRF (Coetzee et al., 1986) and total reflection XRF (Michaelis, 1986). ICP-OES turned out to be better suited for the determination of light elements in soils, plants and environmental samples.

3. ICP mass spectrometry

3.1 Instrumentation

Since the early 1980's inductively coupled plasmas have been used as ion sources for mass spectrometry. This development was made possible by the availability of inexpensive, high-resolution quadrupole filters with high transmission, that did not necessitate high voltages for the extraction of ions. Modern vacuum technology allowed the plasma ion source kept at atmospheric pressure to be combined with the mass spectrometer operated under high vacuum. Such a system (Houk et al., 1980), uses a horizontally-operated ICP (Fig. 1). The plasma impings on a metal cone (sampler) with an aperture between 0.6 and 1 mm made of a metal such as nickel or copper. The

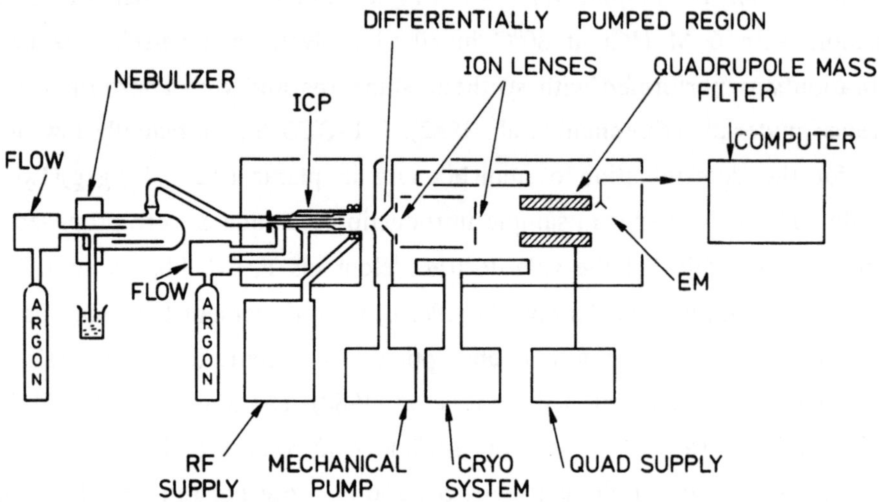

Fig. 1. ICP Mass Spectrometer

region behind the sampler is kept at a pressure of about 1 mbar with the aid of a high-displacement rotary pump. The ion beam passing through the aperture of the sampler, expands, and a part of the beam enters the high vacuum of the mass spectrometer through a second aperture (skimmer). After passage through ion lenses and accelerator plates, the ions enter the quadrupole filter. As a result of the high-frequency voltage, the system transmits ions of a defined charge and mass that are detected by an electron multiplier system. At the frequencies applied the mass range 0 - 300 m/Z can be scanned in 30 ms. The results of repetitive scans can be stored in a multichannel analyzer. The signal at a certain m/Z value can be monitored selectively.

Aerosols can be generated with all devices known from ICP-OES. Pneumatic nebulization is used in most cases. In ICP-MS dry aerosol generation has additional advantages. It decreases the formation of oxide cluster ions, which especially at the low mass range hamper the power of detection and lead to systematic errors as a result of isobaric interferences. The optimization of the working conditions is more complex in ICP-MS than in ICP-OES. The gas flows, of which the nebulizer gas flow is the most important, the RF power, the position of the sampler, and the voltages at the different ion lenses have to be carefully optimized. Because of the small sampling aperture, all circumstances influencing the geometry of the plasma have a larger influence on the performance of ICP-MS than on the performance of ICP-OES.

3.2 Analytical figures of merit

A fraction of the species present in the plasma (argon species, matrix atoms, trace elements to be determined) are ionized and enter the interface. Molecular ions and radicals from the incomplete dissociation of water, acids, or salts and reaction products of these radicals with argon, nitrogen, hydrogen, or metals are present. These products may be formed in the plasma but also in the plasma jet inside the interface. Accordingly, the ICP mass spectra contain:

- signals for singly and doubly ionized elements,
- signals for argon species: Ar^+, Ar_2^+, Ar^{++},
- water and acid species: H_20^+, OH^+, H_30^+, HAr^+, $HAr0^+$,
- reaction products of water or acids with the analyte.

Accordingly, at low atomic mass signals from elements may coincide with signals from argon, water or acid species. Further, signals of doubly-charged ions with mass m may occur at mass m/2 and signals of hydrides and oxides of the analyte elements are found in the spectra. The different isotopes of the elements produce different signals.

In ICP-MS the detection limits are at the 0.02 to 0.1 ng/ml range, which is up to two orders of magnitude lower than in ICP-OES. Moreover, practically all elements can be determined and the detection limits do not differ much from one element to another (Table 1). The detection limits of the halogens, however, are high as their ionization in an argon plasma is poor. The detection limits of the light elements also are higher than those of the heavier elements, which is due to interferences with cluster ions. Single-element optimization of the aerosol gas flow, the RF power, and the sampler position is required to obtain highest power of detection. Only for some of the light elements such as Ca, Mg, and Be and for Fe ($^{56}Fe^{+}$ with $^{40}Ar^{16}O^{+}$), As ($^{75}As^{+}$ with $^{40}Ar^{35}Cl^{+}$), and Se ($^{80}Se^{+}$ with $^{40}Ar^{40}Ar^{+}$) are the detection limits of ICP-MS not better than in ICP-OES. The elements present in the sampler may also cause interferences (Montaser and Golightly, 1987). The power of detection of the halogens was improved (Cl^{+} 5, Cl^{-} 1 ng/ml) by the monitoring of negative ions (Fulford and Quan, 1988).

Interferences due to changes in the matrix composition might result from changes in nebulization, from changes of the ionization in the plasma or in the geometry of the aerosol injection channel, and from changes in the ion energies. Nebulization effects may be minimized by optimizing the aerosol gas flow and by using a peristaltic pump for the transit of the liquid to the nebulizer. "Isobaric" spectral interferences are caused by isotopes of elements with the same mass number. At the low resolution of the quadrupole system these isotopes produce signals at the same m/Z value. These "isobaric" interferences do not noticeably depend on the working conditions and can be corrected with the aid of appropriate software. Spectral interferences due to doubly charged ions and cluster ions stemming from argon species, analyte oxides a.s.o. depend on the aerosol gas flow and on the RF power (Vaughan and Horlick, 1986; Horlick et al., 1985; Tan and Horlick, 1986). The concentrations of the interfering species change with the observation zone in the plasma. Because of the complexity of the interferences in ICP-MS, calibration by standard addition is recommended.

The short-term stability of ICP-MS allows RSDs of better than 1 % to be achieved. A high stability of the nebulizer gas flow is required. By using an internal standard (Thompson and Houk, 1986; Vandecasteele et al., 1988) the precision can be considerably improved. In order to avoid salt depositions at the sampling orifice, at the burner, and at the nebulizer nozzle, which cause long-term drifts, the salt concentration in the analyte should be kept below 1-5 g/L, depending on the chemical composition of the samples.

As with every mass spectrometric technique isotope dilution may be used in ICP-MS. The technique can be applied whenever an element has at least two stable or radioactive long-lived isotopes. With this technique the calibration of the system can be improved. Isotope dilution has been applied extensively for the determination of lead (Dean et al., 1987a). Tracer studies were performed for studies of the metabolism of iron (Ting and Janghorbani, 1987). The precision with which isotopic compositions can be determined is approximately 1 % provided the abundances of the isotopes used do not differ by more than a factor of 10.

Apart from continuous pneumatic nebulization several of the other techniques known from ICP-OES can be applied in ICP-MS. Here they have additional advantages over pneumatic nebulization because a dry aerosol is produced. With the aid of electrothermal evaporation the detection limit for iron can be improved as the intensity of the $^{40}Ar^{16}O^+$ signal which interferes with $^{56}Fe^+$ is considerably reduced (Park et al., 1987). Hydride generation ICP-MS was used for the determination of low concentrations of lead (Wang et al., 1988).

3.3 Environmental applications

The high power of detection of ICP-MS, the easy calibration, the multi-element capacity, the possibility of performing labelling experiments, and the possibility of using ICP-MS as a detector for chromatographs assure that ICP-MS will become a powerful method for environmental analysis.

Thus far only applications to water analysis were reported. With waste waters the same sample decomposition procedures as in AAS and ICP-OES (treatment with H_2O_2/HNO_3) can be used. The analytical figures of merit, systematic errors from signal depressions, and isobaric interferences in ICP-MS for waste water analysis were studied (Herzog and Dietz, 1987). In analyses of waste waters from the Ruhrgebiet factor analysis was applied for the treatment of the large number of data. The high salt concentration in seawater prevents a direct analysis ICP-MS. Solvent extraction and sorption techniques can be used to separate the elements of interest from sodium chloride from the elements of interest. By adsorption of the trace elements on a $Si0_2$ column treated with 8-hydroxyquinoline a preconcentration by a factor of 50 could be obtained, and Ni, Cu, Zn, Mo, Cd, Pb, and U could be determined by ICP-MS in open ocean water (Beauchemin et al., 1988). In river water Na, Mg, K, Ca, Al, V, Cr, Mn, Cu, Zn, Sr, Mo, Sb, Ba, and U could be directly determined and Co, Ni, Cd, and Pb could be measured after preconcentration mentioned (Beauchemin et al., 1987). For As a preconcentration by evaporation was found to be sufficient. The highest accuracy could be obtained with calibration by dilution with stable isotopes (McLaren et al., 1987a). The procedures described have been used for the characterization of a seawater reference sample (Beauchemin et al., 1988). ICP-MS was also applied to the analysis of marine sediments and biological samples. The results of different calibration procedures were compared for the determination of V, Mn, Co, Ni, Cu, Zn, As, Mo, Cd, and Pb at concentrations between 0.2 and 160 $\mu g/g$ in the marine sediment reference material BCSS-1 (McLaren et al., 1987b). The results based on a calibration with external standards differed by up to 30 % from the results of a calibration with standard addition. The differences are caused by signal depressions or enhancements and isobaric interferences. Calibration by dilution with stable isotopes was used in the analysis of lobster (Ridout et al., 1988) and for the determination of Na, Mg, Cr, Fe, Mn, Co, Ni, Cu, Zn, As, Cd, Hg, and Pb in the digested (HNO_3/H_20_2) marine reference samples DOLT-1 ("dogfish liver tissue") and DORM-1 ("dogfish muscle tissue"). It also has been shown that ICP-MS is useful for determining tin in environmental samples Brzezinska-Paudin and Van Loon, 1988). ICP-MS certainly will be a very important technique for hydrological studies (Garbarino and Taylor, 1987) and for studies of the mobility of elements and compounds in the biosphere. The production of the reference samples is only at the very beginning. Innovative techniques will be required to reach the concentration ranges of interest. Wire loop techniques (Boomer et al., 1986) and electrothermal evaporation or direct insertion in the plasma are very useful as shown by the determination of Mn, As, Pd, and Au with electrothermal evaporation ICP-MS and of Fe, Mo, and Ni in the case of direct insertion (Hall et al., 1988).

Determinations of trace elements compounds by ICP-MS can be based on the experience gathered in the coupling of chromatographic techniques with ICP-OES. As already demonstrated by the determination of metals bound to different protein fractions in biological samples (Dean et al., 1987b), ICP-MS allows lower concentration ranges to be reached than attainable with most other methods. The isotopic composition of an element, for instance lead, (Date and Cheung, 1987) can be determined which may be important for source apportionment.

4. Related techniques

In addition to optical emission and mass spectrometry, other types of spectrometry using plasmas were developed and are of potential interest for environmental studies. Atomic fluorescence spectrometry using hollow cathode lamps as primary sources and a plasma as an atom reservoir (Lancione and Drews, 1985; Sanzolone and Meier, 1986) is of interest. This approach allows multielement determinations to be performed. Stray radiation due to high concentrations of calcium does not cause systematic errors in contrast to OES. Atomic fluorescence with an ICP as primary source and a flame as atom reservoir was used for the determination of Cd in lake water (Cavalli et al., 1982). Laser-induced atomic fluorescence with an ICP as an atom reservoir is much more sensitive than ICP-OES especially for elements such as Pb and Cd (Human et al., 1984). Because of the complex laser systems the method is a monoelement method. Similar limitations are encountered with laser enhanced ionization (Turk et al., 1985) and with resonance ionization spectrometry (Hurst, 1981). Metastable transfer spectrometry was applied for the determination of Pb (Melzer et al., 1980) and is also of interest for environmental work.

Literature

Abdullah M, Haraguchi H (1985) Computer-controlled graphite cup direct insertion device for direct analysis of plant samples by ICP AES, Anal Chem 57:2059-2064

Akagi T, Nojiri Y, Matsui M, Haraguchi H (1985a) Zirconium precipitation for simultaneous multielement determination of trace metals in seawater by ICP-AES, Applied Spectrosc 39:662-668

Akagi T, Fuwa K, Haraguchi H (1985b) Simultaneous multi-element determination of trace metals in sea water by ICP AES after coprecipitation with gallium, Anal Chim Acta 177:139-153

Alder JF, Gunn AM, Kirkbright GF (1977) The determination of traces of ammonium-nitrogen in aqueous solution by optical emission spectrometry with a high-frequency inductively coupled argon plasma source, Anal Chim Acta 92:43-48

Aten CE, Bourke JB, Walton JC (1983) Determination of magnesium, iron and zinc in fertilizers by flame atomic absorption and by ICP emission. Comparison of methods, J Assoc Off Anal Chem 66:766-768

Baucells M, Lacort G, Roura M (1985) Determination of cadmium and molybdenum in soil extracts by graphite furnace atomic absorption and ICP spectrometry, Analyst 110:1423-1429

Baucells M, Lacort G, Roura M (1986) ICP emission spectroscopic determination of Ag in tap water, Spectrochim Acta 41B:189-192

Beauchemin D, McLaren JW, Mykytiuk AP, Berman SS (1987) Determination of trace metals in a river water reference material by ICP-MS, Anal Chem 59:778-783

Beauchemin D, McLaren JW, Mykytiuk AP, Berman SS (1988) Determination of trace elements in an open ocean reference material by ICP-MS, J Anal Atom Spectrom 3:305-308

Berman SS, McLaren JW, Willie SN (1980) Simultaneous determination of five trace metals in sea water by ICP AES with ultrasonic nebulization, Anal Chem 52:488-492

Berndt H, Harms U, Sonneborn M (1985) Multielement trace preconcentration from water on activated carbon for the sample pretreatment for atomic spectrometry, Fresenius' Z Anal Chem 322:329-333

Bettinelli M, Baroni U, Pastorelli N (1987) Determination of scandium in coal fly ash and geological materials by graphite furnace AAS and ICP AES, Analyst 112:23-26

Boomer DW, Powell M, Sing RLA, Salin ED (1986) Application of a wire loop direct sample insertion device for ICP MS, Anal Chem 58:975-976

Boumans PWJM (1984) Line coincidence tables for inductively coupled plasma atomic emission spectrometry, Pergamon Press, Oxford, second edition

Boumans PWJM (ed)(1987) Inductively coupled plasma emission spectroscopy, vol I: Methodology, instrumentation and performance, John Wiley & Sons, Inc., New York

Brenner JB, Eldad H, Ehrlich S, Dalman N (1984) Application of ICP-AES with an internal reference to the determination of sulfate and calcium in waters and brines, Anal Chim Acta 166:51-60

Broekaert JAC, Leis F (1979) An injection method for the sequential determination of boron and several metals in waste-water samples by inductively-coupled plasma atomic emission spectrometry, Anal Chim Acta 109:73-83

Broekaert JAC, Leis F (1980) Application of two different ICP-hydride techniques to the determination of arsenic, Fresenius' Z Anal Chem 300:22-27

Broekaert JAC, Tölg G (1987) Recent developments in atomic spectrometry methods for elemental trace determinations, Fresenius' Z Anal Chem 326:495-509

Broekaert JAC, Leis F, Laqua K (1979) Application of an inductively coupled plasma to the emission spectroscopic determination of rare earths in mineralogical samples, Spectrochim Acta 34B:73-84

Broekaert JAC, Leis F, Laqua K (1981) The application of an argon/nitrogen inductively-coupled plasma to the analysis of organic solutions, Talanta 28:745-752

Broekaert JAC, Wopenka B, Puxbaum H (1982) Inductively coupled plasma optical emission spectrometry for the analysis of aerosol samples collected by cascade impactors, Anal Chem 54:2174-2179

Broekaert JAC, Keliher PN, McLaren JW (1986) Selecting a plasma instrument, ICP Inform Newslett 11:689-698

Brotherton T, Caruso J (1987) Evaluation of the grid-type nebuliser for the introduction of high dissolved salt and high solids content solutions into the ICP, J anal atom spectrom 2:695-703

Brzezinska-Paudin A, Van Loon JC (1988) Determination of tin in environmental samples by graphite furnace atomic absorption and ICP MS, Fresenius' Z Anal Chem 331:707-712

Burba P, Cebulc M, Broekaert JAC (1984) Verbundverfahren (Spektralphotometrie, ICP-OES, RFA) zur Bestimmung von Uranspuren in natürlichen Wässern, Fresenius' Z Anal Chem 318:1-11

Butler-Sobel C (1982) Use of an ICP with an extended torch for the determination of nitrogen in aqueous solutions, Applied Spectrosc 36:691-693

Caroli S, Mazzeo AF, Laurenzi A, Senofonte O (1988) Determination of sulphur in coal products by ICP AES, J Anal Atom Spectrom 3:245-248

Cavalli P, Omenetto N, Rossi G (1982) Determination of cadmium at sub-ppm levels in lake sediments by an ICP atomic fluorescence technique, Atom Spectrosc 3:1-4

Cecconie T, Muralidharan S, Freiser H (1988) Characterization of micro-Babington and cross-flow nebulizers for ICP-AES, Applied Spectrosc 42:177-178

Cheng CJ, Akagi T, Haraguchi H (1987) Simultaneous multielement analysis for trace metals in sea water by ICP AES after batch preconcentration on a chelating resin, Anal Chim Acta 198:173-182

Christian GD, Ruzicka J (1987) Flow injection analysis: a novel tool for plasma spectroscopy, Spectrochim Acta 42B:157-167

Cobbold DG (1986) Comments on the analysis of high-solids matrices using the Meinhard concentric nebulizer in ICP-AES, Applied Spectrosc 40:1242-1244

Coetzee PP, Hoffmann P, Speer R, Lieser KH (1986) Comparison of trace-element determination in powdered soil and grass samples by energy-dispersive XRF and ICP AES, Fresenius Z Anal Chem 323:254-256

Cox AG, Cook LG, McLeod CW (1985) Rapid sequential determination of chromium(III)/chromium(IV) by flow injection ICP -AES, Analyst 110:331-333

Cox AG, McLeod CW, Miles DL, Cook JM (1987) Trace enrichment and determination of sulphate by flow injection ICP-AES, J Anal Atom Spectrom 2:553-555

Cox AG, McLeod CW (1986) Preconcentration and determination of trace chromium(III) by flow injection ICP-AES, Anal Chim Acta 179:487-490

Dahlquist RL, Knoll JW (1978) ICP-AES: analysis of biological materials and soils for major, trace, and ultra-trace elements, Applied Spectrosc 32:1-30

Dannecker W, Steiger M, Naumann K (1986) Wind-direction- dependent sampling and multi-element analysis of atmospheric aerosols for improved evaluation of air pollutant concentrations, Fresenius' Z Anal Chem 325:50-56

Davies E, Kempster PL (1986) Spectral interferences in the quantification of arsenic in water samples by ICP emission spectrometry, Spectrochim Acta 41B:1203-1209

Date AR, Cheung YY (1987) Studies in the determination of lead isotope ratios by ICP MS, Analyst 112:1531-1540

Dean JR, Ebdon L, Massey R (1987a) Selection of mode for the measurement of lead isotope ratios by ICP mass spectrometry and its application to milk powder analysis, J Anal Atom Spectrom 2:369-374

Dean JR, Munro S, Ebdon L, Crews HM, Massey RC (1987b) Studies of metalloprotein species by directly coupled high-performance liquid chromatography ICP MS, J Anal Atom Spectrom 2: 607-610

De Galan L (1984) Some considerations on the excitation mechanism in the inductively coupled argon plasma, Spectrochim. Acta 39B:537-550

De Oliveira E, McLaren JW, Berman SS (1983) Simultaneous determination of arsenic, antimony and selenium in marine samples by ICP AES, Anal Chem 55:2047-2050

Dietz F (1986) Symposium 6: Analytik der Wassergüte unter den Aspekten der Umweltbelastung, Fresenius' Z Anal Chem 324:222-223

DIN 38 406 Teil 22 (1987) Deutsche Einheitsverfahren zur Wasser-, Abwasser-, und Schlammuntersuchung - Kathionen (Gruppe E): Bestimmung der 24 Elemente Ag, Al, B, Ba, Cd, Co, Cr, Cu, Fe, K, Mg, Mn, Mo, Na, Ni, P, Pb, Sr, Ti, V, Zn und Zr durch Atomemissionsspektrometrie mit induktiv gekoppeltem Plasma (ICP-AES) (E 22)

Elgersma JW, Maessen FJMJ, Niessen WMA (1986) A low consumption thermospray nebulizer with a fused silica vaporizer for ICP-AES, Spectrochim Acta 41B:1217-1220

Fassel VA, Bear BR (1986) Ultrasonic nebulization of liquid samples for analytical ICP atomic spectrometry: an update, Spectrochim Acta 41B:1089-1113

Fulford JE, Quan ESK (1988) Negative ions in ICP-MS, Applied Spectrosc 42:425-428

Fuwa K (1986) Analysis of environmental materials using spectroscopic methods, Fresenius' Z Anal Chem 324:531-536

Garbarino JR, Jones BE, Stein GP, Belser WT, Taylor HE (1985) Statistical evaluation of an ICP-AES method for routine water quality testing, Applied Spectrosc 39:535-555

Garbarino JR, Taylor HE (1980) A Babington-nebulizer for use in the analysis of natural waters by ICP spectrometry, Applied Spectrosc 34:584-590

Garbarino JR, Taylor HE (1987) Stable isotope dilution analysis of hydrologic samples by inductively coupled plasma mass spectrometry, Anal Chem 59:1568-1575

Gardner WS, Landrum PF, Yates DA (1982) Fractionation of metal forms in natural waters by size extrusion chromatography with inductively coupled argon plasma detection, Anal Chem 54:1196-1198

Gray AL (1985) The ICP as an ion source - origins, achievements and prospects, Spectrochim Acta 40B:1525-1537

Gray AL, Date AR (1983) ICP source mass spectrometry using continuum flow ion extraction, Analyst 108:1033-1050

Goulden PD, Anthony DHJ, Austen KD (1981) Determination of arsenic and selenium in water, fish, and sediments by inductively coupled argon plasma emission spectrometry, Anal Chem 53:2027- 2029

Greenfield S, Jones IL, Berry CT (1964) High-pressure plasmas as spectroscopic emission sources, Analyst 89:713-720

Hall GEM, Pelchat JC, Boomer DW, Powell M (1988) Relative merits of two methods of sample introduction in ICP-MS: electrothermal vaporization and direct sample insertion, J Anal Atom Spectrom 3:791-798

Hartenstein SD, Ruzicka J, Christian GD (1985) Sensitivity enhancements for flow injection analysis ICP-AES using on-line preconcentrating ion-exchange column, Anal Chem 57:21-25

Hawke DJ, Lloyd A (1988) Determination of metals in sewage sludge by ICP-AES using a simplified nitric acid digestion method, Analyst 113:413-417

Herzog R, Dietz F (1987) Anwendung der ICP/MS in der Abwasseranalytik, Gewässerschutz, Wasser, Abwasser 92:109-141

Hiraide M, Ito T, Baba M, Kawaguchi H, Mizuike A (1980) Multielement preconcentration of trace heavy metals in water by coprecipitation and flotation with indium hydroxide for ICP-AES, Anal Chem 52:804-807

Hirata S, Umezaki Y, Ikeda M (1986) Determination of chromium(III), titanium, vanadium, iron(III) and aluminium by ICP-AES with an on-line preconcentrating ion-exchange column, Anal Chem 58:2602-2606

Horlick G, Tan SH, Vaugham MA, Rose CA (1985) The effect of plasma operating parameters on analyte signals in ICP mass spectrometry, Spectrochim Acta 40B:1555-1572

Horvath Z, Barnes RM (1986) Carboxylmethylated polyethylenimine-polymethylene-polyphenylene isocyanate chelating resin preconcentration for ICP spectrometry, Anal Chem 58:1352-1355

Horvath Z, Barnes RM, Murty PS (1985) Iminodiacetic acid - ethylcellulose as a chelating ion exchanger II. Determination of trace metals by ICP-AES, Anal Chim Acta 173:305-309

Houk RS, Fassel VA, Flesch GD, Svec HJ, Gray AL, Taylor ChE (1980) Inductively coupled argon plasma as an ion source for mass spectrometric determination of trace elements, Anal Chem 52:2283-2289

Huang B, Zhang B, Zeng K (1987) New nebulizer - hydride generator system for simultaneous multi-element ICP-AES, Spectrochim Acta 42B:129-137

Human HGC, Omenetto N, Cavalli P, Rossi G (1984) Laser excited analytical atomic and ionic fluorescence in flames, furnaces and inductively coupled plasmas - II. Fluorescence characteristics and detection limits for fourteen elements, Spectrochim Acta 39B:1345-1363

Hurst GS (1981) Resonance ionization spectroscopy, Anal Chem 53:1448A-1456A

Hutton RC, Preston B (1983) A simple versatile hydride generation configuration for ICP, Analyst 108:1409-1411

Ikeda M, Nishibe J, Hamada S, Tujino R (1981) Determination of lead at the ng ml^{-1} level by reduction to plumbane and measurement by ICP emission spectrometry, Anal Chim Acta 125:109 -115

Ishizuka T, Nakajima K (1980) Determination of phosphorus in waste waters by ICP-AES, Anal. Chim. Acta 121:197-203

Jones JB Jr (1982) Simultaneous determination of total boron, calcium, copper, iron, magnesium, manganese, phosphorus, potassium and zinc in fertilizers by ICAP emission spectrometry, J Assoc Off Anal Chem 65:781-785

Jones JW, Capan SG, O'Haver TC (1982) Critical evaluation of a multi-element scheme using plasma emission and hydride evolution AAS for the analysis of plant and animal tissues, Analyst 107:353-377

Kempster PL, Malloch JDR, De Klerk MVDS (1983) Determination of interference correction coefficients for routine trace element analysis of surface water by ICP emission spectrometry, Spectrochim Acta 380:967-972

Koropchak JA, Winn DH (1986) Thermospray interfacing for flow injection analysis with ICP-AES, Anal Chem 58:2558-2561

Krull IS, Bushee D, Savage RN, Schleicher RG, Smith SB (1982) Speciation of Cr(III) and Cr(VI) via reversed phase HPLC with ICP emission spectroscopic detection (HPLC-ICP), Analyt Lett 15:267-281

Krull IS, Bushee DS, Schleicher RG, Smith SB Jr (1986) Determination of inorganic and organomercury compounds by high-performance liquid chromatography - ICP emission spectrometry with cold vapour generation, Analyst 111:345-349

Kuennen RW, Wolnik KA, Frick FL, Caruso JA (1982) Pressure dissolution and real sample matrix calibration for multielement analysis of raw agricultural crops by ICP AES, Anal Chem 54:2146-2150

Kumamaru T, Matsuo H, Okamoto Y, Ikeda M (1986) Sensitivity enhancement for ICP AES of cadmium by suction-flow on-line-exchange preconcentration, Anal Chim Acta 181:271-275

Lafreniere K, Rice GW, Fassel VA (1985) Flow injection analysis with ICP-AES: critical comparison of conventional pneumatic, ultrasonic and direct injection nebulization, Spectrochim Acta 40B:1495-1504

Lancione RL, Drew DM (1985) Evaluation of ICP atomic fluorescence for the determination of mercury, Spectrochim Acta 40B:107-113

Laqua K (1980) Emissionsspektroskopie. In: Ullmanns Encyclopädie der technischen Chemie, Verlag Chemie GmbH, Weinheim, 4. neubearbeitete und erw. Auflage, Bd 5, S 441

Lewin K, Walsh JN, Miles DL (1987) Determination of dissolved sulphide in ground waters by ICP-AES, J Anal Atom Spectrom 2:249-250

Liese T (1985) Zur Bestimmung von Elementen in Pflanzen- und Bodenproben mittels sequentieller ICP-AES, Fresenius' Z Anal Chem 321:37-44

Li-Qiang X, Zhu R (1986) Determination of boron in soils by sequential scanning ICP-AES using side line indexing method, Fresenius'Z Anal Chem 325:534-538

Low GKC, Batley GE, Buchanan SJ (1986) Application of column switching in high-performance liquid chromatography to arsenic speciation analysis with ICAP spectrometric detection, J Chromatogr 368:423-426

Lyons DJ, Rooyafel RL (1982) Determination of molybdenum in plant materials using ICP emission spectroscopy, Analyst 107:331 -335

Matusiewicz H, Horvath Z, Barnes RM (1985) Determination of mercury in drinking water by ICP-AES with electrothermal vaporization, Applied Spectrosc 39:558-560

McLaren JW, Beauchemin D, Berman SS (1987a) Application of isotope dilution mass spectrometry to the analysis of marine sediments, Anal Chem 59:610-613

McLaren JW, Beauchemin D, Berman SS (1987b) Determination of trace metals in marine sediments by ICP mass spectrometry, J Anal Atom Spectrom 2:277-281

McLeod CW, Otsuki A, Okamoto K, Haraguchi H, Fuwa K (1981) Simultaneous determination of trace metals in sea water using dithiocarbamate preconcentration and ICP emission spectrometry, Analyst 106:419-428

McQuaker NR, Kluckner PD, Chang GN (1979) Calibration of an ICP-AES for the analysis of environmental materials, Anal Chem 51:888-895

McQuaker NR, Brown DF, Kluckner PD (1979) Digestion of environmental materials for analysis by ICP-AES, Anal Chem 51: 1082-1094

Melzer JE, Jordan JL, Sutton DG (1980) Determination of trace amounts of lead in water by metastable transfer emission spectrometry, Anal Chem 52:348-349

Michaelis W (1986) Multielement analysis of environmental samples by total-reflection X-ray fluorescence spectrometry, neutron activation analysis and ICP OES, Fresenius Z Anal Chem 324:662-671

Miles DL, Cook JM (1982) The determination of sulphate in natural waters by ICP emission spectrometry, Anal Chim Acta 141: 207-212

Miyazaki A, Barnes RM (1981) Differential determination of chromium(VI)-chromium(III) with poly(dithiocarbamate) chelating resin and ICP-AES, Anal Chem 53:364-366

Miyazaki A, Kimura A, Umezaki Y (1981) Determination of ng ml^{-1} levels of phosphorus in waters by diisobutyl ketone extraction and ICP AES, Anal. Chim. Acta 127:93-101

Miyazaki A, Kimura A, Umezaki Y (1982a) Indirect determination of sub-ng ml^{-1} levels of phosphorus in waters by di-isobutyl ketone extraction of reduced molybdoantimonyl-phosphoric acid and ICP emission spectrometry, Anal Chim Acta 138:121-127

Miyazaki A, Kimura A, Bansho K, Umezaki Y (1982b) Simultaneous determination of heavy metals in waters by ICP AES after extraction into diisobutylketone, Anal Chim Acta 144:213-221

Montaser A, Golightly DW (eds)(1987) Inductively coupled plasmas in analytical atomic spectrometry, VCH Publishers, Inc, New York

Morita M, Uehiro T (1981) Determination of phosphates by liquid chromatography with inductively coupled argon plasma AES detection, Anal Chem 53:1997-2000

Moselhy MM, Vijan PN (1981) Simultaneous determination of trace metals in sewage and sewage effluents by inductively coupled argon plasma atomic emission spectrometry, Anal Chim Acta 130:157-166

Nakata F, Hara S, Matsuo H, Kumamaru T, Matsushita H (1985) Fractional determination of chromium(III) and chromium(VI) via ion chromatography with ICP AES detection, Anal Sci 1:157-160

Park CJ, Van Loon JC, Arrowsmith P, French JB (1987) Sample analysis using plasma source mass spectrometry with electrothermal sample introduction, Anal Chem 59:2191-2196

Que Hee SS, Boyle JR (1988) Simultaneous multielement analysis of some environmental and biological samples by ICP-AES, Anal Chem 60:1033-1042

Ridout PS, Jones HR, Williams JG (1988) Determination of trace elements in a marine reference material of lobster hepatopancreas (TORT-1) using ICP-MS, Analyst 113:1383-1386

Sansoni B, Brunner W, Ruppert H, Dittrich R (1988) Comparative instrumental multi-element analysis I: comparison of ICP source mass spectrometry with ICP-AES, ICP atomic fluorescence spectrometry and AAS for the analysis of natural waters from a granite region, Fresenius' Z Anal Chem 331:154-169

Sanzolone RF, Meier AL (1986) Determination of Ca, Mg, Na, Cd, Cu, Fe, K, Li and Zn in acid mine and reference water samples by ICP atomic fluorescence spectrometry, Analyst 111:645-649

Schlieckmannn F, Umland F (1984) Ein neuartiges Aufschlussverfahren für die Multielementbestimmung von Schwermetallen in Luftstäuben, Fresenius' Z Anal Chem 318:495-497

Schramel P (1988) ICP and DCP emission spectrometry for trace element analysis in biomedical and environmental samples. A review, Spectrochim Acta 43B:881-896

Skogerboe RK, Nanagan WA, Taylor HE (1985) Concentration of trace elements in water samples by reductive precipitation, Anal Chem 57:2815-2818

Sparkes S, Ebdon L (1986) Slurry atomization for agricultural samples by plasma emission spectrometry, Anal Proc 23:410-412

Sugimae A, Mizoguchi T (1982) Atomic emission spectrometric analysis of airborne particulate matter by direct nebulization of suspensions into the ICP, Anal Chim Acta 144:205-212

Sugiyama M, Fujino O, Kihara S, Matsui M (1986) Preconcentration by dithiocarbamate extraction for determination of trace elements in natural waters by ICP AES, Anal Chim Acta 181:159-168

Tan SH, Horlick G (1986) Background spectral features in ICP mass spectrometry, Applied Spectrosc 40:445-460

Taylor P, Dams R. Hoste J (1985) Determination of cadmium, chromium, copper, nickel and zinc in city waste-incinerator ash using ICP AES, Analyt Lett 18:2361-2368

Thompson JJ, Houk RS (1987) A study of internal standardization in ICP-MS, Applied Spectrosc 41:801-806

Ting BTG, Janghorbani M (1987) Application of ICP-MS to accurate isotopic analysis for human metabolic studies, Spectrochim Acta 42B:21-27

Tschöpel P, Tölg G (1982) Comments on the accuracy of analytical results in nanogram and picogram trace analysis of the elements, J Trace Microprobe Techn 1:1-77

Turk GC, Watters RL (1985) Resonant laser-induced ionization of atoms in an ICP, Anal Chem 57:1979-1983

Vandecasteele C, Nagels M, Vanhoe H, Dams R (1988) Suppression of analyte signal in ICP-MS and the use of an internal standard, Anal Chim Acta 211:91-98

Vaughan MA, Horlick G (1986) Oxide, hydroxide and doubly charged analyte species in ICP mass spectrometry, Applied Spectrosc 40:434-444

Vermeiren K (1989) Thermospray verstuiving voor inductief gekoppeld plasma atoomemissiespectrometrie, Ph D Dissertation, University of Ghent

Wang X, Viczian M, Laztity A, Barnes RM (1988) Lead hydride generation for isotope analysis by ICP-MS, J Anal Atom Spectrom 3:821-828

Wendt R, Fassel VA (1965) Induction coupled plasma spectrometric excitation source, Anal Chem 37:920-922

Winge RK, Fassel VA, Kniseley RN, De Kalb E, Haas WJ Jr (1977) Determination of trace elements in soft, hard and saline waters by the ICP multielement atomic emission spectroscopic technique (ICP-MAES), Spectrochim Acta 32B:327-345

Winge RK, Peterson VJ, Fassel VA (1979) ICP-AES: prominent lines, Applied Spectrosc 33:206-219

Winge RK, Fassel VA, Peterson VJ, Floyd MA (1985) Inductively coupled plasma-atomic emission spectroscopy. An atlas of spectral information, Elsevier, Amsterdam

Xiao-Quan S, Tie-Bang W, Zhe-Ming N (1987) Simultaneous determination of major, minor and trace elements in airborne particulates by ICP-AES, Fresenius' Z Anal Chem 326:419-424

PRECONCENTRATION OF REFRACTORY ELEMENTS FOR INDUCTIVELY COUPLED PLASMA ATOMIC FLUORESCENCE SPECTROMETRY

S. Kaya[1], T.M. Durrani, S. Greenfield, and J.F. Tyson

Department of Chemistry, University of Technology, Loughborough, Leicestershire, LE11 3TU, U.K.

1. Introduction

Modern civilisation is now dependent on the large-scale use of a wide range of metals and most of these are naturally present only at trace levels in the biosphere and in biological material. The ability to analyse for trace elements is a pre-requisite for metal speciation studies. Many such studies are concerned with environmental samples including solid waste, waste water, drinking water, atmospheric particulates, particulates from emission sources, and also plant, food and animal tissue.

Tungsten and molybdenum are used in the manufacture of steel. The elements are added at concentrations between 5 and 20% to produce high performance steels in terms of hardness and high temperature strength. The emission of potential pollutants from steel works can be followed by analysing for the refractory elements (present at trace or sub-trace levels) in the environmental samples associated with this industry.

Inductively coupled plasma optical emission spectrometry (ICP-OES) allows a competent analytical chemist or spectroscopist to make an accurate and cost effective determination of minor and trace elements in almost any sample that can be dissolved in a suitable solvent to yield a solution or a slurry of finely divided particles. The earliest environmental applications reported were for particulates from stacks and in ambient air. Greenfield and co-workers (1975) described some environmental applications in which ICP-OES was one of the four instrumental approaches employed to evaluate a potential environmental problem.

There are many similarities between ICP-OES and combustion

---

[1] On leave from İnönü Üniversitesi, Eğitim Fakültesi, Kimya Bölümü, Malatya 44069, Turkey

NATO ASI Series, Vol. G 23
Metal Speciation in the Environment
Edited by J. A. C. Broekaert, Ş. Güçer, and F. Adams

flame spectroscopy techniques of flame atomic emission spectrometry (FAES) and flame atomic absorption spectrometry (FAAS). The ICP possesses several important advantages that make it especially attractive for use as a spectroscopic emission source. These include large linear dynamic range, low detection limits and relative freedom from matrix effects compared with other spectroscopic sources. The singular advantage of the ICP emission source above others is the absence of chemical interferences. This is generally attributed to the extremely high temperature of the central core of the toroidal plasma, reported by Mermet (1975) to be about 6,000 K. This temperature, in a rare gas atmosphere, contains energetic ions and favours complete dissociation of all refractory materials studied to date. It prevents the occurrence of common solute matrix vaporisation interferences that are usually encountered with most other atom reservoirs. The concentration range covered for ICP-OES may be several orders of magnitude.

The use of an ICP as an atomiser in atomic fluorescence spectrometry has been studied in some detail by Montaser and Fassel (1976), who pointed out that atomic fluorescence is, for many elements, a more sensitive technique than ICP-OES with relative freedom from scattering and matrix interferences. The ICP provides a chemically inert environment with minimal quenching of excited states. A modulated light source allows the detector circuits to discriminate against the atomic emission from the atom source and detection limits for several elements by this technique are now the best ever reported for any analytical method.

Conventional line sources, such as hollow cathode lamps have been used as excitation sources for atomic fluorescence spectrometry (AFS). The first use of an inductively coupled plasma as a line source in AFS was demonstrated by Hussein and Nickless (1969). In 1981, work commenced on the development of a dual-plasma atomic fluorescence spectrometer at Loughborough (Greenfield, 1984). The technique is known by the acronym ASIA; atomiser, source, ICPs for AFS. The first version of the instrument could be used in both the fluorescence and the emission mode of operation. The source of exciting radiation is provided by the high power Ar/Ar/Air ICP in a Greenfield Torch which enables solutions of high concentration (up to 50 % m/V) to be aspirated. The atomiser plasma is sustained in an extended-tube torch, which is essentially a Fassel torch with an outer tube extending 6 cm above the load coil minimising the effect of air entrainment (Kosinski *et al.*, 1983) and providing a long tail-flame. This plasma is operated at relatively low powers and gas flow rates and is much less turbulent than a typical plasma used for emission work.

The detection limits obtained by fluorescence for the refractory elements with this version of the instrument were found to be worse than those of non-refractory elements. One possible explanation for this is the occurrence of refractory oxides in the relatively low power atomiser plasma (Greenfield and Thomsen, 1985). A study of the effect of alkane gases on the atomic fluorescence signals of some refractory elements was made and an improvement by a factor of two obtained in the detection limit of tungsten by using ethanolic solutions instead of adding traces of hydrocarbon gases (Greenfield and Thomsen, 1986).

Flow injection (FI) techniques are being used increasingly for sample introduction in ICP-OES, but in most instances the ICP is used simply as another detector, albeit one with multi-element capability (Sharp *et al.*, 1988). The preconcentration of samples by FI techniques involving either liquid-liquid extraction or ion exchange have been reported for several atomic spectrometry determinations.

In this study the on-line preconcentration of tungsten and molybdenum on a miniature column of Amberlite IRA-93 ion exchange resin has been developed for the ASIA instrument. The performance of the new ASIA system in terms of detection limits, calibration curves, and linear range has been determined and compared with the previous system.

2. Experimental

2.1. Apparatus

The previous ASIA instrument is described in detail elsewhere (Greenfield and Thomsen, 1985). The parts of the new ASIA instrument is given in Table 1, and a schematic diagram in Fig. 1. Some modifications have been made to improve the light gathering from the excitation source. A new computer controlled monochromator has replaced the old monochromator and is placed closer to the radiation source. The single lens between the atomiser and monochromator replaces the two lenses used previously and gives an image at the entrance slit with a linear magnification of 1.67.

The flow injection manifolds are shown in Fig.2.

Table 1. Instrumental details of the new ASIA spectrometer

R.f.generators	Radyne RD150, 15 kW, 7 MHz, (source plasma) Radyne SC15, 2.5 kW, 36 MHz, (atomiser plasma)
Torches	Greenfield torch (source plasma) Extended-tube torch (atomiser plasma)
Nebulisers	de Galan with vortex-type cloud chamber (source plasma) Labtest GMK nebuliser and spray chamber
Lens system	2x5.08 cm diam., 5.08 cm focal length, bi-convex quartz lenses (Between source ICP and atomiser) 5.08 cm diam., 5.08 cm focal length lens (Between atomiser ICP and monochromator)
Scanning monochromator	Bentham M300 Monochromator (Computer controlled)
Photo Multiplier	Bentham DH-3 (UV range)
Photomultiplier Supply	Photo pyhsics (0-3 kV)
Current amplifier	Bentham 265HF (Programmable)
Lock-in Amplifier	Bentham 223 (with Bentham 217 Bin/Display)
Light chopper	EG & G Brookdeal 9479
IEEE/488	Bentham 228F, Integrating ADC (Analogue input-100 mV to 10 V, 1 Mohm)
Computer	Opus PC II microcomputer (1 MB RAM, 2x360 kB floppy-disk drive)
Stepping motor drive .	Bentham SMD 3B/IEEE Stepping Motor Drive (with Bentham SMD 2, Power Amplifier)
Peripherals	HP 7470A Plotter, MP-165 Printer
Sample introduction ..	Gilson Minipuls-2, 8 channel peristaltic pump.

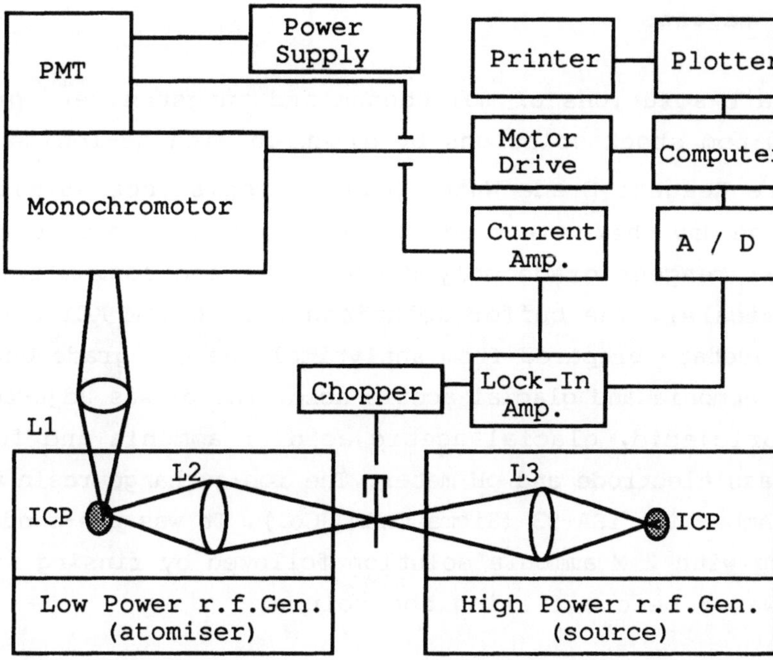

Fig. 1. *Schematic diagram of the Atomiser, Source, ICPs for AFS-ASIA system. L1, L2, and L3 are lenses*

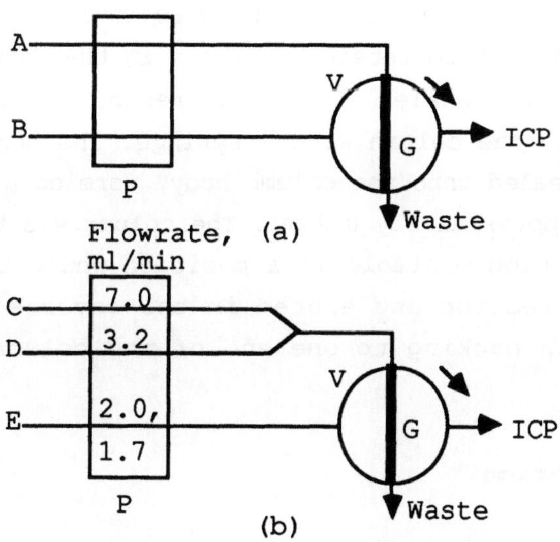

Fig.2.*Preconcentration manifolds with directional valve and peristaltic pump. (a) single line manifold; A, sample mixed with buffer at variable flowrate; B, eluant at variable flowrate; P, pump; V, valve; G, column. (b) dual line manifold; C, sample; D, buffer; E, eluant*

2.2. Reagents

Standart solutions of molybdenum and tungsten were prepared from 1000 ppm stock solutions by dilution with de-ionised water (LiquiPure Reagent Grade Water System, resistance 58 μSiemens. cm^{-1}) on the day they were used. Stock solutions were prepared from analytical reagent grade molybdenum oxide and sodium tungstate (BDH Chemicals). The buffer solutions were 0.1 M KCl and 0.5 M ammonium acetate prepared from analytical reagent grade potassium chloride, ammonia and glacial acetic acid. The pH was adjusted with hydrochloric acid, glacial acetic acid or ammonia and followed with a glass electrode and pH meter. The ion-exchange resin was 16-50 mesh Amberlite IRA-93 (Sigma Chem. Co.). It was pre-conditioned by washing with 2 M ammonia solution followed by rinsing with de-ionised water before filling the column.

2.3. Microcolumn Construction

The body of the column consisted of a 2.3 mm i.d. x 50 mm length of glass tubing. The ends of the column were covered with porous PVC with holes small enough to retain the resin in the column. After one end of the column was sealed with epoxy resin, a slurry of IRA-93 was aspirated into the column with a syringe. The other end of the column was then sealed into the column body, forming a column with a total volume of approximately 0.2 ml. The column was then connected to a rotary injection manifold in a position where the sample was loaded in one direction and eluted in the reverse direction to prevent the resin packing to one end of the column.

2.4. Optimisation

The optimisation parameters were the Ar gas flow rates for injector and plasma, the air flow rate for the source plasma cooling, the RF power of the generators, and the height of atomiser

Table 2. Operating conditions for Mo and W determinations in the fluorescence mode (ASIA)

	Atomiser Plasma		Source Plasma	
	Mo	W	Mo	W
Power (kW)	.58	.62	7.3	7.3
Coolant gas ($l.min^{-1}$)	14	14	30	30
Plasma gas ($l.min^{-1}$)	-	-	30	30
Injector gas ($l.min^{-1}$)	2.5	2.5	2.5	3.0
Height above the load coil (cm)	7.4	7.2	2.0	2.0
Soln. flowrate ($ml.min^{-1}$)	2.7	2.8	3.0	3.0
Slit width (mm)	5.0	5.0	5.0	5.0

torch. The optimisation was carried out using an iterative univariate search method while monitoring the total fluorescence signal (Greenfield *et al.*, 1989). The widest slit width, 5 mm, was used, because of the low fluorescence background. The values of the various operating parameters used for the determination of tungsten and molybdenum are given in Table 2. The linearity of the calibration curves [log (fluorescence intensity) versus log (concentration)] for each element under its optimum conditions were checked. The parameters investigated for the preconcentration manifold included effect of pH, sample flowrates, nature of the eluant and its flowrate, and buffer strengths. Measurements were made on-line for most of the parameters studied in the emission mode of the instrument using manifold (a) in Fig. 2. The calibration runs and the detection limits were determined in the fluorescence mode of the instrument using manifold (b) in Fig.2.

3. Results and Discussion

The retention-pH profile for tungsten and molybdenum are in Fig. 3. Quantitative retention of Mo and W at pH 5 and below was obtained. Both W and Mo gave more or less similar results which indicates the structural similarity of the metal species in the solution. The effect of pH studies indicated that a pH of 2.0 was

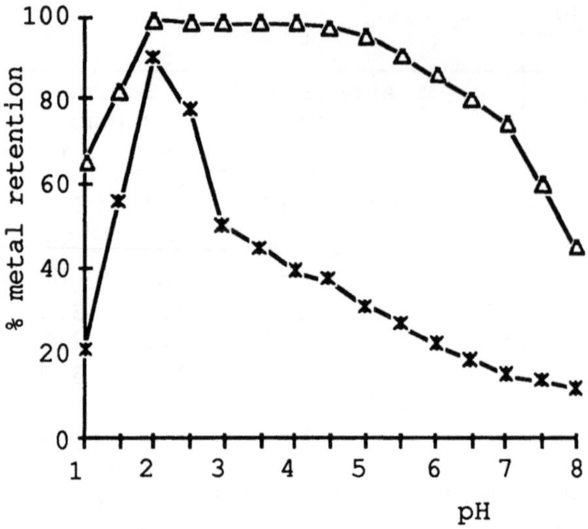

Fig. 3. *Effect of pH on metal retention onto IRA-93 resin:* **Δ**, W; *****, Mo

suitable for maximum retention of W and Mo. However, pH 4.5 was used for the preconcentration of W because of the extended life of the resin column at pH 4.5 compared with that at pH 2.0. The results in Fig.4, show the effect of different sample flowrates through the resin column. The effect for W and Mo is more or less similar due to the reason described earlier. The waste was analysed to calculate the retention efficiency of the column. The results showed an efficiency of >95% for W after loading 3000 μg of W and 90% for Mo after loading 700 μg of Mo on to the column. A number of eluants were studied. Nitric acid and hydrochloric acid up to 4 M concentration were less effective in eluting the retained metal ions from the resin column. Whereas sodium carbonate solution (0.1M) gave a high blank, thus eliminating the possibility of its use. A 2 M ammonia solution was found to be effective in eluting the reatined metal ions with a continuous stream of eluant and the transient

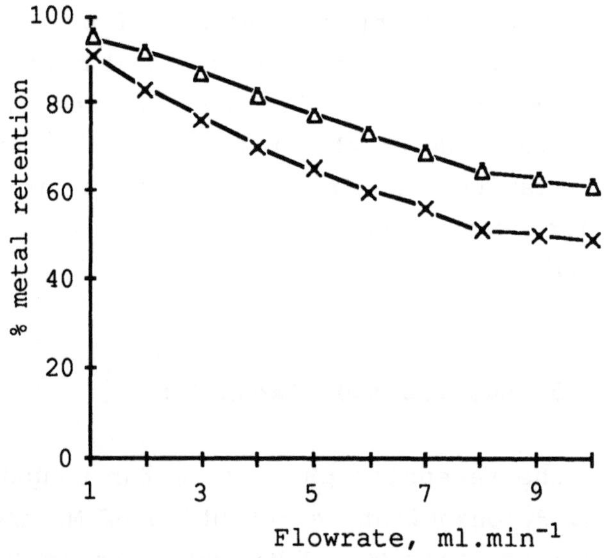

Fig. 4. *Effect of sample flowrates on % metal retention on resin column:* **Δ**, W; X, Mo

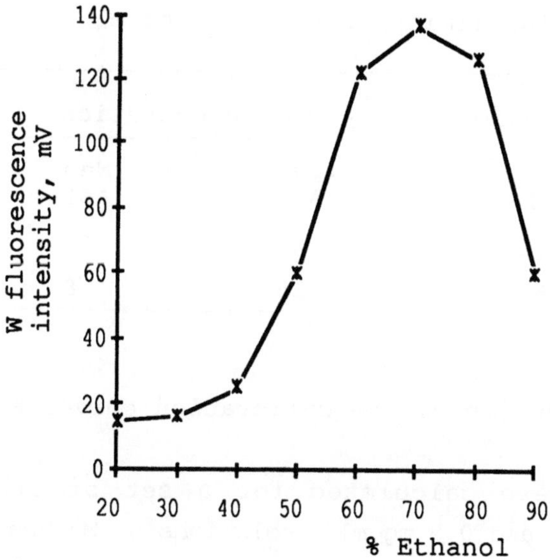

Fig. 5. *The effect of the ethanol percentage on the atomic fluorescence signal of tungsten*

signal appeared as an asymmetric peak returning to the baseline after 40-50 seconds. To apply the method for atomic fluorescence measurements, the sorbed metal ions were eluted with 2 M ammonia solutions in 75% ethanol. The effect of the ethanol percentage on the atomic fluorescence signal of W determined with conventional nebulisation is shown in Fig. 5. The results showed that the fluorescence signal increased as the percentage of ethanol in the solution increased, until a limit was reached after which the signal dropped, probably due to the quenching of the fluorescence signal by carbon species. The on-line procedure developed was as follows. The samples were loaded onto the column using manifold (b) in Fig. 2, for a predetermined time of five minutes and the metal ions were eluted by a stream of ethanolic 2 M ammonia solutions flowing at 2.0 and 1.7

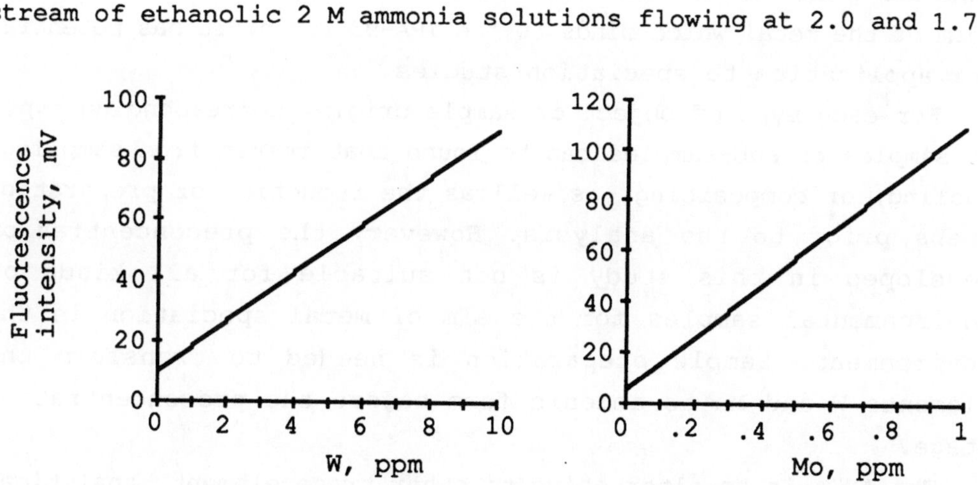

Fig. 6. *The calibration curves for W and Mo*

Table 3. Comparative detection limits (3 s.d.) in ng ml^{-1}

	Conventional nebulisation		Preconcentration	
Element	Previous ASIA	New ASIA	Previous ASIA	New ASIA
W	900	428	65	31
Mo	146	63	15	8

$ml.min^{-1}$ for W and Mo, respectively. The calibration curves are shown in Fig.6.

The detection limits were calculated for a set of five replicate measurements of 0.1 and 0.5 $mg.ml^{-1}$ solutions of Mo and W respectively and are given in Table 3. Precisiens of 5.24 and 4.28% RSD were obtained at these concentrations for Mo and W, respectively.

4. Conclusions

An FI preconcentration manifold can be used to improve the detection limits by ASIA by factor of between 20 and 30 for the elements W and Mo. As the manifold is based on the collection of a form of the metal which binds to the IRA-93 resin, it has potential for application to speciation studies.

For each type of object or sample origin, corresponding types of samples or sub-samples can be found that result from sampling, pooling, or compositing, as well as the reduction or preparation steps prior to the analysis. However, the preconcentration developed in this study is not suitable for all kinds of environmental samples for the aim of metal speciation in the environment. Sample preparation is needed to transform the elements W and Mo to anionic form before the preconcentration stage.

The ASIA is an alternative to other trace-element analytical systems such as ICP-OES, x-ray fluorescence, FAA, neutron activation, and arc and spark emission spectroscopy for environmental as well as other analysis.

Acknowledgements

SK thanks to the İnönü Üniversitesi for study leave and TMD thanks the Ministry of Science and Technology, Government of Pakistan for financial support.

References

Greenfield S (1984) Inductively Coupled Plasmas in Atomic Fluorescence Spectrometry. Anal Proc 21:61-63.

Greenfield S, Jone ILL, McGeachin HMcD, Smith PB (1975) Automatic Multi-Sample Simultaneous multi-element analysis with a H.F. Plasma torch and direct reading spectrometer. Anal Chim Acta 74:225-245.

Greenfield S, Salman MS, Thomsen M, Tyson JF (1989) Comparison of Alternating Variable Search and Simplex Methods of Optimisation for Inductively Coupled Plasma Optical Emission and Atomic Fluorescence Spectrometry. J Anal At Spectrom 4:55-65.

Greenfield S, Thomsen M (1985) Atomiser, source, inductively coupled plasmas, in atomic fluorescence spectrometry (ASIA): Some recent work. Spectrochim Acta 40B:1369-1377.

Greenfield S, Thomsen M (1986) Use of organic additives for the detection of refractory elemnts in plasma atomic fluorescence spectrometry. Spectrochim Acta 41B:677-682.

Hussein AMCh, Nickless G (1969) paper presented at the 2nd ICAS, Sheffield, UK

Kosinski MA, Uchida H, Winefordner JD (1983) Atomic Fluorescence Spectrometry with Inductively Coupled Plasma as Excitation Source and Atomization Cell. Anal Chem 55:688-692.

Mermet JM (1975) Comparison des températures et des densités electroniques mesurees surle gaz plasmagène et sur des éléments excites dans un plasma h.f. Spectrochim Acta 30B:383-396.

Montaser A, Fassel VA (1976) Inductively Coupled Plasmas as atomization cells for Atomic Fluorescence Spectrometry.

Anal Chem 48:1490-1499.

Sharp BL, Barnett NW, Burridge JC, Littlejohn D, Tyson JF (1988) Atomic Spectrometry Update-Atomisation and Excitation. J Anal At Spectrom 3:133R-154R.

Tyson JF (1988) Atomic Spectrometry and flow injection analysis: a synergic combination. Anal Chim Acta 214:57-75.

HPLC-AAS INTERFACES FOR THE DETERMINATION OF IONIC ALKYLLEAD, ALKYLTIN, ARSONIUM AND SELENONIUM COMPOUNDS

W.D. Marshall[1], J.S. Blais[2], and F.C. Adams

Dept. of Chemistry

Universitaire Instelling Antwerpen

Universiteitsplein 1

B-2610 Wilrijk, Belgium

1. Introduction

It has been estimated (Randall, 1984) that only 10 to 15 percent of the known compounds are sufficiently volatile and thermally stable to survive a gas chromatographic separation *intact*. A variety of ingenious derivatization procedures were devised to improve the gas chromatographic behaviour of thermally labile and/or non-volatile analytes. For example, cationic alkyllead species [R_3PbX, R_2PbX_2; R = CH_3, C_2H_5; X = anion] were ethylated, propylated, butylated, phenylated, or converted to their corresponding hydrides either prior to or in conjunction with gas chromatography-atomic absorption spectrometry (GC-AAS) (Van Cleuvenbergen *et al.*, this volume). In our hands the butylation of alkylleads, when present at µg/kg concentrations in fatty tissues, proved to be technically demanding, labour- intensive, and time-consuming. Moreover, under certain conditions, transalkylation (redistribution) reactions were observed as an artefact of the derivatization procedure (Blais and Marshall, 1986). These problems prompted us to investigate the coupling of other chromatographic techniques with quartz tube (QT)-AAS for the determination of selected ionic organometallic species. Relative to GC, high performance liquid chromatography (HPLC) has a lower resolving power which can be improved somewhat (microbore techniques) at the expense of sample capacity. For liquid samples the major limiting factor for virtually all atomic spectroscopic techniques is the process of sample introduction. The conversion of liquid sample to atomic vapour is a notoriously inefficient

---

[1] On leave from [2] Dept. of Food Science and Agricultural Chemistry, Macdonald College of McGill, Ste Anne de Bellevue, Qué., Canada, H9X 1C0.

NATO ASI Series, Vol. G 23
Metal Speciation in the Environment
Edited by J. A. C. Broekaert, Ş. Güçer, and F. Adams

process (Browner and Boorn, 1984a, b). The direct coupling of HPLC with a spectroscopic detector requires that the column eluate be introduced in a continuous and pulseless mode. In flame-AAS (FAAS), efficient atomization of the analyte prior to detection requires the complete and uniform nebulization (and pyrolysis) of the mobile phase. Direct HPLC-AAS couplings have been achieved using a variety of eluate nebulization techniques (Ebdon *et al.*, 1987). However, several problems remain to be solved. The nebulization/atomization process must occur in milliseconds and liquid flows of 1 mL/min become gas flows of liters/min at normal detector operating conditions. On the positive side, the wide latitudes in choosing solvent systems and elution modes have made HPLC a very versatile technique for the separation of both volatile and non volatile analytes. Conventional columns for HPLC are also characterized by high sample capacities (200 μL injections are common).

To avoid the problems of nebulization/atomization, earlier approaches to HPLC-AAS employed post column derivatization. Volatile products were purged from the column eluate in a gas liquid separator (GLS) and swept into the quartz tube (QT) detector and the spent liquid was discarded. Thus, alkyltins in HPLC eluate were converted to their hydrides and detected with picogram limits of detection (Burns *et al.*, 1981). Similarly various arsenic species have been determined by ion chromatography-hydride generation-QT-AAS (Ricci *et al.*, 1981). In contrast to batch hydride generation prior to chromatography, post column hydride generation is attractive because the possibility of molecular rearrangements during the derivatization (Talmi and Bostick, 1975) becomes unimportant provided that there is no post-column mixing of the separated analytes. Several researchers have coupled liquid chromatography with graphite furnace-atomic absorption spectrometry (GF-AAS) (Irgolic and Brinckman, 1984; Ebdon *et al.*, 1987). Whereas in the majority of applications the column eluate was sampled discretely (thus producing a histogram), a modified IC-type carbon rod furnace has been develpoed that can be operated continuously for the detection of alkyltins in column eluate (Nygren *et al.*, 1988).

Although atomic emission techniques offer impressive limits of detection (Broekaert, 1987), the higher costs of acquiring and operating these instruments often makes it difficult to justify their dedicated use to a single chromatographic analysis. Yet biological studies may require a continuing access to a particular speciation procedure. The objectives of the our current studies are to evaluate several HPLC-AAS interfaces for the determination of a variety of ionic organometallic species. It was desirable that the HPLC chromatographic procedure(s) be able to accept the analytes as complexes or as ion pairs (since many of the ionic analytes of interest can be separated from biological matrices in these forms). In this paper our recent progress in the development of interfaces for the determination by HPLC-AAS of ionic akylleads, alkyltins, selenonium, and arsonium compounds is reviewed.

2. Determination of Alkyllead Compounds

A thermospray effect (Blakley and Vestal, 1983) as a means of the flash evaporation of a column eluate appeared to offer a promising approach to couple an HPLC with a quartz T-tube AAS detector. In a preliminary design (Figure 1), a 50 μm i.d. capillary quartz transfer line was positioned within the electrically heated lower portion of the silica T-tube. The eluate [methanol plus 0-30 % (v/v) water] appeared to be completely vapourized if the quartz inlet tube was maintained at 900 °C. The vapour self ignited in the heated upper region of the T-tube (atom keeper). For alkyllead butylates, the approximate limit of detection (as Pb) was 50 ng.

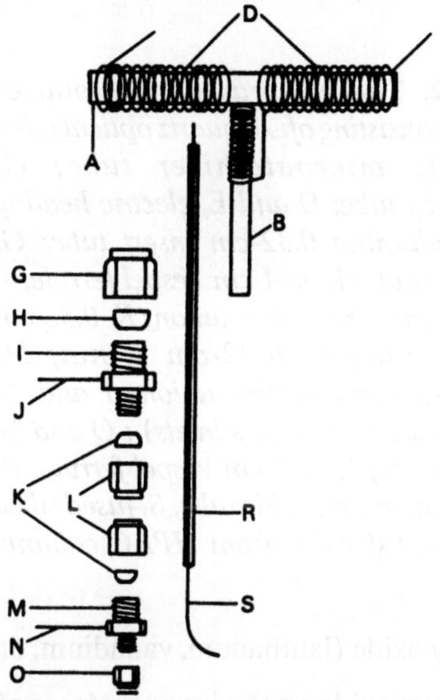

Figure 1. *HPLC-AAS thermospray interface consisting of: A, quartz optical tube; B, inlet tube; D, electric heating coil; G, 0.64-cm stainless steel (ss) nut; H, 0.64-cm Vespel ferrule; I, 0.64-0.32-cm ss reducing union; K, 0.32-cm Vespel ferrule; L, 0.32-cm ss nut; M, 0.32-0.16-cm ss reducing union; J and N, 0.16-cm o.d. ss tubes (gas inlets); O, 0.16-cm ss nut; R, 0.32-cm o.d. silica guide tube; and S, fused-silica transfer line from HPLC column*

Attempts to decrease the detection limit by introducing fuel or oxidant gases (hydrogen, acetylene or oxygen) into the furnace (100-300 mL/min) disrupted the thermospray presumably by cooling the capillary. The thermospray was re-established by lower flow rates and/or increased heating coil temperatures. In contrast to hydrogen or acetylene, oxygen decreased the limit of detection for these analytes appreciably. Oxygen in the thermospray environment resulted in a fireball which apparently sustained the atomization of lead from these analytes. For organic-rich mobile phase (> 90 % MeOH) the fireball overheated the transfer line and caused in-column pyrolysis of the eluate and eventually blocked the transfer line.

A modified interface design (Figure 2) physically separated the thermospray device from the oxygen inlet. The thermospray tube (C) was positioned approximately perpendicular to the lower portion (B) of the T-tube. With this configuration, the heat transfer from the resistant wires to the transfer line remained unaffected by the flow rates of the make-up gases. The fireball formed 0.5 to 1.5 cm from the capillary tip (at the intersection of tubes B and C) where oxygen and eluate vapours meet. A stable thermospray effect could be maintained with methanolic mobile phases containing 0 - 50 % water.

The long-term exposure of the silica tube to aqueous mobile phases resulted in the devitrification of the inner quartz surfaces, a process which was greatly accelerated by the presence of sodium salts of several complexing agents. Sodium atoms are thought to migrate into the hot quartz surfaces forming sodium silicates. On cooling, the modified surfaces contract heterogeneously and destroy the tube. To prolong the lifetime of silica tubes in flames the outer exposed surfaces can be coated with a refractory oxide (lanthanum, vanadium, or aluminum oxide, Brown *et al.*, 1987). A coating of aluminum oxide on the inner surfaces of the silica T-tube was effective against the corrosive action of aqueous methanol containing diphenylthiocarbazone (dithizone) but had little protective effect against sodium dimethyldithiocarbamate (NaDMDTC, 300 μg/ml) in the mobile phase. If NaDMDTC had to be used, the microatomizer assembly had to be kept at 500-600 ^{o}C when not in use. Interestingly, the coating had no detectable effect on the detector response to lead containing analytes suggesting that the silica surface does not play an active role in the atomization of lead. In contrast to the hydrogen radical-mediated atomization observed with the silica T-tube in GC-AAS (Forsyth and Marshall, 1985), the lead atomization in this interface was

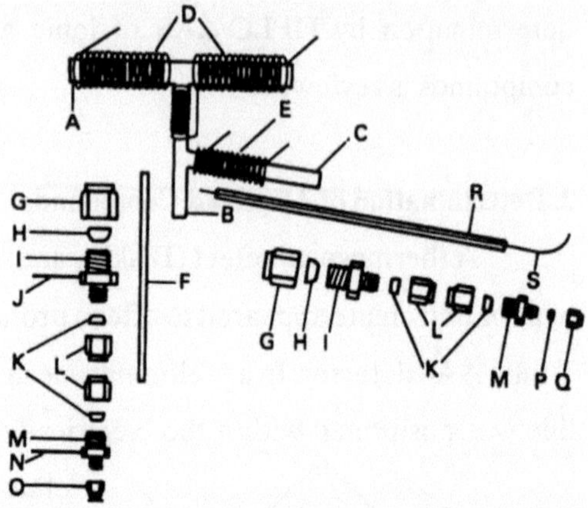

Figure 2. *Thermospray microatomizer assembly consisting of: A, quartz optical tube, B, quartz microatomizer tube; C, thermospray tube; D and E, electric heating coils; F, alumina 0.32-cm insert tube; G, 0.64-cm ss nut; H, 0.64-cm Vespel ferrule; I, 0.64-0.32-cm ss reducing union; K, 0.32-cm Vespel ferrule; L, 0.32-cm ss nut; M, 0.32-0.16-cm ss reducing union; J and N, 0.16-cm o.d. ss tubes (gas inlets) ; O and Q, 0.16-cm ss nut; P, 0.16-cm Vespel ferrule; R, 0.32-cm o.d. quartz guide tube; S, fused silica capillary transfer line from HPLC column*

unaffected by the condition of the quartz surface and was maximized by the presence of oxygen. In this case the atomization appears to be a process mediated by the flame (T = 1,400-1,600 °C) that is supported by the chromatographic eluate (fuel) and oxygen.

Preliminary experiments demonstrated that appreciable portions of Pb^{2+} and Me_2Pb^{2+} in dilute solution (0.5 μg/ml) were sorbed by the walls of borosilicate glass vials used with the autosampler. Repeated chromatography of such a solution from the same vial was characterized by a gradual decrease in response for Me_2Pb^{2+} and Pb^2, whereas the Me_3Pb^+ response remained unchanged. A crystal of NaDMDTC, added to the vial, returned the responses of the divalent analytes to their original values corroborating the analyte adsorption hypothesis. This problem was not observed if the same lead chloride standards were sampled from a polytrifluoroethylene (PTFE) vial.

It was known that microgram amounts of Pb^{2+}, Me_2Pb^{2+}, Et_2Pb^{2+}, Me_3Pb^+ and Et_3Pb^+ could be separated on a Nucleosil C_{18} phase using a methanol/aqueous acetate buffer of pH 4.6 (ionic strength I = 0.2 M) mixture (1:9 v/v) as the mobile phase (Blaszkewicz *et al.*, 1984). The high water concentration of this mobile phase was not compatible with the thermospray interface. However, a water-lean solvent with a complexing agent might be suitable as a mobile phase. A methanolic mobile phase containing dithizone (HDz, 300 μg/ml), water (10 %), and dioxane (15 %) provided a base-line separation of the analytes. (Figure 3, Blais and Marshall, 1989a). Whether the analytes were injected into the HPLC as the chlorides or as the dithizonates did not alter their retention times appreciably.

For water samples spiked with 10 μg Pb/kg of the four lead species, recoveries were excellent (> 86 %). However analogous experiments with a sandy loam soil or a sediment were characterized by unacceptable variations in peak resolution and retention times for replicate samples. Because the chromatographic characteristics of the lead analytes appeared to vary with the concentration (and possibly the nature) of the co-extracted metallics, the

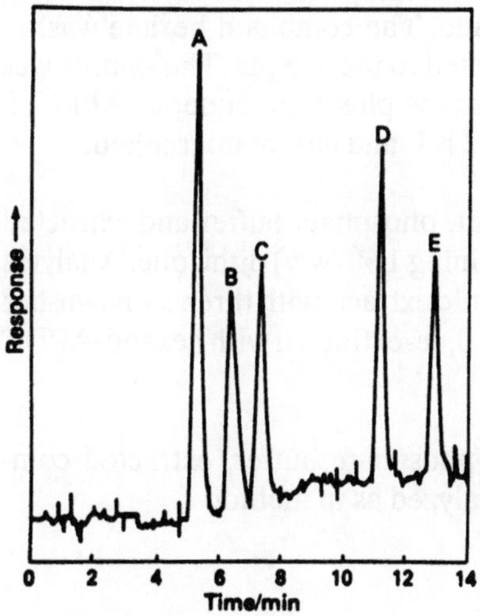

Figure 3. *HPLC-AAS chromatogram of A, Pb^{2+} (20 ng); B, Me_3Pb^+, C, Et_3Pb^+; D, Me_2Pb^{2+}, and Et_2Pb^{2+} (10 ng as Pb of each) obtained with the Nucleosil C_{18} column and dithizone in the mobile phase*

use of dithizone in the mobile phase was abandoned. The five Pb analytes were separated on the Nucleosil C18 column using 3:1 (v/v) CH_3OH/H_2O containing 600 μg/mL of ammonium pyrollidinedithiocarbamate (APDC) delivered at 1 mL/min. (Figure 4).

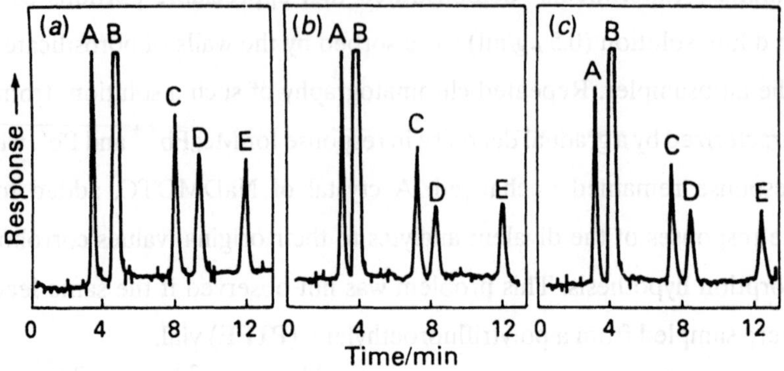

Figure 4. *HPLC-AAS chromatograms of A, Me_3Pb^+; B, Pb^{2+}; C, Et_3Pb^+; D, Me_2Pb^{2+}; and E, Et_2Pb^{2+} which had been spiked (at 10 ng Pb/g) into (a) water, (b) sediment, or (c) soil and reisolated with different procedures*

The applicability of the APDC mobile phase was assessed with water, soil and sediment that had been spiked with 10 μg Pb/kg of each of the four alkylleads. A pre-extraction with hexane followed by a complexometric extraction with hexane-APDC (method 1[3]) resulted in good recoveries from water (Table 1) but not from soil or sediment. The sediment extracts contained a yellowish dissolved material which co-eluted as a large tailing peak with Pb^{2+}. A double extraction procedure (method 2[4]) was then attempted. The

---

[3]Method 1: The spiked matrix (10 g) was combined with 10 mL phosphate buffer (pH 9.0; I = 0.2 M) and pre-extracted twice with 5 mL of hexane. The combined hexane washes were back-extracted with 3 mL of water which was returned to the sample. The sample was then extracted three times with hexane-APDC (5 mL hexane plus 1 mL aqueous APDC, 5 mM). The extracts were combined, dried, reduced to 0.2 mL and chromatographed.

[4]Method 2: The spiked matrix was combined with 10 mL phosphate buffer and extracted three times with 5 mL of 1:1 (v/v) hexane-benzene containing 1 % (w/v) dithizone. Analytes were recovered by back extraction of the combined organic extracts with three 5 mL washes of 0.15 M HNO_3, which were combined, basified to pH 9.0, re-extracted with hexand-APDC and analyzed as in method 1.

[5]Method 3: The spiked soil was slurried with 10 mL phosphate buffer, extracted complexometrically (without hexane pre-extraction) and analyzed as in method 1.

Table 1. Percentage recoveries (± 1 SD[1]) of ionic alkylleads from tap water, a sandy clay loam, and an anoxic sediment using various extraction procedures

Matrix	Method	% Recovery of Analyte			
		Me_3Pb^+	Et_3Pb^+	Me_2Pb^{2+}	Et_2Pb^{2+}
Water	1[3]	96 ± 3	90 ± 5	86 ± 6	91 ± 4
	2[4]	98 ± 4	95 ± 3	92 ± 8	94 ± 4
Sediment	1	U.C.[2]	U.C.	U.C.	U.C.
	2	90 ± .5	86 ± 2	72 ± 4	75 ± 5
Soil	1	50 ± 7	25 ± 4	N.D.[3]	N.D.
	2	66 ± 1	52 ± 4	N.D.	N.D.
	3[5]	83 ± 6	83± 2	73 ± 6	74 ± 3

[1] Standard deviation based on three replicate determinations.
[2] U.C. = unacceptable chromatography
[3] N.D. = none detected

analytes were first recovered from these matrices with dithizone in 1:1 benzene-hexane and then back-extracted into 0.15 M nitric acid. The nitric washes were combined, neutralized and back-extracted with hexane-APDC. The hexane-APDC extracts were concentrated and analyzed. With this procedure analytes were recovered efficiently from water and sediment but not from soil (Table 1).

The low recoveries from soil were surprising; a direct extraction of these analytes from this soil with hexane-APDC, butylation of the extracted species and their determination by GC-AAS gave good recoveries (Blais and Marshall, 1986). A complexometric extraction of spiked soil without hexane pre-extraction (method 3[5]) resulted in recoveries comparable to those reported previously. These observations suggested that this soil contained ligands capable of binding the analytes to form non-polar complexes (hexane soluble) that were surprisingly stable (resistant to the action of dilute acid). Typical chromatograms from the spiked tap water (a), sediment (b), and soil (c) are presented in Figure 4.

The described system has limits of detection between 1 and 2 ng of compound for each of the alkyllead analytes (Blais and Marshall, 1989a). Although minor modifications may have to be made to the extraction procedures depending on the sample matrix, HPLC-AAS offers a promising alternative to GC-AAS for the determination of ionic alkylleads. If sufficient sample (5-10g) is available, it may be the method of choice because sample derivatization is not required. Additionally, the high selectivity of the hexane-APDC extraction system for ionic alkylleads over Pb^{2+} (Blais and Marshall, 1986) facilitates the determination of these analytes in matrices with high levels of Pb^{2+}.

3. Post Column Ethylation

The ethylation of lead(II) and alkyllead cations was considered a promising route for their post-column conversion to volatile species. The derivatization reactions (equations 1-3) are considered to be mediated by alkyl radical transmetalation.

$$2Pb^{2+} + 4NaB(C_2H_5)_4 \longrightarrow [2(C_2H_5)_2Pb(II)] \longrightarrow Et_4Pb + Pb^0 + 4B(C_2H_5)_3 \quad (1)$$
$$R_3Pb^+ + NaB(C_2H_5)_4 \longrightarrow R_3Pb(C_2H_5) + Na^+ + B(C_2H_5)_3 \quad (2)$$
$$R_2Pb^{2+} + 2NaB(C_2H_5)_4 \longrightarrow R_2Pb(C_2H_5)_2 + 2Na^+ + 2B(C_2H_5)_3 \quad (3)$$

The stoichiometry of these reactions is not known because triethylborane can also alkylate lead(II) (Honeycutt and Riddle, 1960) and presumably alkyllead(IV) species. Ionic methylleads have been determined in water in low picogram concentrations by *in situ* ethylation in a closed cell followed by dynamic purging of the products into a trap (Rapsomanikis *et al.*, 1986). The products were then separated by controlled volatilization into a QT-AAS detector. Lead(II) was also recovered but with a relatively poor limit of detection (LOD) (a response of 0.1 to 1.0 % relative to alkyllead cations). On-column ethylation (Ashby *et al.*, 1988) as part of a GC separation procedure has been achieved by addition of $NaBEt_4$ to the head of the GC column. Inorganic lead and $Me_nPb^{(4-n)+}$ (n = 2,3) were determined as Et_4Pb and $Me_nEt_{(4-n)}Pb$ respectively. The desire to determine each of the seven ionic di- or trialkylleads [$R_nR'_mPb^{(4-n-m)+}$; R_n,R'_m = Et, Me; (n+m) = 2 or 3] prompted us to consider post-column ethylation as a volatilization technique prior to detection by electrothermal QT-AAS. This approach might improve the limits of detection observed with the thermospray unit providing that the ethylation procedure could be made continuous and the products could be purged rapidly from the reaction mixture so that the chromatographic efficiency would not be compromised.

The post-column reactor (Figure 5) consisted of a T-union in which column eluate

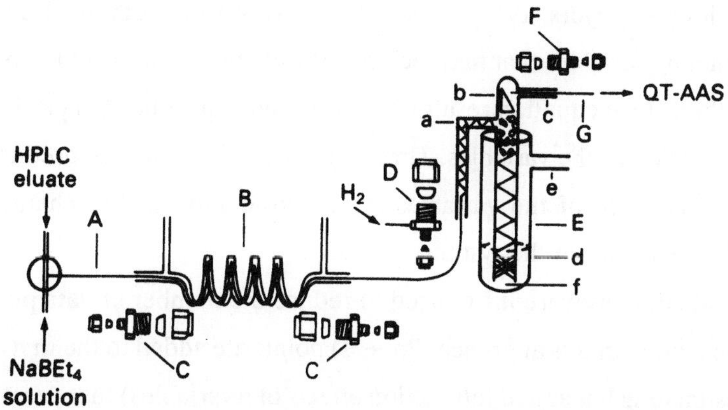

Figure 5. *Post column reactor consisting of a mixing T-union, a 0.3 mm i.d. capillary Teflon tube (A) a heating coil (B) , and a gas liquid separator (GLS, E). Gaseous effluent was channeled from the GLS to the quartz optical tube via a 0.32 mm i.d. fused silica transfer line (G) (Blais and Marshall, 1989b)*

was mixed with $NaBEt_4$, a capillary Teflon tube (A) jacketed with a glass heating coil (B) and a gas liquid separator (GLS). Water (40 - 70 °C) from a thermostated bath was circulated within the heating coil with a centrifugal pump. The volatile tetraalkyllead compounds were purged from the crude reaction mixture and swept up (via tube a) into the QT-AAS in a stream of hydrogen which was added to the GLS via a gas inlet in a modified reducing union (D). Spent liquid effluent was channeled down the inner tube of the GLS into a receiving vessel (d) and discarded to waste via tube (e). The internal volume of the GLS was reduced with coarse sand (symbolized by triangles) overlayered with PTFE chips and held in place with a shaped teflon block (b) at one end and glass wool at the other end.

A constant hydrostatic pressure was maintained within the GLS with a column (f) of liquid mercury (4 mL). Although $NaBEt_4$ does not react with Hg^0 (Honeycutt and Riddle, 1961), it was anticipated that small amounts of Hg^{2+} might be formed at the mercury-water boundary. Because of the possible presence of ethylmercury species in the liquid effluent the waste container was well ventilated in a fume hood.

Preliminary experiments demonstrated that certain of the operator defined variables (to be optimized) interacted with others. Multivariate optimization procedures (Massart *et al.*, 1978) provide a means of optimizing multi-variable processes (in this case the detector response) including variables that interact with each other. Although Simplex optimization procedures (Deming and Morgan, 1973) are less time consuming to perform than factorial designs (Myers, 1976), factorial designs provide additional information on interactions amoung variables that may facilitate the characterization of the process under study. The mathematical equation resulting from the analysis is used to predict the system response as a function of the variables. Three dimensional surfaces of predicted response as a function of any two of the variables are a considerable aid in characterizing and interpreting the behaviour of the system.

Composite designs were introduced to reduce the number of data points required to generate the model. In this approach $(2n + 1)$ points are added to the first-order design (suitable for estimating linear and interaction effects of n variables) that permits the fitting of a second-order model searching for optimal response. To further reduce the number of observations, the 2^n factorial may be replaced by a fractional replicate of 2^n provided that it is chosen so that no main effect is confounded with the grand mean. A half replicate of a 2^5 (2 levels, 5 variables) factorial design (16 observations) was augmented by 11 points to form a composite design, with a total of 84 observations (3 replicate determinations of each data point and three additional replicates of the center point (for the estimation of experimental error).

For the optimization study the column was removed and the injector was connected directly to the mixing T-union of the reactor. The QT skin temperature and the concentration of $NaBEt_4$ solution (0.1 to 1.0 % w/v, added to the mixing T-union) did not influence the detector response significantly ($\alpha < 95$ %). In contrast, the flow rate of the aqueous $NaBEt_4$ solution ($\alpha = 99.99$ %), the temperature of the reaction coil , and the flow rate of the H_2 purge gas ($\alpha = 99$ %) significantly influenced the response. The beneficial influence of the higher reaction coil temperature was considered to reflect the more efficient purging of the derivatives from the heated reaction mixture, whereas the higher flow rate of $NaBEt_4$ decreased the solubility of the derivatives by increasing the water content in the crude reaction mixture. Although a maximal detector response was obtained with H_2 flow rates of 10 ml/min, peak shapes were more symmetrical and peak areas more reproducible at higher flow rates. Optimized values were considered to be: H_2 flow rate, 50 ml/min; $NaBEt_4$ flow rate and concentration, 1 ml/min of a 0.1 % (m/v) aqueous solution; reaction coil temperature, 70 °C; QT temperature 1,000 °C.

With these operating conditions the limits of detection for the ionic alkyllead standards were calculated based on a model of first-order error propagation with background noise distributed normally (Foley and Dorsey, 1984). The LODs of these compounds were: Me_3PbCl, 0.12 ng; Et_3PbCl, 0.15 ng; Me_2PbCl_2, 0.10 ng; and Et_2PbCl_2, 0.13 ng (Blais and Marshall, 1989b). Longer term reproducibility (at 10 x LOD) during ten hours was excellent (3.9 - 5.2 % relative standard deviation).

Injecting more than 3 ng of each of the analytes under conditions of optimum response resulted in appreciable tailing and some overlap of peaks corresponding to Et_3Pb^+ and Me_2Pb^{2+} which was reflected in lower coefficients of correlation for these analytes relative to those observed in the thermospray microatomizer interface (Section 2). Peak tailing at higher concentrations (> 3 ng of analyte) of injected standards was not observed if the H_2 flow-rate was doubled.

Interestingly, Pb^{2+} was derivatized/volatilized only inefficiently (estimated LOD, 100 ng) by the ethylation procedure (Figure 6a) . This situation is desirable since many samples from polluted environments contain very high levels of Pb^{2+} that may contribute to artifact formation during the alkylation procedure. The resolving power of the isocratic solvent system (4:1 methanol/water containing 600 μg/mL ammonium pyrollidinedithiocarbamate) and the Nucleosil C_{18} column is demonstrated by the virtual base line separation of a synthetic mixture containing seven mixed ionic alkylleads (Figure 6b)

In summary, a post-column ethylation coupled with purging the product from the

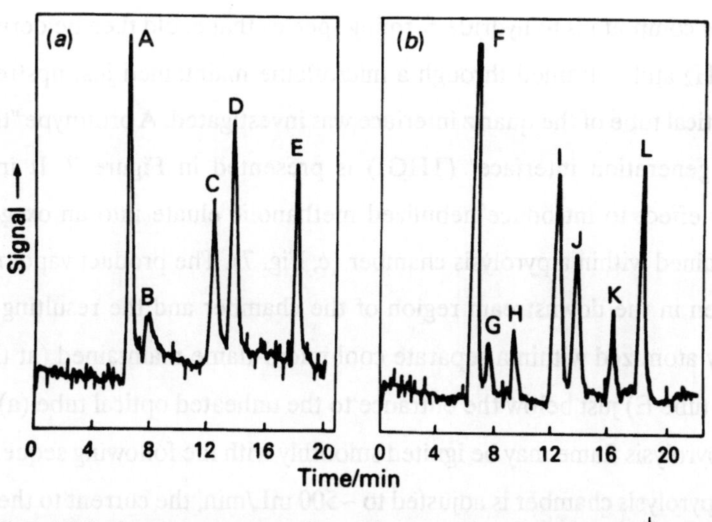

Figure 6. *HPLC-AAS chromatograms of a, 1.0 ng each of A, Me_3Pb^+; C, Et_3Pb^+; D, Me_2Pb^{2+}; E, Et_2Pb^{2+} and B, 100 ng Pb^{2+}. Chromatogram b is a synthetic mixture containing F, Me_3Pb^+; G, $EtMe_2Pb^+$; H, Et_2MePb^+; I, Et_3Pb^+; J, Me_2Pb^{2+}; K, $EtMePb^{2+}$ and L, Et_2Pb^{2+}*

reaction mixture offers a suitable alternative to GC-AAS. Because complexometric extracts containing the analytes can be concentrated without analyte loss, and larger sample volumes can be applied to an LC column, method limits of detection for the two techniques (GC and HPLC) are comparable. The isocratic solvent system also minimizes chromatographic re-equilibration time between samples. One advantage of the ethylation procedure relative to the thermospray-microatomizer interface is that it is compatible with water-rich mobile phases permitting a variety of columns to be employed. Although the ethylation procedure improved the LODs approximately ten-fold, this improvement was achieved at the expense of a more limited linear dynamic range (0.1-3 ng of cation, at maximum sensitivity).

4. Determination of Arsonium and Selenonium Compounds

The application of the direct thermospray-microatomizer interface to the determination of arsonium and selenonium cations was rapidly abandoned. The most intense resonant lines for these elements at 194.3 nm for As and 196.0 nm for Se are in the spectral range of strong emissions by kinetic flame components that decreased, appreciably, the signal to noise ratio of the detector. Deuterium background correction was ineffective. Post-column hydride generation was also somewhat unappealing because these onium compounds do not react readily with reducing agents and acids. On-line pyrolytic degradation of onium compounds to hydride-forming species that could then be derivatized *in situ* with excess H_2 and entrained through a microflame maintained just upstream from the unheated optical tube of the quartz interface was investigated. A prototype "thermochemical hydride generation interface" (THGI) is presented in Figure 7. It incorporates a thermospray effect to introduce nebulized methanolic eluate into an oxygen-supported flame maintained within a pyrolysis chamber (c, Fig. 7). The product vapours are reacted with hydrogen in the downstream region of the chamber and the resulting products are subsequently atomized within a separate cool micro-flame maintained (at the end of the oxygen inlet tube E) just below the entrance to the unheated optical tube (a).

The pyrolysis flame may be ignited smoothly with the following sequence: flow rate of O_2 to the pyrolysis chamber is adjusted to ~500 mL/min, the current to the thermospray heating element is adjusted to heat the outer skin of the thermospray tube (d) to 900-1,000 ^{o}C, the capillary transfer line is inserted half way into the hot thermospray tube and allowed

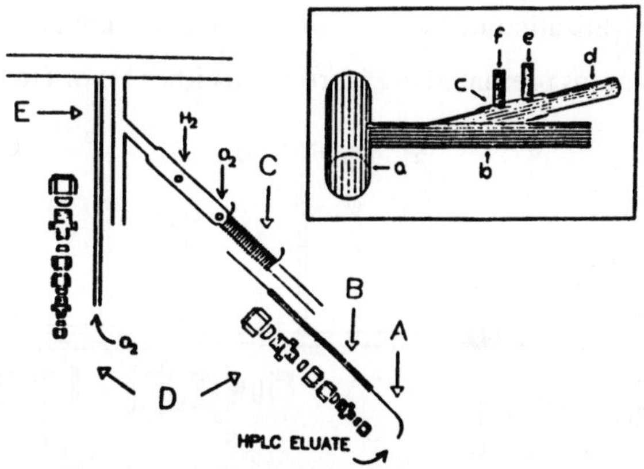

Figure 7. *Thermochemical hydride generation interface. The quartz body consists of an optical tube (a); an analytical flame tube (b); a pyrolysis chamiber (c); a thermospray tube (d) and O_2 and H_2 inlet tubes (e and f). The complete assembly consists of a capillary transfer line (A); a quartz guide tube (B); a heating coil (C) and Swagelok assemblies (D) to position the guide tube within the thermospray tube and the O_2 inlet tube (E) within the analytical flame tube. Oxygen, added via tube E supports an analytical flame maintained just below the the optical tube (Blais and Marshall, 1990)*

to heat for 1-2 min., the HPLC flow rate (100 % methanol) is rapidly adjusted to 0.5 mL/min which will ignite the pyrolysis flame. The accumulation of methanol in the pyrolysis chamber (the result of lower temperatures) must be avoided as it may result in an explosive ignition. Once ignited, the H_2 flow rate is adjusted to 1 L/min and the excess H_2 is ignited manually at both ends of the optical tube. The flow of O_2 to the analytical flame is increased until ignition (~300 mL/min) and subsequently, the flow rate of H_2 is decreased until flameout occurs at both ends of the optical tube (the O_2 analytical flame should remain ignited). The capillary depth, HPLC flow rate, and O_2 flow rate to the pyrolysis tube are then adjusted to obtain a stable thermospray. To extinguish the pyrolysis flame the HPLC flow must first be stopped and the capillary retracted from the heated zone of the thermospray.

The atomization of arsenic and selenium hydrides within silica T-tubes is considered to be mediated by hydrogen radicals (Dedina and Rubeska, 1980; Weltz and Melcher, 1983) that are generated within the reaction zone of the diffuse flame. These free radicals are considered to be formed within a spatially limited cloud which does not extend into the AAS optical beam, thus limiting the background noise.

The postulated thermochemical hydride generation (THG) mechanism is supported by the fact that no AAS signal for Me_3Se^+ or for Me_4As^+ was observed in the absence of a flame maintained just below the unheated optical tube. More direct evidence for the thermochemical derivatization of Me_3Se^+ or of SeO_2 was obtained by trapping experiments using the apparatus presented in Figure 8 and either H_2 or He as the entraining gas.

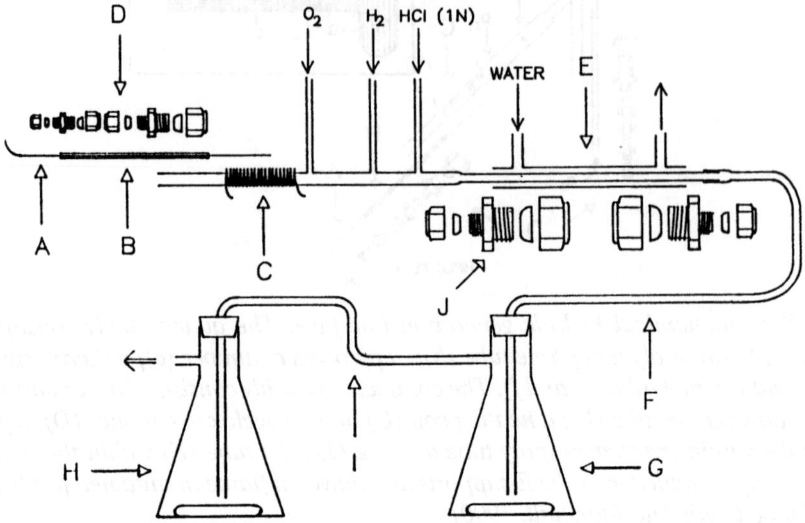

Figure 8. *Pyrolysis apparatus for trapping experiments in which the combustion vapours were reacted with excess H_2, mixed with 1 M HCl, cooled in a condenser (E), entrained through an acidic trap (G) containing 2,3-diaminonaphthalene and into an alkaline trap (H) containing 1-fluoro-2,4-dinitrobenzene*

The product gases emerging from the pyrolysis chamber were mixed with 1M hydrochloric acid, cooled to room temperature in a condenser (E, Fig. 8), and entrained into an acidic solution containing 2,3-diaminonaphthalene (G). Selenium(IV) species entering this trap would be converted to naphthalenepiazselenol. Hydrogen selenide (H_2Se) in the THG product vapours would pass through the acidic solution of trap G and into the alkaline solution in trap H, containing 1-fluoro-2,4-dinitrobenzene. The products, naphthalenepiazselenol in trap G and bis-2,4-dinitrophenyl selenide in trap H, obtained after thermospray pyrolysis of Me_3Se^+ or SeO_2 under reducing (H2) or inert (He) post-pyrolysis atmospheres were isolated and determined by HPLC-THG-QT-AAS. The results are presented in Table 2. Because of the high gas flow-rates in the trapping apparatus, the quantitative recovery of products was not anticipated. These trapping experiments were able to account for some 23 to 31 % of the injected analyte and clearly demonstrated the influence of added

Table 2. Percent recovery of naphthenepiazselenol (NPSe) and of bis-2,4-dinitrophenyl selenide [$(DNP)_2Se$] when Me_3Se^+ or SeO_2 was injected into the pyrolysis apparatus

Analyte	Support Gas	Trap	NPSe	$(DNP)_2Se$
Me_3Se^+	H_2	G	N.D.[1]	N.D.
		H	N.D.	31.2
	He	G	16.7[2]	N.D.
		H	9.3	N.D.
SeO_2	H_2	G	N.D.	N.D.
		H	N.D.	29.6
	He	G	20.4[2]	N.D.
		H	2.9	N.D.

[1] N.D. = none detected
[2] red selenium observed in the transfer line

post-pyrolysis gases. Both SeO_2 and Me_3SeI were converted to H_2Se but only in the presence of H_2.

With the HPLC injector connected directly to the THG interface and operating conditions optimized for the detection of arsonium or selenonium compounds, this detector also provided approximately equal responses to species containing these elements in higher oxidation states (SeO_4^{2-} was a notable exception). Relative responses for different compounds of As and Se are recorded in Table 3. Collectively, these results suggest that the THG interface-detector assembly will detect different oxidation states of both selenium and arsenic efficiently and offers the possibility for the detection arsenic species bound to larger biomolecules.

The operating parameters for the interface-detector were optimized using a multivariate methodology based on a half-replicate 2^5 composite design. These experiments were carried out by recording the average responses (three replicate injections) to 0.4 μg of Me_3SeI or Me_4AsI under different flow rates of O_2 to (1) the pyrolysis chamber, and (2) to the analytical micro-flame; the flow rates of (3) H_2 to the pyrolysis chamber and (4) HPLC column eluate. A fifth parameter was introduced into the model to mimic typical normal- or reverse-phase column eluate composition (0-40 % diethyl ether or 0-40 % water in

Table 3. Relative THG interface responses (± 1 SD)[1] to As and Se compounds

Analyte	Relative Response
tetramethylarsonium iodide	100 ± 1.2
arsenobetaine	98 ± 1.7
arsenocholine	103 ± 1.4
dimethylarsenic acid	106 ± 0.8
arsenic pentoxide	75 ± 2.4
trimethylselenonium iodide	100 ± 1.5
selenocholine tetraphenylborate	104 ± 3.3
selenomethionine	99 ± 2.7
selenium dioxide	101 ± 2.2
sodium selenate	18 + 2.2

[1] Standard deviation based on three replicate injections.

methanol). For these studies, the injector was connected to the interface via a 20-cm length of wide-bore stainless steel tubing (to permit complete mixing of the sample with the mobile phase prior to detection).

Typical predicted response surfaces for Me_4As^+ with the organic eluate (0-40 % diethyl ether in methanol) are presented in Figure 9. Although variations between observed and predicted responses were generally lower than 10 %, the lack of fit tests were significant at α = 95 % for the models. The differences may be attributed, in part, to unmodeled variations in the interface performance under sub-optimal conditions. A rapid accumulation of carbon deposits in the optical tube was observed for 30-40 % ether and low flow rates of O_2 to the pyrolysis chamber (500 mL/min). Additionally, for water rich mobile phases (≥40 % water), a stable thermospray effect was only obtained by repositioning the capillary transfer line within the thermospray heating coil. Despite inadequacies, the accuracy of the model was considered sufficient to determine optimal operating parameters. Calculated limits of detection for selected onium salts of As and Se in the absence of an HPLC column (flow injection technique) are recorded in Table 4.

A cyanopropyl stationary phase and a methanolic mobile phase containing a silanol masking agent (0.05 % triethylamine + 1 % glacial acetic acid) were used to separate

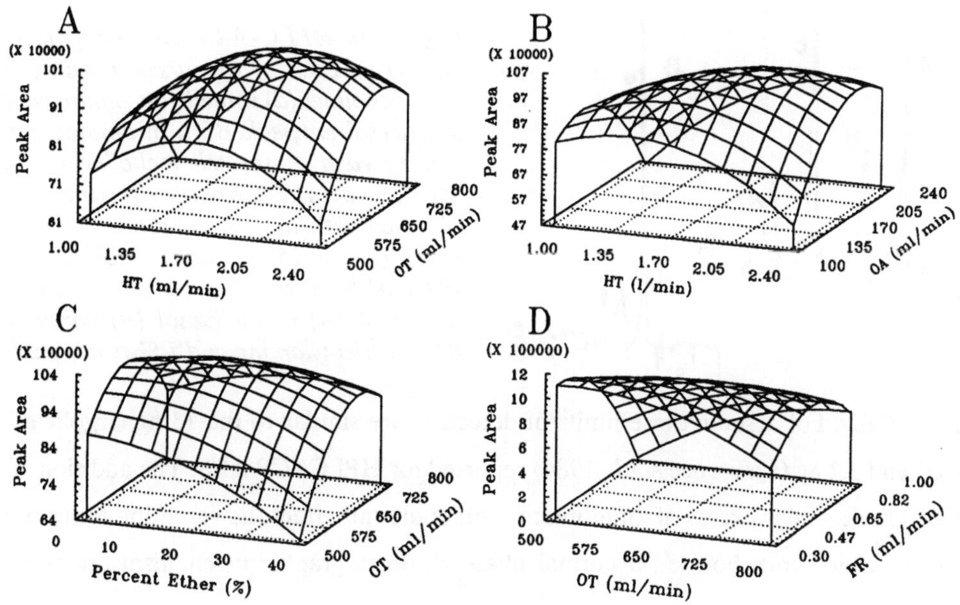

Figure 9. *Predicted variations in detector response (peak area) to Me_4As^+ as a function of: A, H_2 flow rate to the combustion chamber (HT) and O_2 to the combustion chamber (OT); B, HT and O_2 to the analytical flame (OA); C, percent ether in the mobile phase and OT; and D, OT and the flow rate of mobile phase (FR).*

arsonium (arsenobetaine, arsenocholine and tetramethylarsonium) and selenonium (selenocholine and trimethylselenonium) cations using isocratic conditions. Typical chromatograms of standards under conditions of optimal detector operating conditions are presented in Figure 10. Although the two selenonium standards were totally resolved, the Me_3Se^+ was eluted as a relatively broad peak which reduced its limit of detection (Table 4)

Table 4. Calculated limits of detection[1] (ng compound) for selected arsonium and selenonium iodides

	As-betaine	As-choline	Me_4As^+	Se-choline	Me_3Se^+
amount (ng)	22.7(13.3)[2]	26.7(15.1)	14.8(7.6)	59.1(32.4)	88.9(43.9)

[1] Limits of detection were calculated using a model of first order error propagation with noise randomly distributed (Foley and Dorsey, 1984)

[2] Values in brackets are the ng of the corresponding cation.

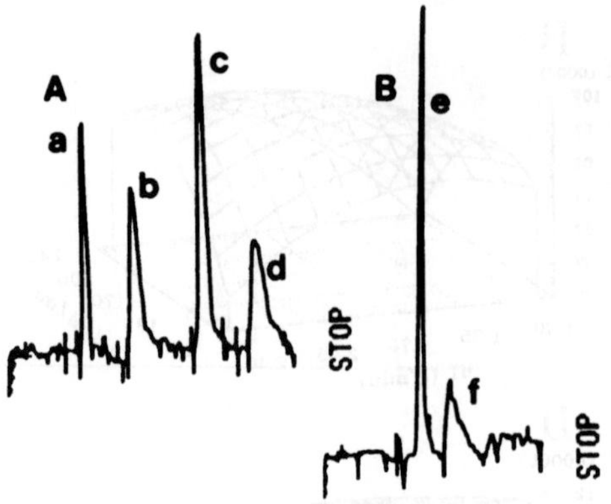

Figure 10. *HPLC-AAS chromatograms of A, 100 ng each of (b), arsenobetaine; (c), arsenocholine; (d) trimethylarsonium cation and (a) solvent peak. B is a chromatogram of 500 ng each of (e) selenocholine and (f) Me_3Se^+ cation. Chromatograms were obtained with a 5μ cyanopropyl stationary phase (0.46 x 15 cm) and a mobile phases (0.65 mL/min) of 30 % (v/v) diethyl ether in methanol (A) or methanol (B) containing 0.05 % triethylamine and 1 % acetic acid*

substantially. For arsenic these limits of detection are similar to the 13 ng (Irgolic *et al.*, 1983) and 3.2 ng (Lawrence *et al.*, 1984) reported for HPLC-ICP-AES. The addition of up to a 500 fold molar excess of ammonium acetate did not affect the retention times of the analytes which corroborated a normal phase chromatographic mechanism under these conditions.

5. Organotin determination

These systems were applied to the determination of alkyltins. The continued intense activity in the area of method development for these analytes and the increased demands for sensitive procedures suggest that alternate procedures are needed. A method that requires no analyte derivatization would be particularly valuable as a screening technique. A method (similar to the approach for alkylleads), in which the analytes would be extracted complexometrically and chromatographed using an ion-pairing technique was envisaged. However, preliminary studies of detector response to tributyltin chloride were not encouraging. Attempts to generate hydride-forming species in the pyrolysis chamber of a THG-interface using O_2, air, or N_2O as the support gas were not successful. Indeed, a maximal response was obtained in the absence of support gases; the analyte was thermosprayed into an electrically heated pyrolysis chamber, which was connected directly to an electrically heated optical tube. With this configuration, LODs approached 2 ng for tributyltin chloride injected into methanol or water/methanol (0.5 mL/min). However, this configuration proved to be especially sensitive to pairing ions in the mobile phase. Either APDC or tropolone added to the water/methanol carrier reduced the response approximately 10-fold whereas the addition of similar concentrations of dithizone or APDC to the water/methanol carrier reduced the response of the direct thermospray interface for lead species by approximately 30 %. Apparently a component(s) of the pyrolyzed mobile phase reduces the concentration of "detectable" Sn species appreciably prior to the atomization. We were unable to improve upon the LODs for ionic alkyltins reported for the HPLC-long silica tube-AAS procedure reported by Kadokami et al. (1988); a post-column derivatization would seem to offer more promise.

Silica tube detection of alkyltin species [hydrides or tetraalkyltin derivatives detected by AAS (Maguire, 1987) or by microwave-induced emission techniques (Besner and Hubert, 1988)] were reported to suffer from varying degrees of peak tailing. Several interface designs were equally susceptible to this phenomenon. The addition of Freon-113 (at 1-4 % in methanol or 30 % aqueous methanol) to the mobile phase (as a possible source of halide radicals) caused only a modest reduction in response but did not perceptibly reduce the tailing phenomenon. Similarly, T-tubes coated with aluminum oxide did not reduce the response and did not alter the severity of the tailing phenomenon.

6. Conclusions

The conditions for optimal detection of trace element species in HPLC eluates will be different for each element. However, by a judicious choice of interface operating parameters, conditions can usually be found that permit the selective determination of trace analyte(s) in biological samples often with little or no sample preparation. The major impediment to further progress remains our lack of understanding of the chemical reactions that occur during the detection process in quartz interfaces. At present, the prediction of optimal interface operating parameters is, at best, an uncertain process. For us, surface response analyses have been especially helpful in designing interfaces and optimizing their performance.

Acknowledgement

Support in the form of a visiting professorship (to WDM) from the National Fund for Scientific Research (N.F.W.O.) is gratefully acknowledged.

References

Ashby J, Clark S, Craig PJ (1988) Methods for the production of volatile organometallic derivatives for application to the analysis of environmental samples. J Anal At Spectrom 3:735-736.

Bayer E, Paulus A (1987) Silanophilic interactions in reverse phase high performance liquid chromatography. J Chromatogr 400:1-4.

Besner A, Hubert J (1988) Effects of dopants on tin emission in a helium microwave-induced plasma. J Anal At Spectrom 3:381-385.

Blais JS, Marshall WD (1986) Determination of alkyllead salts in runoff, soils, and street dusts containing high levels of lead. J Environ Qual 15:255-260.

Blais JS, Marshall WD (1989a) Determination of ionic alkyllead compounds in water, soil and sediment by high performance liquid chromatography - quartz tube atomic absorption spectrometry. J Anal At Spectrom 4:271-277.

Blais JS, Marshall WD (1989b) Post column ethylation for the determination of ionic alkyllead compounds by high performance liquid chromatography - atomic absorption spectrometry. J Anal At Spectrom 4:641-645.

Blais JS, Marshall WD (1990) Determination of arsenobetaine, arsenocholine and tetramethylarsonium by HPLC - thermochemical hydride generation - atomic absorption spectrometry. Anal Chem *in press*.

Blakley CR, Vestal ML (1983) Thermospray interface for liquid chromatography/mass spectrometry. Anal Chem 55:750-754.

Broekaert JAC (1987) Trends in optical spectrochemical trace analysis with plasma sources. Anal Chim Acta 196:1-21.

Brown AA, Roberts DJ, Kahokola KV (1987) Methods for improving the sensitivity in flame atomic absorption spectrometry. J Anal At Spectrom 2:201-204.

Browner RF, Boorn AW (1984a) Sample introduction: the Achilles' heel of atomic spectroscopy. Anal Chem 56:786A-796A.

Browner RF, Boorn AW (1984b) Sample introduction techniques for atomic spectroscopy. Anal Chem 56: 875A-888A.

Burns DT, Glocking F, Harriott M (1981) Investigation of the determination of tin tetraalkyls and alkyltin chloride by atomic absorption spectrometry after separation by gas-liquid chromatography or high performance liquid-liquid chromatography. Analyst 106:921-930.

Dedina J, Rubeska I (1980) Hydride - atomization in a cool hydrogen-oxygen flame using a quartz tube atomizer Spectrochim Acta 35B:119-128.

Deming SN, Morgan SL (1973) Simplex optimization of variables in analytical chemistry. Anal Chem 45:278A-283A.

Ebdon L, Hill S, Ward RW (1987) Directly coupled chromatography - atomic spectroscopy; part 2: Directly coupled liquid chromatography - atomic spectroscopy; a review. Analyst 112:1-16.

Foley JP, Dorsey JG (1984) Clarification of the limit of detection in chromatography.Chromatographia 18:503-511.

Forsyth DS, Marshall WD (1985) Performance of an automated gas chromatograph - silica furnace - atomic absorption spectrometer for the determination of alkyllead compounds. Anal Chem 57:1299-1305.

Honeycutt JB, Riddle JM (1960) Triorganoboranes as alkylating agents. J Amer Chem Soc 82:3051-3052.

Irgolic KJ, Stockton RA, Chakraborti D, Beyer W (1983) Simultaneous inductively coupled argon plasma emission spectrometer as a multi-element-specific detector for high pressure liquid chromatography: the determination of arsenic, selenium and phosphorus compounds. Anal Chim Acta 38B:437-445.

Irgolic KJ, Brinckman FJ (1986) Liquid chromatographic element-specific detection systems for analysis of molecular species. In: Bernhard M, Brinckman FE, Sadler PJ (eds) The importance of chemical "speciation" in environmental processes. Life Sciences Research Reports 33:667-684, Dahlem Konferenzen, Berlin.

Kadokami K, Uehiro T, Morita M, Fuwa K (1988) Determination of organotin compounds in water by bonded-phase extraction and high performance liquid chromatography with long tube atomic absorption detection. J Anal At Spectrom 3: 187-191.

Lawrence KE, Rice GW, Fassel VA (1984) Direct liquid sample introduction for flow injection analysis and liquid chromatography with inductively coupled argon plasma spectrometric detection. Anal Chem 56:289-292.

Maguire RJ (1987) Environmental aspects of tributyltin. Appl Organomet Chem 1:475-498.

Massart DL, Dijkstra A, Kaufman L (1978) Evaluation and Optimisation of Laboratory methods and analytical procedures. Elsevier, Amsterdam.

Myers RH (1976) Response surface methodology. Allyn and Bacon, Boston.

Nygren O, Nilsson CA, Frech W (1988) On-line interfacing of a liquid chromatograph to a continuously heated graphite furnace atomic absorption spectrophotometer for element-specific detection. Anal Chem 60:2204-2208.

Randall (1984) Carbon dioxide based supercritical fluid chromatography - column effencies and mobile phase solvent power. In: Ahuja S (ed) Ultrahigh resolution chromatography, ACS Symposium Series 250:135-169, American Chemical Society, Washington.

Rapsomamikis S, Donard OFX, Weber JH (1986) Speciation of lead and methyllead ions by chromatography/atomic absorption spectrometry after ethylation with sodium tetraethylborate. Anal Chem 58:35-38.

Ricci GR, Colovis G, Hester NE (1981) Ion chromatography with atomic absorption spectrometric detector for determination of organic and inorganic arsenic species. Anal Chem 53:610-613.

Talmi Y, Bostick DT (1975) Determination of alkylarsenic acids in pesticides and environmental samples by gas chromatography with microwave emission spectrometric detection system. Anal Chem 47:2145-2150.

Vlacil F, Hamplova V (1985) A contribution to the description of the retention mechanism of metal chelates during their reverse-phase chromatography. Collect Czech Chem Commun 50:2221-2227.

Weltz B, Melcher M (1983) Investigations on atomization mechanisms of volatile hydride-forming elements in a heated quartz cell. Part 1. Gas phase and surface effects; decomposition and atomization of arsine. Analyst 108:213-224.

Coupling of HPLC and Chemical Reaction Detectors for Trace Analysis of Alkyllead Species

B. Neidhart, M. Blaszkewicz [2], U. Backes [2], and G. Dieckmann [2]

Department of Analytical Chemistry
University of Ulm
Albert-Einstein-Allee 11
D-7900 Ulm, F.R.G.

1. Introduction

Ecological pathways, bioavailability, and the toxicity of a metal are strongly influenced by its chemical binding forms which determine the hydrophilic or lipophilic properties of the compounds. Consequently, metal speciation is indispensable for risk assessment of hazardous and potentially toxic substances. Organolead compounds enter the biosphere mainly due to their worldwide use as antiknock additives in gasoline. Tetraethyl- and tetramethyllead are emitted into the atmosphere during industrial production and handling as well as from car exhausts together with their decomposition products (tri- and dialkyllead species). Environmental degradation takes place, e.g., by photolytic breakdown, reaction with ozone or OH-radicals, and by biodealkylation. The question of biomethylation within the chemical cycle of organolead species in the environment is still controversial.

Although the use of leaded gasoline is restricted by governmental regulations in the United States and in Western Europe, the worldwide annual production of TAL (tetraalkyllead) exceeded 200,000 t in 1988. Therefore it can be expected, that, at least during the next two decades, the "organolead problem" will remain on the agenda, especially concerning ecological pathways, degradation, bioalkylation, and also the effects of a chronic exposure.

Speciation of organolead compounds in the atmosphere, in the aquatic environment, and in biological materials is the only basis on which all the remaining questions can be answered in the future [De Jonghe and Adams, 1982; Fuwa *et al.*, 1982].

[2] Institute of Occupational Health, Central Scientific Unit Analytical Chemistry, Ardeystr. 67, D-4600 Dortmund 1, F.R.G.

NATO ASI Series, Vol. G 23
Metal Speciation in the Environment
Edited by J. A. C. Broekaert, Ş. Güçer, and F. Adams

2. State of the art

All lead and organolead compounds have at least one central lead atom in common, which in many cases determines the detection principle within an analytical procedure. For trace analysis of lead, atomic absorption spectrometry is the method of choice. This fact has led to a multitude of applications for quantitative determination of inorganic lead in various matrices.

The first approaches to the speciation of inorganic and organic lead compounds without chromatographic separation were based on a spectrophotometric detection using specific reagents like dithizone (1,5-diphenylthiocarbazone) [Parker *et al.*, 1961; Henderson and Snyder, 1961; Snyder and Henderson, 1961; Snyder, 1967; Hancock and Slater, 1975; Aldridge and Street, 1981] and PAR (4-(2-pyridylazo)-resorcinol [Pilloni and Plazzogna, 1966; Schmidt and Huber, 1978].

With dithizone for example, trialkyl-, dialkyllead and lead(II) ions can be determined spectrophotometrically by measuring the absorption in chloroform at the different absorption maxima of the three dithizonates. Mathematical treatment of the absorption data allows the quantification of the three groups of compounds in a concentration range of 5 - 25 µg/ml, even if they are coexisting in one sample. This method is advantageous, because three of the four classes of lead compounds in question can be analyzed via complexes with very high molar absorptivities, (60,000 $l \cdot mol^{-1} \cdot cm^{-1}$ for lead(II) and dialkyllead, 23,000 $l \cdot mol^{-1} \cdot cm^{-1}$ for trialkyllead) [Henderson and Snyder, 1961]. Disadvantageous is the low selectivity of dithizone, which also reacts with other metals leading to complex absorption spectra and hereby to difficulties in the mathematical treatment. As the dithizone method works without prior separation steps, speciation within one of the compound classes (e.g. dimethyl-/diethyllead) is impossible. In addition, problems may arise from light sensitivity of the dithizonates and their low solubility in aqueous solutions.

Compared to dithizone, PAR has higher selectivity and specifity for the analysis of lead(II) and divalent organolead compounds and it forms stable, water soluble complexes with these species. The spectrophotometric determination of Pb^{2+} beside R_2Pb^{2+} is possible by masking Pb^{2+} with EDTA. The molar absorptivities are comparably high (40,000 $l \cdot mol^{-1} \cdot cm^{-1}$ for Pb^{2+} and R_2Pb^{2+}) [Shibata, 1967]. However, care must be taken with respect to interferences by other metals. Tri- and tetrasubstituted organolead compounds do not react with PAR.

Further attempts towards the specific determination of organolead compounds were made by combining specific sampling techniques with unspecific detectors as flame AAS [Thilliez, 1967; Harrison *et al.*, 1974; Coker, 1978] or GFAAS [Purdue *et al.*, 1973;

Kashiki *et al.*, 1974; Robinson *et al.*, 1975; De Jonghe and Adams, 1979; Rohbock and Mueller, 1979].

Another possibility for a group specific determination of the four classes of lead compounds is given by a wet chemical extraction procedure [Yamauchi *et al.*, 1981; Yamamura *et al.*, 1981; De Jonghe *et al.*, 1983; Arai, 1986]. This method has the shortcoming that multiple manipulations like extraction, derivatization, reextraction etc. are time consuming and may lead to serious errors.

A "real" speciation of lead(II) and alkyllead compounds is only possible by introducing high performance separation techniques like gas and liquid chromatography.

Tetramethyl- and tetraethyllead, as relatively volatile compounds, can be separated by gas chromatography. Investigations have shown that flame ionization detectors (FID) [Harrison and Laxen, 1978; De Jonghe and Adams, 1983; Du Puis and Hill, 1979] or electron capture detectors (ECD) [Lovelock and Zlatkis, 1961; Potter *et al.*, 1977; Forsyth and Marshall, 1983] have very good sensitivity but little selectivity. A more selective analysis can be achieved by on-line coupling of GC [Nielsen *et al.*, 1981] or LC [Blaszkewicz *et al.*, 1988] with mass spectrometry (MS), plasma AES [Reamer *et al.*, 1978; Sommer and Ohls, 1979; Estes *et al.*, 1980; Estes *et al.*, 1981], or AAS. The combination GC-AAS has turned out to be the most successful of the above mentioned systems [Chau *et al.*, 1976a, b; Robinson *et al.*, 1977; Chau and Wong, 1977; Bye *et al.*, 1978; Chau *et al.*, 1979; Radziuk *et al.*, 1979a, b; De Jonghe *et al.*, 1980a, b; Cruz *et al.*, 1980; Vickrey *et al.*, 1980; Chau and Wong, 1981; Chakraborti *et al.*, 1981; Chau *et al.*, 1983; Andersson *et al.*, 1984; Chau *et al.*, 1984; Chakraborti *et al.*, 1984; Hewitt and Harrison, 1985a, b; Harrison *et al.*, 1985a, b; Van Cleuvenbergen *et al.*, 1985; Hewitt *et al.*, 1986; Radojevic *et al.*, 1986; Van Cleuvenbergen *et al.*, 1986a, b; Rapsomanikis *et al.*, 1986; Blais and Marshall, 1986; Jiang *et al.*, 1987; Chakraborti *et al.*, 1987; Nygren, 1987; Nygren and Nilsson, 1987; Foster *et al.*, 1987; Bai *et al.*, 1987; Allen *et al.*, 1988; Forster and Howard, 1989].

Di- and trialkylated lead compounds can also be separated by GC after off-line pre-column derivatizations; the same is true for Pb^{2+}. The aim of derivatization is the production of tetrasubstituted organolead compounds with sufficient volatility for GC separation. Reagents for derivatization are alkylating substances which can transfer, e.g., propyl or butyl groups to the ionic lead species; hydrogenation is also possible. The named derivatization reagents can be used without any restrictions, because the resulting lead derivates have neither industrial importance nor are they present in the environment. Thus, an unequivocal identification of the different lead species is ensured.

After GC separation the lead species reach the AAS system via an interface and are detected, e.g., in a heated quartz cuvette. The advantages of this procedure are that firstly the whole series of lead species including inorganic lead and alkyllead can be analyzed in one GC run, and secondly that AAS is a very sensitive detection method. Disadvantageous for routine analyses is the apparatus needed, consisting of a complete GC and an AAS system. The pre-column derivatization procedure can possibly disturb the redistribution equilibria and therefore seems to be critical. Additionally uncontrollable reactions can take place in the interface between GC and AAS.

Another approach to lead speciation is LC, especially high performance liquid chromatography (HPLC), without pre-column derivatization [MacCrehan *et al.*, 1977; Koizumi *et al.*, 1979; Vickrey, 1979; Messman and Rains, 1981; Blaszkewicz and Neidhart, 1983; Blaszkewicz *et al.*, 1984; Blaszkewicz, 1984; Ibrahim *et al.*, 1984; Blaszkewicz *et al.*, 1987; Bushee *et al.*, 1987]. For the separation of alkyllead compounds (tetra-, tri-, dialkyllead) reversed phase columns are preferred; but the use of ion exchange materials is also effective [Blaszkewicz *et al.*, 1988]. The problems in HPLC arise on the detector side. The common UV-detector is not selective enough, and consequently it is only useful for extremely clean samples [Bushee *et al.*, 1987]. Additionally, this detector type is not sensitive enough, as it does not allow measurements in the low ng/ml range. These problems can partly be overcome by element specific detectors like flame AAS [Messman and Rains, 1981], GFAAS [Koizumi *et al.*, 1979; Vickrey, 1979], QFAAS [Blais and Marshall, 1989; Blais and Marshall, in press], and AES in plasma [Ibrahim *et al.*, 1984].

However, flame AAS is not sensitive enough. On-line coupling of LC and GFAAS requires a sophisticated apparatus, which is not available commercially. Fractionation and GFAAS measurements are very sensitive but time consuming. The connection of LC to ICP allows continuous measurements with high selectivity but less sensitivity compared to GFAAS. Problems may arise in the stabilization of the plasma if the mobile phase of the LC contains higher amounts of organic solvents.

The combination of HPLC with an electrochemical detector is very sensitive, but not element specific. Due to the reductive operation mode of the detector, this method requires the total exclusion of oxygen from the whole apparatus, which is very laborious to realize [MacCrehan *et al.*, 1977]. In addition, the ElCD shows no response for tetraalkyllead.

Coupling of HPLC with chemical reaction detectors is another possibility for lead speciation [Blaszkewicz *et al.*, 1988; Blaszkewicz and Neidhart, 1983; Blaszkewicz *et al.*, 1984; Blaszkewicz, 1984; Blaszkewicz *et al.*, 1987; Bushee *et al.*, 1987]. One main advantage of this technique is that pre-column derivatization is unnecessary, which means that interferences with the labile chemical equilibria between the alkyllead species, which may occur during the derivatization reaction and which could lead to erroneous results, are excluded.

This continuously operating system reaches a selectivity comparable to GFAAS but is less sensitive. The latter disadvantage can be compensated by suitable enrichment techniques. For environmental measurements enrichment procedures are needed for both techniques.

3. Principles of post-column chemical reaction detectors (PCRD's)

A chemical reaction detector consists of a manifold where the chemical reaction takes place and a physical detector. In principle, the physical detector can be of any type like AAS, ICP, ElCD or other, however, in connection with HPLC, photometers and fluorimeters are mostly used. The chemical reaction system, which is placed between the chromatographic column and the detector can take the form of

i) an open tube (glass) reactor which is run with an air-segmented flow (homogenous reactions),

ii) a packed tube (glass) reactor without air-segmentation (inert packing: homogenous reactions; reactive packing : heterogenous reactions), or

iii) twisted or knitted PTFE capillaries as open tube reactors without air-segmentation.

Increasing selectivity, specifity, and sensitivity are obtained by introducing a chemical reaction system into a detection device. In connection with the chromatographic separation and photometric detection at a single wavelength, continuous isoformation of different chemical species (e.g. tetra-, tri-, and dialkyllead) represents another advantage offered by a PCRD [Frei and Scholten, 1979; Schwedt, 1980; Krull and Lankmayr, 1982; Frei, 1982; Engelhardt and Neue, 1982].

4. Experimental

4.1. Development and optimization of a post-column chemical reaction detector for organolead species

In order to determine lead and organolead compounds by spectrophotometry after LC separation, a colour reagent is needed which reacts with very many lead species under the same conditions, which is inert to other reagents in the reaction system, and which is soluble in the respective mobile phases and solutions of the reaction system. Literature delivers a lot of information on colour reagents which are applicable for the spectrophotometric determination of inorganic or organic lead compounds. For this purpose Table 1 shows the most important analytical reagents, their positive or negative reactivity towards lead species, the absorption maxima, and the suitable solvent for the complex. The 4-(2-pyridylazo)-resorcinol (PAR) reagent best fulfils the demands mentioned. It reacts with inorganic (Pb^{2+}) and organic (R_2Pb^{2+}) ionic lead species (R = alkyl or aryl), and the resulting complexes show very high molar absorptivities. The red complexes are soluble in water, are stable for several hours and their absorption maxima are close enough together to be measured at a single wavelength.

Table 1. *Colour reagents for the speciation of lead compounds. TS_4PP: a,b,c,d-tetra-kis(4-sulfophenyl)porphine [Wang et al., 1980], GHA: glyoxal-bis-(2-hydroxyanil) [Imura et al., 1969], HNN: 1-hydroxy-4-(4-nitrophenylazo)-2-naphthoate [Imura et al., 1971]*

reagent	Pb^{2+}	R_2Pb^{2+}	R_3Pb^+	R_4Pb	λ_{max}/nm	solv.
arsenazo III	+	–	–	–	655	H_2O
TS_4PP	+	–	–	–	464	H_2O
PAR	+	+	–	–	510-520	H_2O
PAN	+	+	–	–	530/555	$CHCl_3$
dithizone	+	+	+	–	540/500/424	$CHCl_3$
GHA	–	+	–	–	680	$CHCl_3$
HNN	–	–	+	–	440	$CHCl_3$

To find out the optimal wavelength for spectrophotometric measurements, the absorbance curves of the PAR - lead species complexes were registered (Figure 1). The evaluation of the curves provides the following maxima (in nm) and the molar absorptivities (in $l \cdot mol^{-1} \cdot cm^{-1}$): for Pb^{2+}-PAR, 512 and 41,000; for Me_2Pb^{2+}-PAR, 514 and 44,000; for Et_2Pb^{2+}-PAR, 518 and 46,000. Most of the following measurements were monitored at 515 nm with a spectrophotometer. However, absorbance can also be measured with a filter photometer on the mercury line at 546 nm.

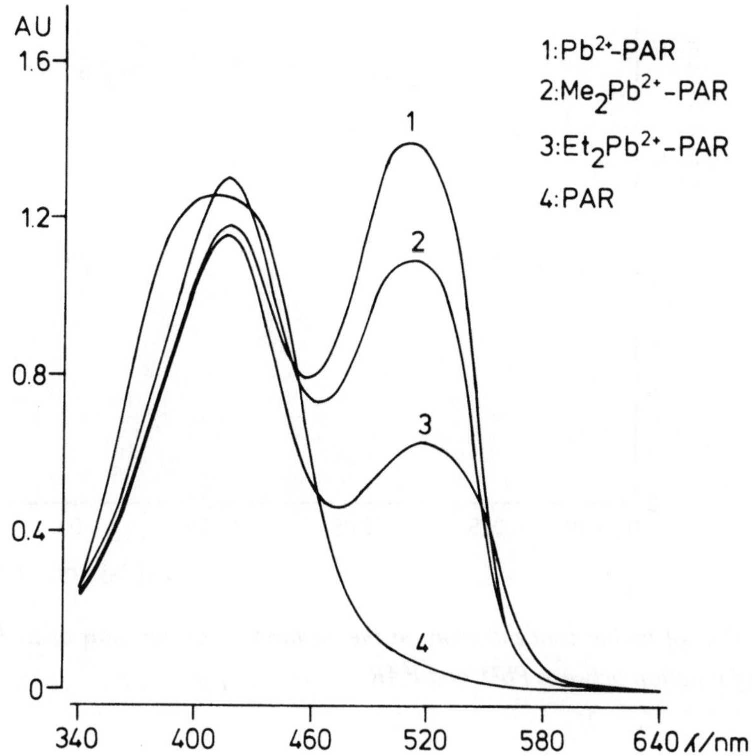

Figure 1. *Absorption spectra of PAR, the Pb^{2+}-PAR-, and the R_2Pb^{2+}-PAR-complex*

4.1.1. Decomposition of organolead compounds

a) Chemical decomposition

The detection of the chromatographically separated lead compounds is performed via the PAR reagent. As PAR forms only complexes with the lead species Pb^{2+} and R_2Pb^{2+}, R_3Pb^+ and R_4Pb have to be transformed into the complexable species. This is performed in the first part of the reaction system, the PTFE reactor TR 1, in which the eluted lead species react with iodine [Blaszkewicz and Neidhart, 1983; Blaszkewicz *et al.*, 1984; Blaszkewicz, 1984]. Iodine as reagent was chosen because of its sufficient reactivity towards organolead compounds, no reactivity towards methanol, and because of its easy handling compared to bromine. The important parameters as iodine concentration, temperature and time for the decomposition reaction were investigated. Figure 2, for example, shows the influence of different iodine concentrations on the signal heights of Pb^{2+} and Et_3Pb^+ species.

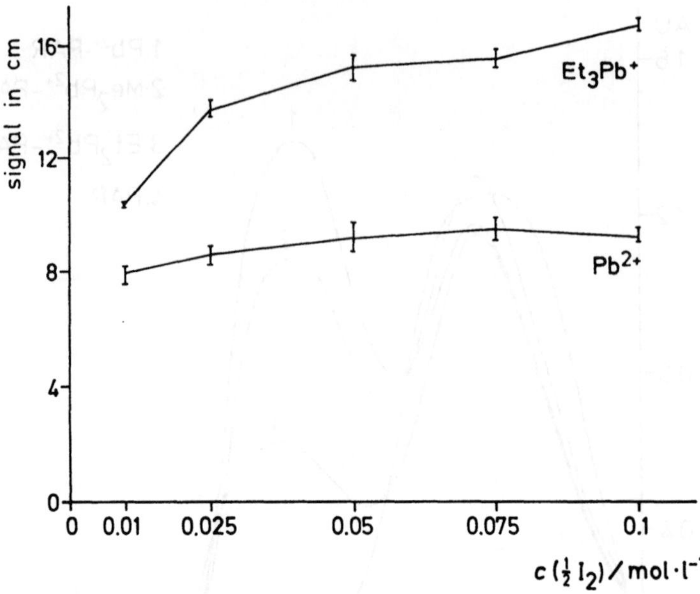

Figure 2. *Effect of iodine concentration on the decomposition reaction of Et_3Pb^+ and on the complex formation between Pb^{2+} and PAR*

The yield of Et_2Pb^{2+} as a decomposition product of Et_3Pb^+ increases with increasing iodine concentrations. A further information from Figure 2 is that high iodide concentrations, produced by the reduction of iodine in the PTFE reactor TR 2, do not interfere with the detection of the Pb^{2+} ion. A precipitation of water-insoluble lead iodide is not noticed in the reaction system. As a reactive halogen iodine reacts with the materials coming into contact with, in this case pump tubing and the peristaltic pump. In order to keep corrosion small, an iodine concentration of c (1/2 I_2) = 0.025 mol/l is used, although the yield of decomposition products is higher at higher concentrations.

The effect of temperature on the decomposition reactions of Et_3Pb^+ and Et_4Pb is shown in Figure 3. A maximum temperature of 50 °C is set as standard condition because the methanol in the mobile phase starts forming gas bubbles in the reaction system at higher temperatures.

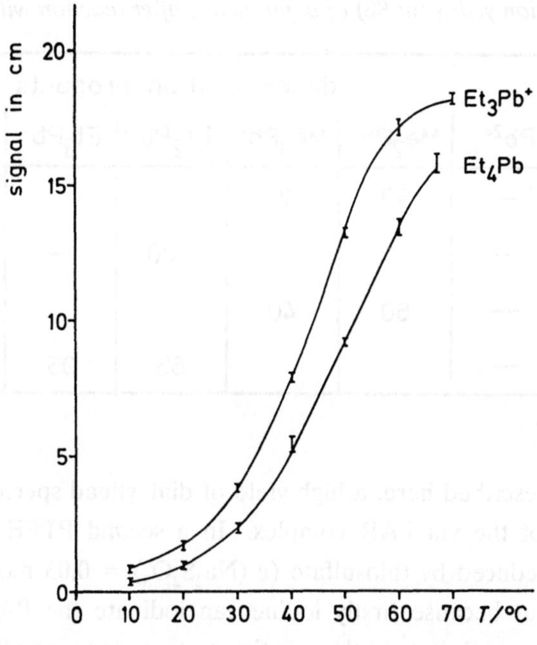

Figure 3. *Effect of temperature on the decomposition reaction by iodine*

The PTFE reactor (TR 1) is such designed that a reaction time of 2.6 min results. An extension of the reaction time is not recommended as the band broadening in the reaction system leads to a decrease of resolution.

Applying the standard conditions of the chemical decomposition by iodine (c (1/2 I_2) = 0.025 mol/l, reaction temperature 50 °C and reaction time 2.6 min), measurements are carried out to investigate the qualitative and quantitative occurrence of different decomposition products. The results are represented in Table 2 for four lead species. In contrast to literature [Hein et al., 1939; Snyder and Henderson, 1961] the decomposition product Pb^{2+} is not found. A complete decomposition of the tetraalkyllead compounds is observed. The reaction stops at the di- and trialkylated derivatives.

Table 2. *Decomposition yields (in %) of organoleads after reaction with iodine*

organolead compounds	decomposition products						
	Pb^{2+}	Me_2Pb^{2+}	Me_3Pb^+	Et_2Pb^{2+}	Et_3Pb^+	Me_4Pb	Et_4Pb
Me_3PbAc	—	93	7				
Et_3PbCl	—			100	—		
Me_4Pb	—	60	40			—	
Et_4Pb	—			65	35		—

For the procedure described here, a high yield of dialkyllead species is important for a sensitive detection of the via PAR complex. In a second PTFE reactor (TR 2) the excessive iodine is reduced by thiosulfate (c ($Na_2S_2O_3$) = 0.03 mol/l). This reduction has to be carried out because firstly iodine can iodinate the PAR reagent, and the reaction products can precipitate in the reaction system, and secondly the deep red color of the iodine solution leads to saturation of the detector at 515 nm.

b) Photochemical decomposition

Beside the use of chemicals, UV absorption by the organolead compounds also leads to a decomposition (photolysis) of the substances. Thus the interaction between radiation and the chemical bonds of the lead species can be used for the detection of organolead species, in analogy to the chemical decomposition. The availability of PTFE reaction coils in photochemical reaction detectors at irradiation wavelengths below 330 nm is based on a diffused radiation transfer and an internal reflectance effect [Scholten *et al.*, 1980a, b]. In order to investigate the photochemical reactions, an irradiation chamber was constructed in which the twisted and additionally coiled PTFE reactor is fixed in a cylindrical form (5.7 m, 0.8 mm inner diameter, 55 mm helix diameter, irradiation time 170 s). A pen-type low-pressure mercury lamp (5 W) is centered into the middle of the PTFE helix. The whole reactor is surrounded by a reflecting aluminum shield. Since the mercury lamp does not develop any excessive heat, even during long operation times, the reactor can be run without cooling. An application of this technique is shown in the chromatogram of Figure 4. Comparing the yields of decomposition products, the iodine reaction is about 30 % more effective than the photochemical reactor. By extending the radiation time and applying lamps with higher intensities, it seems possible to increase the effectivity of the photochemical decomposition.

An aggravating shortcoming of this reactor type in lead speciation is the formation of gas bubbles in the PTFE tubing when the HPLC is run with higher amounts of methanol in the mobile phase, e.g. for elution of tetraalkyllead compounds. These bubbles are transported into the cuvette of the spectrophotometer and lead to great disturbances in the physical detection. In the determination of ionic alkyllead compounds, which are eluted with less methanol in the mobile phase, the photochemical reactor can be applied successfully.

Figure 4. *HPLC chromatogram of organolead compounds after decomposition by UV-light. Nucleosil 5C$_{18}$ (200 x 4), I. 10% MeOH, II. 85% MeOH resp. in 0.1 M acetate buffer, 1 ml/min; PCRD: PAR, pH 9.5, 520 nm, AU 0.1; 1.25 µg Pb^{2+} (1), 1 µg Me_2PbCl_2 (2), 1.5 µg Me_3PbAc (3), 2 µg Et_2PbCl_2 (4), 3 µg Et_3PbCl (7), 3.1 µg Me_4Pb (8), 2.1 µg Me_3EtPb (9), 3.3 µg Me_2Et_2Pb (10), 2.1 µg $MeEt_3Pb$ (11), 3.9 µg Et_4Pb (12) in 250 µl mobile phase II; redistribution products: Me_2EtPb^+ (5), $MeEt_2Pb^+$ (6); solvent peaks: A (injection), B (mobile phase II)*

4.1.2. Color reaction with PAR

4-(2-Pyridylazo)-resorcinol (PAR) is used as the monosodium salt and dissolved in an acetic acid/triethylamine-buffer (pH 10) (c (PAR) = $5 \cdot 10^{-4}$ mol/l). This concentration is sufficient for detecting amounts of Pb^{2+} up to 10 µg (50 nmol)/20 µl injection volume with a linear range of the calibration function when reading the peak height. This range can be extended to 30 µg Pb^{2+}/20 µl when calculating with the peak area. This is also valid for the lead species Me_2Pb^{2+} and Et_2Pb^{2+} and likewise for the tri- and tetraalkyllead compounds, which are detectable only after decomposition to the dialkyllead compounds. The upper dynamic range of the chemical reaction detector including the separation by HPLC is shown in Figure 5 for the Pb^{2+} ion.

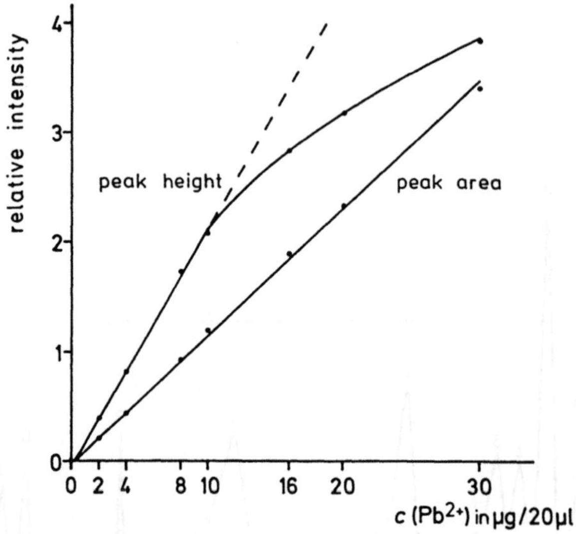

Figure 5. *Dynamic range of the PCRD*

The pH in the reaction system strongly influences the constancy of the baseline during continuous monitoring as well as the formation of the complex. The pH range for the reaction of lead(II) and divalent organolead species with PAR lies between pH 8 and pH 11 with an optimum at pH 9 - 10. The investigations have shown that in this complex mixture, a pH range of 8.7 - 10.2 is suitable to perform measurements with a variation of 5% of the signal intensity. In the reaction system, the pH value is shifted to the optimal range by the addition of alkaline buffer.

Time and temperature are also important parameters for the complexation. The reaction time, which is determined by the dimensions of the PTFE reactors TR 3 and

TR 4 and the flow rate of the reagents, is 30 seconds. This period turned out to be optimal for the formation of the complex.

The influence of temperature on the complexation reaction is different for the Pb^{2+} ion and the organolead species R_2Pb^{2+} (Figure 6). Organolead species are little affected by temperature. The signal intensity slightly declines only at temperatures above 30 °C. The curve of the Pb^{2+} signal shows a maximum at 10 °C. To reach an optimal temperature for the complex formation, the solution, which had to be heated for the decomposition reaction, must be cooled down, e.g. in an ice water bath.

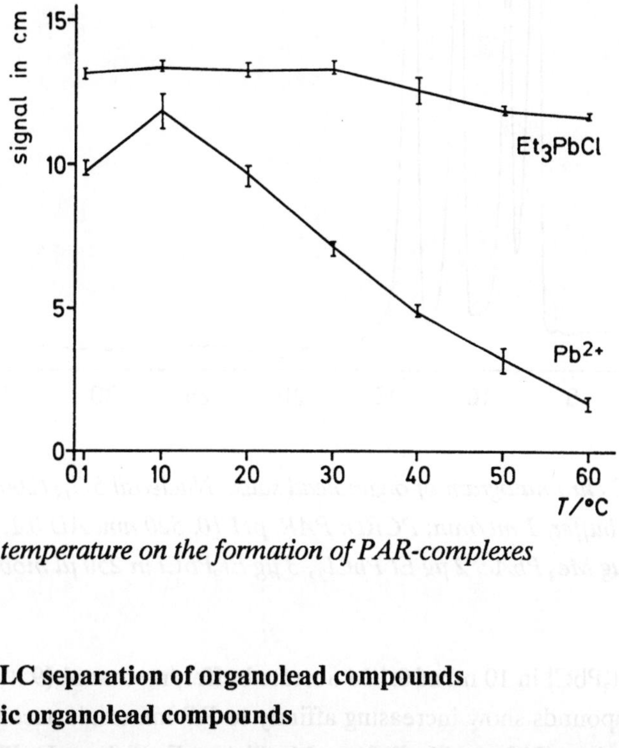

Figure 6. *Effect of temperature on the formation of PAR-complexes*

4.2. HPLC separation of organolead compounds

4.2.1. Ionic organolead compounds

In polar solvents, e.g. aqueous/methanolic buffers, organolead salts dissociate into their respective ions. This is confirmed by chromatography of compounds containing the same organolead but different counter ions. In aqueous buffer solutions with 10 - 20 % of an organic solvent like methanol, ethanol, or acetonitrile, organolead salts are fairly

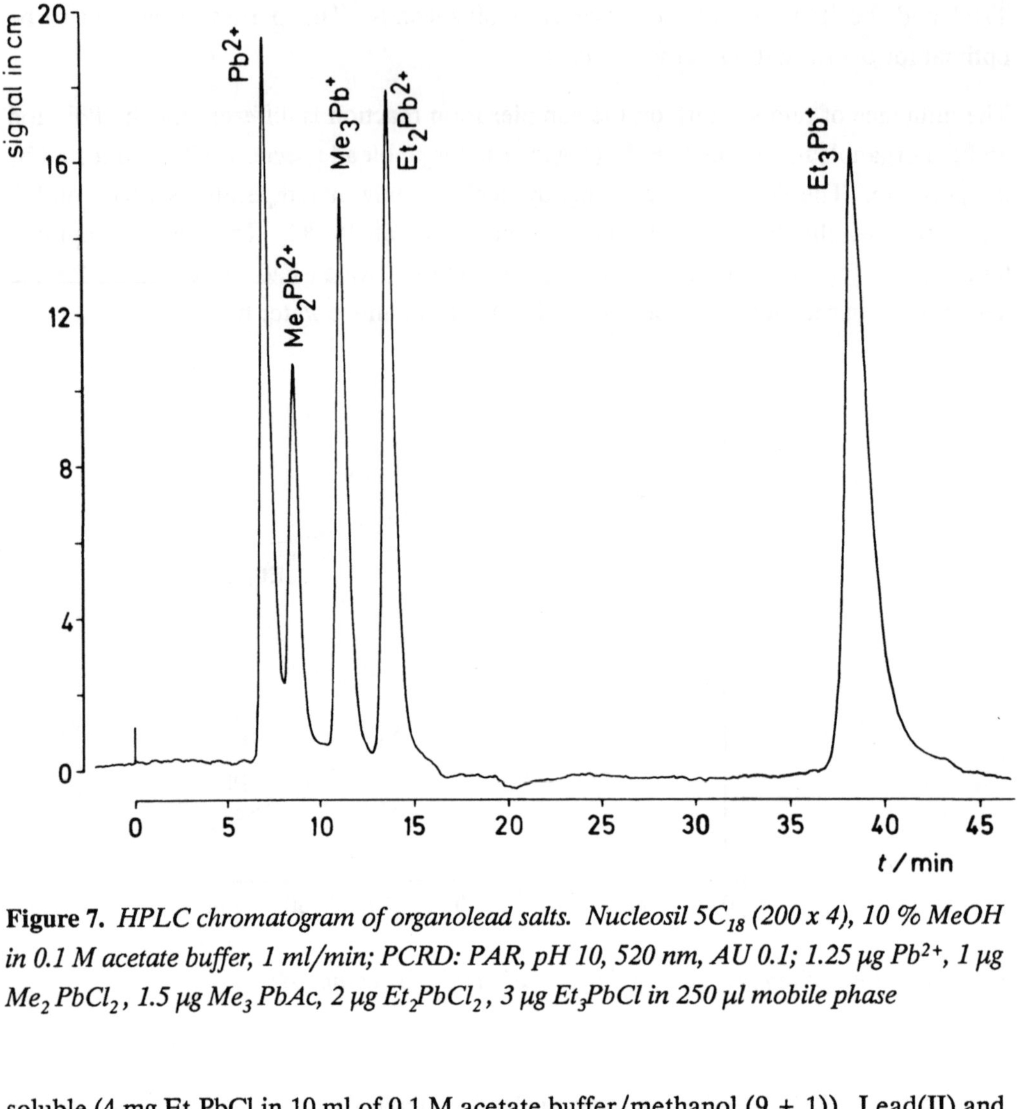

Figure 7. *HPLC chromatogram of organolead salts. Nucleosil $5C_{18}$ (200 x 4), 10 % MeOH in 0.1 M acetate buffer, 1 ml/min; PCRD: PAR, pH 10, 520 nm, AU 0.1; 1.25 µg Pb^{2+}, 1 µg Me_2PbCl_2, 1.5 µg Me_3PbAc, 2 µg Et_2PbCl_2, 3 µg Et_3PbCl in 250 µl mobile phase*

soluble (4 mg Et_3PbCl in 10 ml of 0.1 M acetate buffer/methanol (9 + 1)). Lead(II) and organolead compounds show increasing affinity to RP-materials according to the length of their alkyl chains: (Pb^{2+} < Me_2Pb^{2+} < Me_3Pb^+ < Et_2Pb^{2+} < Et_3Pb^+); the elution of these species from RP-columns is analogous. As an example Figure 7 shows the separation of the 5 species using methanolic acetate buffer. The selectivity of the separation is mainly determined by the methanol content of the mobile phase. The type of RP-material, the dimensions of the column, the ionic strength, and the pH of the mobile phase play a minor role. Increasing methanol concentrations shorten the retention times. At very low salt concentrations their influence on the retention times of Et_3Pb^+ and Me_3Pb^+ becomes obvious (Figure 8).

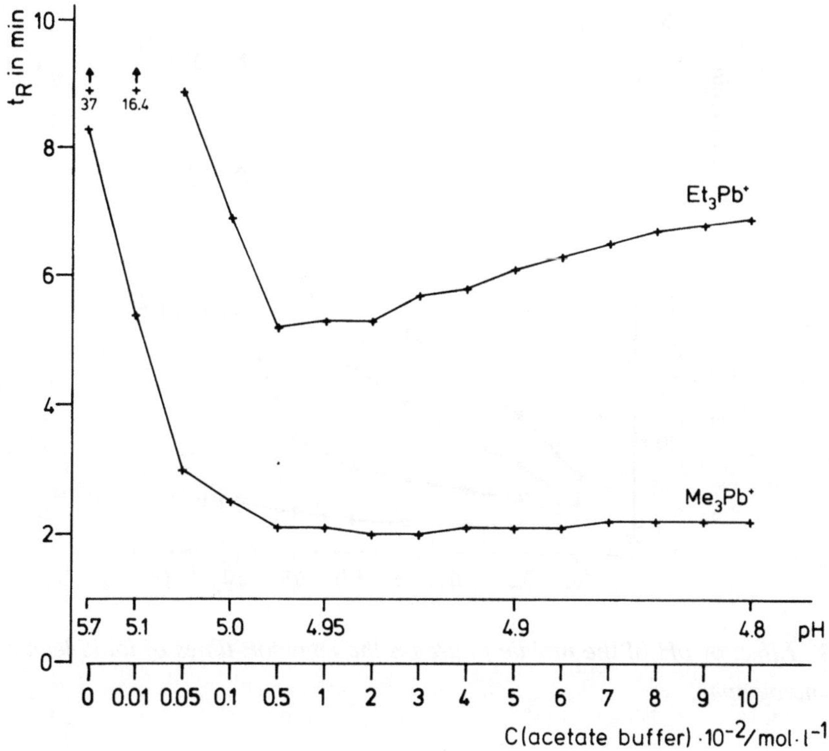

Figure 8. *Effect of concentration of the acetate buffer on the retention times of ionic lead species in RP chromatography*

An even stronger influence can be shown for the pH value (Figure 9). Another important aspect in chromatographic separation of trace amounts of substances is the correlation between resolution (R) and injection volume. It turned out that under standard conditions a volume of 250 µl can be injected under the demand of R = 1. Problems may arise from the Pb^{2+} ion which either elutes as a broad band or together with the solvent front, depending on the composition of the mobile phase.

In general, concerning the whole chromatographic system when analyzing real samples, one should be aware of the fact that slight variations of the methanol content of the mobile phase, temperature changes, and decreasing column efficiency may lead to variations of the retention times.

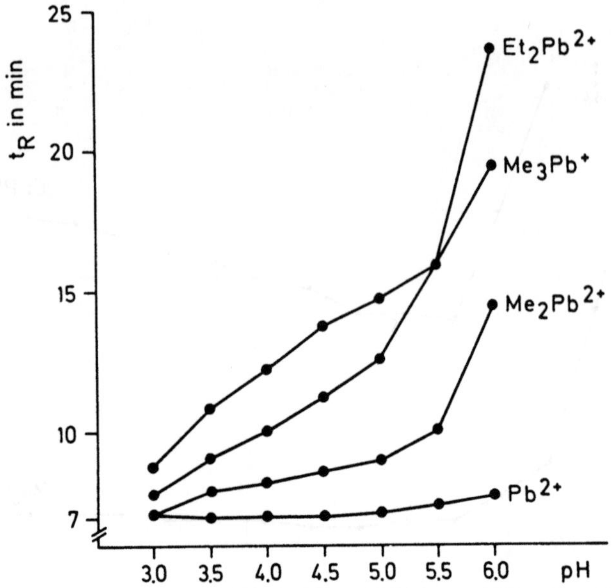

Figure 9. *Effect of pH of the mobile phase on the retention times of ionic lead species in RP chromatography*

4.2.2. Tetraalkyllead compounds

Tetraalkyllead species are nonpolar and thus practically insoluble in water and aqueous buffers. In contrast, they are well soluble in organic solvents like methanol or toluene. With a concentration of 60 % methanol the employed 0.1 M acetate buffer dissolves at least 30 mg of the species of interest in 20 ml. Analogous to the ionic organolead species, the TAL's show an RP-affinity which increases with the length of the alkyl chains. The sequence of elution is Me_4Pb, Me_3EtPb, Me_2Et_2Pb, $MeEt_3Pb$, Et_4Pb. Independently from the type of RP-material used, separation of the TAL's can easily be optimized by variation of the methanol content of the 0.1 M acetate buffer. Very good chromatograms with high peak symmetry can be achieved, e.g., on Spherisorb ODS II (Figure 10).

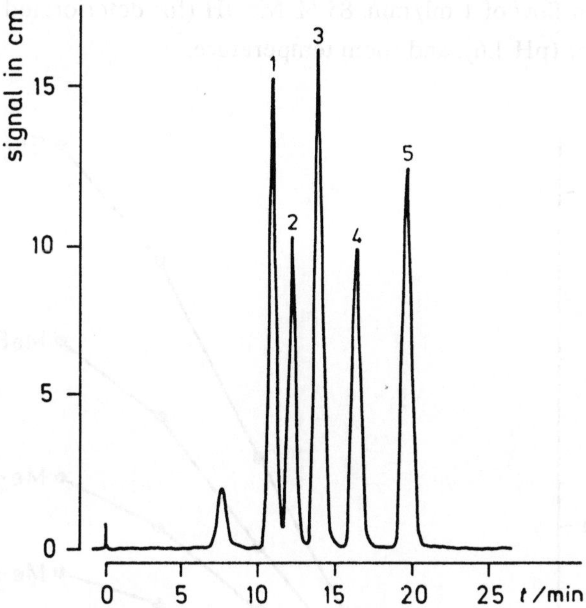

Figure 10. *HPLC chromatogram of tetraalkyllead species. Spherisorb ODS II 5 μm (250x4.6), 90% MeOH in 0.1 M acetate buffer, 1 ml/min; PCRD: PAR, pH 9.5, 515 nm, AU 0.25; 3.6 μg Me_4Pb (1), 2.1 μg Me_3EtPb (2), 3.2 μg Me_2Et_2Pb (3), 2.3 μg $MeEt_3Pb$ (4), 3.9 μg Et_4Pb (5) in 20 μl mobile phase*

In contrast to the ionic lead species, the chromatographic behaviour of TAL's is not very much influenced by the ionic strength of the mobile phase. Instead of methanol, also ethanol, n-propanol, acetonitrile etc. can be used in the mobile phase. The retention times of TAL's are effected by the buffer concentration and the pH value, similar to those of organolead ions, but less extensive.

Separations with sufficient resolution are reached by mobile phases with pH values of 5.5 or higher (Figure 11). Beside the parameters mentioned above, the influence of temperature on the separation of TAL's is also important (Figure 12), and as in the case of the ionic species a maximum volume of 250 μl can be injected without loosing too much in peak resolution. Standard conditions for the separation of TAL's on Nucleosil

$5C_{18}$ (200 x 4) are a flow of 1 ml/min, 85 % MeOH (for deteriorated columns 80 %) in 0.1 M acetate buffer (pH 4.6), and room temperature.

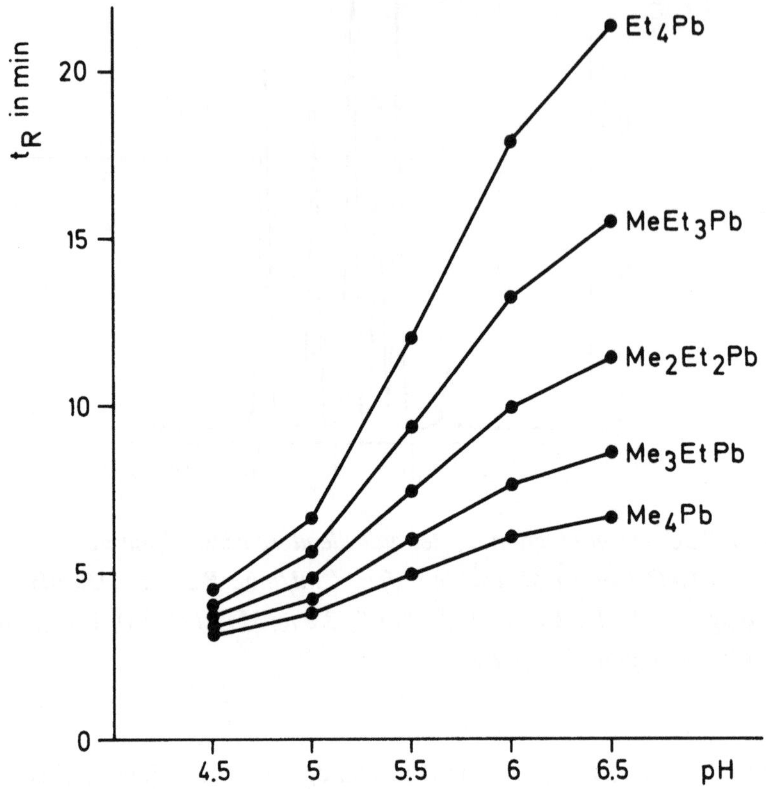

Figure 11. *Effect of pH of the mobile phase on the retention times of tetraalkyllead species in RP chromatography*

4.2.3. Organolead salts and TAL's

Because of the great differences in polarity of all organolead species under investigation, it is not possible to achieve their separation within an acceptable time by isocratic elution. The problem is solved by applying a gradient; however, a continuous gradient interferes with the chemical reaction detector and leads to an instability of the baseline. On the other hand a step gradient produces a peak-like signal in the detector. Under optimized conditions this peak is placed in the chromatogram between the organolead peaks and thus does not influence the evaluation of the signals. An example for the separation of organolead salts and TAL's in one chromatographic run using a step gradient is demonstrated in Figure 13.

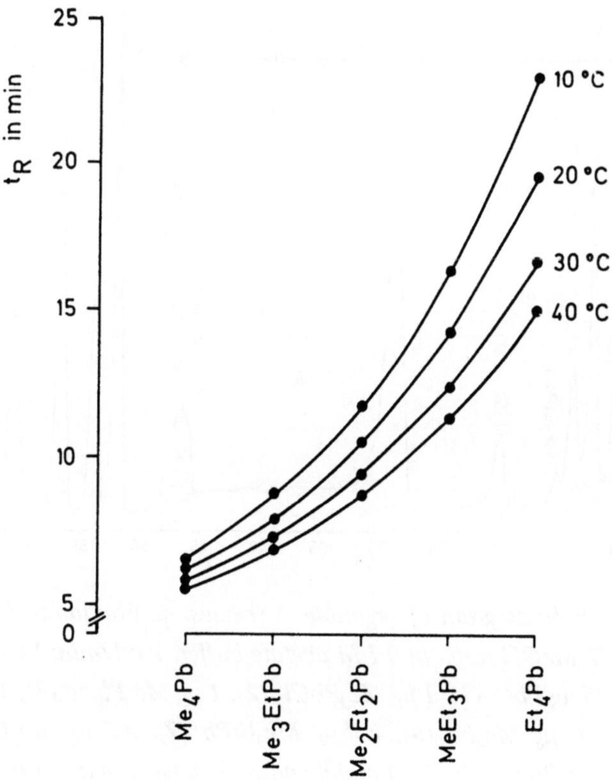

Figure 12. *Effect of temperature on the retention times of tetraalkyllead species in RP chromatography*

4.2.4. Phenyllead compounds

The described HPLC system can also be used for the separation and detection of phenyllead compounds in the presence of all the other alkylleads. Since phenyllead ions are rather unpolar, their separation requires a mobile phase containing at least 60 % methanol in 0.1 M acetate buffer. The separation of TAL's and tetraphenyllead is possible on Nucleosil $5C_{18}$ as well as on Shperisorb ODS II 5 µm, an example of which is shown in Figure 14.

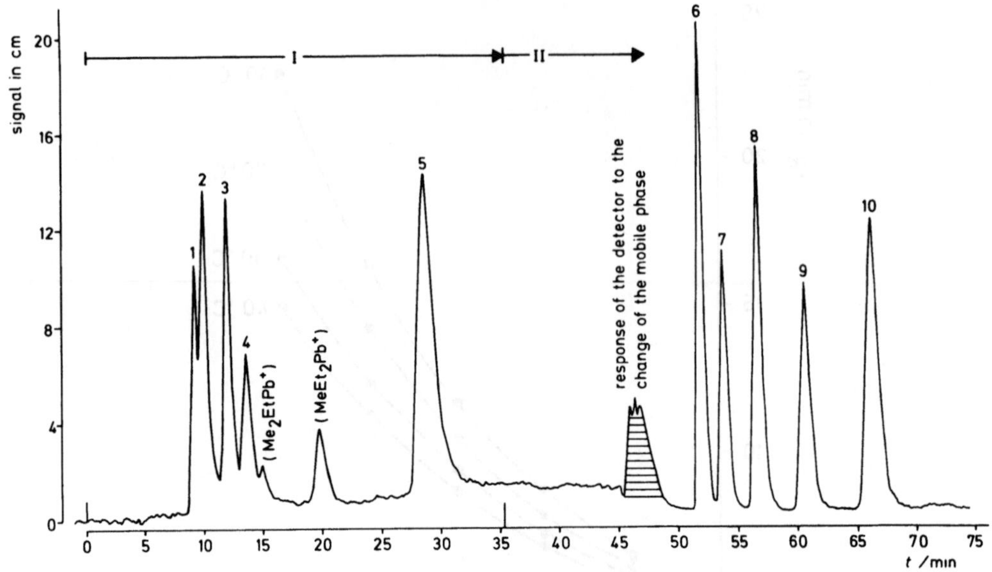

Figure 13. *HPLC chromatogram of organolead species. μ Bondapak $10C_{18}$ (300 x 3.9), I.10 % and II. 80 % MeOH resp. in 0.1 M acetate buffer, 1 ml/min; PCRD: PAR, pH 10, 520 nm, AU 0.1; 1.25 μg Pb^{2+} (1), 1 μg Me_2PbCl_2 (2), 1 μg Me_3PbAc (3), 1 μg Et_2PbCl_2 (4), 3 μg Et_3PbCl (5), 3.7 μg Me_4Pb (6), 2.1 μg Me_3EtPb (7), 3.3 μg Me_2Et_2Pb (8), 2.3 μg $MeEt_3Pb$ (9), 4.7 μg Et_4Pb (10) in 20 μl mobile phase I; solvent peak: A (mobile phase II)*

In this case the sample was prepared from a Ph_4Pb solution in toluene. Although the chemical reaction detector is unsensitive for toluene, a large peak appears in the chromatogram. This fact arises from physical interferences, e.g. changes in density of the mobile phase, as the toluene content of the sample was about 7 %. In contrast to a Nucleosil stationary phase, Ph_4Pb is eluted before Et_4Pb on Spherisorb. This clearly shows the possibilities to obtain different chromatographic selectivities by slightly varying the HPLC parameters.

4.3. Sampling procedures and field measurements

4.3.1. Ionic lead species in water

In order to apply the described analytical procedure to the determination of organolead species in environmental samples, enrichment steps had to be developed because the concentrations of these compounds in the environment are very low. In order to monitore Me_3Pb^+ and Et_3Pb^+ in water samples like rain, melted snow and surface water, a two-step enrichment was used (Figure 15).

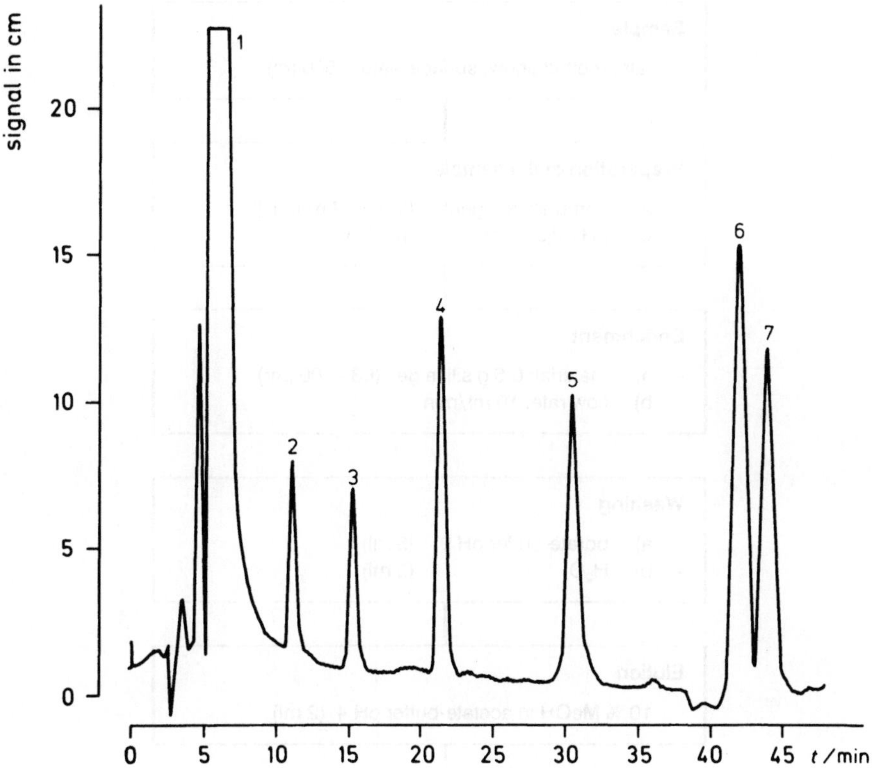

Figure 14. *HPLC chromatogram of alkyl- and phenyllead species. Spherisorb ODS II 5 µm (250 x 4.6), 80 % MeOH and 1 % DMSO in 0.1 M acetate buffer, 1 ml/min; UV-detection: 240 nm, AU 0.05; 17 mg toluene (1), 0.9 µg* Me_4Pb *(2), 0.5 µg* Me_3EtPb *(3), 0.8 µg* Me_2Et_2Pb *(4), 0.6 µg* $MeEt_3Pb$ *(5), 2.5 µg* Ph_4Pb *(6), 1.2 µg* Et_4Pb *(7) in 250 µl mobile phase*

The first part of the procedure is a solid-phase extraction on silica gel, followed by pre-column enrichment. The precolumn, which is integrated in the HPLC system, allows the injection of 5 ml of eluate resulting from the solid-phase extraction. Starting with 500 ml sample volume, this two-step enrichment combined with the HPLC and the chemical reaction detector reaches detection limits of 15 pg/ml for Me_3Pb^+ and 20 pg/ml for Et_3Pb^+. The investigation of several water samples taken from different locations shows the presence of trimethyl- and triethyllead in the concentration range of 20 - 100 pg/ml [Blaszkewicz *et al.*, 1987].

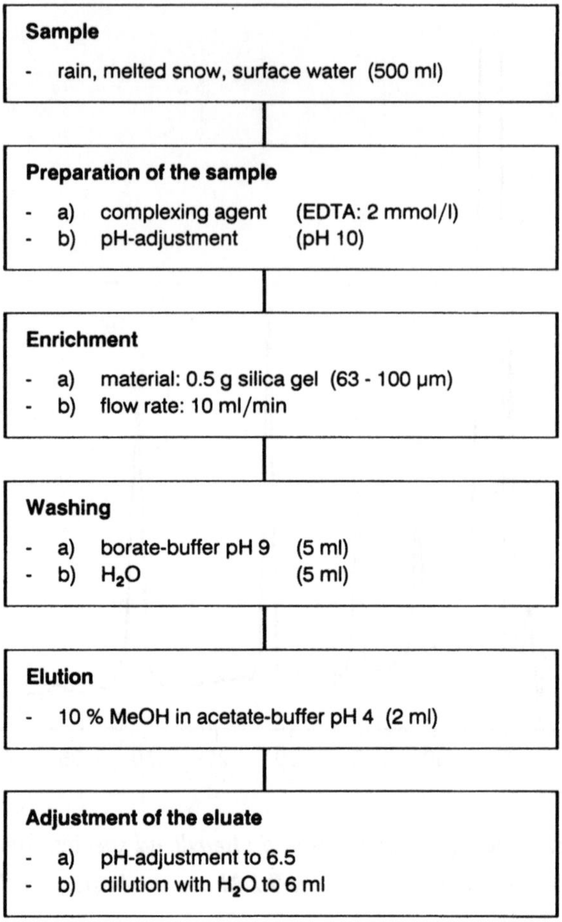

Figure 15. *Trace enrichment of ionic organolead species from water samples*

4.3.2. TAL's in air

The expected low concentrations of TAL's in air made it necessary to develop a special sampling technique, which allows the enrichment of TAL's from some m^3 within an acceptable time. In addition, the enrichment factor should be in the order of $>10^5$ to enable measurements with the above described HPLC system [Backes *et al.*, 1989]. These demands are fulfiled by a two-step enrichment procedure. In the first step the gaseous compounds are absorbed in n-propanol and in the second step they are further concentrated on a reversed phase column. A scheme of the sampling device is shown in Figure 16. It mainly consists of a vertical double glass coil (2) (l = 2.5 m; 34 windings; i.d. = 4 mm), a gas membrane pump, a peristaltic pump, and a gas meter.

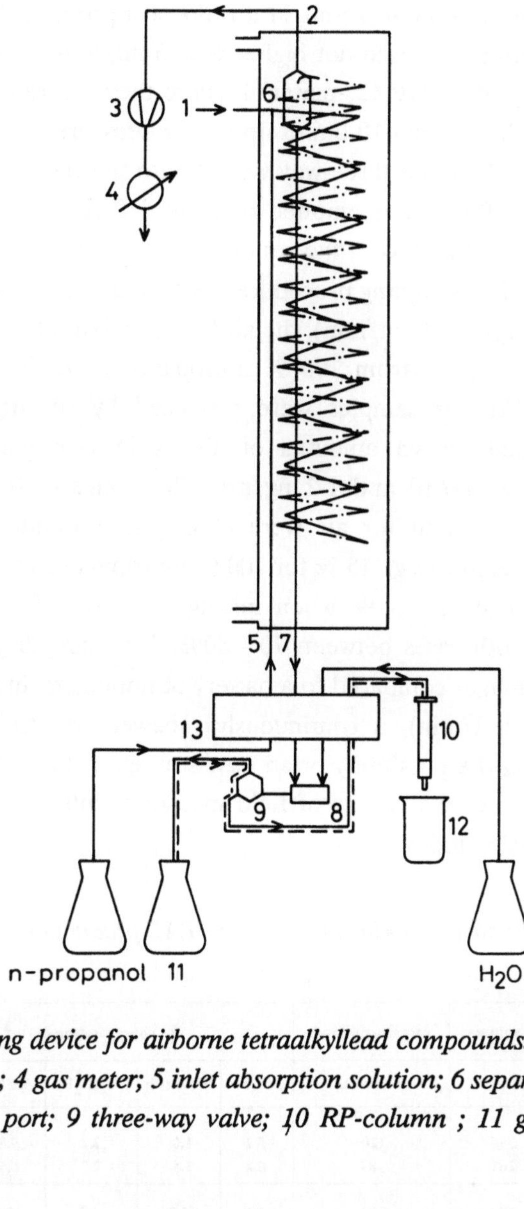

Figure 16. *Sampling device for airborne tetraalkyllead compounds. 1 air inlet; 2 air outlet; 3 membrane pump; 4 gas meter; 5 inlet absorption solution; 6 separator; 7 outlet absorption solution; 8 mixing port; 9 three-way valve; 10 RP-column ; 11 glass flask; 12 waste; 13 peristaltic pump*

The absorption solution is propelled by the peristalic pump 13 through the inlet 5 and forms a thin film when flowing down and up the glass coil. The air is sucked by the membrane pump 3 through inlet 1, then flows parallel to the liquid film, and leaves the glass coil through outlet 2 after having passed the separator 6. From the separator the absorption solution, in which the TAL's are enriched, is pumped by the peristaltic pump through the outlet 7 to the mixing port 8.

Water is added here continuously to result in a ratio of n-propanol:H_2O = 1:5 in the liquid mobile phase. With a flow rate not higher than 3 ml/min the dissolved analytes are pumped over 500 mg of an RP C_{18}-material where they are retained. If, during a longer sampling period, the column 10 builds up a back pressure, the excessive absorption solution is backflushed via the three-way valve 9 into the flask 11. In that case, as the last step of sampling, flask 11 is emptied over the RP-column into the waste 12. Prior to the final analysis of the TAL's, the compounds are eluted from the RP-column with 3 ml of n-propanol. The sampling procedure device was optimized for the flow rate of the absorption solution (0.5-1.0 ml/min), the air flow (2-4 l/min) and the composition of the absorption solution (e.g. from 100 % n-propanol to 70 % n-propanol/30 % methanol). Standard TAL air samples were produced by sucking air through an impinger which contained known amounts of TAL's in n-propanol; the resulting concentrations were 14-60 µg/m³ and 1-2 ng/m³. Flow rates of 0.5 ml/min for the absorption solution and 4 l/min for air were chosen as optimum conditions. The recovery for TML is always low (e.g. 15 % for 100 % n-propanol as absorption solution) but can be increased to about 25 % when mixing n-propanol/methanol 70:30; the recovery for TEL is in both cases between 70 - 80%. The sampling device described here shows several advantages compared to a battery of impingers, in particular high air sampling rates (up to 10 l/min), a continuously renewed absorption solution, high accumulation factors, and the possibility of an easy storage of the final sample (on the RP-column). Table 3 shows the results of field measurements performed at working places of the TAL producing industry.

Table 3. *Tetraalkyllead in air at working places of the TAL production*

location	sample	air volume	sampling period	concentration µg/m³ (as Pb)					
	number	liter	minutes	Me_4Pb	Me_3EtPb	Me_2Et_2Pb	$MeEt_3Pb$	Et_4Pb	Pb^{2+}
gas washer	1	355	120	5.2	1.0	2.2	5.0	10.4	1.4
	2	261	87	6.2	1.8	2.7	6.6	10.4	0.8
mechanical workshop	3	279	103	1.1	<0.3	<0.2	<0.2	0.8	1.3
	4	93	27	5.8	<0.9	<0.6	<0.5	0.6	0.9
	5	95	28	7.5	<0.9	<0.6	<0.5	0.6	1.2
sludge collector	6	230	78	0.7	<0.4	<0.3	<0.2	1.8	0.7
	7	226	72	2.8	<0.4	<0.3	<0.2	1.4	0.1
converter platform	8	215	54	2.7	<0.4	<0.3	<0.2	17.1	2.4
	9	174	56	3.6	<0.5	<0.3	<0.3	46.0	2.2
	10	158	49	3.1	<0.5	<0.4	<0.3	33.8	1.6

MAK-value of Me_4Pb and Et_4Pb: 75 µg/m³ (as Pb)

5. Conclusions

Post-column coupling of a chemical reaction detector to an HPLC system enables the quantitation of organolead species in the lower ng (abs.) region. In combination with an appropriate sampling technique the analytical procedure allows a lead speciation in the environment. As the presented procedure is different from all other methods employed so far, it can be used as an independent method to check the accuracy of alkyllead determinations in environmental samples. Additionally, the procedure, which is simple and easy to perform, should be used for a general monitoring of organolead compounds in the environment, because increasing informations about lead species allow for a better, more differentiated judgement on potential risks and the effectiveness of governmental regulations.

Acknowledgement

The authors kindly acknowledge support by the "Deutsche Forschungsgemeinschaft".

References

Aldridge WN, Street BW (1981) Spectrophotometric and fluorimetric determination of tri- and di-organotin and organolead compounds using dithizone and 3-hydroxyflavone. Analyst 106:60-68

Allen AG, Radojevic M, Harrison RM (1988) Atmospheric speciation and wet deposition of alkyl-lead compounds. Environ Sci Technol 22:517-522

Andersson K, Nilsson C-A, Nygren O (1984) A new method for the analysis of tetramethyllead in blood. Scand J Work Environ Health 10:51-55

Arai F (1986) Determination of triethyllead, diethyllead and inorganic lead in urine by atomic absorption spectrometry. Industr Health 24:139-150

Backes U, Bibernell U, Neidhart B (1989) Air sampling procedure for tetraalkyllead compounds. Fresenius Z Anal Chem 333:706

Bai W, Feng R, Wong H (1987) Study on the species and quantity of alkyl-lead in gasoline by GC in combination with silica furnace AAS. Guangpuxue Yu Guangpu Fenxi 7(2):46-49

Blais JS, Marshall WD (1986) Determination of alkyl-lead salts in run-off, soils and street dusts containing high levels of lead. J Environ Qual 15:255-260

Blais JS, Marshall WD (1989) Determination of ionic alkyllead compounds in water, soil and sediment by high performance liquid chromatography - quartz tube atomic absorption spectrometry. J Anal At Spectrom 4:271-277

Blais JS, Marshall WD (in press) Post column ethylation for the determination of ionic alkyllead compounds by high performance liquid chromatography - atomic absorption spectrometry. J Anal At Spectrom

Blaszkewicz M (1984) Entwicklung eines analytischen Verbundverfahrens zur quantitativen Bestimmung von Organobleiverbindungen im Spurenbereich. Thesis, Wuppertal

Blaszkewicz M, Baumhoer G, Neidhart B (1984) Kopplung von HPLC und chemischem Reaktionsdetektor zur Trennung und Bestimmung von bleiorganischen Verbindungen. Fresenius Z Anal Chem 317:221-225

Blaszkewicz M, Baumhoer G, Neidhart B (1987) Determination of trialkyllead species in water samples. Int J Environ Anal Chem 28:207-214

Blaszkewicz M, Baumhoer G, Neidhart B, Ohlendorf R, Linscheid M (1988) Identification of trimethyllead in urine by high performance liquid chromatography with column switching and chemical reaction detection and by liquid chromatography-mass spectrometry. J Chromatogr 439:109-119

Blaszkewicz M, Neidhart B (1983) A sensitive method for simultaneous determination of airborne organolead compounds. Part 1: Chromatographic separation and chemical reaction detection. Int J Environ Anal Chem 14:11-21

Bushee DS, Krull IS, Smith SB, Schleicher, RG (1987) Determination of organolead compounds by liquid chromatograpy with online extraction and ultraviolet detection. Anal Chim Acta 194:235-245

Bye R, Paus PE, Solberg R, Thomassen Y (1978) Atomic-absorption spectroscopy used as a specific gas-chromatography detector. Comparison of flame and graphite furnace techniques in determination of tetra-alkyl-lead compounds. At Absorpt Newsl 17:131-134

Chakraborti D, De Jonghe WRA, Van Mol WE, Adams FC, Van Cleuvenbergen RJA (1984) Determination of ionic alkyllead compounds in water by gas chromatography/atomic absorption spectrometry. Anal Chem 56:2692-2697

Chakraborti D, Jiang SG, Surkijn P, De Jonghe W, Adams F (1981) Determination of tetraalkyllead compounds in environmental samples by gas chromatography - graphite furnace atomic-absorption spectrometry. Anal Proc 18:347-350

Chakraborti D, Van Cleuvenbergen R, Adams F (1987) Speciation of ionic alkyllead in aerosols by gas chromatography - atomic absorption spectrometry. Int J Environ Anal Chem 30:233-242

Chau YK, Wong PTS (1977) An element-specific technique for the analyses of organometallic compounds. In: Methods and standards for environmental measurement. Proceedings of the 8th IMR Symposium, held Sept. 20-24, 1976, Gaithersburg, Md. National Bureau of Standards, Washington, DC (NBS special publication, Vol. 464)

Chau YK, Wong PTS (1981) Some environmental aspects of organoarsenic, lead, and tin. In: Brinkman FF, Fish RH (eds) Environmental speciation and monitoring needs for trace metal-containing substances from energy-related processes. US Dept. of Commerce, Washington, DC., pp 65-80 (NBS special publication, Vol. 618)

Chau YK, Wong PTS, Bengert GA, Dunn JL (1984) Determination of dialkyllead, trialkyllead, tetraalkyllead and lead (II) compounds in sediment and biological samples. Anal Chem 56:271-274

Chau YK, Wong PTS, Bengert GA, Kramar O (1979) Determination of tetraalkyllead compounds in water sediment, and fish samples. Anal Chem 51:186-188

Chau YK, Wong PTS, Goulden PD (1976a) Gas chromatography-atomic absorption spectrometry for the determination of tetraalkyllead compounds. Anal Chim Acta 85:421-424

Chau YK, Wong PTS, Kramar O (1983) The determination of dialkyllead, trialkyllead, tetraalkyllead and lead II ions in water by chelation/extraction and gas chromatography/atomic absorption spectrometry. Anal Chim Acta 146:211-217

Chau YK, Wong PTS, Saitoh H (1976b) Determination of tetraalkyl lead compounds in the atmosphere. J Chromatogr Sci 14:162-164

Coker DT (1978) A simple sensitive technique for personal and environmental sampling and analysis of lead alkyl vapours in air. Ann Occup Hyg 21:33-38

Cruz RB, Lorouso C, George S, Thomassen Y, Kinrade JD, Butler LRP, Lye J, Van Loon JC (1980) Determination of total, organic solvent extractable, volatile and tetraalkyllead in fish, vegetation, sediment and water samples. Spectrochim Acta 35B: 775 - 783

De Jonghe WRA, Adams FC (1979) The determination of organic and inorganic lead compounds in urban air by atomic-absorption spectrometry with electrothermal atomization. Anal Chim Acta 108:21-30

De Jonghe WRA, Adams FC (1982) Measurements of organic lead in air - a review. Talanta 29:1057-1067

De Jonghe WRA, Adams FC (1983) Gas chromatography with flame ionization detection for the speciation of trialkyl lead halides. Fresenius Z Anal Chem 314:552-554

De Jonghe WRA, Chakraborti D, Adams FC (1980a) Sampling of tetraalkyllead compounds in air for determination by gas chromatography-atomic absorption spectrometry. Anal Chem 52:1974-1977

De Jonghe WRA, Chakraborti D, Adams FC (1980b) Graphite furnace atomic-absorption spectrometry as metal-specific detection system for tetraalkyl-lead compounds separated by gas-liquid chromatography. Anal Chim Acta 115:89-101

De Jonghe WRA, Van Mol WE, Adams FC (1983) Determination of trialkyllead compounds in water by extraction and graphite furnace atomic absorption spectrometry. Anal Chem 55: 1050-1054

Du Puis MD, Hill HH (1979) Analysis of gasoline for antiknock agents with a hydrogen atmosphere flame ionization detector. Anal Chem 51:292-295

Engelhardt H, Neue UD (1982) Reaction detector with dimensional coiled open tubes in HPLC. Chromatographia 15:403-408

Estes SA, Poirier CA, Uden PC, Barnes RM (1980) Gas chromatography with plasma emission spectroscopic detection of Friedel-Crafts catalyzed alkyl group redistribution products among Si, Ge, Sn and Pb atoms. J Chromatogr 196: 265-277

Estes SA, Uden PC, Barnes RM (1981) High-resolution gas chromatography of trialkyllead chlorides with an inert solvent venting interface for microwave excited helium plasma detection. Anal Chem 53:1336-1340

Forster RC, Howard AG (1989) The capillary gas chromatography - atomic absorption spectrometry of organotin and organolead compounds. Anal Proc 26:34-36

Forsyth DS, Marshall WD (1983) Determination of alkyllead salts in water and whole eggs by capillary column gas chromatography with electron capture detection. Anal Chem 55:2132-2137

Foster P, Laffond M, Perraud R, Baussand P, Jacob V (1987) Determination of atmospheric pollutants by gas chromatography - atomic absorption spectrometry or thermodesorption gas chromatography. Int J Environ Anal Chem 28:105-120

Frei RW (1982) Reaction detectors in modern liquid chromatography. Chromatographia 15:161-166

Frei RW, Scholten AHMT (1979) Reaction detectors in HPLC. J Chrom Sci 17:160-167

Fuwa K, Haraguchi H, Morita M, Van Loon JC (1982) A critical review of spectrochemical methods for chemical speciation using element specific detectors with chromatography. Bunko Kenkyu 31:289-305

Hancock S, Slater A (1975) A specific method for the determination of trace concentrations of tetramethyl- and tetraethyllead vapours in air. Analyst 100:422-429

Harrison RM, Laxen DPH (1978) Sink processes for tetraalkyllead compounds in the atmosphere. Environ Sci Technol 12:1384-1392

Harrison RM, Perry R, Slater DH (1974) An adsorption technique for the determination of organic lead in street air. Atmos Environ 8:1187-1194

Harrison RM, Radojevic M (1985) Determination of tetraalkyl and ionic alkyllead compounds in environmental samples by butylation and gas chromatography-atomic absorption. Environ Technol Lett 6:129-136

Harrison RM, Radojevic M, Hewitt CN (1985) Measurements of alkyllead compounds in the gas and aerosol phase in urban and rural atmospheres. Sci Total Environ 44:235-244

Hein F, Klein A, Mesée HJ (1939) Zur titrimetrischen Bestimmung von Blei-Organo-Verbindungen. Fresenius Z Anal Chem 115:177-183

Henderson SR, Snyder LJ (1961) Rapid spectrophotometric determination of triethyllead, diethyllead, and inorganic lead ions, and application to the determination of tetra-organolead compounds. Anal Chem 33:1172-1175

Hewitt CN, Harrison RM (1985a) Total speciation of gas-phase alkyllead in the atmosphere. In: Proceedings of the 5th International Conference on Heavy Metals in the Environment. CEP Consultants, Edinburgh, pp 171-173

Hewitt CN, Harrison RM (1985b) A sensitive specific method for the determination of tetraalkyllead compounds in air by gas chromatography/atomic absorption spectrometry. Anal Chim Acta 167:277-287

Hewitt CN, Harrison RM, Radojevic M (1986) Determination of individual gaseous ionic alkyl-lead species in the atmosphere. Anal Chim Acta 188:229-238

Ibrahim IS, Gilbert W, Caruso JA (1984) Determination of tetraalkyllead by high performance liquid chromatography with ICP detection. J Chromatogr Sci 72:111-115

Imura SI, Fututaka K, Aoki H, Sakai T-N (1971) Spectrophotometric determination of triethyllead in aqueous solution. Japan Analyst 20:704-708

Imura SI, Fututaka K, Kawaguchi T (1969) Spectrophotometric determination of dialkyllead ion in aqueous solution with glyoxal-bis(2-hydroxyanil). Japan Analyst 18:1008-1013

Jiang SG, Chakraborti D, Adams F (1987) Factors influencing sensitivity and accuracy for the determination of alkylselenides and tetraalkyllead compounds by gas chromatography/atomic absorption spectrometry. Anal Chim Acta 196: 271-275

Kashiki M, Yamazoe S, Ikeda N, Oshima S (1974) Determination of lead in gasoline by atomic absorption spectrometry with a carbon rod atomizer. Anal Lett 7:53-64

Koizumi H, McLaughlin RD, Hadeishi T (1979) High gas temperature furnace for species determination of organometallic compounds with a high pressure liquid chromatograph and a Zeeman atomic absorption spectrometer. Anal Chem 51:387-392

Krull IS, Lankmayr EP (1982) Derivatization reaction detectors in HPLC. Int Laboratory 12:12-28

Lovelock JE, Zlatkis A (1961) A new approach to lead alkyl analysis: Gas phase electron absorption for selective detection. Anal Chem 33:1958-1959

MacCrehan WA, Durst RA, Bellama JM (1977) Electrochemical detection in liquid chromatography: Application to organometallic speciation. Anal Lett 10:1175-1188

Messman JD, Rains TC (1981) Determination of tetraalkyllead compounds in gasoline by liquid chromatography-atomic absorption spectrometry. Anal Chem 53:1632-1636

Nielsen T, Egsgaard H, Larsen E, Schroll G (1981) Determination of tetramethyl-lead and tetraethyl-lead in the atmosphere by a two-step enrichment method and gas chromatographic-mass spectrometric isotope-dilution analysis. Anal Chim Acta 124:1-13

Nygren O (1987) High-resolution gas chromatography with graphite-furnace atomic-absorption spectrometry as the detection system. J Anal At Spectrom 2:801-803

Nygren O, Nilsson C-A (1987) Determination of trimethyllead salts in blood using high-resolution gas chromatography - graphite-furnace atomic-absorption spectrometry. J Anal At Spectrom 2:805-808

Parker WW, Smith GZ, Hudson RL (1961) Determination of mixed lead alkyls in gasoline by combined gas-chromatographic and spectrophotometric techniques. Anal Chem 33:1170- 1171

Pilloni G, Plazzogna G (1966) Spectrophotometric determination of diethyllead and diethyltin ions with 4-(2-pyridylazo)-resorcinol. Anal Chim Acta 35:325-329

Potter HR, Jarvie AWP, Markall RN (1977) Detection and determination of alkyllead compounds in natural waters. Water Pollut Contr 76:123-128

Purdue LJ, Enrione RE, Thompson RJ, Bonfield NA (1973) Determination of organic and total lead in the atmosphere by atomic absorption spectrometry. Anal Chem 45:527-530

Radojevic M, Allen A, Rapsomanikis S, Harrison RM (1986) Propylation technique for the simultaneous determination of tetraalkyllead and ionic alkyllead species by gas chromatography/atomic absorption spectrometry. Anal Chem 58:658-661

Radziuk B, Thomassen Y, Butler LRP, Van Loon JC (1979a) A study of atomic absorption and atomic fluorescence atomization systems as detectors in the gas chromatographic determination of lead. Anal Chim Acta 108:31-38

Radziuk B, Thomassen Y, Van Loon JC (1979b) Determination of alkyl lead compounds in air by gas-chromatography and atomic absorption spectrometry. Anal Chim Acta 105:255-262

Rapsomanikis S, Donard OFX, Weber JH (1986) Speciation of lead and methyllead ions in water by chromatography/atomic absorption spectrometry after ethylation with sodium tetraethylborate. Anal Chem 58:35-38

Reamer DC, Zoller WH, O'Haver TC (1978) Gas chromatograph-microware plasma detector for the determination of tetraalkyllead species in the atmosphere. Anal Chem 50: 1449-1453

Robinson JW, Kiesel EL, Goodbread JP, Bliss R, Marshall R (1977) The development of a gas chromatography-furnace atomic absorption combination for the determination of organic lead compounds. Atomization process in furnace atomizers. Anal Chim Acta 92:321-328

Robinson JW, Rhodes L, Wolcott DK (1975) The determination and identification of molecular lead pollutions in the atmosphere. Anal Chim Acta 78:474-478

Rohbock E, Mueller J (1979) Methode zur Messung gasförmiger Bleialkylverbindungen in Außenluft. Mikrochim Acta I:423-434

Schmidt U, Huber F (1978) Spectralphotometrische Bestimmung von Blei (II)-, sowie Dialkylblei- und Trialkylbleiverbindungen in geringen Konzentrationen. Anal Chim Acta 98:147-149

Scholten AHMT, Welling PLM, Brinkman UATh, Frei RW (1980a) Use of PTFE coils in post-column photochemical reactors for liquid chromatography - application to pharmaceuticals. J Chromatogr 199:239-248

Scholten AHTM, Brinkman UATh, Frei RW (1980b) Photochemical reaction detectors in continuous-flow systems - application to pharmaceuticals. Anal Chim Acta 114:137-146

Schwedt G (1980) Chemische Reaktionsdetektoren für die schnelle Flüssigkeits-Chromatographie. Hüthig, Heidelberg, pp 18-26

Shibata S (1967) 2-Pyridylazo compounds in analytical chemistry. In Flaschka HA, Barnard AJ (eds) Chelates in analytical chemistry, Vol. 4. Dekker, New York, pp 116-132

Snyder LJ (1967) Determination of trace amounts of organic lead in air. Composite sample method. Anal Chem 39:591-595

Snyder LJ, Henderson SR (1961) A new field method for the determination of organolead compounds in air. Anal Chem 33: 1175-1180

Sommer D, Ohls K (1979) Emissionsspektroskopie mit induktiv gekoppelter Plasmaflamme als Detektor zur Elementanalyse nach gas-chromatographischer Trennung. Fresenius Z Anal Chem 295:337-341

Thilliez G (1967) Determination of traces of lead in air by atomic absorption spectrometry. Anal Chem 39:427-432

Van Cleuvenbergen RJA, Chakraborti D, Adams FC (1986a) Occurrence of tri- and dialkyllead species in environmental water. Environ Sci Technol 20:589-593

Van Cleuvenbergen RJA, Chakraborti D, Adams FC (1986b) Occurrence of monoalkyllead species during the speciation of organolead. Anal Chim Acta 182:239-244

Van Cleuvenbergen R, Chakraborti D, Van Mol W, Adams F (1985) Ionic alkyllead species in atmospheric fall-out and surface water. In: Proceedings of the 5th International Conference on Heavy Metals in the Environment. CEP Consultants, Edinburgh, pp 153-155

Vickrey TM (1979) Organometallics analyzed by liquid chromatography with GFAA detector. Altex Chromatogram 2(4):2-3

Vickrey TM, Howell HE, Harrison GV, Ramelow GJ (1980) Post column digestion methods for liquid chromatography - graphite furnace atomic absorption speciation of organolead and organotin compounds. Anal Chem 52:1743-1746

Wang WL, Chang C-A, Fan C, Tung S-Y (1980) Use of water soluble porphyrin in analysis. I. Synthesis of meso-tetraphenylpophyrintetrasulfonicacid (TPPS4) and its reaction with lead in spectrophotometric analysis. Fen Hsi Hua Hsueh 8:310-314; cp. (1981) C A 94:202103n

Yamamura Y, Arai F, Yamauchi H (1981) Urinary excretion pattern of triethyllead, diethyllead and inorganic lead in the tetraethyllead poisoning. Industr Health 19:125-131

Yamauchi H, Arai F, Yamamura Y (1981) Determination of triethyllead, diethyllead and inorganic lead ions in urine by hydride generation-flameless atomic absorption spectrometry. Industr Health 19:115-124

SPECIATION OF ORGANOLEAD COMPOUNDS BY GC-AAS

R.J.A. Van Cleuvenbergen, W.D. Marshall[1] and F.C. Adams
Department of Chemistry
University of Antwerp (U.I.A.)
Universiteitsplein 1
B-2610 Wilrijk, Belgium

1. INTRODUCTION : WHY SPECIATE ORGANOLEAD?

The environmental persistence, fate and toxicological impact of trace elements is directly related to the physico-chemical forms in which they occur. From a chemical viewpoint, three classes of trace element containing species may be distinguished: simple inorganic ions (molecules), organically complexed ions (molecules) and element-organic ions (molecules) containing at least one covalent bond between the trace element and carbon. Each of the many physico-chemical forms (species) of a given trace element has its own characteristic environmental distribution and interactive effects - beneficial or toxic - with living organisms. An increasing awareness of this species-specific behaviour over the course of the past two decades has triggered the development of new analytical opportunities - commonly referred to as speciation analysis - and stimulated the creation of techniques combining separation and specific detection which are now well approaching maturity. Thus speciation, in its generally accepted sense, may be defined as the qualitative (identification) and quantitative determination of the individual chemical forms that comprise the total concentration of the trace element in the sample.

It is hardly surprising that lead was one of the first trace metals on which attempts to apply speciation techniques were focused. Starting with the introduction, in 1923, of tetraethyllead as a gasoline additive to prevent premature ignition (knocking) of

[1] on leave from Department of Food Science and Agricultural Chemistry, Macdonald College of McGill, Ste. Anne de Bellevue, Québec, Canada H9X 1C0.

NATO ASI Series, Vol. G 23
Metal Speciation in the Environment
Edited by J. A. C. Broekaert, Ş. Güçer, and F. Adams

the fuel-air mixture in the cylinders of internal combustion engines, a sizable antiknock industry developed. Among the tetraalkyllead compounds particularly tetraethyllead (Et_4Pb) has gained worldwide acceptance. In the seventies the manufacture of tetraethyllead and tetramethyllead reached approximately 3.5 10^5 tonnes Pb annually [Grandjean *et al.*, 1979], which outstripped considerably the production of all but a few other primary chemicals. Yet man and his environment has had to pay a price for this development. It has now been established firmly that the burden of total lead in the environment has been increased dramatically as a result of this one technological application, as evidenced by a variety of geological and archaeological records [Fergusson,1986; Van Cleuvenbergen *et al.*, 1990a]. Moreover this burden has risen globally, despite the fact that the majority of the consumption - and concomitant discharge in the environment - has been restricted to industrialized countries.

The specific form of the compounds entering the biosphere due to the widespread use of organolead was both expected and found to be largely inorganic, reflecting the relative instability of alkyllead species. Nevertheless the assessment of potential hazards from the organic lead fraction which may be discharged during primary manufacture, transfer and filling operations or emitted (along with other combustion products) from the exhaust continues to challenge the environmental scientist. It is now recognized that the acute mammalian toxicity of "organoleads" is from 10 to 100 times greater than that of the inorganic Pb^{2+} ion [Grandjean, 1984]. Moreover these organolead species mediate deleterious changes in biological systems which are distinctly different from classical "plumbism". Thus, although organoleads may comprise only a small fraction of the total lead burden, they may be anticipated to modify the "total lead" toxicity appreciably.

Controversial topics such as the existence and relative importance of lead biomethylation [Wong *et al.*, 1975] or the alleged contribution of the organolead component in wet precipitation to the decline of European forests [Faulstich *et al.*, 1985] have resulted in a continuing interest in the environmental persistence and fate of these products. Nevertheless, many questions remain unresolved concerning the effects of chronic exposure to traces of organoleads or the bioavailability of these compounds to organisms. Several recent reviews have treated the analytical, environmental and toxicological aspects of organolead compounds comprehensively [Grandjean, 1984; Van Cleuvenbergen *et al.*, 1990a].

Table 1 Schematic survey of environmentally important organolead compounds [Van Cleuvenbergen *et al.*, 1990a]

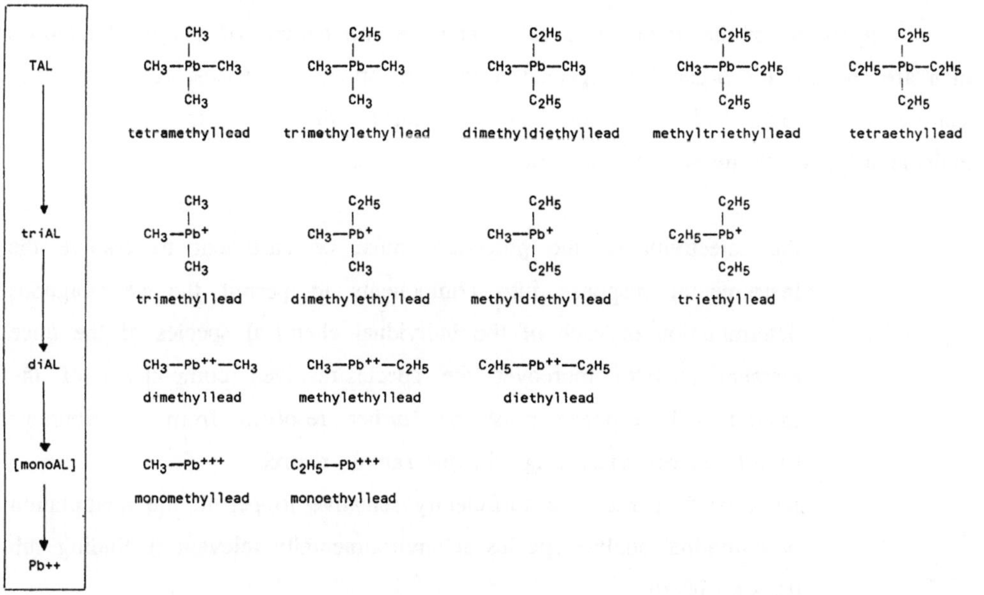

In contrast to other elements, the environmental chemistry of organolead is dominated by a relatively small number of products which are presented in Table 1. These organometallic species may be represented by the general formula $R_nPb^{(4-n)+}(X^-)_{4-n}$ ($R=CH_3$ or C_2H_5, n=1 to 4, X= conventional anionic group). The number of alkyl groups (n) profoundly influences the degree of ionic character of the compound. Virtually all "organolead" speciation studies to date have been restricted to the differentiation of individual species with respect to the number and identity of alkyl substituents bonded to the central lead atom. The identity of the anionic counterion(s) and the characterization of the binding site(s) within the environmental matrix have been ignored. Knowledge of the substituent groups on the lead atom, and their modification without losing the original identity of the species, constitute the major keys to successful speciation.

2. THE METHODOLOGY : STRATEGIC CONCEPTS

A survey of the literature reveals that there is a paucity of analytical methods which are suitable for organolead speciation in environmental matrices. An ideal, truly species-specific analytical detection system must fulfil several requirements simultaneously, which are summarized below:

- the selectivity of the procedure must be sufficient to resolve the instrumental response into components to permit the unambiguous determination of each of the individual chemical species of the trace element analyte; moreover the species-resolved components of the instrumental response must be further resolved from contributions (interferences) originating with the sample matrix
- the procedure must be sufficiently sensitive to permit the quantitation of individual analyte species at environmentally relevant (including sub $\mu g\ kg^{-1}$) levels
- the method should also provide the means of assessing the bioavailability of each chemical species; this implies knowledge of the fraction of the sample matrix to which each of the analyte species was originally associated
- finally, the method should include as few chemical manipulations as possible, thus minimizing the risk of losing the species and/or changing its chemical identity. One must realize that the physical and chemical properties of organometallic compounds have generally not been explored in full detail.

Since the early beginnings of speciation, the use of gas chromatography has proven to be an effective approach, due to its inherently high resolving power and the availability of many selective and sensitive detectors [Ebdon *et al.*, 1986; Cappon,1987]. To be successful, the analyte species must be isolated from the sample in (or converted to) a chemical form amenable to gas chromatographic assay: the analyte or derivative must be sufficiently volatile, thermally stable, resistant to physical (adsorption) and chemical (decomposition) interaction with the stationary phase, and in addition possess suitable physical or chemical properties for selective detection. It follows that the increasing use of liquid chromatography, particularly high performance liquid chromatography, is a logical evolution which considerably expands the type of chemical and physical species that may be studied, often without the need for prior derivatization [Ebdon *et al.*,1987; Marshall, 1988]. Until recently, its use has been somewhat limited by the lower resolving power of this chromatographic technique and especially by difficulties in optimizing the interface with the selective detector systems.

The two types of gas chromatographic or liquid chromatographic detectors which come closest to meeting the idealized characteristics are mass spectrometric and atomic spectrometric systems. Especially the combination of atomic absorption spectrometry with gas chromatography (GC-AAS) has provided a real breakthrough in the general area of speciation chemistry and is, at present, undoubtedly one of the most widely applied techniques; its popularity has to be ascribed to the relatively straightforward interfacing between two instruments which, on their own, are readily available and possess well-known analytical characteristics. In the last five years considerable progress in elucidating the biogeochemical cycle of organolead compounds may be accredited largely to the tandem combination GC-AAS. The current review will summarize and critically discuss the experience gathered while exploring the potential of the technique for alkyllead speciation.

Though often neglected, the sample preparation, which usually serves both the purposes of preconcentration and sample clean-up, is a key component in the speciation procedure. This step is often the most time-consuming to conduct and the most difficult to optimize. This is illustrated by the typical example of a speciation pathway schematized in Figure 1. Usually optimization focuses on the removal of interfering contaminants and stresses the reproducibility and efficiency of analytical recovery, often at the expense of simplicity or speed. Moreover, the specific procedure used is sometimes critically dependent on the complexity of the sample matrix. Thus, the development of a single omnivalent extraction/purification procedure has not yet been realized. The second part of this review article will attempt to outline some remaining problems associated with sample pretreatment, and illustrate the versatility of organolead speciation by discussing a few typical environmental applications from the authors' experience.

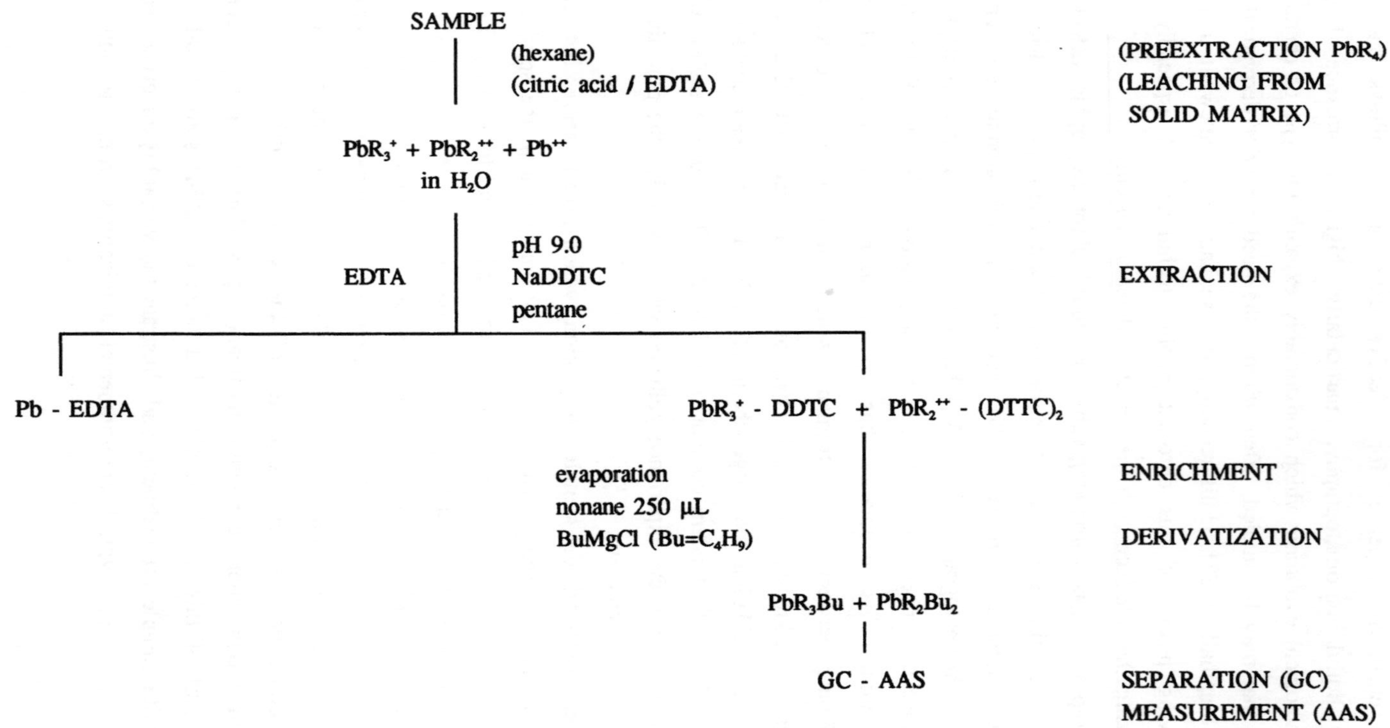

PbR_4 : HEXANE PREEXTRACTION UNDER CONTROLLED CONDITIONS, MEASUREMENT WITH GC-AAS
TOTAL Pb : MEASURED INDEPENDENTLY WITH AAS AFTER WET ACID DIGESTION OF A SAMPLE PORTION

Figure 1 Example of an analytical procedure for speciation of ionic alkyllead in the biosphere [Van Cleuvenbergen *et al.*, 1987]

3. GAS CHROMATOGRAPHY OF ALKYLLEADS: THE ART OF SEPARATING

Initially, organolead speciation was restricted to the tetraalkyllead (TAL) compounds. Their volatility, nonpolar character and sufficient thermal stability make them ideally suited for a gas chromatographic separation. The ionic alkyllead degradation products, on the other hand, being appreciably less volatile, can be chromatographed only with difficulty. A GC-FID (flame ionization detection) chromatogram for a standard mixture of trimethyllead chloride and triethyllead chloride using temperature-programmed conditions is presented in Figure 2. It illustrates well the drawbacks associated with the direct GC separation of this class of compounds: the rather poor peak shape, and the appreciable thermal decomposition leading to TAL redistribution products such as tetramethyllead (TML) and methyltriethyllead (MTEL), are striking and result ultimately in a modest reproducibility. In general, organometallic species suffer from poor peak shapes, displacement and priming effects which tend to become exacerbated at trace levels.

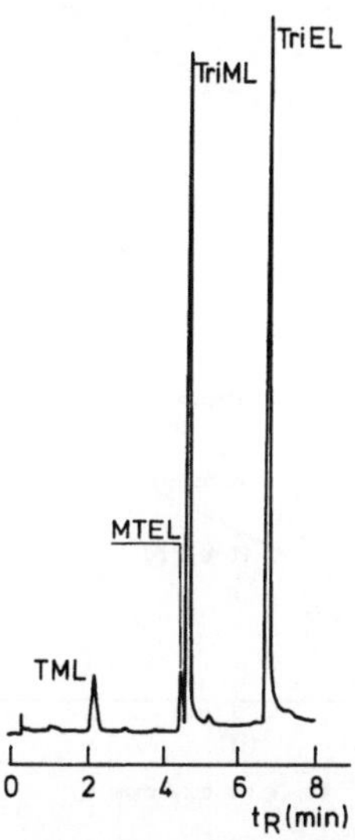

Figure 2 GC-FID chromatogram for a standard mixture of trimethyllead chloride (TriML) and triethyllead chloride (TriEL) using temperature - programmed conditions [De Jonghe *et al.*, 1983a]

A more successful approach is to render the ionic and hydrophilic organolead salts more volatile by derivatization. Of course, the derivative must be one that retains the original integrity of the metal - carbon bonds such that the species identity remains unaltered by the process. Two methods are currently used in the environmental speciation of organometals: hydride formation and alkylation. Taking into account the instability of lead hydrides, particularly those derived from dialkyllead species, and thus the stringent precautions involved in the generation and transport of the hydrides to the detection system, several researchers have experienced alkylation to be a less demanding derivatization technique. Alkylated metals of group IV, $R_nM^{(4-n)+}$, readily accept further alkyl groups from a Grignard or similar reagent to form tetrasubstituted derivatives, $R_nMR'_{4-n}$ which are stable and more volatile than their parent compounds.

The alkylation reaction has been exploited in a variety of ways, including ethylation [Rapsomanikis *et al.*, 1986; Ashby *et al.*, 1988], propylation [Radojevic *et al.*, 1986], butylation [Estes *et al.*, 1982; Chau *et al.*, 1983; Chakraborti *et al.*, 1984] and phenylation [Forsyth *et al.*, 1983], each with its own advantages and disadvantages. Ethylation does not allow complete speciation, as it can transform only methyllead species unambiguously. An interesting development has been the use of an on-column ethylation procedure for the gas chromatographic determination of methylleads [Ashby *et al.*,1988]. Small quantities of sodium tetraethylborate ($NaBEt_4$) applied directly to the head of a GC column were sufficient to derivatize ionic methylleads, resulting in a simplified technique for the determination of these analytes. Propylation and butylation are apparently the best choice at present. As the gas chromatographic behaviour of the tetraalkylated lead derivatives on nonpolar stationary phases appears to be based largely on the molecular weight of the species, retention times can be predicted readily by systematics based on the number of carbon atoms in the alkyl groups sigma bonded to the lead atom [Forsyth *et al.*, 1983; Van Cleuvenbergen *et al.*, 1986b]; an example is presented in Figure 3. If two derivatives have the same total number of carbon atoms, their GC separation is generally difficult, as evidenced by the chromatography of

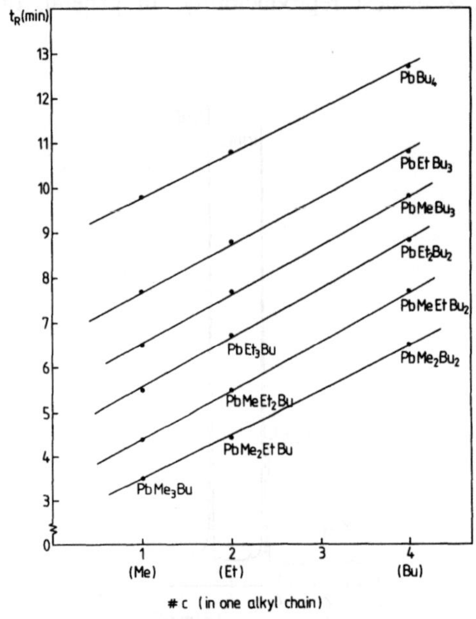

Figure 3 Systematics of gas chromatographic retention time vs. number of carbon atoms in one alkyl chain for butylated ionic alkyllead compounds [Van Cleuvenbergen *et al.*, 1986b]

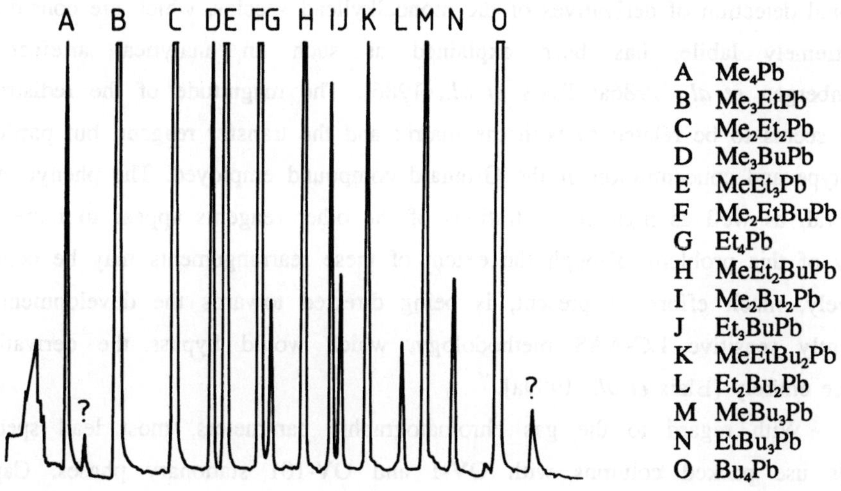

Figure 4 GC-AAS chromatogram for a synthetic mixture of all the tetraalkyllead species containing methyl (Me), ethyl (Et) or butyl (Bu) groups on the lead atom [Blais, 1987]

dimethyldibutyllead and triethylbutyllead (Figure 4). Thus, butylation is likely to be preferred over propylation, as in the latter case several pairs of products (mainly including mixed methyl-ethyl lead species) suffer from this resolution problem. Finally, phenylation has been found to give rise to an increased formation of artefacts [Blais *et al.*, 1986], probably because the intermediate phenyl radical is relatively more stable, thus promoting redistribution reactions.

Although alkylation provides a derivative which is more stable than the corresponding hydride, one disadvantage is that it is not a total sampling technique: the derivatization is carried out in non-aqueous medium of which only a small aliquot is finally injected into the analytical system. The volatility of the tetra-alkylated derivatives restricts the degree to which the final solutions can be concentrated by solvent evaporation; to our knowledge the search for a suitable "keeper" has not been successful. Furthermore there is now ample evidence that during (or after) derivatization minor scale rearrangements (alkyl group redistribution reactions) between the species may occur. The occasional detection of derivatives of the monoalkyllead species, which are considered to be extremely labile, has been explained as such an analytical artefact [Van Cleuvenbergen *et al.*, 1986a; Blais *et al.*, 1986]. The magnitude of the redistribution process seems to be related to both the matrix and the transfer reagent, but particularly to the type and concentration of the Grignard compound employed. The phenyl reagent, in general, as well as high concentrations of the other reagents appear to increase the severity of this problem. Though the extent of these rearrangements may be controlled effectively, much effort, at present, is being directed towards the development of a sufficiently sensitive LC-AAS methodology, which would bypass the derivatization sequence entirely [Blais *et al.*, 1989a].

With regard to the gas chromatographic parameters, most lead speciation methods use packed columns with OV-1 and OV-101 stationary phases. Capillary column (narrow bore fused silica) applications using various detection systems have recently gained popularity [Cappon, 1987; Forsyth, 1987]. Capillary column gas chromatography appears to be fully compatible with the common speciation detectors; the response of modern atomic absorption instruments, for example, is sufficiently rapid to allow an adequate processing of the chromatographic output. Optimal use of a capillary column tends to result in an improved resolution and sensitivity, combined with a considerable gain in analysis time. The main restriction, however, is the limited sample capacity (injection volume) of these columns, a factor which, at present, favours the continued use of packed glass columns for measurements near the detection limit. It may be anticipated that the use of wide bore fused silica columns would be an interesting compromise, with the additional advantage of ease of installation in conventional GC instruments. In our opinion, this controversial approach [Grob *et al.*, 1988] could lead to improved method limits of detection in the future.

4. AAS AS A GC DETECTOR FOR ALKYLLEADS : TO SEE OR NOT TO SEE...

One of the attractive features of interfacing a GC to an AAS detection system is that little or no structural modification of the individual instruments is required. Generally the interface is accomplished by simply connecting a transfer line from the

GC column exit to the atomization source of the AAS, resulting in a species-specific system which is relatively inexpensive and easy to operate. The very high selectivity of such systems over conventional GC detectors is convincingly demonstrated in Figure 5, which compares chromatograms of a gasoline sample with FID and GFAAS detection.

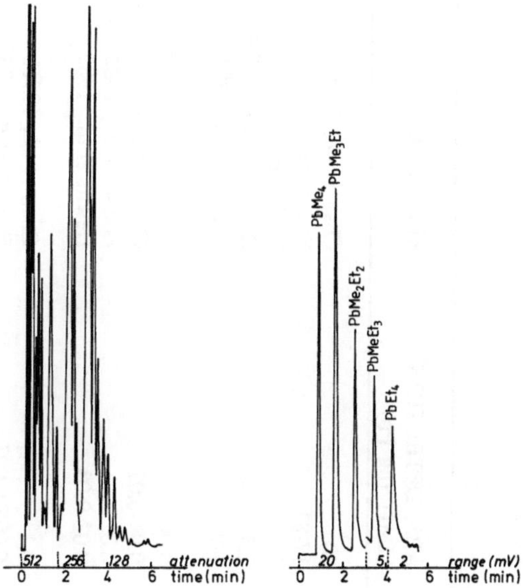

Figure 5 Gas chromatogram of a gasoline sample with FID and GFAAS detection, respectively [De Jonghe *et al.*, 1980a]

Although both the flame and the furnace atomic absorption mode have been successfully used, the resulting configuration is not always suitable for environmental speciation. For coupling a GC to a flame AAS, the transfer line can either be connected to the nebulizer or directly fed into the burner head of a flame AAS. The latter interface achieves increased sensitivity by eliminating dilution of the analyte by the nebulizer gases. Nevertheless, the use of flame AAS remains limited due to its high detection limits, resulting from the short residence time of the analyte atoms within the optical beam of the spectrometer. Limits of detection which typify various atomization interfaces are presented in Table 2.

The first reported GC-AAS coupling, by Kolb *et al.* in 1966, used flame AAS to determine tetraalkyllead compounds in petrol; for environmental speciation, however, this method lacks the requisited sensitivity. A major step forward was the introduction of an electrothermally heated silica tube as an atomizing device [Chau *et al.*, 1976b]. This furnace typically has a T-shape geometry and is encased in a block of preshaped insulating material. It is aligned in the optical path of the spectrometer and heated electrically using resistance wire coiled around the upper tube; the lower tube serves to feed the GC effluent into the atomization cell, and is often modified to allow the

TABLE 2 Survey of reported GC-AAS configurations for speciation of tetraalkyllead

DETECTOR[1]	TRANSFER LINE RUNNING INTO:	ABS. DETECTION LIMIT (ng Pb)	REFERENCE
FAAS	nebulisation chamber	80	Chau *et al.*, 1976a
		17-81	Bye *et al.*, 1978
	hole below burner slot	20	Coker, 1975
		1.5	Radziuk *et al.*, 1979b
	slot of air-propane burner	1-2	Ebdon *et al.*, 1982
F-QFAAS	T tube through side arm	0.09	Radziuk *et al.*, 1979b
		0.04-0.07	Chakraborti *et al.*, 1984
	tube through hole	0.017	Ebdon *et al.*, 1982
		0.02-0.03	Hewitt *et al.*, 1985
ET-QFAAS	T tube through side arm	0.1	Chau *et al.*, 1976b
		0.005-0.007	Forsyth *et al.*, 1985
		0.009-0.011	Rapsomanikis *et al.*, 1986
		0.01-0.03	Radojevic *et al.*, 1986
GFAAS	enlarged injection hole	10	Segar, 1974
	home-made atomizer	0.1	Robinson *et al.*, 1977
	injection hole	0.12-1.1	Bye *et al.*, 1978
	tube via connector	0.04	Radziuk *et al.*, 1979a
	injection hole	0.03	Radziuk *et al.*, 1979b
	inner gas pathway	0.04-0.09	De Jonghe *et al.*, 1980a
	injection hole	0.02	Andersson *et al.*, 1984
	small injection hole	0.008-0.016	Nygren, 1987

[1] FAAS = flame AAS; F-QFAAS = flame heated quartz furnace AAS; ET-QFAAS = electrothermally heated quartz furnace AAS; GFAAS = graphite furnace AAS

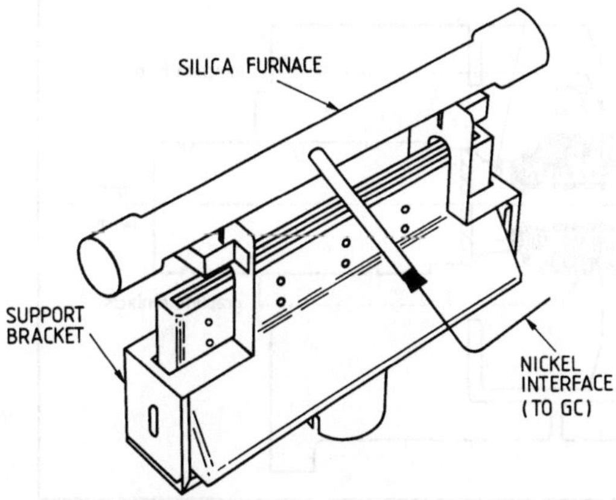

Figure 6 Typical design of a flame heated quartz furnace atomization source used in GC-AAS [Chakraborti *et al.*, 1984]

introduction of one or more atomization support gases. Added hydrogen gas was soon recognized as essential to enhance the sensitivity at the conventional operating temperature of this assembly which is limited to 900-1000 °C, hence considerably less than optimum graphite furnace atomization temperatures for lead. It has been postulated that hydrogen radicals mediate both the formation of lead hydride and its subsequent atomization [Forsyth *et al.*,1985].

An elegant variation on the previous system was pioneered by Radziuk *et al.* [1979]. They experimented with the use of the air-acetylene burner of the flame AAS instrument for heating a suspended tube and were able to achieve a similar increase of atom residence time in the optical path. Though excellent detection systems have been constructed by modifying their approach, automation of the flame heated device remains difficult due to the safety precautions which are necessary when working with open flames. A typical design is illustrated in Figure 6.

The first GC coupling to a commercial graphite furnace was achieved, albeit somewhat crudely, by Segar in 1974, by passing the end of a tungsten transfer line through an enlarged sample introduction hole in the graphite tube. Other successful approaches have used the inert gas purge passageway of the commercial furnace as an entry port for the GC effluent. Thus the effluent may be introduced through a T-fitting from both ends of the furnace and replace the internal gas flow (Figure 7). The latter configuration proved suitable only for sufficiently volatile compounds such as TAL; condensation problems were experienced with the ionic alkyllead derivatives [Chakraborti *et al.*, 1984]. Though either approach can yield favourable absolute detection limits in the low picogram range (Table 2), their environmental application suffers from one major drawback: the limited lifetime of graphite tubes (a few hours) when continuously

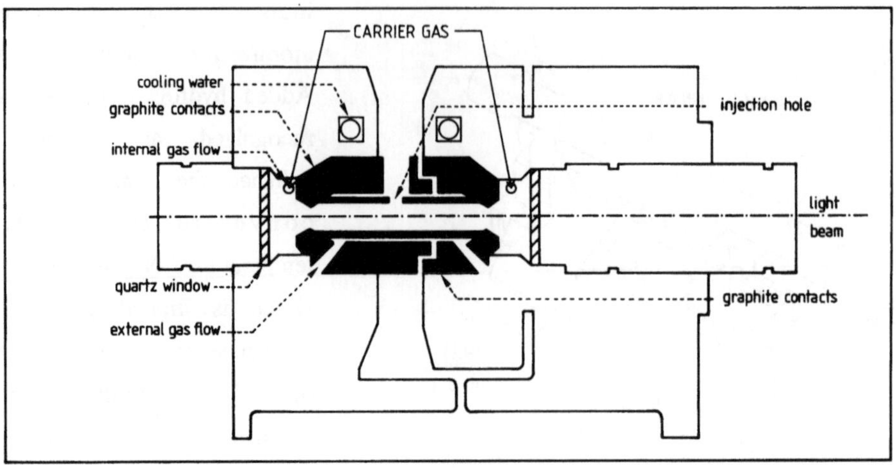

Figure 7 Cross-section of a Perkin-Elmer HGA-74 graphite furnace atomizer using the internal gas flow path for introduction of the GC effluent [modified from De Jonghe *et al.*, 1980a]

held at the atomization temperature during the chromatographic experiment. This not only renders the analysis very expensive, but also gives rise to a gradual change in sensitivity with time which complicates sample throughput.

From the previous survey and the detection limits in Table 2 it has to be concluded that, at present, the detection method of choice for organometal speciation is the electrothermally or flame heated tube, commonly designated as quartz furnace AAS (QFAAS). Different materials, including recrystallized alumina and commercial silica, may be used for the tube assembly. Extensive investigations using this detector have demonstrated that certain factors critically influence the sensitivity of the technique. One already mentioned parameter of utmost importance is the addition of hydrogen gas to support atomization, as graphically illustrated in Figure 8. Without this make-up gas virtually no lead atomization signal is produced, and the metal appears to be deposited on the tube wall, from which it may revolatilize when hydrogen enters the tube. Other gases may also be used, although they are less efficient [Forsyth *et al.*, 1985]. Introduction of air as an extra support gas has a detrimental effect on the lead signal, though it proved useful to enhance the atomization of tin [Chau *et al.*, 1982].

Under continuous reducing conditions, elution of the solvent (usually a hydrocarbon) causes the formation of a black carbon deposit on the silica cell surface, particularly if elevated injection volumes are employed. Its effect (if any) remains largely unknown; it seems not to hamper a sensitive and reproducible lead detection. A timing circuit and pneumatic valve assembly has been described which automatically switches the support gas from hydrogen to air during elution of the solvent front, thus preventing the deposition of carbon [Forsyth *et al.*, 1985].

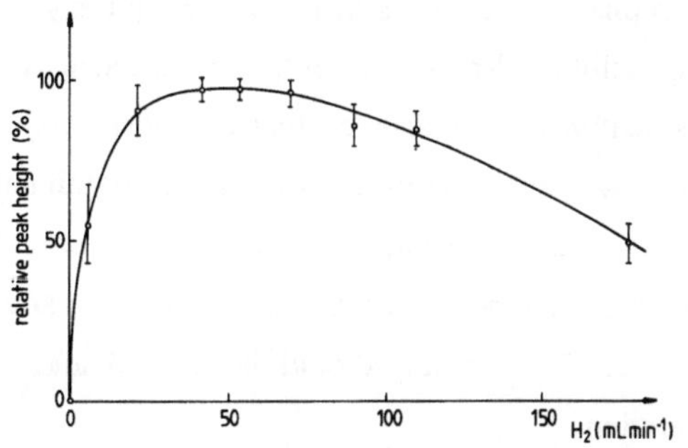

Figure 8 Effect of hydrogen flow rate on the peak height of alkylleads detected by GC-AAS (average of the signal for the five TAL species and standard deviation) [Van Cleuvenbergen *et al.*, unpublished work]

As documented for the hydride AAS determination of arsenic [Welz *et al.*, 1986; Van Cleuvenbergen *et al.*, 1988], the inner wall surface of the detector tube is believed to exert a profound influence on the atomization efficiency. Depending on the history of the silica furnace, sensitivities for lead fluctuate; we have experienced that increasing the hydrogen flow rate may in some cases restore the original (uniform) sensitivity. Probably the surface contaminants play a role in scavenging the hydrogen or other radicals, possibly by promoting radical recombination. For severe contamination problems, a drastic method to modify the condition of the quartz surface is to clean the cell in concentrated hydrofluoric acid [Welz *et al.*, 1981]. Though this may prove valuable in removing excessive contamination by co-chromatographing substances, it deteriorates the furnace rapidly and does not seem to produce consistently a better sensitivity.

When the GC-AAS instrument is brought up to the operating temperature for a series of measurements, the first chromatograms tend to be unreliable (poorer sensitivity and some baseline drift). It is advisable to sacrifice the first two or three injections to prime the system. In this way active sites in the column and the transfer line are efficiently covered. If priming does not result in acceptable analytical characteristics, it becomes more difficult to localize the problem at first sight. Cleaning or replacing of the silica furnace quickly reveals whether this part of the system is responsible. When concentrated extracts from environmental matrices are analyzed frequently, a brownish or black deposit may gradually develop in the upper part of the chromatographic column, due to contaminants in the extracts. At least the first few centimeters of the packing should then be replaced regularly. Silylation will often restore the resolving power of a packed column and can be effected without disconnecting the furnace. The furnace should be cooled and maintained between 200-300 °C during this operation.

A properly designed and optimized interface is a critical part in the experimental setup. It usually consists of a simple transfer line heated either by applying a variable voltage between both ends for conducting materials, or by using heating tape, to avoid any condensation of the organolead species. The temperature can be monitored with strategically placed thermocouples. A wide variety of materials have been recommended for the construction of the interface, including glass, fused silica, several metals (stainless steel, tantalum and nickel) and teflon or glass lined metal tubing. The inherent inertness makes fused silica probably the best choice; it results in a transfer line that is relatively convenient to handle and can pass through metal tubing for heating. To minimize peak broadening, a careful inspection of dead volumes in the connections used and an efficient control of the carrier gas flow and the interface temperature are essential. The length of the interface appears to be less critical, if it is well designed. In Figure 9 an expanded view is presented of a suitable coupling between the transfer line and the T quartz detector tube; depending on the options taken to introduce the atomization support gases, this part of the interface may look rather complicated and is frequently the most demanding to elaborate.

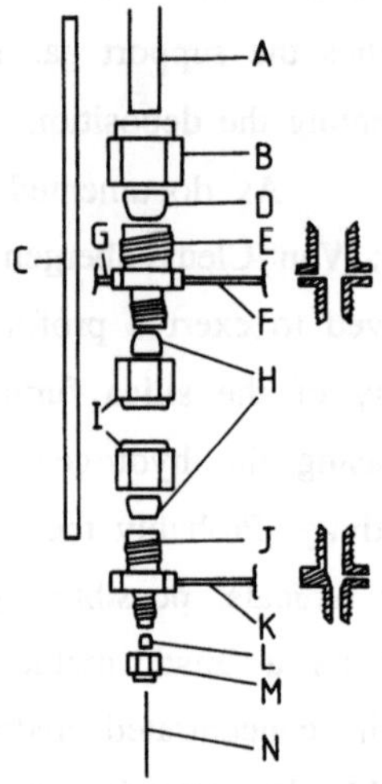

Figure 9 Example of an interface between the lower tube of the quartz T-furnace (A) and a fused silica transfer line (N); F, G and K are inlets for support gases [Forsyth *et al.*, 1985]

5. SAMPLE PREPARATION : THE ACHILLES' HEEL OF THE METHOD?

When speciating organometals in biological matrices such as tissues, body fluids or plant material, sample dissolution without rupturing the chemical structure of the analytes is a first prerequisite. Conventional acid digestion or ashing techniques using oxidizing acids are not suitable. For biological samples with high protein and fat contents, two techniques are currently in use: (1) enzymatic hydrolysis of the sample using mixed crude enzyme preparations containing lipase and proteases [Forsyth *et al.*, 1983], and (2) tetramethylammonium hydroxide as a tissue solubilizer [Chau *et al.*, 1984].

For aqueous samples which contain the organometals in solution, a simple extraction is often sufficient to separate the analytes from the matrix. In case where direct extraction is only partially effective, as often happens when solid matrices like dust, aerosols or sediments are investigated, a leaching procedure using citric acid may improve the recovery [Chakraborti *et al.*, 1986].

The subsequent isolation of particularly tri- and dialkyllead species from the "dissolved" sample has been a challenging task to the analyst, as their ionic characteristics do not allow these compounds to be extracted readily into any organic solvent, in contrast to tetraalkyllead species. Triethyllead and trimethyllead can be transferred as the undissociated chloride to an organic solvent, after saturation of the aqueous phase with NaCl [De Jonghe *et al.*, 1983b]. The procedure, however, is tedious and yields unsatisfactorily low recoveries for dialkyllead species [Noden, 1980]. A major step forward has been realized with the idea of trying out common complexing agents such as diethyldithiocarbamate and dithizone [Barker *et al.*, 1976; Chau *et al.*, 1983; Forsyth *et al.*, 1983]. The resulting organometallic complexes are sufficiently lipophilic so that they can be transferred to an organic solvent and are stable enough to withstand an additional enrichment by subsequent evaporation of the organic solvent. Interestingly, there is an appreciable difference in the polarity of the alkyllead-dithizone or alkyllead-dialkyldithiocarbamate complexes on the one hand and the corresponding Pb^{2+} complexes on the other hand. Thus hexane or a similar solvent can be used to extract alkyllead complexes quite selectively in the presence of large excesses of Pb^{2+}. Alternatively, masking the Pb^{2+} fraction with ethylene diamine tetraacetate (EDTA) results in a more complete discrimination. As mentioned earlier the sample preparation is completed by a derivatization step, during which the compounds are converted to more volatile tetraalkylated lead species without loss of their identity.

One apparently attractive approach to lead speciation is to fractionate the lead present into separate tetraalkyllead, ionic alkyllead and extractable inorganic lead (Pb^{2+}) fractions. Tetraalkylleads are removed from the sample solution/slurry by an initial extraction with an apolar solvent (often hexane). A subsequent reextraction of the sample solution with the same or a similar organic solvent in the presence of a suitable complexing agent - dithiocarbamate or diphenylthiocarbazone (dithizone) - results in the recovery of the ionic alkyllead fraction. Finally, the inorganic Pb^{2+} fraction is recovered using a more polar solvent in the presence of the same complexing agent. Synthetic aqueous mixtures containing alkyllead and inorganic Pb^{2+} standards are fractionated efficiently with this approach and the selectivity of the fractionation can be improved by back-extracting each organic phase with water. However the presence of natural complexing agents in soil and biological tissues can modify the solubilities of ionic alkylleads appreciably if they are present at the $\mu g\ kg^{-1}$ level. An appreciable portion of the trialkyllead fraction can be extracted from these matrices with a hexane wash. Moreover, this fraction may not be recovered completely by back-extraction into water or phosphate buffer. Thus, the preextraction of biological samples prior to complexometric recovery of ionic alkyllead analytes must be approached with caution. The identity of the component(s) of biological materials which interacts with ionic alkylleads and presumably functions as phase transfer reagent has not been studied to our knowledge.

Preconcentration of the organolead species from the air requires a totally different approach and is usually achieved by cryocondensation or by adsorption of the analytes on porous polymers; absorption in organic solvents has also been proposed [Backes *et al.*, 1989]. As currently practiced, cryotrapping suffers from the major drawback of water clogging of the trap at the low temperatures needed for complete collection of tetraalkylleads. As neither chemical drying agents nor a cryogenic pretrap offer a means of efficient water removal from the sample stream, the problem is usually circumvented by adapting either the sampling parameters or the desorption procedure [e.g. De Jonghe *et al*, 1980b]. Due to the resulting complex nature of cryogenic trapping procedures it may be very demanding to assure invariably quantitative recoveries; practical disadvantages preclude semi-automated sampling or use in remote areas. The other major enrichment technique, adsorption on polymers, enjoys continuing interest, but also requires trapping of the desorbing compounds on a cold U-tube prior to injection into the GC-AAS system to ensure a reasonable sensitivity [Hewitt *et al.*, 1985].

There exist several fundamental reasons why sample manipulation should be restricted to a minimum when speciating alkyllead. One of them is the well-documented experience that, especially from an aqueous medium, the covalent TAL species, and to some extent also ionic alkyllead cations, may be adsorbed readily on different types of solid surfaces, including glass recipients [Jarvie *et al.*, 1981; Harrison *et al.*, 1985a]. To date the nature of the adsorption process is not well understood; the best way to overcome these losses in e.g. the determination of alkylleads in natural water is to perform the extraction directly from the sampling bottle, preferably in the presence of excess complexing agent.

In addition, the rapid decomposition of TAL to trialkyllead degradation products may drastically affect the reliability of analyses [Radojevic *et al.*, 1987]. Convincing evidence indicates that in every environmental matrix TAL compounds are eventually converted into inorganic lead via trialkyllead salts; dialkyllead salts are also likely to be involved as intermediates. Information on the stability of the species, and the pathway followed in the transformation reactions, remains fragmentary [Van Cleuvenbergen *et al.*, 1990a]. A typical situation is the recovery problem which arises when dimethyllead is spiked on particulate matter or biological materials; the question whether the species is relatively labile and degrades rapidly - a hypothesis which is contradicted by laboratory simulations of the breakdown process - or becomes tightly bound to unknown components of the matrix, despite the sample dissolution procedure, still awaits for a conclusive answer.

The recovery of the analytes from complicated matrices should in fact be a general point of consideration in speciation techniques. Frequently the recovery efficiency of a procedure is assessed by adding the analytes, usually in the form of soluble salts, to a sample. Since a spiked standard species may behave otherwise than a "natural" one, its recovery may not accurately reflect the removal yield of the analytes originally present in the sample. Basically each attempt at environmental speciation should consider questions such as (1) to what extent species are detected using a certain analytical methodology and (2) which fraction needs to be determined: the (bio)available part, or the total concentration of the species in that matrix (including the part which occurs in tightly bound structures). In addition, it should be stressed that it is essential that recovery trials be conducted at levels which are environmentally relevant.

6. SOME SELECTED APPLICATIONS

It will not be attempted here to cover completely the extensive area of occurrence and fate of organolead in the biosphere. The interested reader is referred to a recent review which has tried to compile the existing data [Van Cleuvenbergen *et al.*, 1990a]. Rather than choosing an exhaustive approach, we have preferred to select few illustrative applications which demonstrate the potential of the GC-AAS technique for lead speciation.

In Table 3 the average concentration of ionic alkyllead and the species composition is presented for a variety of environmental samples [Van Cleuvenbergen *et al.*, 1987; 1990a]. The speciation scheme outlined in Figure 1 was adopted for these analyses. Some representative chromatograms are gathered in Figure 10 [Van

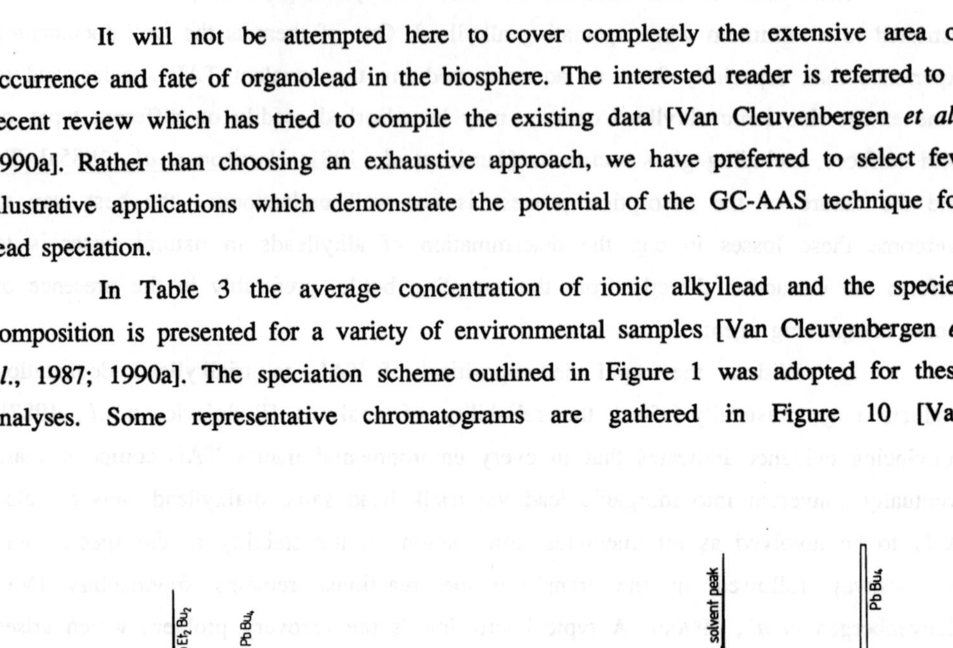

Figure 10 Speciation of ionic alkyllead with extraction / butylation / GC-AAS in rain water (left), sediment (middle) and grass (right, without D_2 correction); the $PbBu_4$ signal results from incompletely masked Pb^{2+}

TABLE 3 Average concentration levels of ionic alkyllead in different environmental matrices, and the distribution (%) of the individual species

SAMPLE TYPE	No.	PERIOD (month/year)	TOTAL IONIC ALKYLLEAD CONTENT (as Pb)	DISTRIBUTION (%)						
				$PbMe_3^+$	$PbMe_2Et^+$	$PbMeEt_2^+$	$PbMe_2^{++}$	$PbEt_3^+$	$PbMeEt^{++}$	$PbEt_2^{++}$
RESIDENTIAL										
aerosol	8	2/86	0.20 ng m^{-3}	2	2	n.d.[1]	2	17	1	77
rain	8	2-5/84	67 ng L^{-1}	51	7	n.d.	15	10	n.d.	17
	57	1/86-7/87	45 ng L^{-1}	44	2	< 1	11	21	1	21
snow	3	1/85	65 ng L^{-1}	46	7	2	9	17	3	16
road runoff	3	11/85	700 ng L^{-1}	24	10	4	n.d.	54	n.d.	8
potable water	1	3/86	5.4 ng L^{-1}	50	11	n.d.	15	10	n.d.	15
lake water	10	12/85-10/86	6.8 ng L^{-1}	60	8	3	4	19	n.d.	7
lake sediment	11	12/85-10/86	0.63 ng g^{-1}	10	4	3	< 1	68	n.d.	15
road dust	3	12/85-1/86	12.8 ng g^{-1}	9	2	1	4	14	2	70
soil	3	5/86	0.47 ng g^{-1}	22	6	2	n.d.	6	n.d.	64
grass	4	6/87	5.9 ng g^{-1}	19	19	n.d.	n.d.	33	n.d.	29
GASOLINE STATION										
road runoff	1	2/86	4300 ng L^{-1}	16	6	3	n.d.	72	n.d.	4
road dust	1	2/86	350 ng g^{-1}	2	< 1	< 1	1	9	1	87
soil	1	3/86	2.8 ng g^{-1}	12	2	1	2	6	n.d.	77
grass	5	11/85-1/86	56 ng g^{-1}	19	6	1	3	19	1	50

[1] n.d. = not detected

Cleuvenbergen *et al.*, 1986b; Chakraborti *et al.*, 1989a; Van Cleuvenbergen *et al.*, 1990b]. The sampling locations, all in the vicinity of our university campus, are typically residential (but in a heavily industrialized and densely populated country) and are characterized by low traffic density roads and car parking areas. The premium brand gasoline distributed by a gasoline station within this suburban area was analyzed (spring 1986) by direct injection of the diluted sample into the GC-AAS system which was calibrated with a mixed TAL standard. A tetraalkyllead content of 0.35 g Pb L^{-1} was found, 88% of which was added as tetraethyllead and the remaining as tetramethyllead. This alkyllead distribution was typical of the average gasoline additive composition around that time in the Antwerp metropolitan area as indicated by a survey of several distributors. The concentrations of ionic alkyllead found in environmental samples collected immediately adjacent to the station (Table 3) clearly reflect this composition by a high relative abundance of the tetraethyllead degradation products.

The high relative abundance of triethyllead in all road runoff water samples agrees with the degradation studies of tetraethyllead in water reported to date. These studies all pointed to a more or less rapid breakdown to triethyllead; diethyllead was occasionally detected in trace quantities, inorganic lead being the final degradation product.

At present we tend to consider a hypothesis of Nielsen *et al.* [1981] to be the most probable explanation for the high diethyllead contribution to the total ionic alkyllead level in street dust, soil as well as in the particulate fraction of the air. Based on the sublimation point of the chloride salts, it was suggested that the dominant proportion of trialkyllead would probably remain gaseous, with a minor fraction condensing on and remaining with particulate matter. Recently the direct detection of gas-phase trialkyllead has been reported; the species were shown to contribute significantly to the total alkyllead level of the atmosphere, their concentration in particles claimed to be only about 1% of that in the vapour phase [Harrison *et al.*, 1985b; Allen *et al.*, 1988]. Our preliminary study of the degradation of ionic alkyllead species in dust and soil does not indicate a much higher stability of diethyllead compared to triethyllead in these matrices [Chakraborti *et al.*, 1989b]; thus a higher persistence seems not to provide an explanation for the elevated relative diethyllead abundance.

As the tri- and dialkyllead compounds present in the atmosphere are quite soluble in water, washout and rainout can be expected to be major sink processes and to constitute a source of these pollutants to the hydrosphere. Our extensive work on this topic [Van Cleuvenbergen *et al.*, 1985; 1986b] points out that the distribution of the species in wet fall-out is intermediate between that found on particulate matter and that expected from the gaseous fraction of the air [Allen *et al.*, 1988], both parts being subject to the removal processes as well as to continuous alkyllead degradation. As apparent from Table 3, we did not observe any marked difference between the average organolead concentration or species composition in rain for the spring of 1984 and for the period 1986-1987. This is consistent with the regulations on gasoline lead in Belgium, which did not change during that time period. To the extent that direct comparisons are valid, it is interesting to consider earlier measurements addressing the speciation of atmospheric tetraalkyllead in the same geographic locate [De Jonghe *et al.*, 1981]. At that time the (most volatile) tetramethyllead species accounted for about 75% of the tetraalkyllead in the gaseous fraction of the atmosphere, a considerably greater proportion than expected from the average gasoline lead composition. This suggests a high contribution of gaseous trimethyllead to the ionic alkyllead concentration levels in rain.

Having established that wet atmospheric fall-out processes transfer ionic alkyllead species into the aquatic environment, we subsequently detected these contaminants in various natural waters. This study revealed that in the lake situated on the university campus trimethyllead was clearly the predominant species, a trend holding true for many other environmental water samples as well. Since the compound accounts for an appreciable part of the alkyllead burden in rain and, moreover, laboratory experiments simulating the degradation indicate that it is the most persistent of the tri- and dialkyllead compounds in the hydrosphere, its high relative abundance can easily be interpreted. In the lake the concentration ratio of triethyllead to trimethyllead was about 20 times higher in the sediments than in the aqueous phase, whereas the same proportion was not observed in estuarine samples [Chakraborti *et al.*, 1989a]. Though the environmental significance of this observation remains unknown, the stagnant character of the lake or some typical chemical constituents are assumed to be involved.

Ionic alkyllead would also appear to be ubiquitous in biological tissues. They have been detected in avian tissues (liver, kidney, brain and breast muscle) from several species, periwinkles, freshwater bivalves and benthos and in marine and freshwater fish. Studies with Herring gulls (*Larus argentatus*, mature and immature birds and eggs) from ten locations within the Great Lakes waterway [Forsyth *et al.*, 1986] and periwinkles (*Littorina irrorata*) from six sites within lower Chesapeake Bay [Krishnan *et al.*,1988a] have confirmed a second source of methyllead salts (other than from direct demethylation of Me_4Pb) which is presumed to be methylation of Pb^{2+}. Further, these results were consistent with an environmentally mediated methylation of ethyllead salts. Yet the environmental burdens of ionic alkylleads would seem to be very low, at least for the compartments which have been examined to date, suggesting that the rates of mineralization of alkyllead species must be at least competitive with the biomethylation of Pb^{2+}. To our knowledge, the rates of conversion of $R_nPb^{(4-n)+}$ to Pb^{2+} have not been measured in environmental matrices although progress has been made with the reported synthesis of [^{210}Pb]-Me_3PbCl [Blais *et al.*, 1989b].

One possible means of assessing the impact of the continued release of organoleads is to monitor residue burdens in appropriate wildlife species. The eggs of certain avian species, if collected at the proper time, are replaced by the birds so that the sampling does not affect the species population adversely. Feeding trials with ethyllead salts to Japanese quail (*Coturnix coturnix japonica*), which were considered to be a relevant model for avian wildlife, demonstrated that ethyllead toxicants are only poorly accumulated in soft tissues (< 1.0%) or transferred to egg of this species [Krishnan *et al.*,1988b]. Methylation (as a metabolic response) was not detected in appreciable amounts (<< 1%). Reequilibration (redistribution) reactions, however, were prevalent; triethyllead was the major species in soft tissues of quail which had received Et_2PbCl_2 in drinking water for 60 days. Moreover, burdens in separate tissues (liver, kidney, brain and breast muscle) were highly correlated suggesting that this toxicant was equilibrated among these tissues. In contrast, Et_2Pb^{2+} was the major toxicant in egg whether Et_3PbCl or Et_2PbCl_2 was provided to the birds. Clearly, the interpretation of exposure to alkylleads based on residue burdens in separate tissues must be approached with caution and tempered with a knowledge of the metabolic fate of these species.

One persistent problem with the determination of ionic alkylleads by GC-AAS has been the low recovery of Me_2Pb^{2+} particularly from biological matrices. For most matrices, it is not known whether the low recovery is the result of instability of this species or whether it is simply bound very strongly to some unknown component(s) of the sample matrix. Recoveries of Me_2Pb^{2+} from blood (from several mammalian species) have been especially disappointing. Interestingly, dialkylleads are potent in vitro inhibitors of the enzyme γ-aminolaevulinic acid dehydratase (γ-ALAD, E.C. 4.2.1.24), a component of human erythrocytes. By monitoring the activity of this enzyme at different pH values, it is possible to differentiate inhibition by Pb^{2+} from inhibition by Me_2Pb^{2+}.

For crude plasma, the inhibition by added Me_2Pb^{2+} was intense and persisted for at least 48 h; appreciable mineralization of the inhibitor to Pb^{2+} was not evident during this time [Hamilton, 1986].

Very little is known concerning the binding of alkylleads to components of other biological materials. Recoveries of ethyllead salts (at the μg kg^{-1} level) by complexometric extraction from chicken egg homogenate were improved considerably by digesting the homogenate for 24 h with a crude mixture of lipase and protease [Forsyth *et al.*, 1983]. In contrast, the recovery of Me_3Pb^+ was less influenced by the native homogenate. That the binding to components of the matrix may be appreciable is suggested by bioassays for inhibition of anaerobic nitrogen transformations in soil or sediment. No appreciable inhibition of denitrification or N-fixation was observed for up to 10 μg g^{-1} of alkyllead chloride ($Me_nPbCl_{(4-n)}$, $Et_nPbCl_{(4-n)}$; n = 2, 3) added to soil or sediment whereas partial or complete inhibition was observed if the same toxicants were

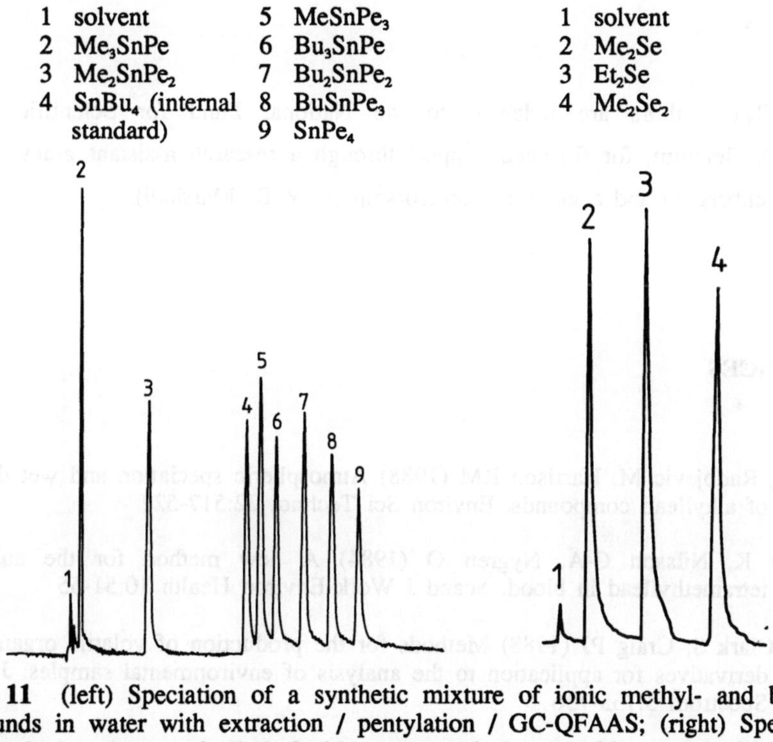

Figure 11 (left) Speciation of a synthetic mixture of ionic methyl- and butyltin compounds in water with extraction / pentylation / GC-QFAAS; (right) Speciation of a synthetic mixture of alkylselenide compounds in air with cryotrapping / GC-GFAAS

added to supernatants from these preparations [Blais *et al.*, 1988; in press]. It is also interesting to note that the microbial population appeared to adapt to the presence of these toxicants.

As a final remark, it is worth noting that the speciation methodology developed for lead is not restricted to this element only. Minor procedural modifications permit the species-specific determination of related elements such as tin [Dirkx *et al.*, 1988] and selenium [Jiang *et al.*, 1982] (Figure 11). The experience gathered when speciating lead may thus prove valuable for monitoring the impact of other trace metals on the biosphere. At a moment when the consumption of organic lead is declining in response to more stringent vehicle emission standards the use of several other organometallic compounds continues to increase.

ACKNOWLEDGEMENTS

The authors are indebted to the National Fund for Scientific Research (N.F.W.O.), Belgium, for financial support through a research assistant grant (to R.J.A. Van Cleuvenbergen) and a visiting professorship (to W.D. Marshall).

REFERENCES

Allen AG, Radojevic M, Harrison RM (1988) Atmospheric speciation and wet deposition of alkyllead compounds. Environ Sci Technol 22:517-522

Andersson K, Nilsson C-A, Nygren O (1984) A new method for the analysis of tetramethyllead in blood. Scand J Work Environ Health 10:51-55

Ashby J, Clark S, Craig PJ (1988) Methods for the production of volatile organometallic derivatives for application to the analysis of environmental samples. J Anal At Spectrom 3:735-736

Backes U, Bibernell U, Neidhart B (1989) Air sampling procedure for tetraalkyllead compounds. Fres Z Anal Chem 333:706

Barker AJ, Ellis SRM, Clarke AB (1976) The removal of organic lead from aqueous effluents by a combined chemical complexing - solvent extraction technique. Advances in Chemistry Series 155:381-395

Blais JS, Marshall WD (1986) Determination of alkyllead salts in runoff, soils, and street dusts containing high levels of lead. J Environ Qual 15:255-260

Blais JS (1987) Determination of alkyllead compounds and synthesis of alkyllead radiotracers. M Sc thesis, Dept Food Science and Agricultural Chemistry, McGill University, Montreal, Canada

Blais JS, Marshall WD, Knowles R (1988) Toxicity of alkyllead salts to anaerobic nitrogen transformations in sediment. J Environ Qual 17:457-462

Blais JS, Marshall WD (1989a) Determination of ionic alkyllead compounds in water, soil, and sediment by high performance liquid chromatography - quartz tube-atomic absorption spectrometry. J Anal At Spectrom 4:271-276

Blais JS, Momplaisir GM, Marshall WD (1989b) Synthesis, storage and transfer of $[^{210}Pb]$-$(CH_3)_3PbCl$ across tomato fruit cuticle. Appl Organomet Chem 3:89-98

Blais JS, Doige CA, Marshall WD, Knowles R (in press) Persistence and toxicity of alkyllead salts to anaerobic nitrogen transformations in soil. Arch Environ Contam Toxicol

Bye R, Paus PE, Solberg R, Thomassen Y (1978) Atomic absorption spectroscopy used as a specific gas chromatography detector; comparison of flame and graphite furnace techniques in the determination of tetraalkyllead compounds. Atom Abs Newslett 17:131-134

Cappon CJ (1987) GLC speciation of selected trace elements. LC-GC 5:400-418

Chakraborti D, De Jonghe WRA, Van Mol WE, Van Cleuvenbergen RJA, Adams FC (1984) Determination of ionic alkyllead compounds in water by gas chromatography/atomic absorption spectrometry. Anal Chem 56:2692-2697

Chakraborti D, Van Cleuvenbergen R, Adams F (1986) Determination of ionic alkyllead compounds in road runoff water and road dust by butylation and gas chromatography - atomic absorption spectrometry. In: Lester JN, Perry R, Steritt RM (eds) Chemicals in the environment. Selper Ltd, London, p 298-304

Chakraborti D, Van Cleuvenbergen R, Adams F (1987) Speciation of ionic alkyllead in aerosols by gas chromatography - atomic absorption spectrometry. Int J Environ Anal Chem 30:233-242

Chakraborti D, Van Cleuvenbergen RJA, Adams FC (1989a) Ionic alkyllead compounds in environmental water and sediments. Hydrobiologia 176/177:151-158

Chakraborti D, Dirkx W, Van Cleuvenbergen R, Adams F (1989b) Speciation of ionic alkyllead in potable water and soil. Sci Tot Environ 84:249-257

Chau YK, Wong PTS, Saitoh H (1976a) Determination of tetraalkyllead compounds in the atmosphere. J Chromatog Sci 14:162-164

Chau YK, Wong PTS, Goulden PD (1976b) Gas chromatography - atomic absorption spectrometry for the determination of tetraalkyllead compounds. Anal Chim Acta 85:421-424

Chau YK, Wong PTS, Bengert GA (1982) Determination of methyltin (IV) and tin (IV) species in water by GC/AAS. Anal Chem 54:246-249

Chau YK, Wong PTS, Kramar O (1983) The determination of dialkyl-, trialkyl-, tetraalkyllead and lead (II) species in water by chelation/extraction and gas chromatography - atomic absorption spectrometry. Anal Chim Acta 146:211-217

Chau YK, Wong PTS, Bengert GA, Dunn JL (1984) Determination of dialkyllead, trialkyllead, tetraalkyllead, and lead (II) compounds in sediment and biological samples. Anal Chem 56:271-274

Coker DT (1975) Determination of individual and total lead alkyls in gasoline by a simple rapid gas chromatography/atomic absorption spectrometry technique. Anal Chem 47:386-389

De Jonghe W, Chakraborti D, Adams F (1980a) Graphite-furnace atomic absorption spectrometry as a metal - specific detection system for tetraalkyllead compounds separated by gas-liquid chromatography. Anal Chim Acta 115:89-101

De Jonghe WRA, Chakraborti D, Adams FC (1980b) Sampling of tetraalkyllead compounds in air for determination by gas chromatography - atomic absorption spectrometry. Anal Chem 52:1974-1977

De Jonghe WRA, Chakraborti D, Adams FC (1981) Identification and determination of individual tetraalkyllead species in air. Environ Sci Technol 15:1217-1222

De Jonghe W, Adams F (1983a) Gas chromatography with flame ionization detection for the speciation of trialkyl lead halides. Fres Z Anal Chem 314:552-554

De Jonghe WRA, Van Mol WE, Adams FC (1983b) Determination of trialkyllead compounds in water by extraction and graphite furnace atomic absorption spectrometry. Anal Chem 55:1050-1054

Dirkx W, Van Mol W, Van Cleuvenbergen R, Adams F (1988) Speciation of organotin compounds in surface water. In: Astruc M, Lester JN (eds) Heavy metals in the hydrological cycle. Selper Ltd, London, p 441-446

Ebdon L, Ward RW, Leathard DA (1982) Development and optimisation of atom cells for sensitive coupled gas chromatography - flame atomic absorption spectrometry. Analyst 107:129-143

Ebdon L, Hill S, Ward RW (1986) Directly coupled chromatography - atomic spectroscopy; part 1: Directly coupled gas chromatography - atomic spectroscopy; a review. Analyst 111:1113-1138

Ebdon L, Hill S, Ward RW (1987) Directly coupled chromatography - atomic spectroscopy; part 2: Directly coupled liquid chromatography - atomic spectroscopy; a review. Analyst 112:1-16

Estes SA, Uden PC, Barnes RM (1982) Determination of n-butylated trialkyllead compounds by gas chromatography with microwave plasma emission detection. Anal Chem 54:2402-2405

Faulstich H, Stournaras C (1985) Potentially toxic concentrations of triethyllead in Black Forest rainwater samples. Nature (London) 317:714-715

Fergusson JE (1986) Lead: petrol lead in the environment and its contribution to human blood lead levels. Sci Tot Environ 50:1-54

Forsyth DS, Marshall WD (1983) Determination of alkyllead salts in water and whole eggs by capillary column gas chromatography with electron capture detection. Anal Chem 55:2132-2137

Forsyth DS, Marshall WD (1985) Performance of an automated gas chromatograph - silica furnace- atomic absorption spectrometer for the determination of alkyllead compounds. Anal Chem 57:1299-1305

Forsyth DS, Marshall WD (1986) Ionic alkylleads in Herring gulls from the Great Lakes region. Environ Sci Technol 20:1033-1038

Forsyth DS (1987) Design and evaluation of a capillary gas chromatography - atomic absorption spectrometer interface. Anal Chem 59:1742-1744

Grandjean P, Nielsen T (1979) Organolead compounds: environmental health aspects. Residue Rev 72:97-148

Grandjean P (ed)(1984) Biological effects of organolead compounds. CRC Press, Boca Raton Florida

Grob K, Frech P (1988) 0.53-mm i.d. vs 0.32-mm i.d. GC columns: are large diameter columns necessary? Int Lab 18,8:18-23

Hamilton JD (1986) Experimental evaluation of porphobilinogen synthase (PBGS, E.C. 4.2.1.24) as a physiological index of lead body burden. M Sc thesis, School of Occupational Health and Safety, McGill University, Montreal, Canada

Harrison RM, Radojevic M (1985a) Determination of tetraalkyl and ionic alkyllead compounds in environmental samples by butylation and gas chromatography - atomic absorption. Environ Technol Lett 6:129-136

Harrison RM, Hewitt CN, Radojevic M (1985b) Environmental pathways of alkyllead compounds. In: Lekkas TD (ed) Proc Int Conf Heavy metals in the environment, Athens 1985, vol 1. CEP Consultants Ltd, Edinburgh, p 82-84

Hewitt CN, Harrison RM (1985) A sensitive, specific method for the determination of tetraalkyllead compounds in air by gas chromatography/atomic absorption spectrometry. Anal Chim Acta 167:277-287

Jarvie AWP, Markall RN, Potter HR (1981) Decomposition of organolead compounds in aqueous systems. Environ Res 25:241-249

Jiang S, De Jonghe W, Adams F (1982) Determination of alkylselenide compounds in air by gas chromatography - atomic absorption spectrometry. Anal Chim Acta 136:183-190

Kolb B, Kemmner G, Schleser FH, Wiedeking E (1966) Elementspezifische Anzeige gaschromatographisch getrennter Metallverbindungen mittels Atom-absorptions-spektroskopie (AAS). Fres Z Anal Chem 221:166-175

Krishnan K, Marshall WD, Hatch WI (1988a) Ionic alkylleads in salt marsh Periwinkles (Littorina irrorata). Environ Sci Technol 22:806-811

Krishnan K, Marshall WD (1988b) Avian tissues as bioindicators of exposure to alkylleads: metabolism of ethyllead salts by Japanese Quail. Environ Sci Technol 22:1038-1043

Marshall WD (1988) Chromatographic approaches to trace element speciation. In: Kramer JR, Allen HE (eds) Metal speciation: theory, analysis and application. Lewis Publ, Chelsea, p 155-182

Nielsen T, Egsgaard H, Larsen E, Schroll G (1981) Determination of tetramethyllead and tetraethyllead in the atmosphere by a two-step enrichment method and gas chromatographic - mass spectrometric isotope dilution analysis. Anal Chim Acta 124:1-13

Noden FG (1980) The determination of tetraalkyllead compounds and their degradation products in natural water. In: Branica M, Konrad Z (eds) Lead in the marine environment. Pergamon, Oxford, p 83-91

Nygren O (1987) High resolution gas chromatography with a graphite furnace atomic absorption spectrometer as detector. J Anal Atom Spectrom 2:801-803

Radojevic M, Allen A, Rapsomanikis S, Harrison RM (1986) Propylation technique for the simultaneous determination of tetraalkyllead and ionic alkyllead species by gas chromatography/atomic absorption spectrometry. Anal Chem 58:658-661

Radojevic M, Harrison RM (1987) Concentrations, speciation and decomposition of organolead compounds in rainwater. Atmos Environ 21:2403-2411

Radziuk B, Thomassen Y, Van Loon JC, Chau YK (1979a) Determination of alkyllead compounds in air by gas chromatography and atomic absorption spectrometry. Anal Chim Acta 105:255-262

Radziuk B, Thomassen Y, Butler LRP, Van Loon JC, Chau YK (1979b) A study of atomic absorption and atomic fluorescence atomization systems as detectors in the gas chromatographic determination of lead. Anal Chim Acta 108:31-38

Rapsomanikis S, Donard OFX, Weber JH (1986) Speciation of lead and methyllead ions in water by chromatography/atomic absorption spectrometry after ethylation with sodium tetraethylborate. Anal Chem 58:35-38

Robinson JW, Kiesel EL, Goodbread JP, Bliss R, Marshall R (1977) The development of a gas chromatography - furnace atomic absorption combination for the determination of organic lead compounds; atomization processes in furnace atomizers. Anal Chim Acta 92:321-328

Segar DA (1974) Flameless atomic absorption gas chromatography. Anal Lett 7:89-95

Van Cleuvenbergen R, Chakraborti D, Van Mol W, Adams F (1985) Ionic alkyllead species in atmospheric fall-out and surface water. In: Lekkas TD (ed) Proc Int Conf Heavy metals in the environment, Athens 1985, vol 1. CEP Consultants Ltd, Edinburgh, p 153-155

Van Cleuvenbergen RJA, Chakraborti D, Adams FC (1986a) Occurrence of monoalkyl-lead species during the speciation of organolead. Anal Chim Acta 182:239-244

Van Cleuvenbergen RJA, Chakraborti D, Adams FC (1986b) Occurrence of tri- and dialkyllead species in environmental water. Environ Sci Technol 20:589-593

Van Cleuvenbergen R, Chakraborti D, Adams F (1987) Speciation of ionic organolead compounds in the biosphere. In: Angeletti G, Restelli G (eds) Proc 4th European Symposium on Physico-chemical behaviour of atmospheric pollutants, Stresa 1986 (Commission of the European Communities Air Pollution Research Report 2). D Reidel Publ Comp, Dordrecht Holland, p 571-578

Van Cleuvenbergen RJA, Van Mol WE, Adams FC (1988) Arsenic speciation in water by hydride cold trapping - quartz furnace atomic absorption spectrometry: an evaluation. J Anal At Spectrom 3:169-176

Van Cleuvenbergen RJA, Adams FC (1990a) Organolead compounds. In: Hutzinger O (ed) Handbook of Environmental Chemistry, vol 3 part E. Springer, Weinheim, p 97-153

Van Cleuvenbergen R, Chakraborti D, Adams F (1990b) Speciation of ionic alkyllead in grass and tree leaves. Anal Chim Acta 228:77-84

Welz B, Melcher M (1981) Mutual interactions of elements in the hydride technique in atomic absorption spectrometry. Anal Chim Acta 131:17-25

Welz B, Schubert-Jacobs M (1986) Investigations on atomization mechanisms in hydride-generation atomic absorption spectrometry. Fres Z Anal Chem 324:832-838

Wong PTS, Chau YK, Luxon PL (1975) Methylation of lead in the environment. Nature (London) 253:263-264

Analysis of Mercury - Species in Sediments

R.-D. Wilken and H. Hintelmann

GKSS-Research Center
Institute for Chemistry
Max-Planck-Straße
D-2054 Geesthacht, Germany

1. Introduction

Mercury emission is estimated to be about 50,000 t a year. Progressive industrialization means that the anthropogenic share meanwhile amounts to one fifth of this amount [Tölg and Lorenz, 1977]. Most of the mercury release is from coal burning in coal fired power plants. Hg^0 is the form which reaches the atmosphere. Mercury is distributed extensively in this way although no mercury accumulation is found in the surroundings of coal power plants. Thus mercury is an ubiquitous element. The second main emission source of mercury is the chlor-alkali industry. The resulting mercury waste leads to high local mercury contaminations of the adjacent rivers. This segment of industry can be considered to be the main source of the high mercury pollution of the river Elbe. The natural input of mercury in the environment is chiefly observed at tectonically active locations in the earth's crust. Where continental plates collide, volcanos, geysers and thermal springs are the main mercury emission sources. Whereas natural input seldom leads to accumulation and therefore is not an environmental hazard, industrial mercury emmission can lead to environmental disasters. Meanwhile the japanese Minamata Bight has, sadly, become renowned. Catalyst wastes containing mercury, which were dumped in the Minamata river, reach the human level of food chain via fish and this has lead to disasterous poisonings. Methylmercury chloride was the main mercury species responsible for these poisonings [Fujiki, 1972].

NATO ASI Series, Vol. G 23
Metal Speciation in the Environment
Edited by J. A. C. Broekaert, Ş. Güçer, and F. Adams

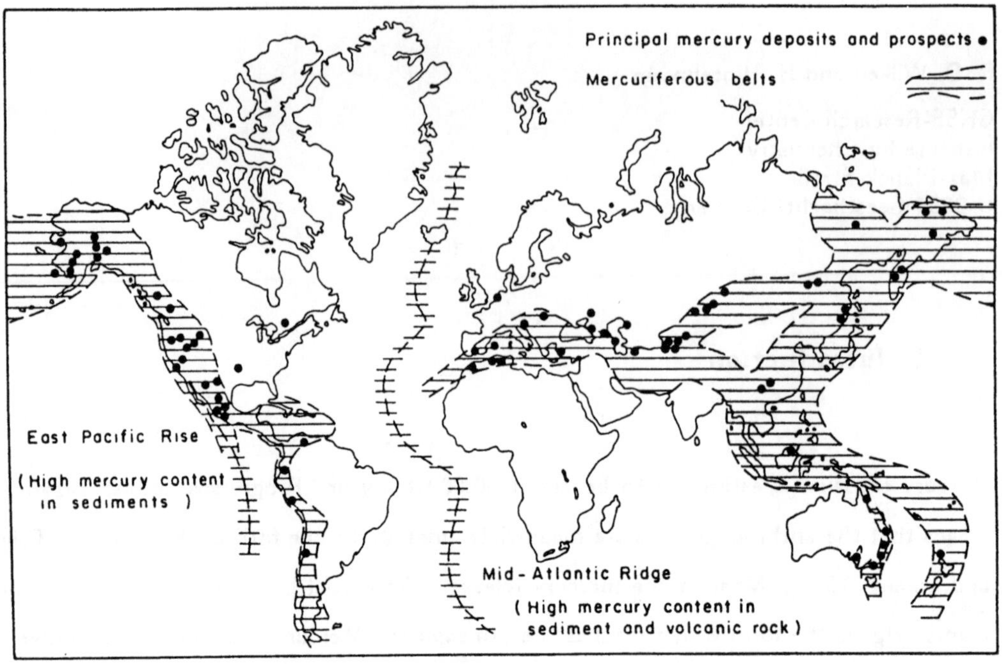

Figure 1. Natural Hg emission sources [Ramamoorthy and Moore, 1984]

Mercury belongs to those metals which undergo many transformation reactions in nature. The chemical equilibrium

$$Hg_2^{2+} \rightleftharpoons Hg^0 + Hg^{2+}$$

is easily shifted from right to left and vice versa. For example the readily volatilised Hg^0 may be removed from the equilibrium, or microorganisms may transform soluble Hg^{2+} into other mercury species. The multitude of possible chemical or biochemical reactions of mercury make the already intricate mercury chemistry even more complicated.

One of the suggested mercury cycles is given in figure 2 [Jonasson and Boyle, 1972].

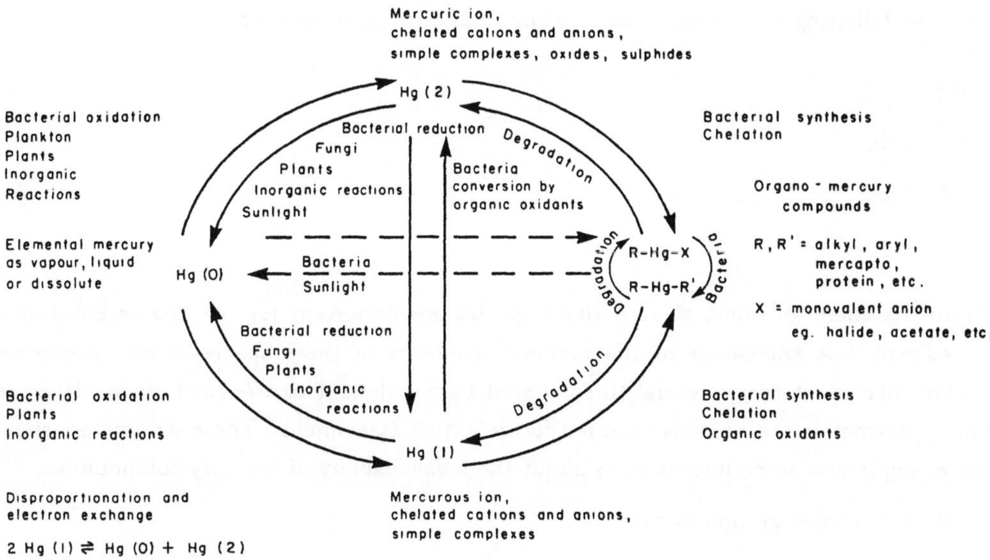

Figure 2. Cycle of mercury conversions in nature [Jonasson and Boyle, 1972]

An exact knowledge of the mercury compounds allows statements on the toxic state of a sediment. The total mercury content alone does not suffice, because the different mercury species clearly have different toxicities. In general the mercury species may be divided into the following three groups, with increasing toxicity

- inorganic mercuric salts
- mercury vapour (Hg^0), aryl - or methoxymercuric - compounds
- alkyl mercuric - compounds.

2. Analysis

In order to investigate the metabolism, transport und toxicity of pollutants in ecosystems it is important to determine their exact physico-chemical forms. From the analytical point of view the following well defined mercury compounds must be examined:

- Hg^0
- Hg_2^{2+}, Hg^{2+}
- CH_3-Hg-X
- CH_3-Hg-CH_3

From the chemical point of view these species are important for the characterization of a sediment. A knowledge of the sediment contents of these species allows preliminary statements about the potential threat posed by a sediment to the food chain. However, more parameters are possibly required for definitive assessment. These parameters should for example give some informations about the bioavailability of mercury compounds.

With this in mind groupings such as

- "dissolved inorganic" mercury
- "ion-exchangeable" mercury
- "total organic" mercury

may be used.

The current state of refinement of analytical methods will be discussed below in relation to the compounds mentioned above.

The basis of all procedures for the analysis of mercury and its compounds is the determination of total mercury.

2.1. Total mercury

The determination of total mercury in sediments is at present easy to carry out, as intercomparison tests show; it is however not as good as for other heavy metals. A wet digestion with oxidizing agents in open or closed systems is the usual method implemented. Of course there are some methods used, with different detection limits. The digestion solution is often analysed using cold vapour atomic absorption spectrometry (CVAAS) [Jonasson and Boyle, 1972]. Neutron activation analysis (NAA) is another sensitive and often used reference method [Van Delft and Vos, 1988]. But NAA is only available as a routine method in a few laboratories.

2.2. Group separation

The analytical methods discussed below are methods for the determination of individual and well defined mercury species. As previously mentioned, a variety of interconversion reactions involving a number of chemical compounds of mercury take place in nature. Another point to consider is the continuous biological production and degradation of mercury species which is particularly important for methylmercury. It is therefore obvious that even the best analysis can only describe the steady state in the environment. Even slight variations in pH, redox potential or salt content may change the ratio of the individual species drastically. The total concentration of invidual mercury compounds may not correlate well with their bioavailability. Mercury species dissolved in pore water are probably more toxic than compounds which are strongly attached to particles in the sediment. For this reason methods have frequently been reported to determine group parameters of mercury and not the individual species. These group parameters represent several mercury compounds with different chemical structure [Robertson *et al.*, 1985; Di Giulio and Ryan, 1987; Schmidt and Freimann, 1984; Jackson *et al.*, 1982].

A differentiation between inorganic and organic mercury compounds is possible with the atomic absorption technique. A procedure was developed which allows a sequential reduction of the different mercury species [Magos, 1971]. Thus, inorganic mercury can be easily reduced by tin(II)-chloride. Organic compounds are more resistant and are reduced by $SnCl_2$ only with the catalytic aid of certain heavy metal cations such as Cu^{2+} or Cd^{2+}. Another way to reduce the organic mercury compounds is with sodiumborohydride [Oda and Ingle, 1981]. With this procedures the organic compounds are not unambiguously identified, because only the total of the organomercurials are determined.

2.2.1. "Total dissolved inorganic" mercury

The above definition should encompass all the mercury which is dissolved in the aqueous phase and which could be measured by CVAAS after reduction with $SnCl_2$. Thus, this measurement represents a combination of Hg^0, Hg_2^{2+} and Hg^{2+}. The aqueous phases can be obtained either by extracting the sediment with distilled water or by measuring the pore water concentrations. Because this fraction is easily reducible it is sometimes referred to as the "reactive" mercury fraction [Schmidt and Freimann, 1984].

2.2.2. "Ion-exchangeable" mercury

This fraction can be estimated after leaching the sediment with various solutions e.g. solutions of ammonium acetate or potassium chloride. These extraction procedures should release (exchange) the mercury associated with organic matter in sediment.

2.2.3. "Total organic" mercury

The measurement of organic forms of mercury in sediments has received much less attention and has frequently involved the analysis of "total organic" mercury. This would include alkyl and aryl mercuric species as well as mercury bound to natural humic substances dissolved in the aqueous phase. This "total organic" mercury is determined by first measuring the amount of "total dissolved inorganic" mercury. The water sample is then oxidized by the addition of an oxidizing reagent e.g. BrCl [Robertson *et al.*, 1987] in order to convert all forms of mercury to Hg^{2+}. The organically bound mercury which has thus been liberated is reduced and analysed by CVAAS. The difference between these measurements provides a measure of "total organic" mercury concentration. With a more detailed protocol, the "organic" mercury can be divided into a humic acid and a fulvic acid fraction. This can be achieved by extracting the sediment with solutions of different pH because humic and fulvic acids have varying solubilities in alkaline and acid solutions [Di Giulio and Ryan, 1987]. One disadvantage of these group separations is the difficulty of working always under the same experimental conditions. Another more serious weakness is that the binding sites of the mercury compounds may be changed due to the leaching conditions. So the obtained results might not represent the "real" situation in the sediments. This restriction must be considered particularly for the sequential leaching methods where the sediments come in contact with a number of chemicals. Similar fractionation studies have revealed that mercury was present in sediments in relatively immobile and non-bioavailable forms. In most cases the mercury compounds are associated with organic matter of the sediments. The water soluble and the ion exchangeable fraction contains almost no mercury at all. Due to the origin or composition of the sediments the majority of the mercury species are found in the "dissolved organic" fraction or in the residual, unsoluble fraction.

2.3. Individual inorganic species

2.3.1. Hg^0

The analysis of mercury in its elemental state in environmental samples is very difficult and, to date, no satisfactory procedure has been found. To begin with, there is the problem of isolating the Hg^0 from the sediment. It is possible to volatilise Hg^0 from the sediment because of its high vapour pressure. But it must be remembered that Hg^0 is not the only volatile mercury species. Dimethylmercury and, depending on the experimental conditions, methylmercury may also be volatilised from sediments. Hence the evaporation of Hg^0 out of a suspension of sediment by a stream of nitrogen and selective collection of the Hg^0 on silver wool is recommended [Robertson *et al.*, 1987]. Subsequently, the collected mercury is transfered to a gold net and determined by CVAAS. In this manner it is possible to obtain the amount of Hg^0 in the sediment directly. Floyds group determines Hg^0 indirectly. All the mercury species which can be volatilised by steam are distilled out of the sediment and captured in an acidic $K_2S_2O_8$ - solution. All organomercurial compounds and Hg^0 are oxidized to Hg^{2+} and then determined by CVAAS. A modified procedure measures organomercurial compounds only. Elemental mercury will be removed by a reaction with zinc granules, placed in front of the trapping solution. From the difference between these measurements the Hg^0 content is obtainted. Both of these procedures have a critical weakness: it is only Hg^0 which is selectivly adsorbed from silver or stannium or is it also possible to collect organomercurial species? There is evidence in the literature that dimethylmercury and methylmercury can be adsorbed on gold or silver surfaces [Robertson *et al.*, 1987; Dumarey, 1985].

Tests in our laboratory confirm that dimethylmercury and methylmercury were adsorbed on gold/platinum nets, as they are used for CVAAS [Kuhn, 1989]. The kinetics of adsorption is much slower than for Hg^0. Up to four nets must be connected serially for a quantitative absorbance of dimethylmercury which was volatiliseed out of a water solution. It has also been observed that Hg^0 evaporates quantitatively out of a water solution within a very short period of time, and that for dimethylmercury and methylmercury this process takes much longer. Another source of uncertainty with the Hg^0 analysis is sample handling and the sample preservation. The usual addition of oxidizing agents is obviously impossible and the addition of acids is problematic, because the change in pH may lead to a reduction or a release of Hg^0. The danger of mercury evaporation makes it advisable to analyse the samples immediately after collection or at least to isolate Hg^0 from the sediment right at the sampling site rather than conducting the analysis in the laboratory.

No reference materials are available for Hg^0 and the author's opinion is that the production of certified materials is impossible. For this reason the quality and the comparability of analytical methods will even in the future be restricted.

2.3.2. HgS

The solubility of mercurysulphide in water is extremely low ($1.1 \times 10^{-16} \mu g/l$), therefore mercury which is fixed as HgS is a lesser evil than other mercury species. However microorganisms can oxidize HgS to $HgSO_4$ and reintegrate it into the mercury - cycle [Fagerström and Jernelöv, 1971].

Until recently, there was little interest in the determining HgS in sediments, but it is now possible to determine HgS using a sequential extraction scheme [Revis *et al.*, 1989]. In the first extraction all organic, inorganic and elemental mercury was removed from sediment with concentrated HNO_3. Subsequently, HgS is removed from the sediment using a second extraction with saturated sodium sulfide solution. This solution is then digested and the mercury is detected by CVAAS. Using this method, 74-100 % of the total mercury in a soil sample was in the form of HgS.

2.4. Individual organic species

2.4.1. Alkylmercury compounds

With regard to the organomercurials, it is the alkyl compounds which are more toxic than the aryl compounds. Methylmercury is especially dangerous. Since the pioneer work of Westøø in Sweden and Sumino in Japan [Westøø, 1968; Sumino, 1968] several investigations in this field were carried out.

To begin with, a suitable strategy for the isolation of methylmercury from sediments must be developed. It must be taken into consideration, that this compound is a strong thiophilic reagent. Many compounds, which contain sulfydryl (-SH) end groups are present in sediments. Proteins, in particular, contain this functional group because of their content of cystein as a component amino acid. These end groups can also be found in humic acids.

The situation is even more complex, because mercury also has an affinity for other complexing agents, which occur in organic matrices, namely organic acids and the amino group ($-NH_2$). The sediments are therefore acidified before any further treatment. Using halogenic acids has two advantages: with a high acid concentration the complexing groups such as $-S^-$, $-NH_2$ and $-COO^-$ are protonated and the methylmercury cation CH_3Hg^+ is set free. The halogen anion then forms the covalent unpolar methylmercury halogenide with this cation, which can be separated by different procedures from the acid solution.

The steam distillation can be mentioned here, because there is evidence in the literature that methylmercury chloride is volatilised by steam [Floyd and Sommera, 1975; Nagase *et al.*, 1980]. The methylmercury in the steam can be trapped with an organic solvent, an oxidizing $K_2S_2O_8$-solution or in a concentrated hydrochloric acid solution and could be analysed afterwards.

Another possibility is the separation of methylmercury chloride from the Hg^{2+}, which could also be in solution, by means of an acid ion-exchange column. Here the Hg^{2+} is fixed on the column as an $HgCl_4^{2-}$-complex, whereas the methylmercury is eluted as methylmercury chloride. The separated methylmercury chloride is analysed with this procedure after an UV-digestion or a wet digestion with CVAAS [Stoeppler *et al.*, 1987a].

The extraction of methylmercury from the acid solution with organic solvents is much more common. In a first extraction step methylmercury is extracted with toluene although benzene has also been used. For the further clean up, subsequent back extraction into an aqueous solution is recommended. This is achieved by a thiosulphate or cystein solution [Newsome, 1971; Uthe, 1971]. If the alkylmercury compounds are to be separated by high pressure liquid chromatography (HPLC), the aqueous extract can be injected directly. If the separation is to be carried out by gas chromatography (GC), another step is needed to transfer the methylmercury back into an organic phase. By acidifying the cystein extracts the methylmercury chloride is set free and can be extracted with toluene.

Membrane separation of Hg species

In our laboratory a procedure is under investigation, in which the alkylmercury is separated by a membrane from interfering agents [Grabau, 1989]:

Several different membranes were investigated for their separation behaviour regarding Hg^{2+} and CH_3Hg^{2+}. The best results were obtained with a pore-free silicone- and a microporous Celgard membrane, which is transversed only by the methylmercury. The diffusion through the silicon membrane was appreciably quicker than through the other membranes under investigation. An example is given in figure 3.

It has been shown in laboratory dialysis experiments with methylmercury spiked sediments, that after 24 hours only 4 % of the spiked amount of methylmercury penetrated a silicone membrane. Thus the recovery of methylmercury using this procedure is less than quantitative. Presumably, only the labile fraction of the total methylmercury is captured in this manner, which is possibly in connection with the bioavailibility of this compound under these circumstances.

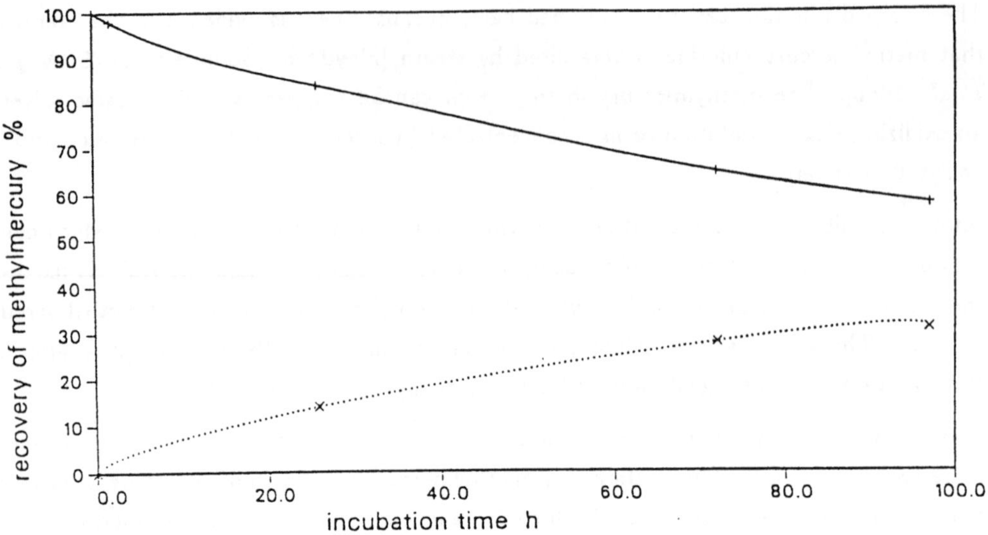

Figure 3. Methylmercury chloride passing a Celgard membrane (— feed, - - - product)

Gas Chromatography: separation and detection

The separation and quantitative determination of organic mercury compounds was, in the past, performed mainly with GC in combination with a suitable detector.

The electron capture detector (ECD) has been used frequently and responds efficiently to the halogen in alkylmercury compounds [Cappon and Smith, 1977; Filipelli, 1987; Cappon and Toribara, 1987]. On the other hand, the response to dialkylmercury compounds is, as expected, poor. Another disadvantage of the ECD is, that the extracts must be very thoroughly cleaned due to the sensitivity of the detector. For this reason many other element specific detectors have also been investigated and, frequently, these detectors are not available commercially.

An exception to this is the coupling of a mass-spectrometer as a detector to the GC or another chromatographic system; GC/MS results and spectra have been reported frequently [Johansson *et al.*, 1970; Frimmel and Winkler, 1975]. Very good detection efficiencies are shown by detectors for example, which use a microwave induced plasma (MIP-AES) [Bache and Lisk, 1971; Talmi, 1975; Chiba *et al.*, 1983]. Another possibility in an element specific detection lies in using an atomic absorption spectrometer after a chromatographic separation. Here the organomercurials are eluted from a column, degraded pyrolitically in a hot glas tube and the resulting Hg^0 is measured in the measuring cell [Dumarey *et al.*, 1982; Guy-Bin *et al.*, 1989; Paudyn and Van Loon, 1985; Robinson and Wu, 1985].

The biggest problem lies not in the detection system, but in the chromatographic behaviour of the mercury compounds. In the past often unwelcomed interactions between the column material and the mercury compounds or about thermal instabilities were reported [O'Reilly, 1981; Rodriguez-Vazquez, 1977]. In particular the injector block of the GC is seen as critical, where thermal degradation can occur [Dressman, 1972]. Because metal surfaces, under certain circumstances, can initiate catalytical degradation, the separation should be performed, if possible, in the absence of metal surfaces; glass columns should be used wherever possible.

The best chromatography has been reported with the so called DEGS (diethyleneglycolsuccinate) and BUDS (butanediolsuccinate) - column materials. A series of problems can still arise:

- peak tailing
- irreproducible retention on the column, which is mainly due to strong interactions with the column material and to column degradation.
- when extracts from real environmental samples are injected, the peak height and -area decreases significantly with successive injections, whereas standards give good chromatograms [O'Reilly, 1982].
- the chromatographic system must be conditioned for extended periods before and between analytical runs.

The conditioning is achieved by injecting alkali halide (KCl, KI) or by injection of mercury as $HgCl_2$. The conditioning can also be called a passivation, because the interactions between the column material and the mercury compounds are reduced.

The successful separation of organomercury compounds with capillary columns has been reported [Cappon and Toribara, 1987; Guy-Bin *et al.*, 1989].

HPLC: separation and detection

Another procedure for the separation of organic mercurials is by HPLC. This method has developed into a powerful analytical method in the last years. Chromatography on non polar phases, especially with octadecyl modified silica gel, has been applied successfully to mercury analysis. These so called "reverse phases" provide a wide scope with regard to eluent composition. Mixtures of methanol/water or acetonitrile/water are commonly used.

In the analysis of organometallics, the derivatisation of the compounds plays an important role. Without derivatisation, the interactions of small non polar molecules (methylmercury chloride) with the stationary phase is only small. As a consequence, organomercury compounds are eluted from the column together with the solvent front and there is no separation of the analytes.

The "charge neutralisation chromatography", which has been developed especially for mercury compounds [MacCrehan *et al.*, 1977], is in opposition to the expensive chelatisation of the metalorganyles by dithiocarbamates or dithizone [Langseth, 1986]. In this method mercaptoethanol is added to the eluent and the complexation takes place "in situ" during the sample injection. The reaction occuring can be described as follows:

$$RHgCl + HS(CH_2)_2OH \rightleftharpoons RHgS(CH_2)OH + H^+ + Cl^-$$

The resulting mercaptoethanol complex is retained on the column, so that a chromatographic separation is possible.

A variety of detectors have been used for the detection of liquid chromatographically separated mercury compounds. Electrochemical detection [Evans and McKee, 1988; MacCrehan and Durst, 1978] should be mentioned first. This method is characterized by its selectivity and high sensitivity, but a disadvantage is the costly preparation of the sample and the eluents, because the solvents used must be free of oxygen, which interferes with the detection process. With other detectors used, the coupling to the HPLC is not readily performed. In a recent procedure, the detection by atomic emission follows the excitation of the mercury species by microwave plasma [Tölg *et al.*, 1984] which is contained in a laboratory assembled device. A similar measuring principle is the atomic emission spectroscopy in an inductively coupled plasma (ICP/AES) [Krull *et al.*, 1986]. The inductively coupled plasma coupled with a mass-spectrometer is a relatively new instrumentation. A coupling of a liquid chromatographic system to such an instrument is commercially available. The determination by atomic absorption has also been reported. The HPLC column eluate can be collected in fractions and subsequently the fractions can be assayed automatically by graphite furnace AAS (GFAAS) [Brinckman *et al.*, 1977]. It is also possible, to pass the eluant through a heated glass tube, so that the mercury compounds are degraded to Hg^0 which is passed into the measuring chamber of the AAS [Holak, 1982]. Both procedures are characterized by a very costly experimental design.

We developed an HPLC procedure, which allows the detection of the separated mercury compounds with the commonly used UV detector [Wilken *et al.*, 1990]. Although the extinction maxima of the organomercurials are below 210 nm (a wavelength, at which the adsorption of the elutant begins), the detection is also successful at 230 nm.

The procedure developed is described in table 1.

Table 1. HPLC-procedure for the analysis of methylmercury

1.)	Weigh 10 g wet sediment, add 50 ml 6 M HCl, shake overnight, centrifuge at 2000 · g
2.)	Extract supernatant solution with 100 ml toluene, reduce to ca. 5 ml (rotary evaporator, 50°C)
3.)	Extraction with 1 ml 0.1 mM $Na_2S_2O_3$, 0.05 M CH_3COONH_4 (adjusted to pH = 5.5 with acetic acid), centrifuge to hasten phase separation
4.)	HPLC – chromatography of the aqueous phase eluent: 20% methanol, 1 mM CH_3COONH_4, pH = 5.5, 0.5 mM 2-mercaptoethanol column: LiChrospher RP-18, 150-3 (5 μm) glass cartridge
5.)	Detection: UV-photometric at 230 nm

2.4.2. Dialkylmercury compounds

From a toxicological point of view, the most important dialkylmercury compound in the environment is dimethylmercury. It is probably formed by microbiological methylation of Hg^{2+}. This compound is extremely volatile and is easily degraded by light or chemical processes. Nevertheless this compound must also be considered because it is as acutely toxic as monomethylmercury.

If dimethylmercury is to be separated from sediments, the sample should not be acidified. In acidic media the CH_3-Hg-CH_3 is degraded quickly to CH_3Hg^+ and could no longer be detected [Udelnow and Rautschke, 1986; Frimmel and Winkler, 1975]:

$$CH_3\text{-}Hg\text{-}CH_3 + HX \rightleftharpoons CH_3\text{-}Hg\text{-}X + CH_4$$

This facile dealkylation should be taken into consideration if monomethylmercury values are to be interpreted. The results obtained could possibly be a result of the sum of dimethylmercury and monomethylmercury; but it must also be said, that the concentrations of dimethylmercury in the environment are considered to be very low.

Thus it is most favorable to extract the sediment directly with organic solvents. The dimethylmercury contained in the organic phase could be analysed afterwards in different ways: one possibility is to convert it into methylmercury chloride [Udelnow and Rautschke, 1986; Frimmel and Winkler, 1975] and then to determine it by the methods described for methylmercury, or another possibility is to determine it directly as dimethylmercury. In the latter approach a gas chromatographic separation of the extract and detection with a suitable detector is recommended. The ECD, often used in gaschromatography, is not so well suited for the determination of dimethylmercury. Other combinations are therefore described. Good detection possibilities are found in coupling GC/AAS [Paudyn and Van Loon, 1985; Guy-Bin *et al.*, 1989; Dumarey *et al.*, 1982] and GC/MIP [Bache and Lisk, 1971; Talmi, 1975], which both allow the element specific determination of mercury compounds.

Another approach to $(CH_3)_2Hg$ isolation is the water steam distillation [Floyd and Sommera, 1975]. Because dimethylmercury is a very volatile compound, it could be volatilised from sediment by gas or water steam. Afterwards the dimethylmercury must be trapped in a suitable solvent and analysed. The critical point in this method is the discrimination from other volatile mercury compounds, where especially Hg^0 must be considered. In the literature the selective trapping of Hg^0 on Zn [Floyd and Sommera, 1975] or Ag [Robertson *et al.*, 1987] is recommended.

Dimethylmercury is possibly the only mercury compound which could be extracted from sediments with organic solvents only without direct acidification; other species are insoluble or are complexed by sediment components. We conducted therefore experiments to digest dimethylmercury in organic solvents to Hg^{2+} with following determination by CVAAS [Kuhn, 1989].

Digestion in organic solvents

As a digestion method we used a UV-digestion with a Hg vapour high pressure lamp (1000 W). With 10-15 min of irradiation of the organic solvents the maximum digestion was achieved. Thus up to 85 % of dimethylmercury dissolved in methanol was digested, 93 % dissolved in ethanol and 78 % dissolved in hexane. The digested solutions could be analysed directly with the CVAAS and $SnCl_2$ as a reduction agent. No interferences from the solvents were observed.

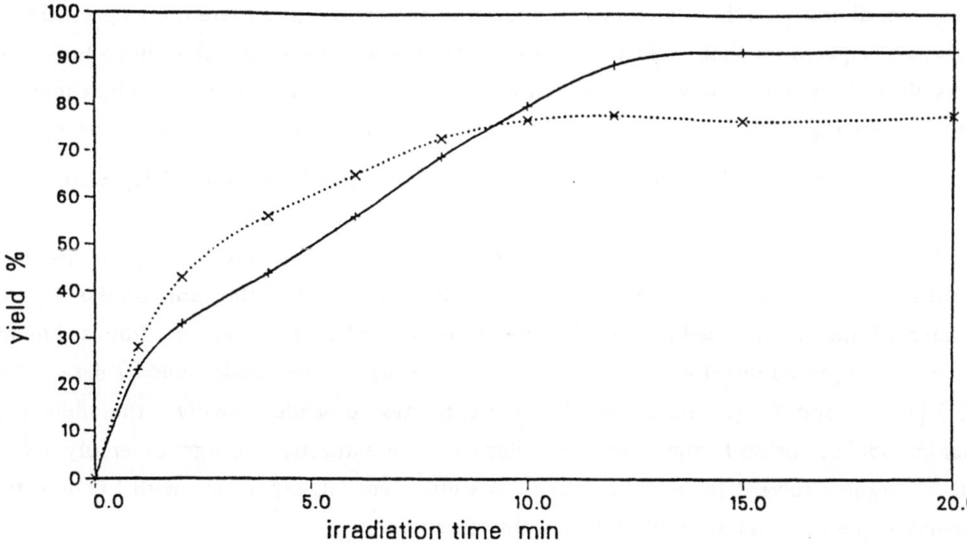

Figure 4. CH_3-Hg-CH_3 digestion (— in ethanol, - - - in heptane)

Sample preparation

Another unsolved problem in the analysis of dimethylmercury is the sample preparation. Because of the instability of this compound an acidification of the sample is not possible. As the best conservation method immediate shock freezing with liquid nitrogen is proposed, in spite of the fact, that only small amounts of samples could be conserved in this manner. As for the Hg^0 analysis no certificated reference materials are available.

2.4.3. Other organomercurial compounds

A series of organomercurials were developed for agriculture and fruit growing applications in which these products were used principally as seed dressings and fungicides. Because of their toxicity they are banned in most countries and have been replaced by other pesticides. Commonly used were phenylmercury and alkoxyethylmercury salts. Therefore some efforts were also made to develop methods for the determination of such compounds in sediments.

There are, of course, other mercury compounds in sediments, and therefore true speciation analysis is required for that problem. For the determination of mercury containing fungicides in disinfected seeds or in harvested fruits, it may be sufficient to determine the whole mercury content after digestion of the sample, but in sediments an isolation seems to be neccessary.

An extraction method often used is with an organic dithizone [Tatton and Wagstaffe, 1969; Tobschall and Rubischung, 1980]. This extract can be investigated by thin layer chromatography (TLC) [Tobschall and Rubischung, 1980]. The identification is done by the Rf-values of the single compounds and the quantitative analysis is possible if the spots on the layer are scraped off and are analysed in the AAS after digestion. The organic extract from sediments could also be placed directly on a GC column and the organomercurials could be detected by ECD [Tatton and Wagstaffe, 1969]. Another extraction procedure involves the addition of sodium iodide solution to the sediment, followed by an extraction of organomercury iodides into an organic solvent [Ealy *et al.*, 1973]. A subsequent GC separation with EC detection permitted the determination of methoxyethylmercury.

3. Biological Aspects

In the biosphere mercury and its compounds undergo numerous interconversions which could be induced by chemical, photochemical or biochemical means. In soils, sediments and aquatic systems microbial reactions are of great significance. The formation of methylmercury by microorganisms probably is the most important biological process. Mercury methylation is not restricted to a limited number of bacteria, but there is some evidence that sulphate reducing bacteria might play an important role in mercury methylation [Compeau and Bartha, 1987]. These microorganisms can produce large quantities of methylmercury with increasing salinities. The methylation of mercury can be regarded as a detoxification mechanism [Wood, 1974] for bacteria. On the other hand, however, it must be noted that microorganisms are at the beginning of the aquatic food chain which ends at man.

Concentrations of mercury in fish of 20-50 μg/kg with methylmercury making up 90-99% of the total mercury have been reported [Stoeppler *et al.*, 1987b]. It has been reported that fish or crayfish are not able to convert inorganic mercury to methylmercury and that food is the dominant pathway for methylmercury accumulation. This hypothesis was corroborated by experiments with caged crayfish which were exposed to different contaminated and uncontaminated waters and to varied diets containing known mercury concentrations [Parks *et al.*, 1988].

A dynamic equilibrium exists between methylation and demethylation in aquatic systems and sediments. The greatest potential for mercury methylation was observed in surfacial sediments which are in contact with the overlying water phase and decreased with depth into sediment cores. This profile of methylating activity correlates well with the known depth distribution of overall microbial activity in sediments and implies that the greatest potential for mercury methylation in sediments occurs where microbial activity is high. In the water column the mercury methylation was low but detectable [Korthals and Winfrey, 1987]. The demethylation activity was also greatest in surface sediments but decreased only slightly with sediment depth. The result was that the methylation/demethylation ratio was < 1 in the water column, exhibited a sharp peak in surfacial sediment and decreased in deeper sediment [Korthals and Winfrey, 1987]. Reduction of pH can increase methylation activity at the sediment surface. This indicates that the net rate of methylmercury production in the water column and at the sediment water interface can increase as a result of lake acidification, and that this may explain, at least partially, why the mercury concentration of fish appears to increase during lake acidification [Xun *et al.*, 1987].

The methods for the determination of methylation and demethylation activities rely mainly on radioactive tracers. The sediments are incubated with $^{203}Hg^{2+}$ and the methylmercury formed during incubation is extracted afterwards with organic solvents. Finally the radioactivity in the extracts is measured and may be regarded as the amount of methylmercury formed. These methods give only relative production rates and no absolute rates. The rate of demethylation can also be determined by measuring the methane which is produced during demethylation [Spangler *et al.*, 1973]. The disadvantage of these methods is that the sediments must be incubated with large quantities of mercury compounds. So it is only possible to simulate the conditions of a highly contaminated sediment. These conditions may not be identical to conditions at uncontaminated sites. The methylation and demethylation activities in these sediments are probably different.

Acknowledgement

We greatfully acknowledge the analytical work of Jörg Grabau and Andreas Kuhn.

References

Bache CA, Lisk DJ (1971) Gas chromatographic determination of organic compounds by emission spectrometry in a helium plasma. Application to the analysis of methylmercuric salts in fish. Anal Chem 43:950-952

Brinckman FE, Blair WR, Jewett KL, Iverson WP (1977) Application of a liquid chromatograph coupled with a flameless atomic absorption detector for speciation of trace organometallic compounds. J Chromatog Sci 15:493-503

Cappon CJ, Smith JC (1977) Gas-chromatographic determination of inorganic mercury and organomercurials in biological materials. Anal Chem 49:385-369

Cappon CJ, Toribara TY (1987) Mercury speciation by capillary GLC. LC & GC 4:1010-1014

Chiba K, Yoshida K, Tanabe K, Haraguchi H, Fuwa K (1983) Determination of alkylmercury in seawater at the nanogram per liter level by gas chromatography/atmosheric pressure helium microwave-induced plasma emission spectrometry. Anal Chem 55:450-453

Compeau GC, Bartha R (1987) Effect of salinity on mercury-methylating activity of sulfate-reducing bacteria in estuarine sediments. Appl Environ Microbiol 53:261-265

Di Giulio RT, Ryan EA (1987) Mercury in soils, sediments and clams from a North Carolina peatland. Water Air Soil Poll 33:205-219

Dressman RC (1972) The conversion of phenylmercuric salts to diphenylmercury and phenylmercuric chloride upon gas chromatographic injection. J Chromatog Sci 10:468-471

Dumarey R, Dams R, Sandra P (1982) Interfacing an atomic absorption spectrophotometer with a fused silica capillary gas chromatographic system for the determination of trace organomercury(II) compounds. Journal of HRC & CC 5:687-689

Dumarey R, Dams R, Hoste, J (1985) Comparison of the collection and desorption efficiency of activated charcoal, silver and gold for the determination of vapor-phase atmospherical mercury. Anal Chem 57:2638-2643

Ealy JA, Shults WD, Dean JA (1973) Extraction and gas chromatographic determination of methyl-, ethyl- and methoxymethylmercury(II) halides. Anal Chim Acta 64:235-241

Evans O, McKee GD (1988) Determination of mercury(II) and organomercury compounds by reversed-phase liquid chromatography with reductive electrochemical detection. Analyst 113:243-246

Fagerström T, Jernelöv A (1971) Formation of methylmercury from pure mercuric sulphide in aerobic sediment. Water Res 5:121

Filipelli M (1987) Determination of trace amounts of organic and inorganic mercury in biological materials by graphite furnace atomic absorption spectrometry and organic mercury speciation by gas chromatography. Anal Chem 59:116-118

Floyd M, Sommera LE (1975) Determination of alkylmercury compounds in lake sediments by steam distillation-flameless atomic absorption. Anal Letters 8:525-535

Frimmel F, Winkler HA (1975) Differenzierte Bestimmung verschiedener Quecksilberverbindungen in Wasser und Sediment. Vom Wasser 45:285-298

Fujiki M (1972) The transitional condition of Minamata Bay and the neighbouring sea polluted by factory waste water containing mercury. 6th Int Conf Water Poll Res Paper No. 12

Grabau J (1989) GKSS, personal communication

Gui-Bin J, Zhe-ming N, Shun-rong W, Heng-bin H (1989) Organic mercury speciation in fish by capillary gas chromatography interfaced with atomic absorption spectrometry. Fresenius Z Anal Chem 334:2730

Holak W (1982) Determination of methylmercury in fish by high-performance liquid chromatography. Analyst 107:1457-1461

Jackson TA, Parks JW, Jones PD, Woychuk RN, Sutton JA, Hollinger JD (1982) Dissolved and suspended mercury species in the wabigoon river (Ontario, Canada): Seasonal and regional variations. Hydrobiol 92:473-487

Johansson B, Ryhage R, Westøø G (1970) Identification and determination of methylmercury compounds in fish using combination gas chromatography-mass spectrometer. Acta Chem Scand 24:2349-2354

Jonasson IR, Boyle RW (1972) Geochemistry of mercury and origins of natural contamination of the environment. The Canadian Mining and Metalurgical Bulletin 1-8

Korthals ET, Winfrey MR (1987) Seasonal and spatial variations in mercury methylation and demethylation in an oligotrophic lake. Appl Environ Microbiol 53:2397-2404

Krull IS, Bushee DS, Schleicher RG, Smith SB (1986) Determination of inorganic and organomercury compounds by high-performance liquid chromatography - inductivly coupled plasma emission spectrometry with cold vapour generation. Analyst 111:345-349

Kuhn (1989) GKSS, personal communication

Langseth W (1986) Determination of organic and inorganic mercury compounds by reverse-phase high-performance liquid chromatography after extraction of the mercuries as their dithizonates. Anal Chim Acta 185:249-258

Langseth W (1986) Determination of organic and inorganic mercury compounds by reverse-phase high-performance liquid chromatography after extraction of the mercuries as alkyldithiocarbamate chelates. Fresenius Z Anal Chem 325:267-271

MacCrehan WA, Durst RA, Bellama JM (1977) Electrochemical detection in liquid chromatography: Application to organometalic speciation. Anal Letters 10:1175-1188

MacCrehan WA, Durst RA (1978) Measurement of organomercury species in biological samples by liquid chromatography with differential pulse electrochemical detection. Anal Chem 50:2108-2112

Magos L (1971) Selective atomic-absorption determination of inorganic mercury and methylmercury in undigested biological samples. Analyst 96:847-853

Nagase H, Ose Y, Sato T, Ishikawa T, Mitani K (1980) Differential determination of alkylmercury and inorganic mercury in river sediment. Intern J Environ Anal Chem 7:285-293

Newsome WH (1971) Determination of methylmercury in fish and in cereal grain products. J Agr Food Chem 19:567-569

O'Reilly JE (1982) Gas chromatographic determination of methyl and ethyl mercury: "Passivation" of the chromatographic column. J Chrom 238:433-444

Oda CE, Ingle JD (1981) Speciation of mercury by cold vapor atomic absorption spectrometry with selective reduction. Anal Chem 53:2305-2309

Parks JW, Sutton JA, Hollinger JD, Russel DD (1988) Uptake of mercury by caged crayfish. Appl Organomet Chem 2:181-184

Paudyn A, Van Loon JC (1985) Determination of organic forms of mercury and arsenic in water and atmospheric samples by gas chromatography-atomic absorption. Fresenius Z Anal Chem 325:369-376

Ramamoorthy S, Moore JW (1984) Heavy metals in natural waters, 1st edn. Springer, Berlin Heidelberg New York

Revis NW, Osborne TR, Sedgley D, King A (1989) Quantitativ method for determining the concentration of mercury(II) sulphide in soils and sediments. Analyst 114:823-825

Robertson DE, Sklarew DS, Olsen KB, Bloom NS, Crecelius EA, Apts CW (1987) Measurement of bioavailable mercury species in fresh water and sediments. Research Project 2020-3, Final Report, Batelle Pacific Northwest Laboratories

Robinson JW, Wu JC (1985) Speciation of mercury compounds in biological samples using GC-AAS. Spect Letters 18:47-69

Rodriguez-Vazquez JA (1977) Gas-chromotographic determination of organomercury(II) compounds. Talanta 25:299-310

Schmidt D, Freimann P (1984) Determination of very low levels of mercury in sea water (North Sea, Baltic Sea, Northern North Atlantic Ocean). Fresenius Z Anal Chem 317:385-387

Spangler WJ, Spigarelli JL, Rose JM, Flippin JS, Miller HH (1973) Methylmercury: Bacterial degradation in lake sediments. Science 180:192-193

Stoeppler M, Ahmed R, May K (1987a) Wet deposition of mercury and methylmercury from the atmosphere. Sci Total Environ 60:249-261

Stoeppler M, May K, Reisinger K (1987b) Studies in the ratio total mercury/methylmercury in the aquatic food chain. Tox Environ Chem 13:153-159

Sumino K (1968) Analysis of organic mercury compounds by gas chromatography, Part II. Kobe J Med Sci 14:131-148

Talmi Y (1975) The rapid sub-picogram determination of volatile organo-mercury compounds by gas chromatography with a microwave emission spectrometer detektor system. Anal Chim Acta 74:107-117

Tatton JO'G, Wagstaffe PJ (1969) Identification and determination of organomercurial fungicide residues by thin-layer and gas chromatography. J Chromatogr 44:284-289

Tobschall HJ, Rubischung P (1980) Identification and determination of some environmental organomercurials in recent fluvatile sediments by means of thin-layer chromatography. Chem Erde 29:239-275

Tölg G, Lorenz I (1977) Quecksilber - ein Problemelement für den Menschen. Chem uns Zt 5:150-156

Tölg G, Kollotzek D, Oechsle D, Kaiser G, Tschoepel P (1984) Application of a mixed-gas microwave induced plasma as an on-line element-specific detector in high-performance liquid chromatography. Fresenius Z Anal Chem 318:485-489

Udelnow C, Rautschke R (1986) Determination of mercury species in wastewater. Z Chem 26:297-298

Uthe JF (1971) Determination of total and organic mercury levels in fish tissue. In: Westley B (ed) Int Symp Identification Meas Environ Poll [Proc.] p 207-212

Van Delft W, Vos G (1988) Comparison of digestion procedures for the determination of mercury in soils by cold vapour atomic absorption spectrometry. Anal Chim Acta 209:147-156

Westøø G (1968) Determination of methylmercury in various kinds of biological material. Acta Chem Scand 22:2277-2280

Wilken RD, Hintelmann H, Ebinghaus R (1990) Biologische Quecksilberumsetzungen in der Elbe. Vom Wasser 74:383-392

Wood JM (1974) Biological cycles for toxic elements in the environment. Science 183:1049-1052

Xun L, Campbell NER, Rudd JWM (1987) Measurement of specific rates of net methylmercury production in the water column and surface sediments of acidified and circumneutral lakes. Can J Fish Aquat Sci 44:750-757

STABILITY STUDY OF Cr(III) AND Cr(VI) IN WATER FOR PRODUCTION OF AN AQUEOUS CHROMIUM REFERENCE MATERIAL

Steen DYG, Rita CORNELIS, Ben GRIEPINK* and Paul VERBEECK**
Laboratory of Analytical Chemistry,
Rijksuniversiteit Gent
I.N.W., Proeftuinstraat 86
B-9000 Gent, Belgium

1 INTRODUCTION

The differences in biological availability and in toxicity of hexavalent chromium compared with trivalent chromium is of great scientific and public health concern. However little is known about the fate of the different chromium species in the aquatic environment. Further studies are needed to investigate the environmental impact of chromium(III) and (VI) discharges.

To perform such studies successfully an aqueous Cr(III) and Cr(VI) reference material is needed to assess the quality of measurement and obtain comparability (Griepink, 1984; WHO, 1988). Such material would need to have long term stability of both species in water at concentrations similar to those occurring in waters and in a chemical environment which, though inevitably artificial, resembles the natural one.

In this study, different parameters have been investigated such as the contents of Cr(III) and Cr(VI) as a function of time, different solutions, possible losses due to adsorption, temperature dependency and choice of the material of the container.

---

* CEC, BCR, Directorate-General for Science, Research and Development, Wetstraat 200, B-1049 Brussel, Belgium

** Laboratory of Astronomy, Geophysics, Mathematical Statistics, Rijksuniversiteit Gent, Krijgslaan 281, B-9000 Gent, Belgium

NATO ASI Series, Vol. G 23
Metal Speciation in the Environment
Edited by J. A. C. Broekaert, Ş. Güçer, and F. Adams

Requirements on water quality with regard to dissolved hexavalent chromium in drinking water and solubility considerations of Cr(III) have been taken into account when choosing a representative concentration 25 µg/l of Cr(III) and 25 µg/l of Cr(VI). After preliminary experiments it was concluded that a 50 mmol/l HCO_3^-/H_2CO_3 buffer (typically HCO_3^- content in river water is approx. 1 mmol/l) at pH 6.4 under CO_2 blanket could be considered as a suitable medium to achieve stable solutions of the Cr(III) and Cr(VI) species. In choosing hydrogencarbonate as a buffer and an agent to prevent hydrolysis of Cr(III), a matrix very close to that of real waters is achieved.

Both Cr(III) and Cr(VI) were detected using their respective label ^{51}Cr. The method used to differentiate between Cr(III) and Cr(VI) is based on the cationic behaviour of the Cr(III), and the anionic one of Cr(VI) species.

2 CHROMIUM CHEMISTRY IN WATER AND ANALYSIS

2.1 Chromium chemistry

Chromium in natural waters exists essentially in the oxidation states 3+ and 6+. At pH values below approx. 9 it forms anionic species when in the 6+ state and cationic species when in the 3+ state (Hem, 1977).

The diagram of electric potential vs pH shows that the hydrated $Cr(H_2O)_6^{3+}$ ion cannot exist above pH 4.0 because of hydrolysis which behaviour is characterized by the slow kinetics of its polymerization reactions (= formation of hydroxide or oxide bridges) (Nriagu and Nieboer, 1988). Cr(III) with its d^3-configuration exhibits a tendency to form inert complexes. The reaction kinetics of such complexes are very slow, and can take a few minutes up to several days. The extent of hydrolysis increases with increasing temperature and pH. Taking advantage of its kinetic inertness such hydrolysis can be prevented by forming complexes with strong coordinating agents in sufficiently high concentration.

In the absence of chelating agents, $Cr(OH)^{2+}$ and $Cr(OH)_2^+$ are the predominant species at pH between 6 and 7 (Nriagu and Nieboer, 1988).

The low solubility of Cr(III) is likely to be the major reason why it rarely occurs at concentrations above the drinking water standard of 50 μg/l (Hem, 1977).

Chromium(VI) is only present in aqueous solutions as anionic chromate except H_2CrO_4 which occurs at very low pH (Tandon et al., 1984).

The value of the redox potential of chromates indicates the strongly oxidizing properties of Cr(VI). Therefore it is unstable in acid solutions in the presence of electron donors (Nriagu and Nieboer, 1988).

Cr(VI) reduction is inhibited to a considerable extent by storing the sample at 4 °C (Stollenwerk and Grove, 1985) whereas Pavel et al., (1985) conclude that cooling does not remarkably prolong the lifetime of Cr(VI) in pure water solutions. This disagreement should be solved.

Finally it should be emphasized that chromates are anions and in fact can behave as ligands to metal ions such as e.g. $Cr(H_2O)_6^{3+}$ and $Cr(OH)_n^{3-n}$ (Nriagu and Nieboer, 1988).

2.2 Chromium analysis

There are many procedures published for the determinations of Cr(III) and Cr(VI) in aqueous samples. These involve precipitation, adsorption, solvent extraction, voltammetry, volatilization, colorimetry with 1.5 diphenylcarbazide, as Cr(VI) complexing agent, and various forms of chromatography (MacCarthy et al., 1989).

The present work relies on the separation of labelled Cr(III) and Cr(VI) through anion exchange extraction of Cr(VI) with Amberlite LA-1 or LA-2 diluted in MIBK (methyl isobutyl ketone) and stripped in 6 Mol/l HCl. MIBK has a relatively large solubility in water which could make the extraction troublesome if it were not that the extraction of Cr(III) with MIBK is poor (Griepink, 1984). The measurement of ^{51}Cr(III) and ^{51}Cr(VI), both in a separate phase, is done by ^{51}Cr counting in a NaI well type scintillation detector, coupled to a single channel analyzer. This method assumes that Cr(VI) solely exists as anionic and Cr(III) exclusively as cationic species. This assumption is justified by preventing hydrolysis of Cr(III) with a complexing agent (here HCO_3^- ions) which allows the Cr(III) complex to remain cationic.

The good repeatability of the results over months also demonstrates that irreversible sudden changes do not occur.

The performance of the Amberlite anion exchange procedure has been checked twice with the Chelex-100 cation exchanger.

3 DESIGN AND PROCEDURE

Complex formation is known to enhance or prevent the polymerization of Cr(III) that accompanies hydrolysis.

In trying to find an agent to prevent hydrolysis, several compounds may be considered. Complexes are known with ligands for which the point of attachment is the oxygen atom (e.g. H_2O, OH^-, carboxylic acid, phenols, pyrophosphate), the nitrogen atom (e.g. NH_3, amines) or the sulfur atom (e.g. sulfhydryl groups) (Nriagu and Nieboer, 1988).

Cr(VI) oxydizes carboxylic acids (Connett and Wetterhahn, 1983); they can therefore not be considered as complexing agents. Pyrophosphate will in the long run hydrolyse to orthophosphate which will precipitate with Cr(III). Complexation with lactate, hydrogencarbonate or compounds involved in the Krebs Cycle (e.g. citrate) prevents hydrolysis of Cr(III). However lactate cannot be considered because it reduces Cr(VI) slowly (Wallaeys, 1987). So hydrogencarbonate became the obvious choice.

After experiments at different pH-values, it was concluded that a 50 mmol/l HCO_3^-/H_2CO_3 buffer (pH 6.4) would stabilize Cr(III) and Cr(VI) species. The stability of the species in this medium has been investigated in this study.

To maintain a good buffer capacity over months and therefore a constant pH value, loss of CO_2 from the system has to be prevented. Samples were stored under a CO_2 blanket at atmospheric pressure and constant temperature.

Three different solutions were prepared :

a) 50 mmol/l HCO_3^-/H_2CO_3 buffer pH 6.4;

b) 50 mmol/l HCO_3^-/H_2CO_3 pH 6.4 containing acetone (1% v/v) to increase the solubility of CO_2 ; (preliminary experiments had not shown a reduction of Cr(VI) by acetone in this buffer);

c) 50 mmol/l HCO_3^-/H_2CO_3 buffer pH 6.4 containing F^- (50 mmol/l) as a complexing agent with Cr(III) (Baes and Mesmer, 1976).

All three series were stored under CO_2 at atmospheric pressure.

Legislation on dissolved hexavalent chromium in drinking water in the FRG, USA, WHO and EEC states an allowable maximum of 50 µg/l (Smart and Sherlock, 1985). To develop a representative reference material, this value was taken into account as well as the solubility of Cr(III) which leads to the concentrations chosen: for Cr(III) 25 µg/l and Cr(VI) 25 µg/l.

The adsorption and reduction characteristics of Cr(VI) in diluted solutions depend on the container material. Wallaeys (1987) investigated such behaviour of Cr(III) and Cr(VI) in glass, polyethylene, plexiglas, nylon, PVC and PTFE. There was no loss of Cr(III) in any of these materials, but for Cr(VI) substantial losses were observed owing to adsorption and reduction during storage. Only PTFE was promising for storage for a limited period of time, whereas quartz was not studied at all, so in this work PTFE containers were used as well as quartz vials. In fact preliminary experiments showed good stability with quartz. Moreover it offers advantages as compared to PTFE such as the possibility to seal and to prevent losses of CO_2 and H_2O by evaporation through the container walls.

Table 1 summarizes the investigations done in the two container materials at different temperatures.

TABLE 1 . Solutions used in the experiments with Cr(III) and Cr(VI) concentrations of 25 μg/l at pH 6.4 (both species as labelled ^{51}Cr compounds)

Medium / Material	50 mmol/l HCO_3^-/H_2CO_3	50 mmol/l $H_2CO_3^-/H_2CO_3$ +1% acetone (v/v)	50 mmol/l HCO_3^-/H_2CO_3 +50 mmol/l F^-
Quartz	5 °C	5 °C	5 °C
PTFE	5 °C and 20°C	5 °C and 20°C	5 °C and 20°C

4 EXPERIMENTAL

4.1 Radiotracer technique

The 320 keV γ-ray of ^{51}Cr ($t_{1/2}$ = 27.71 d) was measured with a 3" x 3" NaI well type scintillation detector coupled to a single channel pulse height analyzer. The counting system was provided with an auto-sample changer.

4.2 Reagents and materials

^{51}Cr labelled solutions were obtained in concentrations of 14.8 mg/l Cr(III) (94.72 MBq/ml) as $CrCl_3$ in HCl 0.1 mol/l and 3.55 mg/l Cr(VI) (37.00 MBq/ml) as Na_2CrO_4 in water from Amersham International plc, Buckinghamshire, England. For use in tracer experiments these solutions were further diluted with water and/or chemically identical unlabelled solutions to achieve the desired concentrations and counting rates.

Millipore Milli-Q water as well as acids purified by sub-boiling distillation in quartz were used throughout.

The liquid anion exchanger Amberlite LA-1/LA-2 was obtained from Sigma Chemical Company, St. Louis, MO, USA and the chelating cationic resin Chelex-100 (200-400 mesh, sodium form) from Bio-Rad Laboratories, Richmond, CA, USA. All solvents and reagents were p.a. grade.

The $NaHCO_3$ was analysed by NAA (Neutron Activation Analysis). Irradiation time was 14 h at a flux of $5 \cdot 10^{12}$ neutrons/$cm^2.s$. The Cr concentration measured was below the limit of detection (LOD = 3 x S)(S = standard deviation). As LOD = 6 ng Cr/g $NaHCO_3$, less than 25 ng Cr/l water sample was introduced. Stock LA-1/LA-2 extraction solution : Add 100 ml LA-1/LA-2 to 50 ml 6 M HCl, stirr continuously and dilute to 250 ml with MIBK (methyl isobutyl ketone) in a calibrated flask (2:1:2) (Minoia et al., 1985).

After long term storage (≥ 1 month) this mixture has to be stirred thoroughly and time given for the 2 phases to separate. This restores the binding capacity for Cr(VI).

- The extractions were performed in PE test tubes from Kartell Milano, Italy.
- The container material was Teflon-PFA from Savillex Corporation, Minesota, USA and Suprasil quartz vials from Heraeus Quarzschmelze GmbH, W.Germany.

The PTFE vials used were leached with HCl, HNO_3 and washed 3x with distilled water and 3x with Milli-Q water.
The quartz tubes were cleaned by treatment with boiling acid (3 Mol/l HNO_3 + 3 Mol/l HCl, 1:1), rinsed 3x in distilled water and 3x in Milli-Q water, further treated for 8h with HNO_3 vapours and finally steam treated 8h with Milli-Q water.

- 50 mmol/l HCO_3^-/H_2CO_3 buffer solution (pH 6.4) : 4.2 g $NaHCO_3$ + 250 ml 0.1 Mol/l HCl + Milli-Q H_2O up to 1 liter, CO_2 blanket.
- 50 mmol/l HCO_3^-/H_2CO_3 + 1% acetone (v/v)buffer solution (pH 6.4) : 4.2 g $NaHCO_3$ + 250 ml 0.1 Mol/l HCl + 10 ml acetone + Milli-Q H_2O up to 1 liter, CO_2 blanket.
- 50 mmol/l HCO_3^-/H_2CO_3 + 50 mmol/l F^- buffer solution (pH 6.4) : 4.2 g $NaHCO_3$ + 250 ml 0.1 Mol/l HCl + 2.1 g NaF + Milli-Q H_2O up to 1 liter, CO_2 blanket.

4.3 Methods

All solutions were sampled twice and all measurements were performed in duplicate. The recovery of the labelled species was always verified by comparison with the initial radioactivity.

The reliability of the method was checked with solutions containing all the ingredients described in table 1, but labelled with a single radiotracer, either $^{51}Cr(III)$ or $^{51}Cr(VI)$.

The separation of Cr(VI) in the organic phase and Cr(III) in the aqueous phase by using Amberlite LA-1/LA-2 anion exchange was performed by mixing 1 ml water with 2 ml stock LA-1/LA-2 extraction solution, shaking 2 min and separating the phases by centrifugation for 10 min at 2500 rpm.

The separation of Cr(III) from the sample was done using Chelex-100 chelating cationic exchange resin and the transfer of 1 ml sample to a column containing approx. 0.5 g Chelex-100 soaked in ~ 50 mmol/l HCO_3^-/H_2CO_3 buffer, adjusted to pH ~ 5. The resin retains the Cr(III) and the eluent contains the Cr(VI).

5 RESULTS AND DISCUSSION

5.1 Propagation of errors

In order to interpret the results obtained they must be supported by a priori error estimates.

The overall random error was estimated from the measurements of 10 ion exchange separations from a single batch. The total standard deviation (= S) of the procedure was calculated to be 2.2%. The standard deviation on the counting statistics was found to be 0.9% (n=10). By analysis of variance the standard deviation, of the overall chemical separation was estimated to be 2.0% (n=10).

$$S^2_{total} = S^2_{counting} + S^2_{chemical\ sep.}$$

$$2.2^2 = 0.9^2 + S^2_{chemical\ sep.}$$

The standard random error of the mean for $S_{chemical\ sep.}$ is :

$$\frac{S_{chemical\ sep.}}{\sqrt{n}} = \frac{2.0}{\sqrt{10}} = 0.63\ \%$$

The error at any given time is considered well defined by these 10 repetitive measurements (d.f.= 9). Each value is then compared to its starting value. As this implies that the same object is involved, an estimate of the covariance between the 2 measurements becomes necessary. Calculation of the covariance according to Kendall and Stuart (1977) shows that none can be shown to exist, considering that at each time 2 samples were taken and the number of d.f.= 1.

A confidence interval has to be established around the starting value, which is arbitrarily put at 100%. In order to prove the likelihood of the subsequent sampling a 0.90 C.I. of the mean is considered appropriate. A larger C.I. (95%) would only guarantee a much more reduced probability to discard a possibly correct value.

The error bar admitted within this 0.90 C.I. is computed as following :

$$\bar{\Delta}_t = (\bar{x}_0 - x_t) \pm t^{d.f.}_{0.05} \cdot \sqrt{s_0^2 + s_t^2}$$

Because $S_0 = S_t$ (respecting identical counting statistics)

and $S_0 = S_t = \sqrt{S^2_{counting} + S^2_{chemical\ sep}}$

and due to the fact that the standard deviation on the counting statistics amounts to 2.2%

$$s_0 = s_t = \sqrt{2.2^2 + 2.0^2} = 3.0\%$$

the difference allowed

$$d = t^{9}_{0.05} \cdot \sqrt{3.0^2\% + 3.0^2\%}$$

$$= 1.833 \cdot 3.0 \sqrt{2} = 8\%$$

This implies that when Δ=0 falls within the Δ-d and Δ+d interval the recorded value at time t cannot be said to differ significantly from the value at t=0.

To conclude a recovery is interpreted as follows :

≥ 92% → plausibly no losses of Cr(III) and/or Cr(VI) ;

< 92% → losses of Cr(III) and/or Cr(VI).

5.2 Results

The Amberlite LA-1/LA-2 anion exchange method, checked respectively with labelled $^{51}Cr(III)$ and $^{51}Cr(VI)$ solutions (total Cr concentration of 50 μg/l) gave the following results.

The % recovery of Cr(III) in the aqueous phase amounted to 97.9 ± 2.3 (S%)(n=3) and the activity of $^{51}Cr(III)$ in the organic phase was below the limit of quantitation (LOQ = 10 x S = 0.9 μg/l).

The % recovery of Cr(VI) in the organic phase was 98.7 ± 1.0 (n=4) and in the aqueous phase 2.7 ± 0.6. The latter was an impurity of the Cr(VI) label.

Chelex-100 cation exchange experiments, with a pH adjustment to approximately 5, necessary to achieve complete recovery, yielded the following results :

The % recovery of Cr(III) on the resin amounted to 99.5 ± 1.4 (n=3) and the activity of $^{51}Cr(III)$ in the eluent was below the limit of quantitation (LOQ = 10 x S).

The % recovery of Cr(VI) in the eluent was 96.7 ± 1.0 (n=3) and on the resin 2.5 ± 0.8.

Measurements of $^{51}Cr(III)$ and $^{51}Cr(VI)$ in water samples after Amberlite LA-1/LA-2 or Chelex-100 treatment resulted statistically in the same recovery.

All data of the experiments outlined in the scheme (Table 1) are plotted in figures 1 to 10. The individual standard deviation ($S_{r,t}$) is indicated as well. The recovery at t=0 has been arbitrarily put to 100% (full line) and the confidence interval is indicated by a dotted line (see paragraph 5.1).

In the cases with the PTFE material, stored at 5 °C, the results for Cr(III) in the hydrogencarbonate buffer (Fig. 1 and 2) are without significant losses over a 140 day period. The apparent exception shown in Fig. 1 is probably an analytical artifact as the mean value after 120 days is within the 92% confidence interval. On the contrary significant losses of Cr(III) occurred in the PTFE experiments, at 20 °C, as given in Fig. 3. The stability of Cr(VI) in the PTFE material at 5 °C is sufficient after a 140 day period (Fig. 2) but two significant losses of Cr(VI) occurred in the experiments described in Fig. 1.

The stability of Cr(III) in hydrogencarbonate + acetone (1% v/v) buffer in PTFE (Fig. 5 and 6) was not sufficient. Although Cr(VI) was statistically recovered after 140 days at 5 °C a significant loss was previously observed after 76 days

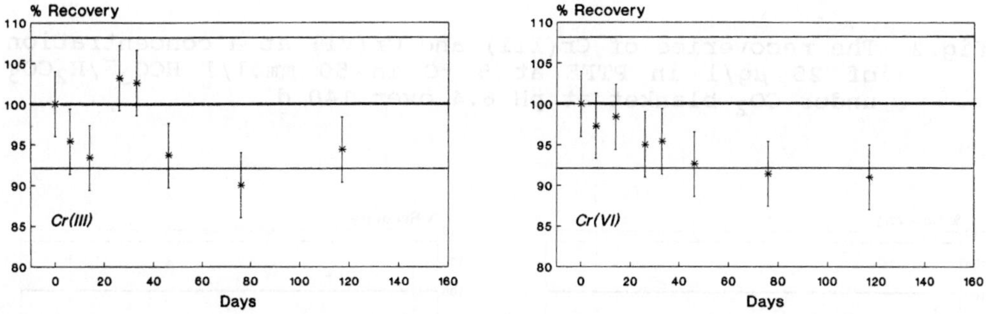

Fig.1 The recoveries of Cr(III) and Cr(VI) at a concentration of 25 µg/l in PTFE at 5 °C in 50 mmol/l HCO_3^-/H_2CO_3 under CO_2 blanket at pH 6.4 over 120 d

(Fig. 5). At 20 °C significant losses of Cr(VI) were observed (Fig. 6). The hydrogencarbonate + NaF (50 mmol/l) buffer (Fig. 8 and 9) proved to cause instability of both Cr(III) and Cr(VI) in PTFE.

As a consequence acetone and NaF addition to the matrix when stored in PTFE containers cannot be considered for use as a possible reference material.

The content of Cr(VI) in quartz proved to be stable in the three media studied (Fig. 4, 7 and 10). In all three media in quartz vials Cr(III) showed a satisfactory stability up to 76 days, but significant losses were found after 140 days (Fig. 4, 7 and 10).

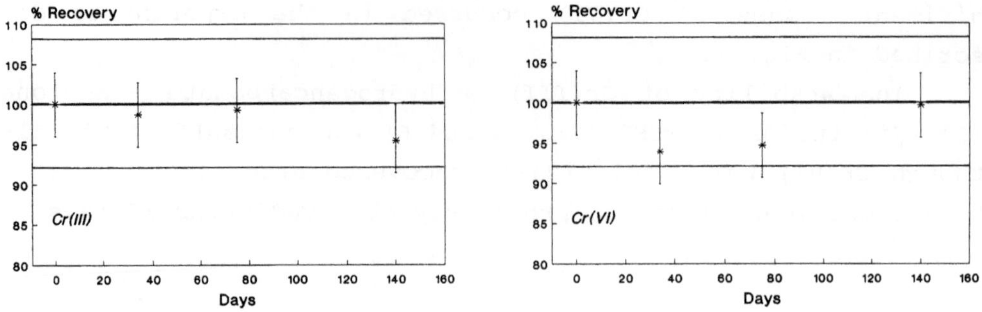

Fig.2 The recoveries of Cr(III) and Cr(VI) at a concentration of 25 μg/l in PTFE at 5 °C in 50 mmol/l HCO_3^-/H_2CO_3 under CO_2 blanket at pH 6.4 over 140 d

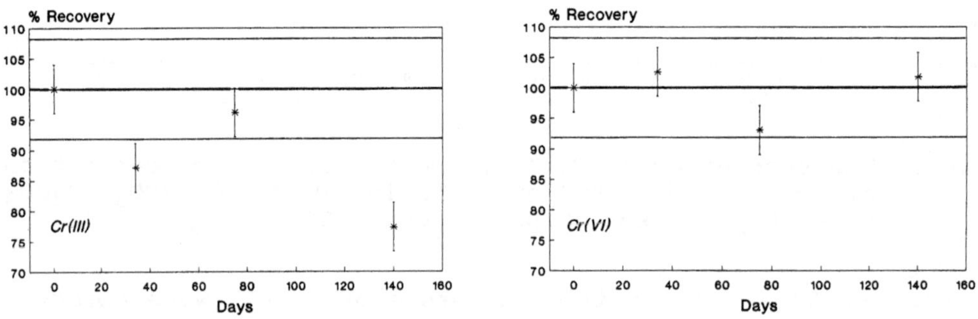

Fig.3 The recoveries of Cr(III) and Cr(VI) at a concentration of 25 μg/l in PTFE at 20 °C in 50 mmol/l HCO_3^-/H_2CO_3 under CO_2 blanket at pH 6.4 over 140 d

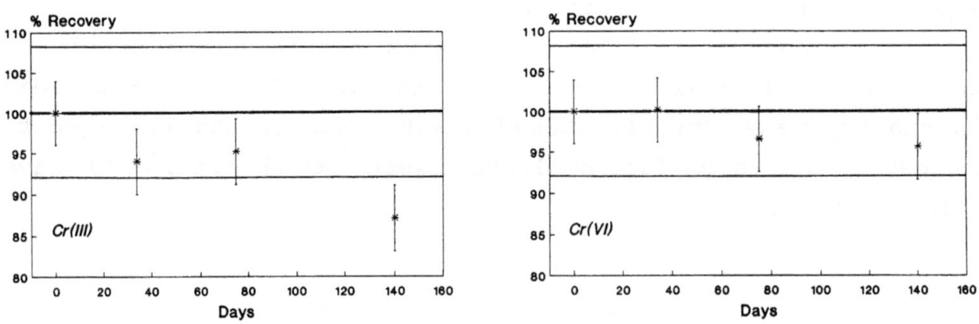

Fig.4 The recoveries of Cr(III) and Cr(VI) at a concentration of 25 μg/l in quartz at 5 °C in 50 mmol/l HCO_3^-/H_2CO_3 under CO_2 blanket at pH 6.4 over 140 d

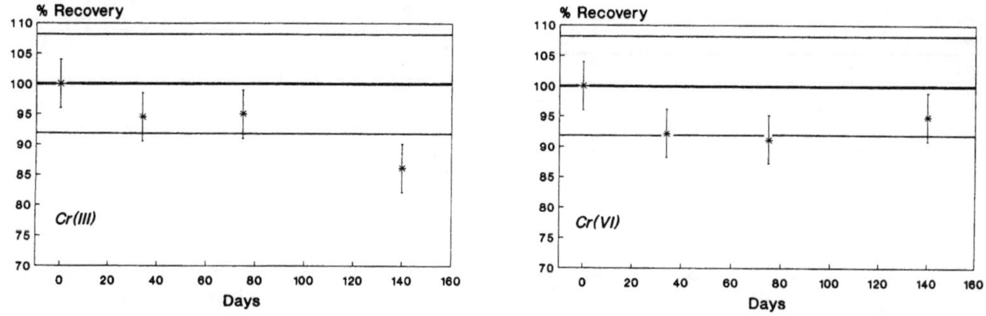

Fig.5 The recoveries of Cr(III) and Cr(VI) at a concentration of 25 μg/l in PTFE at 5 °C in 50 mmol/l HCO_3^-/H_2CO_3 + 1% acetone (v/v) under CO_2 blanket at pH 6.4 over 140 d

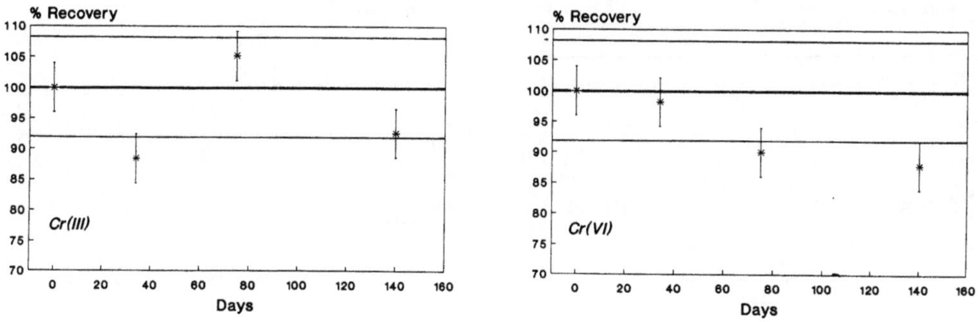

Fig.6 The recoveries of Cr(III) and Cr(VI) at a concentration of 25 μg/l in PTFE at 20 °C in 50 mmol/l HCO_3^-/H_2CO_3 + 1% acetone (v/v) under CO_2 blanket at pH 6.4 over 140 d

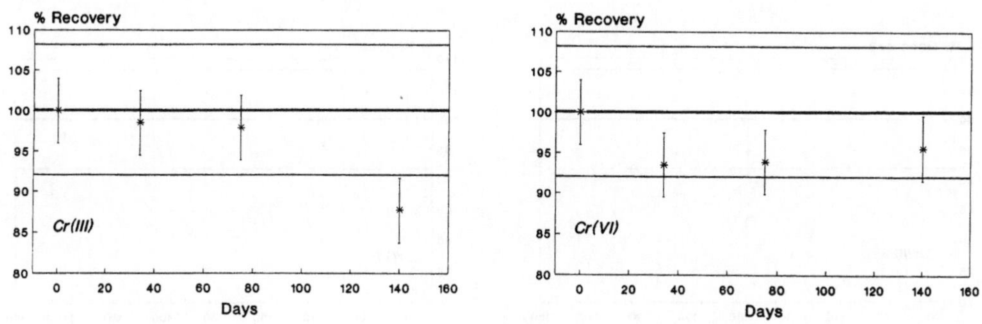

Fig.7 The recoveries of Cr(III) and Cr(VI) at a concentration of 25 μg/l in quartz at 5 °C in 50 mmol/l HCO_3^-/H_2CO_3 + 1% acetone (v/v) under CO_2 blanket at pH 6.4 over 140 d

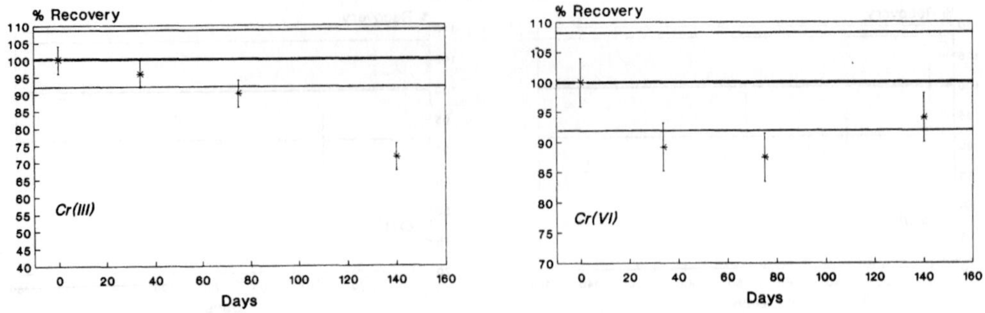

Fig.8 The recoveries of Cr(III) and Cr(VI) at a concentration of 25 μg/l in PTFE at 5 °C in 50 mmol/l HCO_3^-/H_2CO_3 + 50 mmol/l F^- under CO_2 blanket at pH 6.4 over 140 d

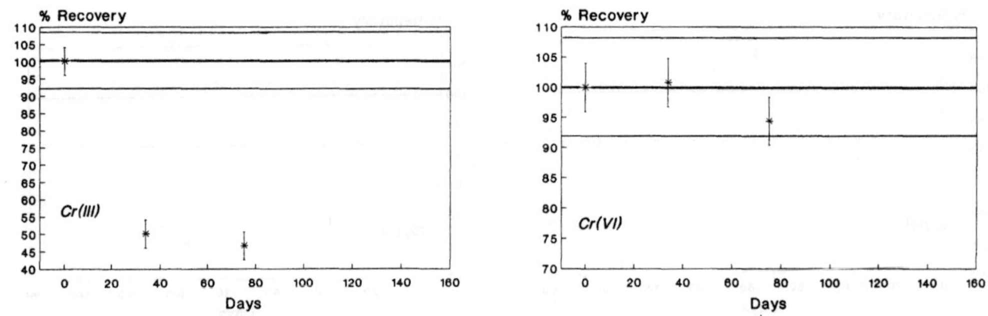

Fig.9 The recoveries of Cr(III) and Cr(VI) at a concentration of 25 μg/l in PTFE at 20 °C in 50 mmol/l HCO_3^-/H_2CO_3 + 50 mmol/l F^- under CO_2 blanket at pH 6.4 over 76 d

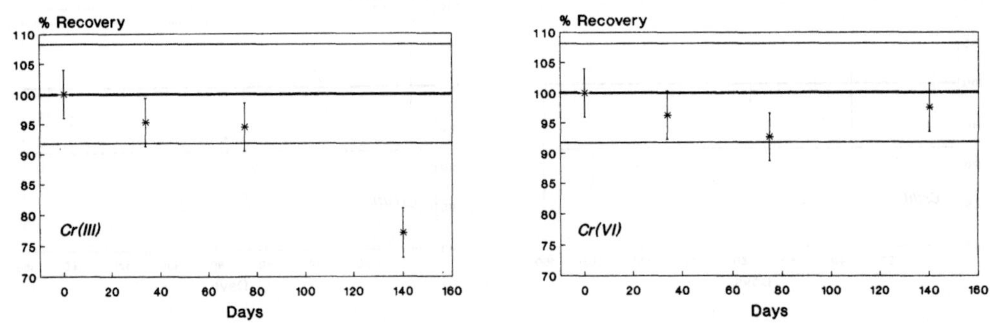

Fig.10 The recoveries of Cr(III) and Cr(VI) at a concentration of 25 μg/l in quartz at 5 °C in 50 mmol/l HCO_3^-/H_2CO_3 + 50 mmol/l F^- under CO_2 blanket at pH 6.4 over 140 d

6 CONCLUSION

The result of these first experiments indicates that stable solutions of Cr(III) and Cr(VI) at levels of 25 μg/l can be obtained in a 50 mmol/l HCO_3^-/H_2CO_3 buffer in PTFE if kept at 5 °C under a CO_2 blanket.

7. ACKNOWLEDGEMENTS

The authors acknowledge with gratitude the interest shown throughout their research by Professor Dr. R. Dams, Director of the Institute. They express their thanks to Mr. L. Mees for his valuable technical assistance.

S.Dyg obtained a sectoral grant from the CEC through the intermediary of the Danish Institute of Occupational Health, Copenhagen. R. Cornelis is a research director of the National Fund for Scientific Research, Belgium.

REFERENCES

Baes CF Jr, Mesmer RE (1976) The hydrolysis of cations. John Wiley & Sons, New York

Connett PH, Wetterhahn KE (1983) Metabolism of the carcinogen chromate by cellular constituents. In : Clarke MJ (ed) Structure and bonding, vol 54. Springer, Berlin Heidelberg, New-York, Tokyo

Griepink B (1984) Trace element analysis in surface water Pure Appl Chem 56 : 1477-1498

Hem JD (1977) Reactions of metal ions at surfaces of hydrous iron oxide. Geochim Cosmochimica Acta 41 : 527-538

Kendall M, Stuart A (1977) The advanced theory of statistics, 1 : 247. Charles Griffin & Co Ltd, London

MacCarthy P, Klusman RW, Rice JA (1989) Water Analysis. Anal Chem 61 : 269R-304R.

Minoia C, Mazzucotelli A, Cavalleri A, Minganti V (1983) Electrothermal atomisation atomic-absorption spectrophotometric determination of Cr(VI) in urine by solvent extraction separation with liquid anion exchangers. Analyst 108 : 481-484

Nriagu JO, Nieboer E (1988) Chromium in the natural and human environments, vol. 20. John Wiley & Sons, New York

Pavel J, Kliment J, Stoerk S, Suter O (1985) Preservation of traces of chromium VI in water and waste water samples. Fresenius Z Anal Chem 321 : 587-591

Smart GA, Sherlock JC (1985) Chromium in foods and the diet, Food Add Contaminants, 2 : 139-147

Stollenwerk KG, Grove DB (1985) Reduction of hexavalent Chromium in water samples acidified for preservation. Environm Qual 14 : 396-399
Tandon RK, Crisp PT, Ellis J (1984) Effect of pH on Chromium-(VI) species in solution. Talanta, 31 : 227-228
Wallaeys B (1987) Determination of trace elements by continuous ambulatory peritoneal dialysis, Ph.D thesis, Rijksuniversiteit Gent, Belgium
WHO (1988) Chromium. Environmental Health Criteria vol 61 World Health Organisation, Geneva

Organic Arsenic Compounds in Petroleum and Natural Gas

Kurt J. Irgolic and Bal K. Puri*
Institut für Analytische Chemie, Karl-Franzens-Universität Graz, A-8010 Graz Austria and Department of Chemistry, Texas A&M University, College Station, Texas 77843, U.S.A.

1. Introduction

Arsenic, a monoisotopic element with mass number 75, and several of its compounds have been known since ancient times. The average concentration of arsenic is estimated to be 2.5 mg/kg in the earth's crust and 1.5 µg/kg in ocean water. Thus, arsenic is about as abundant - or better - scarce as the elements uranium, tungsten, molybdenum, cesium, and most of the rare-earth elements (Cox, 1989). Arsenic is ubiquitous. Arsenic concentrations of several thousand mg/kg can be found in soils - even in those not influenced by human activities that dispersed arsenic compounds into the environment (U.S. National Academy of Sciences, 1977; Tanaka, 1988). Ocean water with 1.5 µg arsenic per kg has one of the lowest arsenic concentrations among environmental compartments. In contrast, marine organisms have arsenic concentrations several thousand-fold higher than their aqueous environment (Hanaoka et al., 1988). Because petroleum and natural gas were formed during geological time periods from arsenic-rich marine organisms, these hydrocarbon materials should be blessed with significant concentrations of arsenic compounds. Very little information is available in the literature about the concentrations of total arsenic in crude petroleum, in liquid hydrocarbons derived from petroleum, and in natural gas.

Arsenic is generally considered to be one of the most potent poisons. This arsenophobia has its roots in history and lives on in the minds of many people. Arsenic - or more accurately - its compounds can

* On leave from Department of Chemistry, Indian Institute of Technology, Hauz Khas, New Delhi-110 016, India

NATO ASI Series, Vol. G 23
Metal Speciation in the Environment
Edited by J. A. C. Broekaert, Ş. Güçer, and F. Adams

certainly influence life processes in undesirable ways, interfere with catalytic reactions of industrial importance, and diminish the quality of environmental compartments, if present at excessive concentrations. At the same time, arsenic compounds may be necessary for the proper functioning of biochemical reactions that are essential for life. To prevent measures from being instituted that regulate arsenic concentrations in waters, soils, air, and hydrocarbon materials in an inappropriate and too restrictive manner, and to ensure, that life of all types is properly protected, the total arsenic concentrations in pertinent environmental samples must be known. Then the arsenic compounds present in these samples must be identified and quantitatively determined. The fact that trace elements exert their beneficial or toxic influences through inorganic and organic trace element compounds is now well documented for arsenic (Irgolic, 1988). Because large quantities of petroleum and natural gas are produced, refined, chemically changed, or burned, the concentrations of arsenic and the types of arsenic compounds present in these materials are of academic, environmental, industrial, and toxicological importance.

2. Determination of Total Arsenic Concentrations

Because of the great interest of environmental scientists in the effects of arsenic, a large variety of methods were developed for the determination of total arsenic. Examples of such methods are instrumental and radiochemical neutron activation analysis; X-ray fluorescence with X-ray excitation; proton-induced X-ray emission; atomic fluorescence spectrometry; flame atomic absorption and emission spectrometry; electrothermal atomic absorption spectrometry; inductively-coupled, direct current, and microwave plasma emission spectrometries; electrochemical methods; colorimetric methods using molybdenum-blue or silver diethyldithiocarbamate; and the many variants of the hydride generation technique (Irgolic, 1990a). Some of these methods work best, when all organic material in the sample has been mineralized and others may require separation of the analyte from the matrix. A few of these methods can be used with solid and liquid samples, but most of them require the sample to be an aqueous solution. Samples consisting of hydrocarbon liquids or hydrocarbon gases generally cannot be directly analyzed for arsenic with these methods, although a few direct analyses are reported in the literature. For instance, arsenic was determined in crude oils, fuel oils, and lubricating oils with neutron activation analysis, X-ray fluorescence, and atomic absorption spectrometry (Braier and Eppolito, 1975), with inductively-coupled argon plasma atomic emission spectrometry (Decocq et al., 1986; Carr and Borst, 1978) and with inductively-coupled argon plasma-mass spectrometry (Hausler, 1987). For plasma emission spectrometries viscous hydrocarbons must be dissolved in a solvent such as xylene to overcome viscosity problems. Rather volatile hydrocarbons must also be diluted with toluene to prevent the plasma from being extinguished. Electrothermal atomic absorption spectrometry ashes the samples and thus removes the organic matrix. However, loss of analyte and interferences during the atomization stage are always a real threat to the quality of the analytical data. Most of the instruments thus far employed for the direct determination of arsenic in liquid hydrocarbons are expensive, are not easily available, and often have rather high detection limits for arsenic. None of the methods were used for the arsenic determinations in natural gas.

The best detection limits for arsenic were achieved by reduction of arsenite or arsenate with sodium borohydride to arsine, collection of the generated arsine in a liquid-nitrogen-cooled trap, and transportation of the arsine into a microwave plasma (Cappon, 1987). Amounts of arsenic in the low picogram range were determined in this manner. Although other detectors for hydride generation systems such as DC-helium plasmas or quartz-tube atomizers have detection limits of only 0.1 to 1 nanogram, they are - because of their robustness - much more frequently found in analytical laboratories than microwave plasma units (Irgolic, 1990b). Hydride generation systems convert the analyte present in the aqueous phase into a gas that is then swept by a helium/hydrogen stream into the detector. The arsenic compounds in natural gas are already in the gaseous state. Therefore, introduction of the natural gas into one of the detectors would appear to be the most direct way to determine arsenic. However, the DC-helium and microwave plasmas - although stable to helium/hydrogen flows - become unstable, when the gas stream is loaded with hydrocarbon gases. The plasma may even be extinguished. The more powerful inductively coupled or DC plasmas may tolerate analytically acceptable concentrations of hydrocarbon gases in the argon stream. Exploratory investigations with these instruments produced promising results.

For routine determinations of arsenic in liquid and gaseous hydrocarbons that should be adaptable to use in the field extraction of the arsenic from the gas into an aqueous phase followed by hydride generation recommends itself because of the excellent detection limits and commercial availability of hydride generation systems in combination with an atomic absorption spectrometer. A gas chromatograph coupled to a microwave plasma emission detector, a system introduced by Hewlett Packard last year, might be capable of accepting hydrocarbon gases. As the quickest path to a routine determination of arsenic in hydrocarbon matrices the extraction of the arsenic compounds from the hydrocarbon into an aqueous or nitric acid phase followed by mineralization of the extracts, reduction of the arsenate to arsine, and detection of arsenic with a He plasma emission system was chosen.

3 .Determination of Total Arsenic in Liquid Hydrocarbons

In principle, a liquid hydrocarbon sample could be mineralized with hot mixtures of oxidizing acids and the digest analyzed for arsenic. In practice, these mineralizations are difficult to carry out, because the reaction mixtures might catch fire when performed in open vessels and might produce dangerous pressures generated by the products of the reaction such as carbon dioxide, water, and nitrogen oxides when carried out in closed vessels. To circumvent these difficulties, experiments were conducted to evaluate water and nitric acid as agents for the transfer of the arsenic compounds from the hydrocarbon into the aqueous phase. The arsenic in crude oil samples that are not too viscous was quantitatively extracted, when one milliliter of the oil was boiled in an open beaker with 100 milliliters of distilled water for one hour. Very viscous oils such as the NIST Residual Fuel Oil 1634b could not be extracted in this manner. Dissolution of the fuel oil with xylene (2 g oil, 25 mL xylene) produced a solution, from which the arsenic could be extracted with boiling water. The oil-water mixtures were then filtered through a Whatman No. 1 filter. The filter was washed with distilled water. The filtrate and the washings were combined and mixed with 30 millilters of concentrated nitric acid and 15 milliliters of concentrated sulfuric acid. The mixture was heated until white fumes of sulfur trioxide appeared. The cooled solution was carefully diluted with distilled water and heated again until sulfur trioxide fumes became noticable. The mineralization must be carried out with exact adherence to the recipe (Puri and Irgolic, 1990) to ensure the quantitative conversion of the methylated arsenic compounds to arsenate (Ihnat and Thompson, 1980). The digest now containing only inorganic arsenic was appropriately diluted with distilled water and an aliquot of this solution analyzed for total arsenic by the hydride generation technique with a DC-helium plasma detector (Clark et al.,1980).

Samples of refined hydrocarbons with a boiling point much lower than crude oil samples are best extracted with boiling dilute nitric acid (10 mL conc. nitric acid, 90 mL water). This extraction procedure gave reliable arsenic concentrations with the standard fuel oil diluted with xylene and with a commercial atomic absorption standard for arsenic in an organic matrix (triphenylarsine in xylene).

The method was tested with the NIST Residual Fuel Oil. The arsenic concentration found by the extraction/mineralization/hydride generation method (0.121 μg/g) agreed with the certified concentration (0.12 μg/g) within experimental error. The arsenic concentrations in the 53 crude oil samples investigated ranged from 0.04 to 514 mg/L with a median of 0.84 mg/L. These results were obtained with a single extraction. A series of experiments, in which the oily residue from the first extraction was boiled again with water showed, that 97 percent of the arsenic are transferred by one extraction from the oil to the aqueous phase (Puri and Irgolic 1990).

4. Determination of Total Arsenic in Natural Gas

The determination of arsenic in natural gas can be accomplished as described for liquid hydrocarbons as soon as the arsenic is transferred from the gas into the liquid phase. This task is efficiently performed by vigorously shaking one liter of the gas sample with 10 milliliters of concentrated nitric acid on a platform shaker at the highest speed. The shaking time required for quantitative transfer of the arsenic is determined by the efficiency, with which the gas and the nitric acid are mixed and thus depends on the geometry of the flask and the construction of the shaker. Shaking times between one-half to one hour were found to be adequate, when a one-liter round-bottomed flask and a platform shaker operating at 300 rpm were used. The collection of a gas sample is simple, when a gas pipeline is the source of the gas. A dry one-liter, one-necked flask with a side arm carrying a Teflon stopcock is connected with a Teflon tube to a gas-outlet-valve on the pipeline. The neck is kept open and the stopcock on the side arm is opened. Gas is passed through the flask at a reasonably fast rate until all the air is displaced. A dropping funnel containing ten milliliter concentrated nitric acid is inserted into the open neck of the flask and at the same time the gas valve on the pipeline and the Teflon stopcock are closed. The stopcock is quickly turned through the open position to equilibrate the gas pressure in the flask to atmospheric pressure. After the flask has been agitated on the shaker the liquid phase is quantitatively transferred into a volumetric flask, the flask is filled to the mark with distilled water, an appropriate aliquot is mineralized, and the digest used to determine the total arsenic concentration. In collecting gas from a pipeline care should be taken, that the temperature changes on expansion do not lead to condensation of hydrocarbons in the flask. In addition, the flask should be visually inspected during the flushing process to check on particulates that might be blown into the flask. Should particles be seen, the flask must be cleaned and dried. To prevent contamination by particulates, the valve on the pipeline should be kept fully open for a few minutes to dislodge particles in the line before filling the flask.

When the collection of a gas sample at a pipeline, where the gas supply is unlimited, is not possible and the sample must be collected in rather small pressure bombs, the extraction procedure can still be used. In this case the flask cannot be flushed with the gas, because this operation would exhaust the sample. However the flask can be evacuated with an oil pump to less than one Torr, the evacuated flask connected to the bomb with a Teflon tube, and the flask carefully filled with the gas to a pressure slightly above atmospheric pressure. The gas sample thus collected is treated as described for the sample obtained from the pipeline. Severe problem were encountered with steel bombs. The arsenic concentrations found in gas collected in small 304-stainless steel bombs were always lower than the concentrations obtained in samples not stored in bombs. Ten-fold concentration differences were not uncommon. Sorption of arsenic compounds by the walls of the steel bombs is the likely cause for these discrepancies. The small steel bombs had a volume of approximately one liter. When gas collected in a large cylinder at a pressure of 850 psi was analyzed, the arsenic concentration was the same as in the sample collected directly. The smaller surface to volume ratio in the large cylinder probably is one of the reasons that a small and negligible percentage of the arsenic in the gas was sorbed onto the walls. Sorption will become less important as the gas pressure and concomitantly the quantity of arsenic in the cylinder increases. A gas sample kept in the large cylinder at 850 psi and a directly collected sample had the same arsenic concentration of 0.53 µg/L, whereas the same gas collected at the same time but kept in a small steel bomb had 0.33 µg/L. Representative concentrations for gas from another well are 2.06 (direct) and 0.05 µg/L (small bomb, 250 psi). Additional investigations of the pressure dependence of the concentration differences and of the influence of the material from which the collection vessels are made on the sorption of arsenic compounds are needed.

The hydride generation system for these determinations of arsenic is calibrated in terms of nanograms of arsenic in the aliquot placed into the reduction vessel. The concentration of arsenic in the natural gas is then calculated from the quantity of arsenic found, the appropriate multipliers connected with the dilution of the digest, and the volume of the flask, in which the gas sample was collected. The detection limit of the hydride generation system for arsenic is approximitaly one nanogram of arsenic. When the entire digest from an extraction of one liter of gas is transferred into the reduction vessel, the detection limit of the entire procedure is also one nanogram. This detection limit increases with decreasing volume of the aliquot taken from the digest for analysis. In most instances, the detection limit for the entire procedure (extraction of the gas sample, mineralization of the extract with a mixture of concentrated nitric/sulfuric acid, hydride generation, trapping of the arsine, detection by helium-DC plasma emission spectrometry) is not the limiting factor in the determination of arsenic in natural gas samples. The blank, the amount of arsenic present in all the reagents used in the process and introduced from the environment, fixes the quantity of arsenic that can be reliably determined in a gas sample. In a laboratory, in which much work with arsenic is carried out, the blank was found to be 10 nanogram arsenic. This high blank prevents arsenic concentrations lower than 5 ng/L to be determined with acceptable precision. If lower concentrations of arsenic must be determined reliably, a laboratory free of contamination with arsenic compounds has to be used.

5. Identification of Arsenic Compounds in Petroleum and Natural Gas

The hydride generation system can be used to identify arsenic compounds present in the aqueous or dilute-nitric-acid extracts obtained from liquid hydrocarbons and natural gas. To obtain information about arsenic compounds, the extracts must not be mineralized. Only those arsenic compounds can be detected that are reducible by sodium borohydride to sufficiently volatile arsines. The generated arsines are collected in a trap cooled with liquid nitrogen, then volatilized in sequence of increasing boiling points, and transported into the detector. An oil sample with a total arsenic concentration of 1.67 µg/mL contained per milliliter 23 ng arsenic as inorganic arsenic (arsenite and/or arsenate), 104 ng in monomethylated form, and 520 ng in trimethylated form. Some dimethylated arsenic was also present but was difficult to quantify, because the signals for di- and trimethylated arsenic overlapped. Approximately 1 µg As/mL remained unidentified (Puri and Irgolic, 1990). The arsenic compounds not detectable by the hydride generation system (compounds that are not reducible or reducible compounds that are not sufficiently volatile) could be tetramethylarsonium compounds, trialkylated arsenic compounds with alkyl groups larger than methyl, or complex organic arsenic compounds similar in structure to compounds found in marine organisms (Irgolic, 1986). To identify these recalcitrant arsenic compounds, methods based on liquid chromatographs coupled to element-specific detectors (Irgolic and Brinckman, 1984) must be employed.

The extracts obtained from natural gas were similarly investigated. Fresh extracts introduced into the hydride generation system produced only signals indicative of trimethylated arsenic. As the extracts aged at room temperature, signals for inorganic, monomethylated, and dimethylated arsenic increased in intensity, whereas the signal for trimethylated arsenic decreased. In the oxidizing environment of concentrated nitric acid the trimethylated arsenic compounds lose methyl groups stepwise and are ultimately transformed into arsenate. One of the arsenic compounds in natural gas, therefore, is a trimethylated arsenic compound, for which trimethylarsine, a liquid boiling at 53°C at atmospheric pressure, trimethylarsine oxide, a white, very hygroscopic solid melting at 192°C, and trimethylarsine sulfide, a yellowish solid melting at 183°C are candidates. The oxide and sulfide are unlikely to be in the gaseous state at room temperature. Because trimethylarsine was identified as one of the arsenic compounds in natural gas by gas chromatography-mass spectrometry (Irgolic et al., 1990a), a large fraction of the arsenic in natural gas must be in form of trimethylarsine. Additional support for the presence of trimethylated arsenic and perhaps of trimethylarsine in natural gas is the identification of trimethylarsine sulfide by mass spectrometry and NMR spectroscopy as the major component of white solids found in the pipelines (Irgolic et al., 1990b). The mass spectra clearly showed, that dimethylethylarsine sulfide, diethylmethylarsine sulfide, and triethylarsine sulfide at concentrations that decreased with increasing molecular mass of the compounds were also present. These results suggest, that the gas contains the corresponding trialkylarsines. The reactions of these arsines with elemental sulfur or with a compound that is a sulfur donor may be responsible for the conversion of the trialkylarsines to the trialkylarsine sulfides.

6. Genesis of the Arsenic Compounds Present in Petroleum and Natural Gas

Although the presence of trialkyarsenic compounds in petroleum and natural gas may be surprising initially, the consideration of the composition of the marine source materials for these hydrocarbon mixtures inevitably leads to a reasonable hypothesis for the origin of these arsenic compounds. Zooplankton, phytoplankton, macroalgae, and higher forms of marine life are known to accumulate arsenic to concentrations several thousand-fold above the arsenic concentration in seawater. The predominant arsenic compounds in the marine organisms are trimethyl(carboxymethyl)-arsonium zwitterion (arsenobetaine) and dimethyl(ribosyl)arsine oxides (Edmonds and Francesconi, 1988). These methylated arsenic compounds are the likely precursors of the arsenic compounds in petroleum and natural gas. During the diagenetic and subsequent processes that converted at elevated temperatures and high pressures the buried marine biomass into liquid and gaseous hydrocarbons, arsenobetaine could have been transformed into trimethylarsine as outlined in equations 1 and 2. The decarboxylation of arsenobetaine to a tetramethylarsonium compound (eqn. 1) should proceed

$$[(CH_3)_3As^+CH_2COOH]^+ X^- \longrightarrow [(CH_3)_4As]^+ X^- + CO_2 \qquad (1)$$

$$[(CH_3)_4As]^+ X^- \longrightarrow (CH_3)_3As + CH_3X \qquad (2)$$

without any difficulties under concitions, under which marine biomass was transformed into oil and gas. The reductive demethylation of the tetramethylarsonium compound (eqn. 2) should produce trimethylarsine that can then partition between the liquid petroleum and the natural gas. This hypothesis nicely explains the presence of trimethylarsine in natural gas that was ascertained by gas chromatography-mass spectrometry and suggests, that the trimethylated arsenic compound in crude petroleum is also trimethylarsine. A conciderable fraction of the arsenic in petroleum is trimethylated, but the valency of arsenic in this compound is not known. Even if the arsenic compound in petroleum were trimethylarsine oxide or sulfide, reactions could be envisaged that would produce these pentavalent arsenic compounds from trimethylarsine. The ethylmethylarsines and the triethylarsine that probably are also present in natural gas demand an addition to the hypothesis. The reaction of trimethylarsine with an alkyl halide is well known and produces a trimethylalkylarsonium halide. The arsonium compounds with several alkyl groups preferentially lose methyl halide upon thermal demethylation. Repeated ethylation-demethylation reactions could produce the ethylmethylarsines and the triethylarsine (eqn. 3). For these reactions to proceed, ethyl halides must be present during

$$(CH_3)_3As + C_2H_5X \longrightarrow \left[(CH_3)_3AsC_2H_5\right]^+ X^- \xrightarrow{\text{heat}} (CH_3)_2AsC_2H_5 + CH_3X$$

$$(CH_3)_2AsC_2H_5 \xrightarrow{+\,C_2H_5X} \left[(CH_3)_2As(C_2H_5)_2\right]^+ X^- \xrightarrow[-\,CH_3X]{\text{heat}} (CH_3)As(C_2H_5)_2$$

$$(CH_3)As(C_2H_5)_2 \xrightarrow{+\,C_2H_5X} \left[(CH_3)As(C_2H_5)_3\right]^+ X^- \xrightarrow[-\,CH_3X]{\text{heat}} (C_2H_5)_3As \qquad (3)$$

the period between burial of the marine biomass and the extraction of the petroleum or gas from the reservoir. That ethyl groups are not scarce during this period is indicated by the presence of ethane in natural gas. Experiments to verify this hypothesis are now in the planning stage.

The recent biological origin of trimethylarsine must also be considered. Microorganisms are capable of transforming inorganic arsenic to methylarsenic compounds such as trimethylarsine (Challenger, 1945; Pickett et al., 1988). The formations, in which petroleum and natural gas were formed and are now found, certainly contain inorganic arsenic at the mg/kg level as a minor constituent of the mineral assemblies forming the rocks. This inorganic arsenic could have been used by microrganisms as the source of arsenic for the production of trimethylarsine. Whether microorganisms can survive in a petroleum or natural gas deposit at pressures and temperatures prevalent several thousand meters below the earth's surface, is an interesting question. Alternatively, microbial action on the rocks exposed in the bore hole could and probably does generate some trimethylarsine. A detailed consideration of the rate of production achievable under these conditions (microbial population, surface area of exposed rocks, arsenic concentration in the rocks, permeability of the rocks for microbial intrusion) will probably establish, that microbial generation of trimethylarsine cannot produce sufficient trimethylarsine to reach an arsenic concentration of one milligram per cubic meter at a daily extraction of 100 million cubic feet of natural gas. Although a small contribution to the trimethylarsine in natural gas from recent biological activities cannot be excluded, the majority of the trimethylarsine is very likely the product of reactions that transformed marine biomass to hydrocarbons and concomitently the organic arsenic compounds present in it to trimethylarsine.

The presence of monomethylated and dimethylated arsenic compounds in crude petroleum is compatible with recent biological activity and the formation from geologically ancient marine arsenic compounds. Microarganisms are known to form mono- and dimethylarsenic compounds from inorganic arsenic (McBride and Wolfe, 1971; Braman and Foreback, 1973). Dimethylribosylarsine oxides could be the ancient marine precursors of dimethylated arsenic compounds. These ribosylarsine oxides are known to be cleaved to dimethyl(hydroxyethyl)arsine oxide in an anaerobic environment (Edmonds and Francesconi, 1988). Whether this degradation product can be converted chemically into dimethylarsine and monomethylarsenic compounds under the conditions prevalent during the formation of hydrocarbons is not known. Much additional work is needed to clarify the genesis of the arsenic compounds in crude petroleum and in natural gas. This work will also be very helpful to understand how mercury and selenium compounds known to be present in natural gas and other compounds of metals and metalloids not yet identified in these matrices were formed.

Acknowledgement The financial support of this work by the Robert A. Welch Foundation of Houston, Texas and NATO support for travel and accomodations that made possible the participation in the NATO Advanced Study Institute at Cesme, Turkey is gratefully acknowledged.

References

Braier HA, Eppolito J (1975) Determination of trace metals in petroleum: instrumental methods. In "The Role of Trace Metals in Petroleum." Yen TF (ed), Ann Arbor Science Publ., Ann Arbor, Michigan, USA, pp 65-87.

Braman RS, Foreback CC (1973) Methylated forms of arsenic in the environment. Science 182: 1247-1249.

Cappon CJ (1987) GLC speciation of selected trace elements. LC-GC 5: 400-418, and references therein.

Carr CD, Borst JE (1978) Use of inductively coupled plasma emission spetroscopy for the analysis of undigested organic materials and oils. Applied Research Laboratory, Sunland, California, USA; presented at the 1978 Pittsburgh Conference.

Challenger F (1945) Biological methylation. Chem Rev 36: 315-361.

Cox PA (1989) "The Elements." Oxford University Press, Oxford, UK.

Decocq A, Brocas AA, DeWindt J, Druon C (1986) Automatic wear metals control in diesel lubricating oil by ICPES. Am Lab November 1986: 128-133.

Edmonds JS, Francesconi KA (1988) The origin of arsenobetaine in marine animals. J Appl Organomet Chem 2: 297-302.

Hanaoka K, Yamamoto H, Kawashima K, Tagawa S, Kaise T (1988) Ubiquity of arsenobetaine in marine animals and degradation of arsenobetaine by sedimentary micro-organisms. J Appl Organomet Chem 2: 371-376.

Hausler D (1987) Trace element analysis of organic solutions using inductively coupled plasma-mass spectrometry. Spectrochim Acta 42B: 63-73.

Ihnat M, Thompson BK (1980) Acid digestion, hydride evolution atomic absorption spectrometric method for determination arsenic and selenium in foods: Part II. Assessment of collaborative study. J Assoc Off Anal Chem 63: 814-839.

Irgolic KJ, Brinckman FE (1984) Liquid chromatography-element-specific detection systems for molecular speciation studies. In "The Importance of Chemical Speciation in Environmental Processes." Bernhard M, Brinckman FE, Sadler PJ (eds), Springer Verlag, Berlin, pp 667-684.

Irgolic KJ (1986) Arsenic in the environment. In "Frontiers in Bioinorganic Chemistry." Xavier AV (ed). VCH Publishers, Weinheim, Germany (FRG), pp 399-408.

Irgolic KJ (1988) Arsenic compounds in marine and terrestrial organisms: analytical, chemical, and biochemical aspects. J Appl Organomet Chem 2: 303-307.

Irgolic KJ (1990a) Determination of organometallic compounds in environmental samples with element-specific detectors . In "Trace Metals Analysis and Speciation." Krull IS (ed). Elsevier, Amsterdam, in press.

Irgolic KJ (1990b) The determination of arsenic and arsenic compounds in environmental samples excluding human materials. In "Hazardous Metals in the Enviroment." Stöppler M (ed). Elsevier, Amsterdam, in press; and references therein.

Irgolic KJ, Spall D, Puri BK, Ilger D, Zingaro RA (1990a) Determination of arsenic and arsenic compounds in natural gas samples. J Appl Organomet Chem, in press.

Irgolic KJ, Spall D, Puri BK (1990b) Trialkylarsine sulfides from natural-gas pipelines. J Appl Organomet Chem, in preparation.

McBride BC, Wolfe RS (1971) Biosynthesis of dimethylarsine by Methanobacterium. Biochem 10: 4312-4317.

Pickett AW, McBride BC, Cullen WR (1988) Metabolism of trimethylarsine oxide. J Appl Organomet Chem 2: 479-482.

Puri BK, Irgolic KJ (1990) Determination of arsenic in crude petroleum and liquid hydrocarbons. Environ Geochem Health, in press.

Tanaka T (1988) Distribution of arsenic in the natural environment with emphasis on rocks and soils. J Appl Organomet Chem 2: 283-295.

US National Academy of Sciences (1977) Arsenic. Washington, DC.

APPLICATION OF POLAROGRAPHIC AND VOLTAMMETRIC TECHNIQUES IN ENVIRONMENTAL ANALYSIS

Günter Henze

Abteilung für Anorganische und Analytische Chemie,
Universität Trier
Fachbereich Chemie, Universität Kaiserslautern
Federal Republic of Germany

INTRODUCTION

Polarography and voltammetry are suited for the trace analysis of numerous ecotoxic metals, metalloids and various organic pollutants. Moreover, these methods are useful for the differentiation of oxidation states and the characterization of metal complexes (Henze and Neeb 1986). Therefore, they are of interest in environmental research, particularly for metal speciation, and a number of papers appeared on this subject in recent years (Florence 1986, Florence and Batley 1980, Nürnberg 1983, Mizuike 1987).

In this paper, a survey on the possibilities and limitations of polarography and voltammetry in environmental analysis and further aspects of research in this field are presented.

SPECIATION STUDIES

The highly sensitive anodic-stripping-voltammetry in the differential pulse mode (DPASV) is applicable for trace metal speciation in natural waters. This technique can be used to distinguish between "labile" and "non -labile" or "bound" metal species.

The labile metal species take part in the electrode reaction and include free hydrated metal ions and inorganic metal complexes. In natural waters mainly simple sulfato-, chloro-, carbonato-, hydroxo - complexes etc. are existing. These are

NATO ASI Series, Vol. G 23
Metal Speciation in the Environment
Edited by J. A. C. Broekaert, Ş. Güçer, and F. Adams

weak complexes and upon reduction at the electrode during the voltammetric determination, the different states of equilibria are disturbed and are subject to a continuous change. When released from the complexes in the double layer of the working electrode, the free cations of this kind of species are reduced, analogous to the free cations of the hydrated metal ions. In this way the total content of the labile metal species is determined without differentiation.

The non-labile metal species comprise soluble metal-organic complexes with humic acid, fulvic acid, NTA or EDTA, etc. as ligands and metal species associated with colloids. From this type of species only these can be determined individually which can be isolated from the water sample, because organic components present in natural waters may interfere with the electrode process in the voltammetric determination. In the practice of electrochemical analysis, the differentiation between weakly and strongly bound metal-organic complexes is useful, where the weak complexes are defined as labile organic complexes.

In DPASV the metals of the labile species may be deposited at constant potentials on the surface of a mercury drop or film electrode, without prior sample decomposition. In this procedure, performed at the natural pH of the water sample or in mildly acidic conditions, all labile species of each metal are concerned, which will dissociate in the double layer of the electrode. It may be mentioned that the deposition potential, the supporting electrolyte and the pH-value of the sample, influence the quantitation of the labile species.

The non-labile or bound metal species are thus determined as the difference between the total element concentration and their labile part in water samples.

Before DPASV determination of the total dissolved elemental content in polluted waters, such as river, lake or costal waters, the sample has to be filtrated and decomposed, which is done in practice mostly by oxidative UV-irradiation of the acidified sample. By irradiation of the unacidified sample, only the weakly bound organic complexes were decomposed and changed in labile states.

Another group, the lipid-soluble metal species, may be determined by DPASV after extraction and UV-irradiation. The sample residue contains that part of bound species which is associated with colloids and suspended particulate materials. For their decomposition low temperature ashing in an oxygen plasma or wet digestion is useful (Table 1).

In evaluation of the potentiality of DPASV for the analysis of metal species, it has to be considered that this technique is not particularly species-specific, nor it is for labile metal species.

Table 1. Speciation scheme for metals in water samples (sample treatment by membrane filtration)

Filtrate Analysis

Aliquot	Operation	Interpretation
1.	ASV at natural pH (for seawater) or at pH 4.6 (acetate buffer) for fresh water	"labile" metal species free (hydrated) and labile inorganic metal complexes
2.	UV irradiate of the natural sample (8 h) +0.1 % H_2O_2; then ASV	labile (weak) organic metal complexes
3.	Acidify to 0.05 M HNO_3 + 0.1 % H_2O_2; UV irradiate (8 h); then ASV	total metal
4.	Extract with hexan - 20 % butan-1-ol; ASV on acidified, UV-irradiated aqueous phase	lipid soluble metal species

Residue Analysis

	low temperature ashing in an oxigen plasma or wet digestion; then ASV	metal content, associated with inorganic and organic colloids and suspended particulate materials

More information on the composition of labile species can be derived from the shift of the polarographic half-wave potential, which depends on kind and degree of complexation. For this purpose the so-called pseudopolarograms can be used. (Nürnberg et al. 1976).
Pseudopolarograms are obtained by plotting stripping peak current as a function of the deposition potential for different ligand concentrations. From the resulting voltammograms, which are analogous to dc-polarograms, the stability constants and coordination numbers can be evaluated. As this kind of speciation studies requires a great number of measurements, computer assisted systems for generating the desired pseudopolarograms are useful. On the basis of such measurements, it was reported, that in marine waters for example, copper and lead would be present to a great extent as carbonato-complexes, cadmium as chloro-complexes and zinc as hydroxo-complexes (Florence 1986).

In general, it is difficult to directly determine a specified species of a trace metal in aquatic samples, because the selectivity of the voltammetric technique is not satisfactory. Therefore, the separation and concentration techniques are required prior the determinations of metal species. (Mizuike 1987).
An example of special interest is the stripping voltammetric determination of lead species in seawater after preceding membrane filtration. The polarographic behaviour of lead(II) and trimethyllead(IV) is different, so that with different deposition potentials in anodic stripping voltammetry, a differential determination of the species is possible (Bond et al. 1983)

REDOX STATE

Electrochemical methods can, however, be sucessfully applied to the determination of elements at different oxidation states.

The differentiation of arsenic(III) and arsenic(V) is realized by anodic, as well as by cathodic, stripping voltammetry. In the electrochemical preconcentration step, only arsenic(III) was reduced. The electroinactive arsenic(V) may be determined after reduction to arsenic(III) by sodium-sulfite or gaseous sulfur dioxide. The determination is highly sensitive and selective by differential-pulse-cathodic-stripping-voltammetry (DPCSV). In this procedure, arsenic(III) is preconcentrated on a mercury drop as an intermetallic compound with copper and redissolved by cathodic stripping on the basis of a further reduction to minus three valency state. This method is capable of detecting 0.2 ng/ml arsenic and is applicable to water analysis (Henze et al. 1980)

Accordingly, selenium(IV) and selenium(VI) can be simultaneously determined by DPCSV (Henze 1979, 1981). Moreover, polarography and/or voltammetry have been used to distinguish between iron(II/III), chromium(III/VI), tin(II/IV), uranium(IV/VI), manganese(II/IV) and thallium(I/III).

TRACE METAL ANALYSIS

Only in a few cases are the total element concentrations in aquatic ecosystems so high that polarographic methods can be applied. In most cases however, as already mentioned, concentrations are very low, so that stripping voltammetric methods have to be used. For a determination in the sub-ng/g range and upper-pg/g range only, the highest sensitive stripping technique with adsorptive accumulation of a surface active metal chelate has to be applied.

This method has been useful in practice for the estimation of nickel and cobalt in water samples, based on the formation and preconcentration of their dimethylglyoxime complexes at the electrode, followed by reduction of the accumulated chelate by a cathodic voltage scan (Wang 1985).

Similar highly sensitive techniques for water analysis have been developed for uranium, titanium, gallium, aluminum and molybdenum (Table 2). Interferences in these determinations caused by surface active organic compounds, which lower the sensitivity owing to adsorption, are removed by oxidative UV-irradiation.

The previously mentioned interferences by various organic compounds have special effect in the metal analysis of solid environmental samples like soils, sewage sludges and biological materials. All of these samples necessarily have to be decomposed prior to the analysis, which inevitably entails the real binding forms of the metals. Therefore, only the total element concentrations can be measured after such a decomposition procedure, generally by dry or wet ashing techniques. Investigations for the polarographic or voltammetric determination of element species in extracts of solid environmental samples up until now are unsatisfactory and result in distribution pattern for the organically bound elements.

Table 2. Adsorptive stripping measurements of metals ions in water samples

Analyte	Ligand	Electrolyte	Detection Limit	Ref.
Gallium	Solochrome Violet RS	Acetate Buffer (pH 4.8)	1×10^{-9} M	Wang et al. 1986
Molybdenum	Eriochrome Blue Black R	Acetic Acid + Ammonia (pH 1.6)	5×10^{-9} M	Willie et al. 1987
Titanium	Mandelic Acid	Mandelic Acid + Ammonia (pH 3.3)	3×10^{-11} M	Li et al. 1989
Uranium	Pyrocatechol	Acetate Buffer (pH 4.7)	2×10^{-9} M	Lam et al. 1983
Platinum	Formazone	Formaldehyde + Hydrazine + H_2SO_4	4×10^{-10} M	van den Berg et al. 1988
Aluminium	Solochrome Violet RS	Acetate Buffer (pH 4.5)	5×10^{-9} M	Wang et al. 1985

For the determination of the total element concentrations in soils and sewage sludges, digestions with nitric acid and hydrochloric acid are used. This serves to completely release the environmentally relevant heavy metals from the dried sample. In this process sparingly soluble oxides mainly of aluminium, chromium and titanium, cannot be determined.

Today the determination of, e.g. zinc, cadmium, lead and copper directly in the digested solution can be performed continuously and automatically with microprocessor-controlled devices. These computerized polarographs mostly work according to the principle of segmentation of the voltage run, where into each individual segment only the conditions of one element are entered. The polarogram resp. voltammograms recorded for the selected operation sequence in the determination of these metals in sewage sludges are shown in Figure 1 (Meyer et al. 1987).

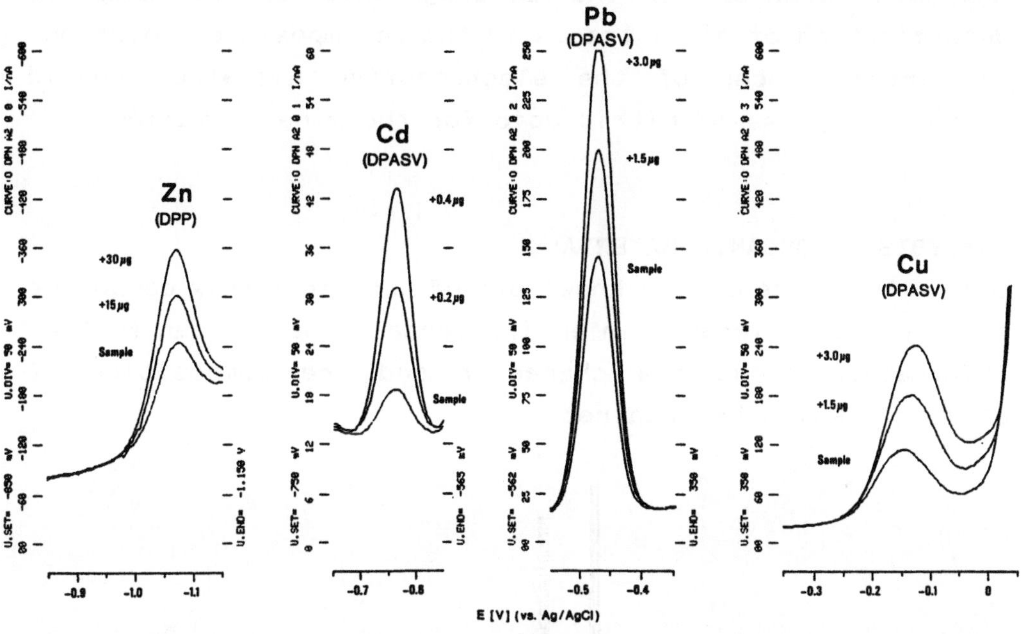

Fig. 1. Determination of heavy metals in sewage sludge in the denounced operation sequence (registrated with the Metrohm VA-Processor 646)
Operation Sequence:
Sweep 1: -850 ... - 1150 mV for Zn^{2+} by DPP;
Sweep 2: -750 ... - 565 mV for Cd^{2+} by DPASV;
Sweep 3: -562 ... - 350 mV for Pb^{2+} by DPASV;
Sweep 4: -350 ... + 100 mV for Cu^{2+} by DPASV

By far more extended are the sample preparations for the polarographic or voltammetric analysis of the total element concentrations in biological materials.

Wet and dry decomposition in open systems are normally beset with systematic errors due to contamination or volatilization, which falsify the analytical results if the determinands lie in the ng/g-range. In the mineralization of organic matrices in pressure bomb systems, the blank depends to a great degree on the chemicals used and the material of the inserts.

In any cases, the main disadvantage of the pressurized decomposition is that biological samples are not completely mineralized in pressure bombs, unless high temperatures are applied which require special apparatus.

The main problem in the polarographic or voltammetric determination of metal traces in the decomposition solutions are interferences of the electroactive nitrated organic compounds by use of nitric acid for the mineralization.

ANALYSIS OF ORGANIC MATERIALS

Only with comprehensive knowledge of organic matrix compounds and organic pollutants in environmental samples, can further information about the character and the composition of element species be obtained.

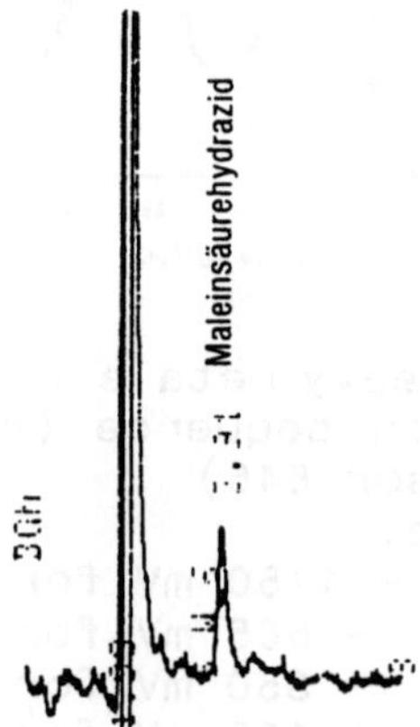

Fig. 2. Chromatogram of 275 pg maleic hydrazide in 20 µl of the injected sample. Eluent: Water/Methanol (90:10) + KNO_3 (pH 3); Working Electrode: Glassy Carbon; Detection Potential: + 1,3 V (vs Ag/AgCl)

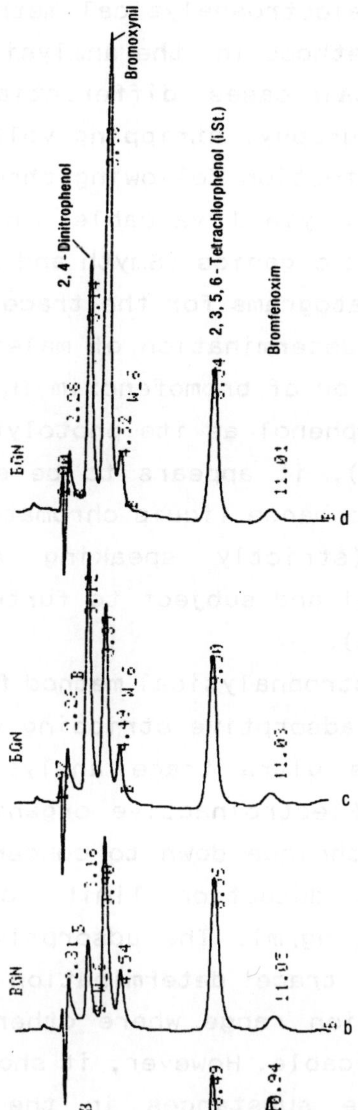

Fig. 3. Chromatograms of bromofenoxim and its photolytic decomposition products bromoxynil and 2.4-dinitrophenol detected at a glassy carbon electrode by + 1,3 V (vs. Ag/AgCl); Injected Volume 20 µl; Column: Nucleosil C_{18}, 10 µm; Eluent: Water-Methanol (22 : 78) + KNO_3 (pH 3). a. 14 ng bromofenoxim + 1 ng 2,3,5,6-tetrachlorophenol; b. 120 min exposure; c. + 2.5 ng 2,4-dinitrophenol; d. + 4 ng bromoxynil

It may be mentioned that electroanalytical methods are not unconditionally standard methods in the analysis of organic compounds. However, in certain cases, differential pulse and alternating current polarography, stripping voltammetry and on-line electrochemical detection following chromatographic separation techniques, may yield valuable information on environmentally significant organics (Smyth and Smyth 1987). When considering the chromatograms for the trace analysis of pesticides, e. g. for the determination of maleic hydrazide in the pg-range (Figure 2) or of bromofenoxim in presence of bromoxynil and 2.4-dinitrophenol as its photolytic decomposition products (Figure 3), it appears to be evident that particularly the high performance liquid chromatography with voltammetric detection (strictly speaking amperometric detection) is very powerful and subject to further development (Meyer and Henze 1989).
Another very promising electroanalytical method for traces of organic pollutants is the adsorptive stripping voltammetry, already mentioned for the ultra trace analysis of heavy metals in water samples. Electroinactive organic compounds can be studied by this technique down to concentrations of about 0,1 μg/ml and the detection limit for reducible organics is as low as 0.1 ng/ml. The adsorptive stripping voltammetry thus permits trace determination of organic compounds at a concentration range where other analytical methods are often not applicable. However, it should be noted that other surface active substances in the sample can interfere with the determination.
In environmental analysis, the possibility of determining trichlorobiphenyl by adsorptive stripping voltammetry is of great interest. Because the conventional method for quantifying polychlorinated biphenyls (PCBs) is the very expensive combination of gas chromatography with mass spectrometry, a less extended electrochemical method would be desirable; but intense investigations are necessary to reach this goal. Up until now the method was applied to the analysis of waste and natural waters (Lam and Kopanica 1984).

A further important ecological example is the determination of surfactants, which are probably the most common organic pollutants in the aqueous environment. At low concentrations they are usually determined electrochemically or by spectrophotometry.

Electrochemical techniques for the determination of surface active substances are based, in general, on adsorption phenomena at the electrode/electrolyte interface. The polarographic adsorption analysis is done by measurements of the capacity depression or the desorption peaks of the capacity-potential curves with the alternating current mode.

A further method for studying the content of surface active substances in solution is based on suppression of polarographic maxima by adsorbable molecules. This method is in practical utilization not so useful because it lacks specifity and depends on a great number of parameters, such as the composition of the supporting electrolyte, properties of the electrode, etc. (Kalvoda 1987).

Along with the already mentioned organic harmful substances, the aromatic nitro- and nitroso-compounds, the polycyclic aromatic hydrocarbons, as well as all kinds of organohalogen compounds, are most prominent in ecological considerations. The electrochemical behaviour of aromatic nitro- and nitroso-compounds has been extensivly studied and sensitive polarographic methods often in combination with a chromatographic separation, have been developed.

Very interesting are also the clues to the relations between redox behaviour and the carcinogenic properties of these compounds. Compounds, which are reduced at relatively positive potentials, are likely to be most toxic, especially those which are reduced directly to the amin (Smyth and Osteryoung 1980).

Polycyclic aromatic hydrocarbons can be electrochemically determined with good sensitivity by anodic differential pulse voltammetry with a glassy carbon working electrode. This method is suitable for the analysis of individual polycyclic hydrocarbons (e.g. 3,4-benzpyrene, anthracene) down to

2×10^{-8}M with acetonitrile or sulpholane as solvent (Smyth and Osteryoung 1980).

Polarographic investigations of the simple C_1- and C_2-halogenated hydrocarbons served for the characterization of the reduction of carbon-halogen-bounds, in general.

Otherwise, these considerations were important for a judgement of the redox-properties, because in this way the slowly proceeding degradation of the chlorinated hydrocarbons can be explained (Meyer and Henze 1987, 1988).

In Table 3 the ranges for the peak-potentials of the investigated compounds are shown, obtained by the differential-

Table 3. Peak potentials of the C_1- and C_2-chlorinated and -brominated hydrocarbons in different solvents (ethanol, dimethylsulfoxide, dimethylformamide, propylencarbonate, acetonitrile) and supporting electrolytes (tetraalkyl-ammonium-salts)

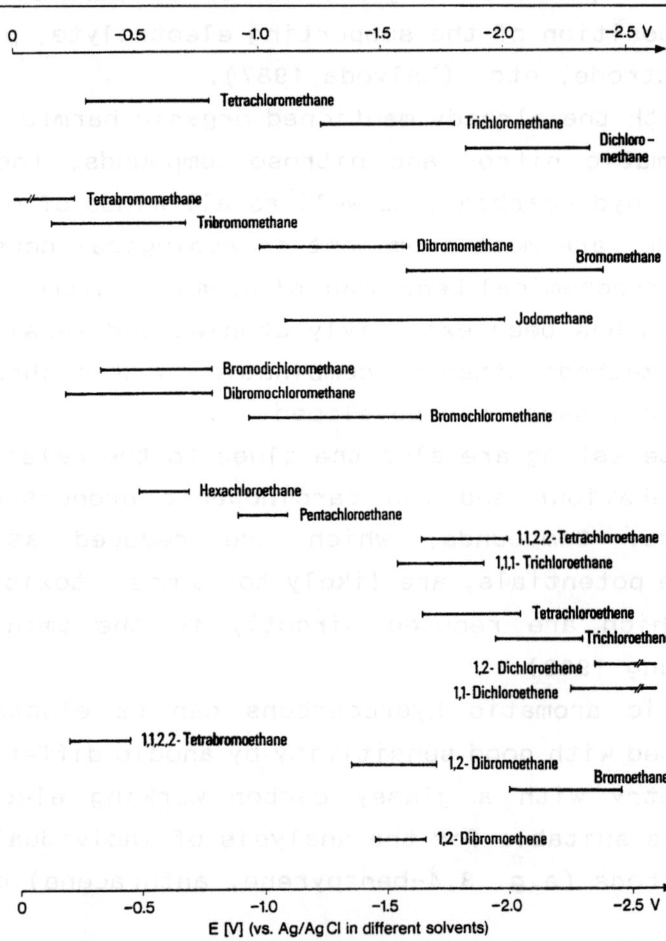

pulse-polarography, registrated in different solvents and supporting electrolytes. It is seen that the brominated substances are generally more easily reduced than the chlorinated ones.

In all cases the irreversible reduction process of the C_1- and C_2- halogenated hydrocarbons are discussed as radical mechanisms, combined with a stepwise transfer of two electrons.

For the C_1- compounds the electrode process corresponds to:

Starting reaction: $CX_4 + e^- \rightarrow [CX_4^{\cdot -}] \rightarrow \cdot CX_3 + X^-$.

Consecutive reaction: $\cdot CX_3 + e^- \rightarrow [CX_3^-] \xrightarrow[-\,OH^-]{+\,H_2O} CHX_3$

Likewise halogen substitutes are released stepwise from the C_2-halogenated hydrocarbons and are reduced in the case of their geminal position by developing one peak to the corresponding ethane.

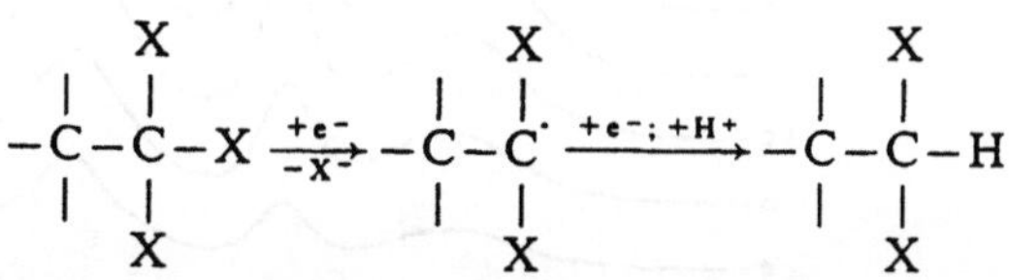

In case of a vicinal position of the halogens, the compounds are reduced to the corresponding ethene or ethine via biradicals

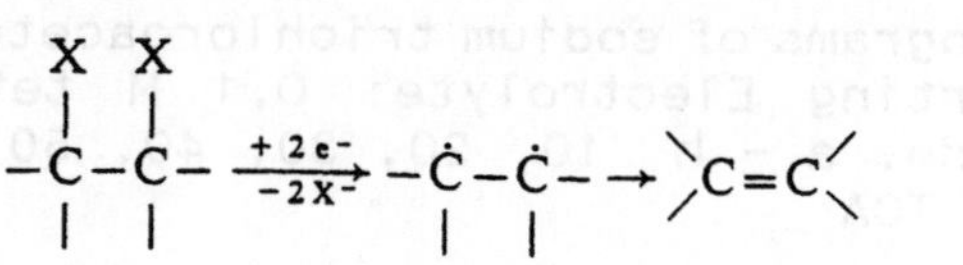

or carbanions.

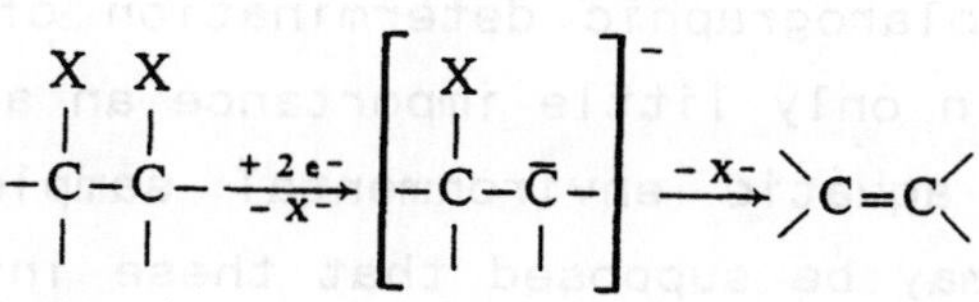

Similar electrode processes are behind the polarographic reduction of various organohalogen pesticides with an aliphatic-bound halogen (Jehring et al. 1989).

Fig. 4 shows dp-polarograms of trichloroacetic acid, used as a herbicide. The three peaks correspond to the stepwise reduction to the unsubstituted acid, via the di- and mono-chloroacetic acid.

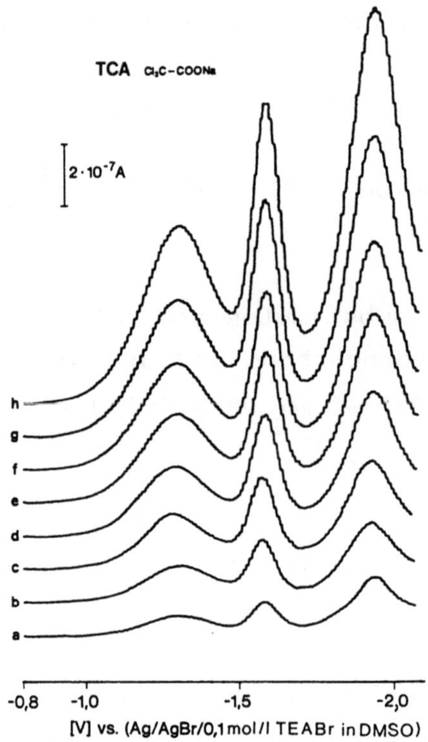

Fig. 4 Polarograms of sodium trichloroacetate (TCA) in DMSO. Supporting Electrolyte: 0.1 M tetraethylammonium-bromide. a - h 10, 20, 30, 40, 50, 60, 70, 80, 100 µl/ml TCA

The direct polarographic determination of pesticides will certainly gain only little importance an account of the low contents in aquatic environmental samples or in soils. However, it may be supposed that these investigations have smoothed the way for some compounds to be detected very sensitive by amperometry followed high performance liquid chromatography.

CONCLUSIONS

Electroanalytical methods, especially the highly sensitive stripping voltammetric techniques, are suited to trace analysis of metals and organic compounds in environmental samples, chiefly in aquatic systems.

Moreover, stripping voltammetry is the most applied technique for trace element speciation in waters. However, it is not possible to distinguish by this technique between the different labile species of an individual metal. Only a single stripping peak is obtained for a mixture of labile species, which may coexist for unique metals in natural waters. Consequently, without sample treatments, the voltammetric procedures are limited in the characterization of groups of species. For more information of labile species, pseudopolarograms may be used.

Polarography and voltammetry are very powerful in the determination of the redox states of elements in solutions. For the analysis of organic pollutants, the amperometric detection following the seperation by high performance liquid chromatography is very sensitive and selective. This technique may be applicable for trace analysis of pesticides in environmental samples.

References

1. Berg van den CMG, Jacinto GS (1988) The determination of platinum in sea water by adsorptive cathodic stripping voltammetry. Anal Chim Acta 211 : 129

2. Bond AM, Bradbury JR, Howell GN, Hudson HA, Hanna PJ, Strother S (1983) Electrochemical reduction of the trimethyllead(IV) cation in seawater. J Electroanal Chem 154 : 217

3. Florence TM (1986) Electrochemical approaches to trace element speciation in waters (a review) Analyst 111 : 489

4. Florence TM, Batly GE (1980) Chemical speciation in natural waters. Crit Rev Anal Chem 9 : 219

5. Henze G, Monks P, Tölg G, Umland F, Weißling E (1979) Über die simultane Bestimmung von Selen und Tellur im unteren ppb-Bereich durch Cathodic-Stripping-Voltammetry. Fresenius Z Anal Chem 295 : 1

6. Henze G, Joshi AP, Neeb R (1980) Bestimmung von Arsen im sub-ppb Bereich durch Differential-Pulse-Cathodic-Stripping-Voltammetry. Fresenius Z Anal Chem 300 : 267

7. Henze G (1981) Determination of toxic elements by cathodic-stripping-voltammetry. Mikrochim Acta (Wien) II 343

8. Henze G, Neeb R (1986) Electrochemische Analytik. Springer, Berlin, Heidelberg, New York

9. Jehring H, de la Chevallerie-Haaf U, Meyer A, Henze G (1989) Polarographisches Verhalten von Pesticiden mit Organohalogenbindungen. Fresenius Z Anal Chem 332 : 890

10. Kalvoda R (1987) Polarographic determination of adsorbable molecules. Pure & Appl Chem 59 : 715

11. Lam NK, Kalvoda R, Kopanica M (1983) Determination of uranium by adsorptive stripping voltammetry. Anal Chim Acta 154 : 79

12. Lam NK, Kopanica M (1984) Determination of trichlorobiphenyl by adsorptive stripping voltammetry. Anal Chim Acta 161 : 315

13. Li H, Berg van den CMG (1989) Determination of titanium in sea water using adsorptive cathodic stripping voltammetry. Anal Chim Acta 221 : 269

14. Meyer A, Henze G (1987) Polarographisches Verhalten der C_1-Halogenkohlenwasserstoffe. Fresenius Z Anal Chem 327 : 123

15. Meyer A, de la Chevallerie-Haaf U, Henze G (1987) Determination of zinc, cadmium, lead and copper in soils and sewage sludges by microprocessor-controlled voltammetry in comparison with atomic absorption spectrometry. Fresenius Z Anal Chem 328 : 565

16. Meyer A, Henze G (1988) Polarographisches Verhalten der C_2-Halogenkohlenwasserstoffe. Fresenius Z Anal Chem 329 : 764

17. Meyer A, Henze G (1989) Untersuchungen zur HPLC-Bestimmung von Pesticiden mit amperometrisch detektierbaren Hydroxylgruppen. Fresenius Z Anal Chem 332 : 898

18. Mizuike A (1987) Recent developments in trace metal speciation in fresh water. Pure & Appl Chem 59 (4) : 555

19. Nürnberg HW, Valenta P, Mart L, Raspor B, Sipos L (1976) Applications of polarography and voltammetry to marine and aquatic chemistry. Fresenius Z Anal Chem 282 : 357

20. Nürnberg HW (1983) Investigations on heavy metal speciation in natural waters by voltammetric procedures. Fresenius Z Anal Chem 316 : 557

21. Smyth MR, Osteryoung J (1980) Electroanalysis of environmental carcinogens. In: Electroanalysis in hygiene, environmental, clinical and pharmaceutical chemistry, Elsevier Scientific Publishing Company, p. 423

22. Smyth WF, Smyth MR (1987) Electrochemical analysis of organic pollutants. Pure & Appl Chem 59 : 245

23. Wang J, Farias PAM, Masmoud JS (1985) Stripping voltammetry of aluminium based an adsorptive accumulation of its solochrome violet RS complex at a SMD electrode. Anal Chim Acta 172 : 57

24. Wang J (1985) Stripping Analysis. VCH Publishers, Inc.

25. Wang J, Zadeii JM (1986) Determination of traces of gallium based on stripping voltammetry with adsorptive accumulation. Anal Chim Acta 185 : 229

26. Willie SN, Berman SS, Page JA, van Loon GW (1987) The voltammetric determination of Mo in seawater after adsorptive accumulation of the eriochrome blue black R complex. Can J Chem 65 : 957

Metal Speciation in Flue Gases, Work Place Atmospheres and Precipitation

D. Klockow, R.D. Kaiser, J. Kossowski, K. Larjava[1], J. Reith, and V. Siemens
Institute of Spectrochemistry and Applied Spectroscopy (ISAS), and University of Dortmund, Dept. of Chemistry, Dortmund, F.R.G.

1. Introduction

The determination of metals in technical and ambient aerosols is a field of analytical chemistry related, for instance, to emission control, investigation of atmospheric processes, or to effects research. In this context not only the total content of a certain metallic element in a volume of air or in a dust sample, respectively, is of interest, but also its distribution between different phases or over different size ranges and its occurence in definite chemical forms (species).

In aquatic chemistry the determination of dissolved and particulate metal species has been given considerable attention to for many years (see, e.g., the contributions by Buffle, Förstner et al., Frimmel, Klöppel et al., Lund, Tessier et al., Wilken et al., this volume). In emission measurements or air quality studies, however, attempts towards metal speciation are scarce and have concentrated mainly on gaseous metal compounds (Irgolic et al., Neidhart et al., van Cleuvenbergen et al., this volume).

Flue gases, work place atmospheres and ambient air are complex colloids, so called aerosols, in which droplets and/or solid particles are suspended in a gas mixture. Metal compounds may also be present in gas, liquid (dissolved) or solid phase in such aerosols, their partitioning being a function of the physical and chemical properties of the metals under study and of the aerosol generating processes involved (see Table 1). Examples for the latter are sea spray, soil dust raise by winds, volcanic eruptions, or

[1] On leave from the Technical Research Centre of Finland (VTT), Laboratory of Heating and Ventilation, Espoo, Finland

NATO ASI Series, Vol. G 23
Metal Speciation in the Environment
Edited by J. A. C. Broekaert, Ş. Güçer, and F. Adams

technical processes such as cement manufacturing, which typically produce so called primary particles. In these cases material, through mechanical action, is emitted directly in particulate (droplet) state (predominantly ≥ 1 μm diameter) (Whitby et al., 1980; Klockow, 1982). In contrast processes like fossil fuel combustions or ferrous and nonferrous metallurgical operations generate secondary particles (≤ 1 μm diameter), besides primary ones, by condensation of vapour after its escape from high temperature sources. By this means large fractions of emitted heavy metals (Pb, Cd, Zn) are accumulated in the submicron size range (Natusch et al., 1980a; Dannecker, 1982; Müller, 1982); minor amounts may even stay in a "non-filtrable" state (Bruckmann et al., 1990). Often the metals are enriched at surfaces of primary particles such leading to their compositional heterogeneity (Nießner et al., 1987a). In a similar way as condensation of gaseous matter also homogeneous or heterogeneous chemical transformation of trace gases in natural and technical aerosols leads to secondary particulates (gas-to-particle conversion). Such a "reactive phase transfer" is of considerable relevance for the tropospheric chemistry of compounds such as SO_2, NO_2, NH_3, gaseous organic compounds and most probably also metal alkyl species (Calvert et al., 1985; van Cleuvenbergen et al., 1990).

aerosol generation process	classification of aerosol particles	particle size range	examples
dispersion (mechan. generation)	**primary** solid particles and liquid droplets	coarse (≥ 1 μm diam.)	sea spray cement manuf. metal grinding
condensation (of vaporized or chemically generated matter)	**secondary** solid particles and liquid droplets	fine (≤ 1 μm diam.)	homog.chem. transform. fossil fuel burning welding
	primary particulates converted to **secondary** particles or droplets through heterog. nucleation	coarse (≥ 1 μm diam.)	heter. chem. transform. fossil fuel burning fog

Table 1. Aerosol generation through technical and natural processes

2. Metal Speciation in Aerosols

A speciation scheme for characterization of a metal containing aerosol may be devided into a physical and a chemical part (see Fig. 1). The physical part includes investigation of the partitioning of the metal under study between different phases or determination of the metal in different particle (droplet) size ranges. The chemical part deals with identification and quantification of well defined chemical forms of the respective metal in gas, liquid and solid phase.

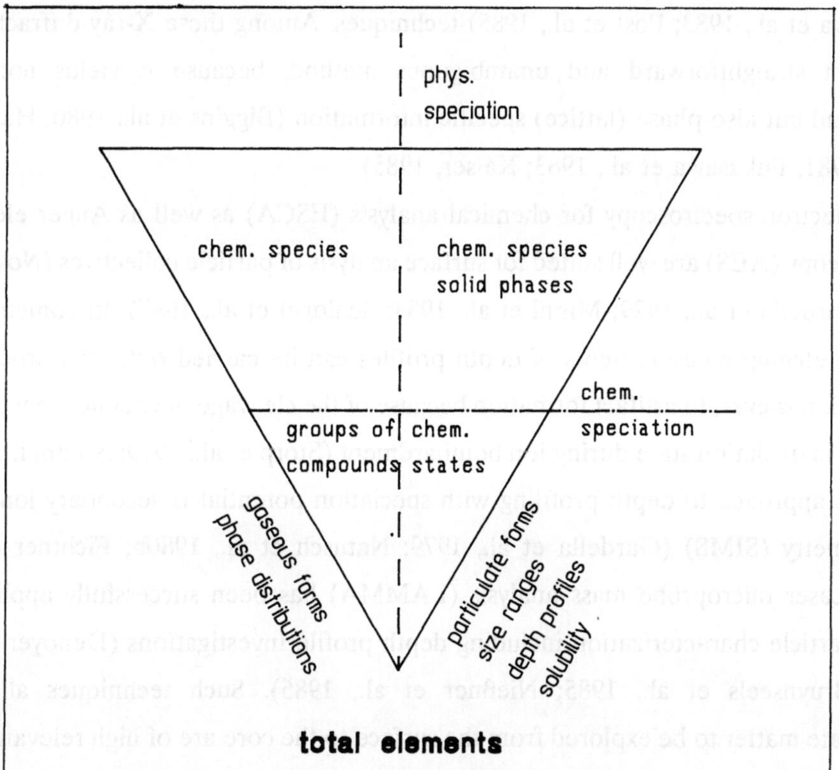

Figure 1. Hierarchy of trace metal speciation in aerosols

Metal speciation in aerosols first of all requires proper sampling techniques which allow reliable separation and preconcentration of gaseous and particulate metal compounds without artifact formation. In principal such techniques are based on the interaction of inertial and viscous forces (impactors, filters), electrical forces (electrostatic separaters, electrical mobility analyzers) and diffusion (diffusion batteries, denuders, filters) (Klockow, 1982; Nießner et al., 1987a). Complementary to traditional filtration and impactor sampling diffusion processing has found increasing application in aerosol collection during the past years (Klockow, 1982; Nießner et al., 1987a,b), and it will be discussed further in this paper that it is also a promising approach to metal collection and speciation in hot flue gases (Larjava et al., 1990).

In particulate matter collected from flue gases, workplace air or ambient air on a suitable substrate metal speciation may be achieved by thermal (Prack et al., 1983; Kaiser, 1985) or X-ray (Grasserbauer, 1975; Biggins et al., 1980; Harrison et al., 1981; Fukasawa et al., 1983; Post et al., 1985) techniques. Among these X-ray diffraction is the most straightforward and unambiguous method, because it yields not only compound but also phase (lattice) specific information (Biggins et al., 1980; Harrison et al., 1981; Fukasawa et al., 1983; Kaiser, 1985).

Electron spectroscopy for chemical analysis (ESCA) as well as Auger electron spectroscopy (AES) are well suited for surface analysis of particle collectives (Novakov, 1977; Gardella et al., 1979; Minni et al., 1984; Schlögel et al., 1987). In combination with ion etching measurements of depth profiles can be carried out. This procedure may lead, however, to artifact formation because of the cleavage of chemical bonds and changes in oxidation state during ion bombardment (Storp et al., 1979; Nießner, 1981). A direct approach to depth profiling with speciation potential is secondary ion mass spectrometry (SIMS) (Gardella et al., 1979; Natusch et al., 1980b; Fichtner et al., 1990). Laser microprobe mass analysis (LAMMA) has been successfully applied to single particle characterization including depth profile investigations (Denoyer et al., 1983; Bruynseels et al., 1985; Nießner et al., 1985). Such techniques allowing particulate matter to be explored from the surface to the core are of high relevance for risk assessment in inhalation toxicology, since the effects caused by inhaled particles in the respiratory tract are related to a large extent to the species enriched at their surfaces. Recently attempts towards continuous surface metal speciation in aerosols, based on the photoelectric effect, have been made (Nießner et al., 1989).

A chemical alternative to merely physical depth profile investigation, which somehow reflects the fate of aerosol particles in a natural aqueous environment, is time resolved solvent leaching (TRSL) (Natusch, 1978; Natusch et al., 1980a) in combination with, for instance, ESCA. Though TRSL has never been fully exploited, it offers an interesting and inexpensive way to elucidate the composition and possible chemical and physiological behavior of particle surfaces. Also bulk analysis of microgram amounts of collected dust as to metal species can be carried out by employing phase selective leaching procedures (Klockow et al., 1985); this will be shown later in this contribution.

Since wet deposition is the major removal process of many metals from the atmosphere, the chemical forms in which these metals are present in atmospheric condensed water (rain, snow, fog etc) are of particular interest from the point of view of effects on terrestrial ecosystems. For such purpose speciation methods as known from aquatic chemistry (see other contributions in this volume) may be employed. However, to just quantify which fraction of a certain metal is present in precipitation in dissolved form and which fraction in suspended particulates, a simple filtration approach as described further below is sufficient.

3. Diffusion Processing for Metal Speciation in Flue Gases

Many metals escaping from high-temperature processes have detrimental effects to man and environment. Because of this, great efforts are made to drastically reduce such emissions, e.g. from combustion sources (Vogg, 1988), and to provide proper measurement techniques for emission survey.

Metals which volatilize easily at process temperatures are the most difficult ones to control because of their appearance in both, particulate and vapor phase. Thermodynamic model calculations have shown (Mojtahedi et al., 1987), that during combustion of municipal solid waste (combustion temperatures sometimes are as high as 1.400 °C) metals such as mercury, cadmium, lead and zinc may appear in elemental form, as halides or as oxides and may partially be present in gaseous phase. The situation is similar in fluidised-bed combustion and gasification of coal (Mojtahedi, 1989). As to the gaseous fraction it is only known by now, that mercury remains almost totally in vapor phase under control unit (filter) operating conditions (e.g. 200 °C) (Braun et al., 1986; Mojtahedi et al., 1987; Germani et al., 1988), and that different mercury species show very different deposition behavior in scrubbers (Braun et al., 1986). For these reasons it appears to be very important to distinguish between gaseous and particulate forms of the metal under study and, at least in case of mercury, also between different chemical species in the two phases.

Conventionally, metal aerosols are collected through filter media under flue gas conditions. New methods have been developed to assure also collection of metals present in vapor phase (Germani et al., 1988; Larjava et al., 1986; Larjava et al., 1990; Jockel et al., 1987). Metzger and Braun (1987) have used a solid absorption method for speciation of elemental mercury (Hg^0) and mercuric chloride ($HgCl_2$) in municipal incinerator flue gases. The chloride is first separated from the sample gas stream by using an anion exchanger (Dowex 1 X8). Elemental mercury is passing through and is collected with an activated carbon/iodine scrubber.

These methods described in the literature are quite complicated and only in one case allow distinction between different compounds. Therefore, a denuder sampling system was developed for collection of volatile metal species released from high-temperature processes, and for separating them from particulate forms. The system (see Fig. 2) consists of three parts:

- a cyclone used as a pre-separator to remove coarse particles $D_{p,ae} > 2\ \mu m$,
- a coated glass denuder for the collection of gaseous metal species,
- a quartz filter for the collection of fine particles, $D_{p,ae} < 2\ \mu m$

The developed denuder sampling technique was tested for elemental cadmium (Cd^0), elemental mercury (Hg^0) and mercuric chloride ($HgCl_2$) in the laboratory and for mercury also under real flue gas conditions.

Elemental silver (Ag) and potassium chloride (KCl) were used as denuder coatings. Coating procedures, as well as the laboratory devices used in these investigations, are described elsewhere (Siemens, 1989; Larjava et al., 1990). The collected metal species together with the coating material were eluted from the denuders with strong nitric acid (HNO_3, suprapur). Analyses were carried out by using atomic absorption spectrometry (cold vapor technique for Hg and graphite furnace technique for Cd).

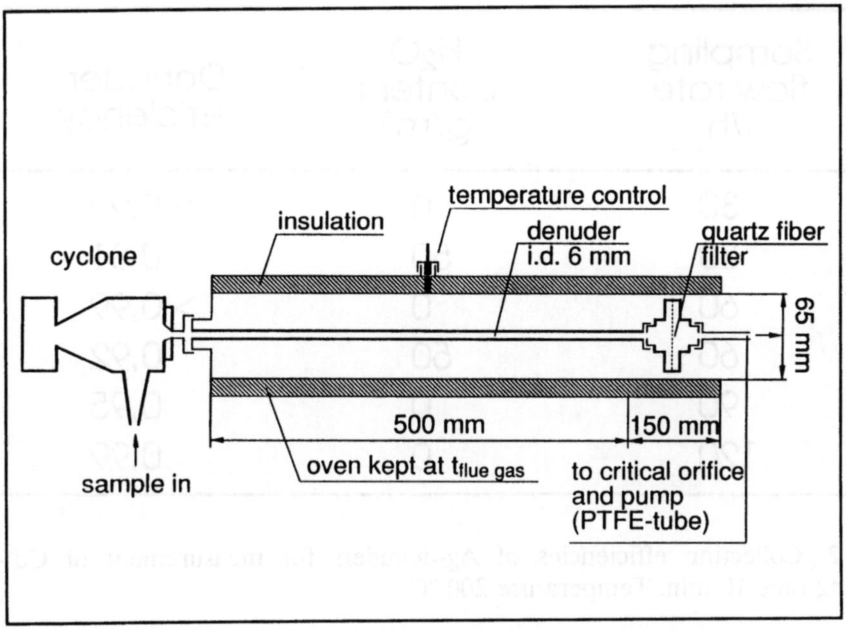

Figure 2. Denuder sampling system for flue gas measurements

Collection efficiencies of Ag-denuders for the measurement of gaseous elemental cadmium are presented in Table 2. The tests were carried out in the laboratory, two of them by using humidified test aerosol (Larjava et al., 1990). Water vapor seems to suppress Cd^0 deposition to some extent, the efficiencies observed, however, are still acceptable. The sampling flow rate was varied between 30 and 120 L/h, but in this range didn't show any significant influence on the collection efficiency.

Ag- and KCl-denuders were investigated with respect to the measurement of elemental mercury and mercuric chloride. Tests were carried out in the laboratory at different sampling temperatures. It could clearly be seen that the sampling temperature should not exceed 120 °C. At least elemental mercury is so volatile that it will be evaporated from a Ag-coating even at these relatively low temperatures.

Sampling flow rate l/h	H_2O content g/m3	Denuder efficiency
30	0	> 0,99
30	50	0,81
60	0	> 0,99
60	50	0,92
90	0	0,95
120	0	0,99

Table 2. Collection efficiencies of Ag-denuders for measurement of Cd^0-vapor. Sampling time 10 min. Temperature 200 °C

Tables 3 and 4 show that KCl-denuders don't collect Hg^0 at all, but retain almost 100 % of $HgCl_2$. Ag-denuders, at the other hand, collect perfectly both mercury species when the sampling temperature is low enough. From this a possibility for mercury speciation by using the denuder technique arises: if two denuders are employed in series, so that the first is coated with KCl and the second with Ag, then mercury halides may be deposited in the first tube and elemental mercury in the second. Water vapor and gaseous hydrochloric acid (up to 400 mg/m^3) did not influence the deposition of the mercury species in the denuders.

The linearity of the denuder sampling system was also investigated for Cd^0, Hg^0 and $HgCl_2$. The amounts of these species found in the tubes were always a linear function of sampling time. The detection limits have been estimated to be a few tenths of a microgram per cubic meter when sampling times of 30 min and flow rates between 40 and 60 L/h are employed.

Average temperature °C	Collection efficiency	
	KCl-denuder	Ag-denuder
30	0	0,997
60	0	0,997
90	0	0,998
130	0	0,588

Table 3. Collection efficiencies of Ag- and KCl-denuders for measurement of Hg^0-vapor at different temperatures. Sampling time 15 min. Test aerosol sampling flow rate 40 L/h

Average temperature °C	Collection efficiency	
	KCl-denuder	Ag-denuder
30	0,994	0,994
60	0,994	0,994
90	0,994	0,993
130	0,914	0,809

Table 4. Collection efficiencies of Ag- and KCl-denuders for measurement of $HgCl_2$-vapor at different temperatures. Sampling time 15 min. Test aerosol sampling flow rate 40 L/h

Ag-denuders were also tested under real stack conditions with respect to the collection of total gaseous mercury. In these experiments mercury was vaporized during light fuel oil burning of mercury containing waste. To compare the denuder approach with an established high-efficiency method, measurements were carried out in parallel with a liquid scrubber technique (4 % $K_2Cr_2O_7$ in 20 % HNO_3) (Dumarey et al., 1981).

Two denuder samples were taken during one scrubber collection, the sampling flow rate being 60 L/h. Although the average mercury concentrations in flue gas found by the two methods only differ by about 10 % (see Table 5), the concentration variations with time are much more pronounced in case of the denuder technique. This is the result of the short sampling time and consequently of a good time resolution achievable with denuders.

Estimated output (Calc. from input data)	Hg mg/m³ Ag-denuder t_s = 10 min	$K_2Cr_2O_7/HNO_3$ t_s = 60 min
≈ 3	7,0	4,1
	1,5	
	3,2	2,3
	2,5	
Average ≈ 3	3,6	3,2

Table 5. Comparison of Ag-denuder and $K_2Cr_2O_7/HNO_3$-absorption method for collection of Hg^0-vapor under real flue gas conditions. Flue gas temperature: 90 °C, working temperature of Ag-denuder: 90 °C, working temperature of dichromate-scrubber: 15 °C, sampling flow rate in both systems: 60 L/h

4. Microchemical Phase Analysis for Metal Speciation in Workplace Air

The investigations described below were carried out in order to examine the potential of selective leaching procedures for speciation of nickel, cobalt and chromium compounds in airborne particulate matter. For this purpose artificially generated aerosols (Kaiser, 1985; Klockow et al., 1985; Reith, 1990) and, in case of nickel, also electrofilter dust of a nickel refinery and dust collected at a thermal metal plating workplace (Kaiser, 1985) were used as test materials.

Extraction procedures have been employed for phase analysis in metallurgical samples already in the past century (Meinecke, 1871). The separation of metallic and oxidic phases in such samples is possible by selective dissolution of the metallic matter through $CuSO_4$ (Frerichs, 1908) or $KCuCl_3$ (Riott, 1941) solutions. It could be shown, that the $KCuCl_3$ reagent is also well suited for the selective separation of Ni° and Co° from other nickel and cobalt species in dust collected on filters (Kaiser, 1985; Reith, 1990). In order to achieve a selective dissolution of metal sulfides in the presence of Ni°, Co° and the corresponding oxides an aqueous solution containing hydrogen peroxide and di-ammonium hydrogen citrate (Katsnel'son et al., 1960) was used (see Fig. 4).

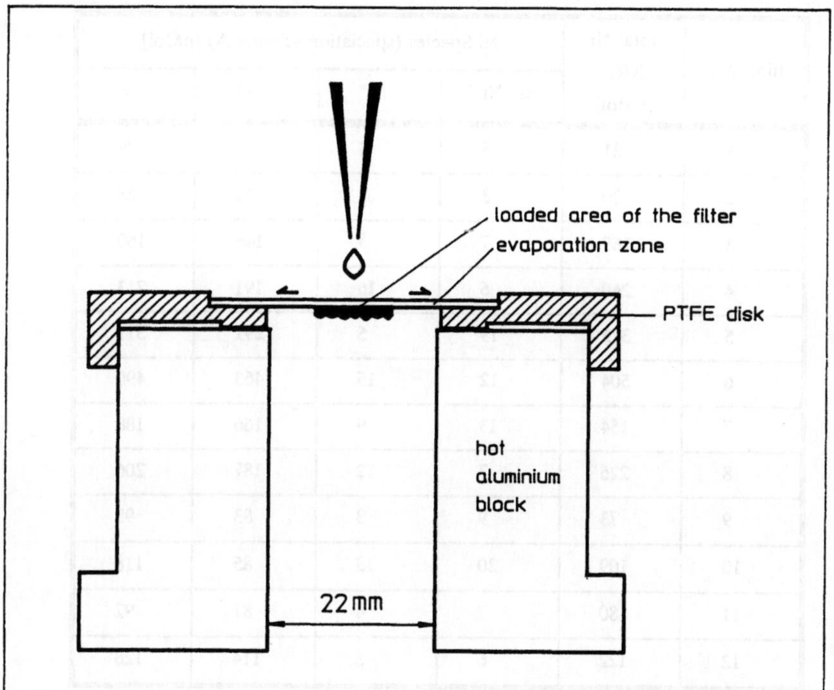

Figure 3. Ring oven and its working principle

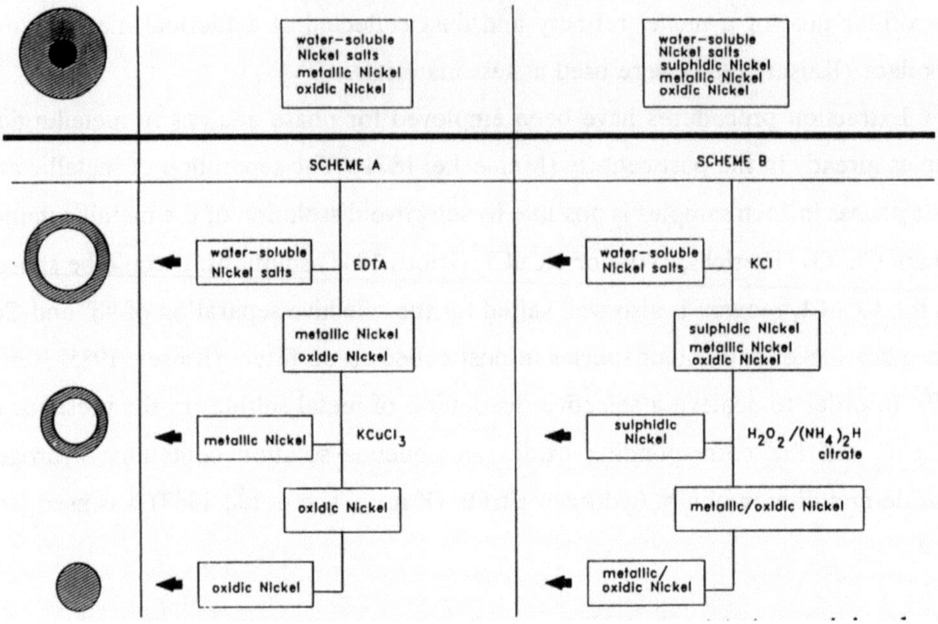

Figure 4. Ring oven speciation schemes for investigation of nickel containing dust

filter No.	total Ni XRF [nMol]	Ni-Species (speciation scheme A) [nMol]			
		sol. Ni^{2+}	Ni^{o}	NiO	Σ
1	21	5	4	20	29
2	20	3	3	22	28
3	189	7	5	148	160
4	260	6	16	191	213
5	373	19	5	291	315
6	504	12	15	463	490
7	154	13	9	166	188
8	226	7	12	187	206
9	73	9	3	83	95
10	109	20	13	85	118
11	80	7	4	81	92
12	122	6	8	114	128

Table 6. Comparison of XRF analysis (total Ni) and microchemical speciation by the ring oven technique (sol. Ni^{2+}, Ni^{o}, NiO). Dust collected at a thermal metal plating workplace

Speciation of chromium in aerosols is complicated by the fact, that Cr(VI) compounds may be reduced to Cr(III) already during the sampling step (Reith, 1990; Neidhardt et al., 1990). Possible means to overcome such difficulties are drying of the aerosol under study during collection (Reith, 1990) or employing collection media (impregnated filters) of high alkalinity (Neidhardt et al., 1990). Taking into consideration the measures necessary to avoid Cr(VI) reduction a speciation scheme could be developed (see Fig. 5), which allows the separate determination of soluble Cr(VI), sparingly soluble Cr(VI), soluble Cr(III) and insoluble chromium in aerosols (Reith, 1990). Most relevant for the performance of the scheme is the application of cation exchange to the separation of Cr(III) and Cr(VI).

To apply the extraction procedures mentioned to metal speciation in particulate matter collected on a suitable filter, the Weisz ring oven (Weisz, 1970) was used (see Fig. 3). The ring oven basically consists of a cylindrical, heatable block of aluminium carrying a central bore-hole of 22 mm diameter. It allows the washing of soluble components out of material in the center of a filter, which is placed horizontally on top of the block, and concentration of the solutes in the heated ring zone (due to evaporation of the solvent used). The bore-hole may be reduced to a smaller diameter by means of an auxiliary adaptor, such making possible two-step separations on one filter. Alternatively, PTFE covers with apertures larger than 22 mm may be put over the aluminium block (see Fig. 3) in order to enlarge the diameter of the "heat wall" and of the corresponding rings containing the eluted material. These features make the ring oven a highly versatile device for speciation purposes, because it allows the different operations to be carried out directly on the collection filter using only microgram amounts of material. The only requirement to be fulfilled is, that the filterholder employed for sampling forces the dust to deposit in a spot not larger than about 10 mm in diameter. Depending on the reagent used for leaching, the species either stay in the center of the filter or are concentrated in an "outer" (~ 28 mm diameter) or "inner" (~ 22 mm diameter) ring (see Figs. 4 and 5). After the separation procedure the filter is concentrically cut into annular parts containing the separated fractions, and the amounts of metals present in these parts are determined by, for instance, flameless atomic absorption spectrometry after nitric acid digestion of the filter material.

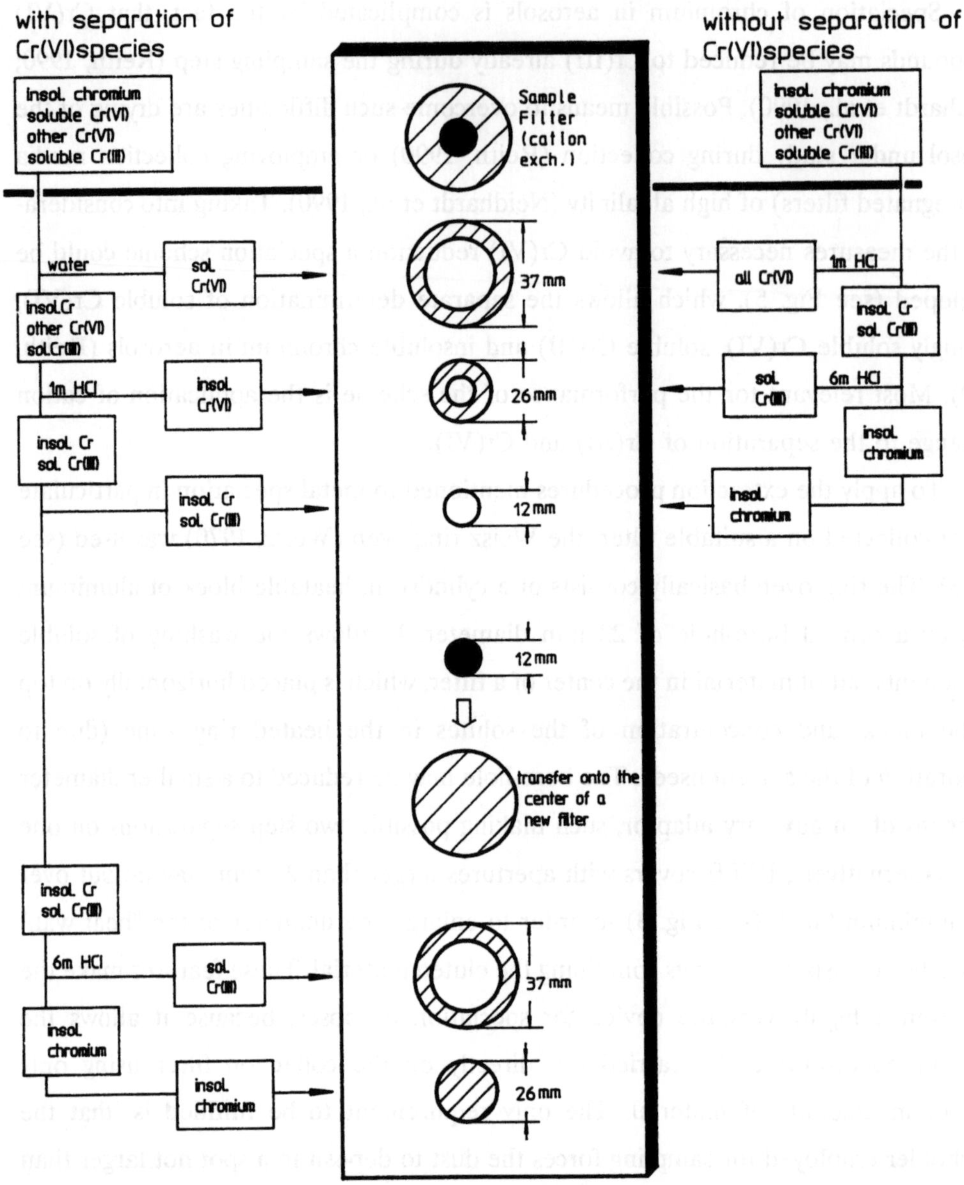

Figure 5. Ring oven speciation schemes for investigation of chromium containing dust

So far selective solvent leaching using the ring oven technique has been practically applied to nickel speciation in dust. One example is given by Table 6, which shows results of analyses of airborne particulate matter collected at a thermal metal plating workplace. As can be seen, the total amount of nickel per filter as determined by X-ray fluorescence analysis compares to some extent with the amount of nickel, which has been calculated from those nickel species (Ni^{2+} sol., Ni^{o}, NiO) determined through application of speciation scheme A (see Fig. 4).

5. In-Situ Separation of Dissolved and Particulate Metal Species in Precipitation

The cycling of water vapour through the atmosphere is inevitably connected to its transfer into the liquid or solid phase and its deposition as rain (snow, fog, etc.). In this removal of condensed water other airborne matter is involved through a number of physico-chemical processes (Pruppacher et al., 1980; Hales, 1986), so that precipitation contains quite a number of dissolved and suspended substances being carried to the ground (wet deposition).

Some metals present in precipitation, in particular the so called heavy metals, may be detrimental to terrestrial ecosystems (Merian, 1984). Therefore the knowledge of the chemical forms in which these metals are deposited, i.e. in readily bioavailable dissolved form or as slowly resorbable or non resorbable particulate matter, is of considerable practical interest with respect to effects research.

In order to distinguish between dissolved and suspended metal species, a wet-only collector (Nguyen et al., 1979) was equipped with filtration units which allow the separation and preconcentration of particulate matter and dissolved metal ions in precipitation during rain events (see Figs. 6 and 7). The rain water collected in the funnel (274 mm inner diameter) is pumped at a rate of 4 ml/min through a filter holder (Fig. 7) containing two filters in series. The first one, a cellulose nitrate filter (0.45 μm pore size), retains suspended particulate matter, and the second, a strongly acidic cation exchange filter of high capacity, dissolved metal ions. The exchange efficiency is rather insensitive as to variations of the pH value beween 3 and 7.

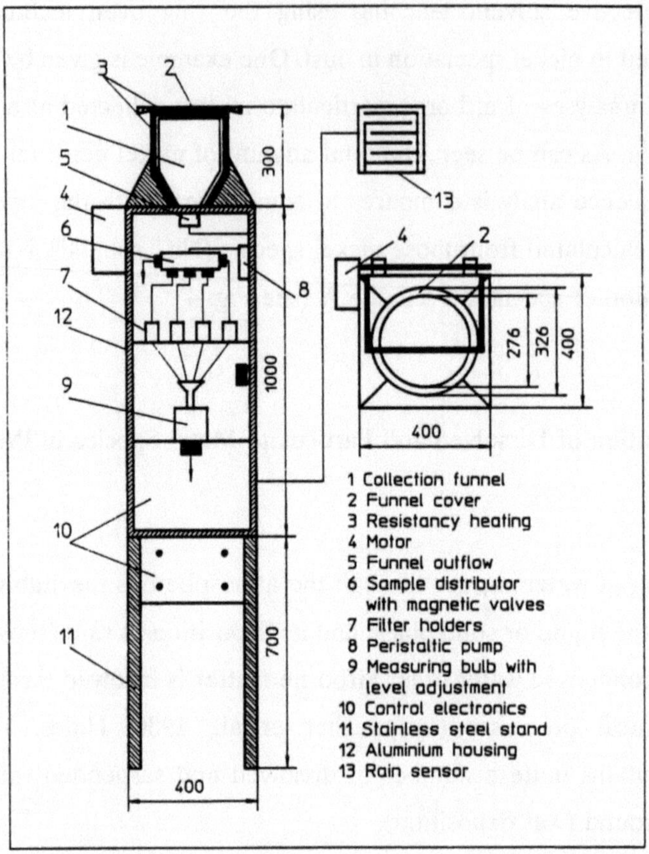

Figure 6. Wet-only rain collector with filtration units. Dimensions in mm

After filtration of 500 ml of rain water a measuring bulb with level adjustment and overflow automatically switches the pumped liquid to another filter holder through operation of the corresponding magnetic valve. By this means 4 sets of filters may be loaded without attendance of the rain collector. The used up filters are replaced by new ones, dried and analysed by X-ray fluorescence (XRF) without further treatment. Under the conditions given (500 ml of rain water for preconcentration) detection limits below 1 μg/L may be achieved for some of the metals under study.

Measurements of this type have been carried out for longer periods at the campus of the University of Dortmund. Examples of results are shown in Figs. 8 and 9. Although the combination of preconcentration through ion exchange filters with

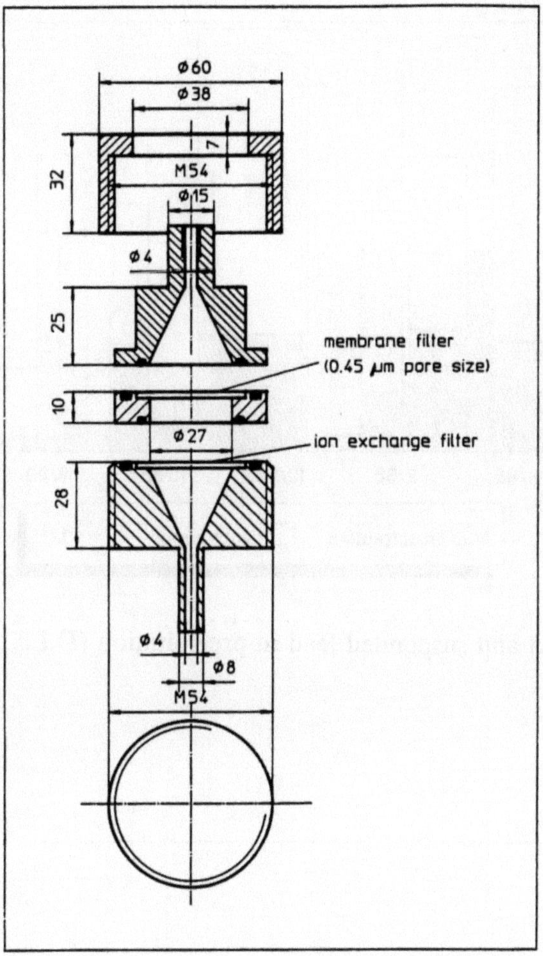

Figure 7. Filtration unit. Dimensions in mm

XRF is well established in trace metal analysis (Toussaint et al., 1977; Gendre et al., 1977; Lieser et al., 1977; Wegscheider et al., 1981), it has not been applied so far to wet deposition measurements. The advantages, however, the proposed technique offers for this purpose, are obvious: unattended preconcentration of the material of interest during rainfall, distinction between particulate and dissolved species, multielement analysis without sample manipulation and danger of contamination.

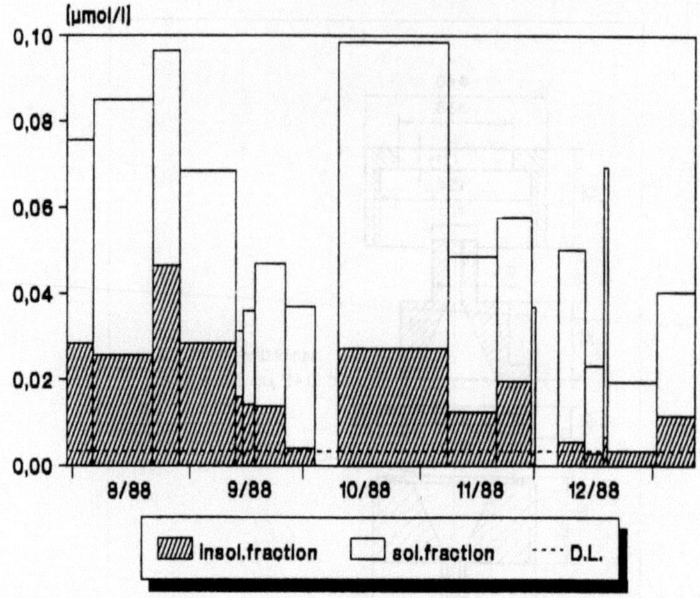

Figure 8. Dissolved and suspended lead in precipitation (D.L.: detection limit)

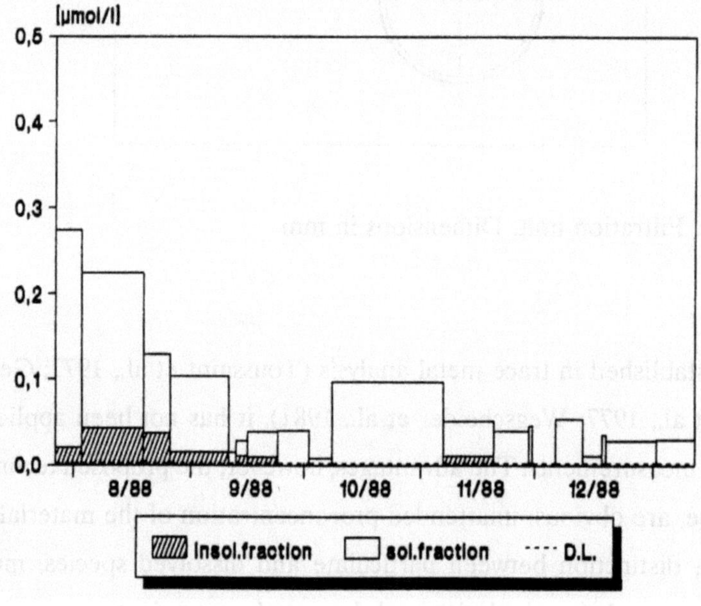

Figure 9. Dissolved and suspended copper in precipitation. (D.L.: detection limit)

6. Conclusions

Metal speciation in technical and ambient aerosols requires a rather extensive set of analytical tools able to provide not only compound but also phase selective information. Consequently the use of proper sampling techniques has to be given special attention to.

As could be shown, denuders are an unconventional but very efficient means for selective extraction of vapour phase metals from hot flue gases. For in-situ separation and preconcentration of dissolved and particulate metal species in rain, simple filtration units may be mounted in a wet-only collector. For collection of airborne particulate matter filters or impactors are most frequently used. Metal speciation in the deposited dust may be carried out by means of the physical methods briefly discussed in paragraph 2. These methods are unique and highly complementary as to the information they provide (lattice, surface, single particle composition, etc.). However, to obtain mass concentrations of certain metal species in collected dust, selective solvent leaching as described in this contribution is a useful microchemical alternative to existing physical approaches.

Literature

Biggins, P.D.E., Harrison, R.M. (1980) Chemical speciation of lead compounds in street dusts. Environ. Sci. Technol. 14:336-339

Braun, H., Metzger, M., Vogg, H. (1986) Zur Problematik der Quecksilberabscheidung aus Rauchgasen von Müllverbrennungsanlagen, 2. Teil. Müll und Abfall 3/86:89-95

Bruckmann, P., Bonitz, P., Reich, Th., Haß, I. (1990) Messungen des filtergängigen Anteils von Schwermetall- und Metalloid-Verbindungen in Außenluft. VDI-Ber. 838:467-483

Bruynseels, F., Tavares, T.M., van Grieken, R. (1985) Characterization of individual particle types in coastal air by laser microprobe mass analysis. Intern. J. Environ. Anal. Chem. 23:1-14

Calvert, J.G., Lazrus, A.L., Kok, G.L., Heikes, B.G., Walega, J.G., Lind, J.G., Cantrell, C.A. (1985) Chemical mechanisms of acid generation in the troposphere. Nature 317:27-35

Dannecker, W. (1982) Untersuchung der Inhaltsstoffe von Stäuben und Aerosolen einer kommunalen Müllverbrennungsanlage. VDI-Ber. 429:265-271

Denoyer, E., Natusch, D.F.S., Surkyn, P., Adams, F.C. (1983) Laser microprobe mass analysis (LAMMA) as a tool for particle characterization: a study of coal fly ash. Environ. Sci. Technol. 17:457-462

Dumarey, R., Heindryckx, R., Dams, R. (1981) Determination of mercury emissions from a municipal incinerator. Environ. Sci. Technol. 15:206-209

Fichtner, M., Lipp, M., Goschnick, J., Ache, H.J. (1990) Massenspektrometrie neutraler und ionischer Sekundärpartikel zur chemischen Analyse von Salzen. Kernforschungszentrum Karlsruhe, Report KfK 4684 (33 pp)

Frerichs, G. (1908) Determination of iron in ferrum reductum. Arch. Pharm. 246:190-205

Fukasawa, T., Iwatsuki, M., Tillekeratne, S.P. (1983) X-ray diffraction analysis of airborne particulates collected by an Anderson sampler. Compound distribution vs. particle size. Environ. Sci. Technol. 17:596-602

Gardella, J.A., Hercules, D.M. (1979) Surface spectroscopic examination of Diesel particulates - a preliminary study. Intern. J. Environ. Anal. Chem. 7:121-136

Gendre, G., Haerdi, W., Lindner, H.R., Schreiber, B., Frei, R.W. (1977) Trace metal enrichment on chemically modified filter papers. Intern. J. Environ. Anal. Chem. 5: 63-75

Germani, M.S., Zoller, W.H. (1988) Vapor-phase concentrations of arsenic, selenium, bromine, iodine and mercury in the stack of a coal-fired power plant. Environ. Sci. Technol. 22:1079-1085

Grasserbauer, M. (1975) Die Bedeutung der Valenzbandspektren in der Elektronen strahl-Mikroanalyse IV. Mikrochim. Acta 1975 II:55-68

Hales, J.M. (1986) The mathematical characterization of precipitation scavenging and precipitation chemistry. In: O. Hutzinger (ed.) The Handbook of Environmental Chemistry, Vol. 4, Part A; Springer-Verlag, Berlin, p. 150-217

Harrison, R.M., Williams, C.R., O'Neill, I.K. (1981) Characterization of airborne heavy metals within a primary zinc-lead smelting works. Environ. Sci. Technol. 15:1197-1204

Jockel, W., Körper, D. (1987) Untersuchung der Emissionsrelevanz flüchtiger Schwermetalle und ihrer Verbindungen. Technischer Überwachungsverein Rheinland e.V., Forschungsbericht 87-10404154 (156 pp)

Kaiser, R.D. (1985) Erzeugung, Charakterisierung und Trennung verschiedener Nickelverbindungen in Luft. - Ein Beitrag zur Speziation von Metallen in Aerosolen. Doctoral Thesis, University of Dortmund (145 pp)

Katsnel'son, E.M., Osipova, E.Y. (1960) Determination of nickel sulfide. Obogashchenie Rud. 5:24-26 (C.A. 55:26854 (1961))

Klockow, D. (1982) Analytical chemistry of the atmospheric aerosol. In: Georgii, H.W., Jaeschke, W. (eds.) Chemistry of the unpolluted and polluted troposphere; Reidel, Dordrecht, p. 57-90

Klockow, D., Kaiser, R.-D. (1985) Generation of various types of nickel containing aerosols and detection of nickel species in these aerosols. In: Brown, S.S., Sunderman Jr., F.W. (eds.) Progress in nickel toxicology; Blackwell Sci. Publ., Oxford, p. 235-238

Larjava, K., Kauppinen, E. (1986) Development of the volatile metal aerosol sampler based on vaporization/condensation. J. Aerosol Sci. 17:516-519

Larjava, K., Reith, J., Klockow, D. (1990) Development and laboratory investigations of a denuder sampling system for the determination of heavy metal species in flue gases at elevated temperatures. Intern. J. Environ. Anal. Chem. 38:31-45

Lieser, K.H., Röber, H.M., Burba, P. (1977) Abtrennung von Spurenelementen aus Wasser mit Hilfe von Celluloseaustauscher-Filtern und ihre Bestimmung durch Röntgenfluoreszenzanalyse mit Radionuklidanregung. Fres. Z. Anal. Chem. 284:361-368

Meinecke, K.(1871) Bestimmung des Phosphor- und Schwefelgehaltes in Eisen. Fres. Z. Anal. Chem. 10:280-284

Merian, E. (ed.) (1984) Metalle in der Umwelt. Verteilung, Analytik und biologische Relevanz. Verlag Chemie, Weinheim (722 pp)

Metzger, M., Braun, H. (1987) In-situ mercury speciation in flue gas by liquid and solid sorption systems. Chemosphere 16:821-832

Minni, E., Gustafsson, T.E., Koponen, M., Kallionaki, P.-L. (1984) A study of the chemical structure of particles in the welding fumes of mild and stainless steel. J. Aerosol Sci. 15:57-68

Mojtahedi, W., Backman, R., Larjava, K. (1987) Fate of some trace elements in fluidised-bed combustion and gasification processes. Publications of the Technical Research Centre of Finland, Espoo, Nr. 42 (44 pp)

Mojtahedi, W. (1989) Trace metals volatilisation in fluidised-bed combustion and gasification of coal. Combust. Sci. and Tech. 63:209-227

Müller, J. (1982) Größenverteilung und atmosphärische Verweilzeit von schwebstaubgebundenen Stoffen. VDI-Ber. 429:231-234

Natusch, D.F.S. (1978) Potentially carcinogenic species emitted to the atmosphere by fossil-fueled power plants. Environ. Health Perspect. 22:79-90

Natusch, D.F.S., Taylor, D.R. (1980a) Environmental effects of western coal combustion. Part IV. Chemical and physical characteristics of coal fly ash. Report EPA-600/3-80-093 (218 pp)

Natusch, D.F.S., Denoyer, E.R., Keyser, T.R., Kirton, S.E., Taylor, D.R., Zeller, M.V. (1980b) Surface analysis of particles emitted to the atmosphere. AIChE Symp. Series 76 (201):127-133

Neidhardt, B., Backes, U., Herwald, S., Lipmann, Ch., Straka-Emden, B. (1990) Chromspeziesanalyse in Schwebstaubproben. VDI-Ber. 838:485-498

Nguyen, V.D., Valenta, P., Nürnberg, H.W. (1979) Voltammetry in the analysis of atmospheric pollutants - The determination of toxic trace metals in rain water and snow by differential pulse stripping voltammetry. Sci. Total Environ. 12:151-167

Nießner, R. (1981) Ein neuer Weg zur Bestimmung starker Säuren und ihrer Salze in der Atmosphäre. Doctoral Thesis, University of Dortmund (299 pp)

Nießner, R., Klockow, D., Bruynseels, F., van Grieken, R. (1985) Investigation of heterogeneous reactions of PAH's on particle surfaces using laser microprobe mass analysis; Intern. J. Environ. Anal. Chem. 22:281-295

Nießner, R., Klockow, D (1987a) Sampling and analysis of air constituents: a survey. In: E.E. Pickett (ed.) Atmospheric pollution; Hemisphere Publ. Corp., Washington, p. 191-213

Nießner, R., Malejczyk, M., Schilling, M., Klockow, D. (1987b) Die Diffusion als Probenahmeprinzip zur Gas/Partikel-Trennung. VDI-Ber. 608:153-180

Nießner, R., Hemmerich, B., Panne, U. (1989) Possibilities and limitations of the photoelectric aerosol sensor array applied to heavy metal aerosols. Fres. Z. Anal. Chem. 335:728-737

Novakov, T., Chang, S.G., Dod, R.L. (1977) Application of ESCA to the analysis of atmospheric particulates. In: Hercules, D.M., Hieftje, G.M., Snyder, L.R., Evenson, M.A. (eds.) Contemporary topics in analytical and clinical chemistry, Vol. 1; Plenum Press, New York, p. 249-286

Post, J.E., Busek, P.R. (1985) Quantitative energy dispersive analysis of lead halide particles from the Phoenix urban aerosol. Environ. Sci. Technol. 19:682-685

Prack, E.R., Bastiaans, G.J. (1983) Metal speciation by evolved gas/inductively coupled plasma atomic emission spectrometry. Anal. Chem. 55:1654-1660

Pruppacher, H.R., Klett, J.D. (1980) Microphysics of clouds and precipitation. D. Reidel Publ. Comp., Dordrecht (714 pp)

Reith, J. (1990) Beiträge zur Speziation von Kobalt und Chrom in Arbeitsplatz-Aerosolen. Doctoral Thesis, University of Dortmund (257 pp)

Riott, J.P. (1941) Determining metallic iron in iron oxides and slags. Ind. Eng. Chem. (Anal. Ed.) 13:546-549

Schögl, R., Indlekofer, G., Oelhafen, P. (1987) Mikropartikelemissionen von Verbrennungsmotoren mit Abgasreinigung - Röntgen-Photoelektronenspektroskopie in der Umweltanalytik. Angew. Chem. 99:312-322

Siemens, V. (1989) Versuche zur diffusionskontrollierten Voranreicherung von Quecksilberspezies in Abgasen. Diploma Thesis, University of Dortmund (80 pp)

Storp, S., Holm, R. (1979) ESCA (electron spectroscopy for chemical analysis) investigations of ion beam effects on surfaces. J. Electron Spectrosc. Relat. Phenom. 16:183-193

Toussaint, C.J., Aina, G., Bo, F. (1977) X-ray fluorescence determination of traces of dissolved and suspended metals in water. Anal. Chim. Acta 88:193-196

Van Cleuvenbergen, R.J.A., Adams, F.C. (1990) Organolead compounds. In: O. Hutzinger (ed.) The Handbook of Environmental Chemistry, Vol. 3, Part E; Springer-Verlag, Berlin, p. 97-153

Vogg, H. (1988) Von der Schadstoffquelle zur Schadstoffsenke - Neue Konzepte der Müllverbrennung. Chem.-Ing. Techn. 60:247-255

Wegscheider, W., Knapp, G. (1981) Preparation of chemically modified cellulose exchangers and their use for the preconcentration of trace elements. Crit. Rev. Anal. Chem. 11:79-102

Weisz, H. (1970) Microanalysis by the ring oven technique. Second Ed., Pergamon Press, Oxford (167 pp)

Whitby, K.T., Sverdrup, G.M. (1980) California aerosols: their physical and chemical characteristics. In: Hidy, G.M., Mueller, P.K., Grosjean, D., Appel, B.R., Wesolowski, J.J. (eds.) The character and origins of smog aerosols; Wiley & Sons, New York, p. 477-517

EVALUATION OF NEW ORGANIC PHASE EXTRACTION PROCEDURES FOR STUDYING THE ROLE OF TERRESTRIAL HUMIC SUBSTANCES IN THE SPECIATION OF IRON AND PLUTONIUM

Robert A Bulman, Gyula Szabo[1] and Angela J Wedgwood
National Radiological Protection Board
Chilton
Didcot OX11 0RQ, ENGLAND

1. INTRODUCTION

Although the movement through the environment of the transuranic elements plutonium and americium has been extensively investigated for many years (adequately summarized elsewhere (Bulman, 1984; Choppin and Allard, 1985)), there remains considerable uncertainty of the long term influence of humic substances upon their movement through soils and in groundwater. In addition, the role played by these biogeopolymers on the uptake into plants is not understood.

Detailed investigations of the nature of the complexing a sites which bind the actinides pose obvious problems in studying the speciation of the actinides as the elements are spectroscopically silent for most laboratory instrumentation procedures. Their only attribute which facilitates investigation of their speciation is their detection at exceedingly low levels. Detection of ^{239}Pu at 10^{-14} to 10^{-15} g (ca. 1 10^{7} atoms, 0.01 mBq) is undertaken routinely in some laboratories investigating its environmental behaviour.

The processes used to isolate humic substances from soils have used: (i) strong alkali which is then followed by acidifi-

---

1 On study leave from "Frederic Joliot-Curie" Research Instiute for Radiobiology and Radiohygiene, H-1775 Budapest, Hungary.

NATO ASI Series, Vol. G 23
Metal Speciation in the Environment
Edited by J. A. C. Broekaert, Ş. Güçer, and F. Adams

cation to separate humic and fulvic acids (ii) extraction with sodium pyrophosphate which is a weak complexing agent (Bremner and Lees, 1948), (iii) extraction with a variety of chelating agents (Kanwar, 1954; Martin and Reeve, 1957; Coffin and DeLong, 1960). It is debatable if any of these methods can be considered suitable when studying the speciation of metal ions bound by humic substances because of the strong likelihood that weakly bound cations will be hydrolyzed by the strong base or competitively bound by the metal-binding agents. If weakly bound cations of Pu(IV) are hydrolyzed the resulting polymeric Pu(IV) hydroxides could be adsorbed or entrapped within the humic substances and thus it would be impossible to determine the potential mobility of the Pu for entry into ecosystems. To overcome these limitations we have sought mild, non-destructive procedures for isolation of Pu(IV) - and Am(III) - humates from soils and sediments into which these elements had been introduced by anthropogenic processes. Gel filtration on controlled pore glass of organic materials extracted from esturine sediment by molten N-methyl-morpholine-N-oxide (melting point ca 70 °C) showed a co-elution of ^{239}Pu and humic substances (Bulman *et al*, 1984). However, with phases richer in humic substances gel permeation chromatography was not possible using this novel extractant and consequently alternative extraction procedures have been sought. In an earlier publication we have reported a comparison of in situ silylation of humic acids with other extraction procedures for the determination of ^{239}Pu in soils (Szabo *et al*, 1990). In this paper we extend these studies of in situ silylation of humic substances by investigating the speciation of Fe and ^{239}Pu in the extracts by using non-aqueous gel chromatographic fractionation .

2. MATERIALS and METHODS

Chemicals: Chromatographic grade silica (Merck 0.040-0.063 mm) was obtained from BDH, trimethylchlorosilane (TMCS) and humic acid were obtained from Aldrich Chemical Co. Ltd. and controlled pore glass (PG-240-200) was obtained Sigma Chemical Co Ltd. Polystyrenes of known molecular weights were obtained from Polymer Laboratories, Ltd, Church Stretton, Shropshire. All other chemicals were obtained from normal commercial suppliers.

Soil, known to contain ^{239}Pu and originating from a marine input, was collected from beneath sea-washed turf 7 km to the south of Sellafield, a commercial nuclear fuel reprocessing facility operated by British Nuclear Fuels plc. A freshly composted garden soil was used to evaluate the extraction procedures described below.

Sodium hydroxide extraction: The procedure of Cameron *et al* (1970) was used to extract humic substances. Essentially the soil (10g) was suspended in 0.1M NaOH solution (100 ml), stirred at 60 °C for 4 h and clarified by centrifugation at 3000 rpm in a bench top centrifuge (IEC *Centra-3*). The alkaline solution was adjusted to pH 5 with 6 M HCl and freeze dried. The dried residue was then silylated with TMCS using the basic procedure outlined below. The extracted material is subsequently identified as "TMSi-NaOH -humate".

In situ silylation with TMCS: Soil (10g) was dried to constant weight at 60 °C, suspended in a solution of dimethylformamide (DMF) (100 ml) containing TMCS (10 ml) and stirred for 4 h at 60 °C. The clarified residue was concentrated *in vacuo* at 60 °C and stored in an anhydrous environment. The extracted material is subsequently identified as "TMSi-humate".

Modification of silica chromatographic media: To minimize any interaction of the polar groups of humic substances with silanolic groups of the chromatographic media, the media was silylated with TMCS - a process which also gave a mild hydrophobic character to the column.

Non-aqueous gel permeation chromatography: Fractionation of TMSi-humate and TMSi-NaOH-humate was achieved by elution with DMF through: (i) Sephadex LH20 (90 x 1.8 cm); (ii) a column (90 x 1.8 cm) of trimethylsilylated chromatographic grade silica and by subsequent passage through a tandem column (20 x 1.6 cm) of the trimethylsilylated PG-240-200. Sample introduction, typically 1 ml of DMF containing ca. 500 µg Fe and 1 Bq ($4.4\ 10^{-9}$ g) of ^{239}Pu, was injected via a three-way valve (Wright-Amicon Ltd, Stroud, Gloucs.) Chromatographic retention data were measured by using an LKB 2150 solvent delivery system and LKB Diode-array detector. The eluent was pumped at 1.25 ml/min and samples collected every 60 seconds from Sephadex LH20 or 120 seconds from the silica columns. Chromatograms were recorded on an Olivetti M24 personal computer and the collected data analyzed by using commercially available software (*WAVESCAN*, LKB).

Assignment of relative molecular masses of eluted fractions: Estimates of the molecular weights of the eluted fractions were attempted by using a calibration plot of the molecular weight (MW) against retention time of polystyrenes of 1, 4, 50 and 400 kDa.

Elemental analysis: Aliquots of the solutions to be chromatographed and the collected fractions were digested with nitric acid until free of organic material and divided into two fractions of equal size. The iron content of one series of fractions was analyzed by atomic absorption spectroscopy and the other series analyzed for ^{239}Pu by α-spectrometry after isolation by ion exchange (Talvitie, 1971).

3. RESULTS

The recovery of Fe and ^{239}Pu from the columns is presented in Table 1. The elution profiles off Sephadex LH20 of Fe in the TMSi-humate and TMSi-NaOH-humate isolated from garden soil are presented in Fig 1. The elution profile off Sephadex LH20 of ^{239}Pu and Fe in the TMSi-humate from the saltmarsh soil are presented in Fig 2. The elution profile off the silica media columns of ^{239}Pu and Fe in TMSi-humate extracted from the saltmarsh soil are presented in Fig 3.

Only approximate assignment of molecular weights to the fractions eluted from the silica columns could be achieved as it was impossible to construct a straight line plot of log molecular weight versus retention time for the polystyrenes of known molecular weight.

Table 1. Percentage recovery of ^{239}Pu and Fe TMSi-humate and TMSi-NaOH humate from non-aqueous gel permeation chromatography columns

	TMSi-NaOH-humate	TMSi-humate		
	Garden Soil	Saltmarsh Soil		Garden soil
	Sephadex LH20	Silica	Sephadex LH20	Sephadex LH20
Fe	85	70	61	75
^{239}Pu	--	60	64	--

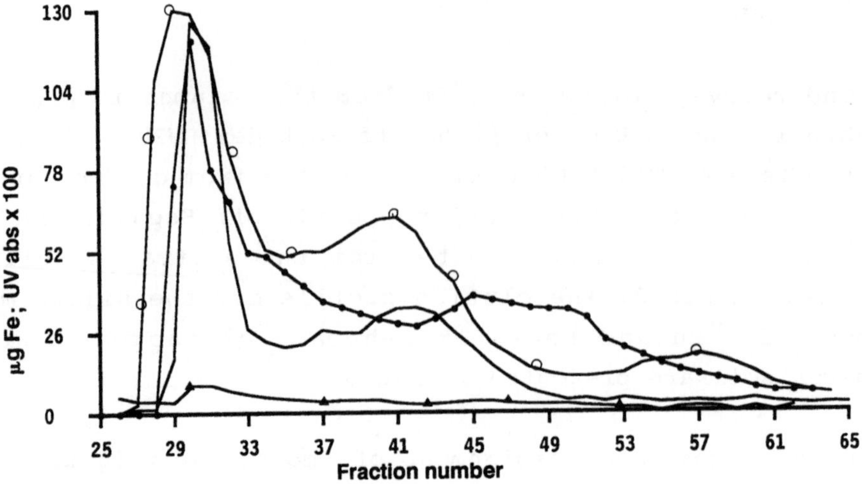

Fig. 1. Elution profile of materials originating from garden soil and chromatographed on Sephadex LH20: (i) TMSi-humate, Fe (—) and UV absorption (o-o); and (iii) TMSi-NaOH-humate, Fe (▲-▲) and UV absorption (•-•)

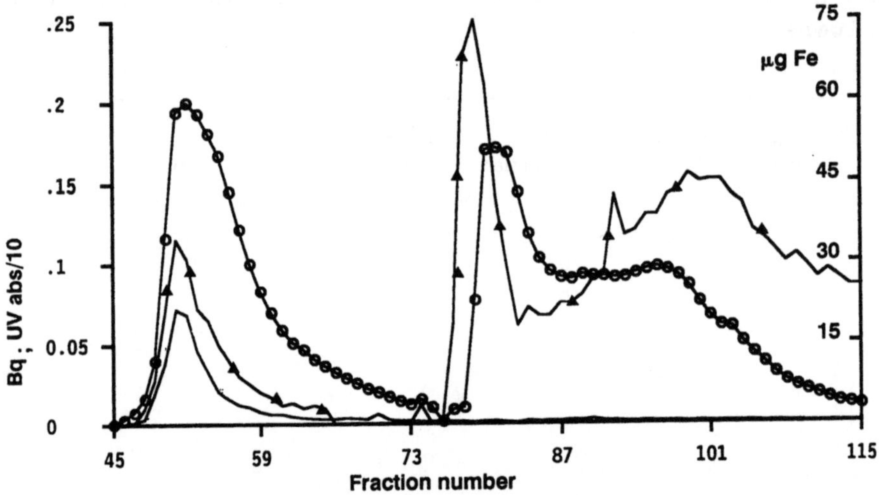

Fig 2. Elution profile of TMSi-humate from salt marsh soil and chromatographed on Sephadex LH20: Fe (▲-▲), ^{239}Pu (—) and UV absorption (o-o)

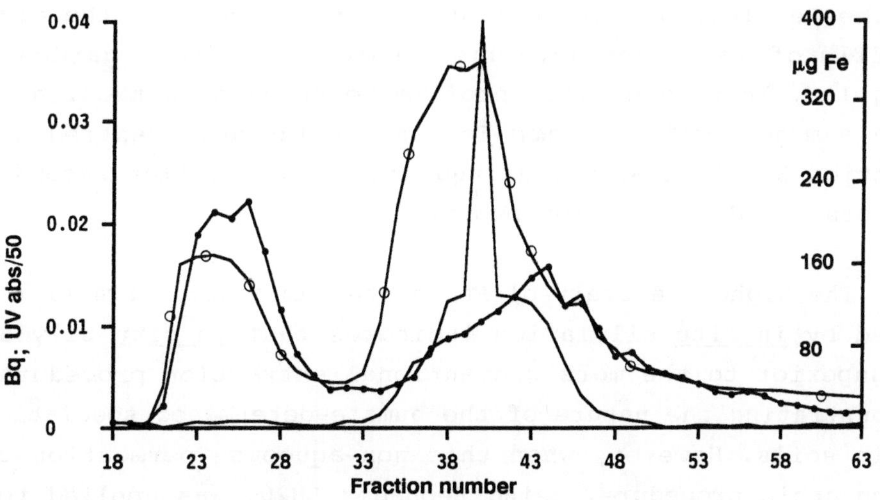

Fig 3. Elution profile of TMSi-humate from silica columns: Fe (▲—▲), ^{239}Pu (•-•) and UV absorption (o-o)

4. DISCUSSION

Because of the levels of ^{239}Pu in the salt marsh soil (3.9 Bq/g) and the labour-intensive analytical chemistry required to determine ^{239}Pu in eluted fractions, the merits of <u>in situ</u> trimethylsilation of humates over alkaline extraction have, at the moment, been restricted to an assessment of the extraction procedure on the speciation of humates binding Fe.

Fig 1 shows the elution profile of Fe in from Sephadex LH 20, of TMSi-NaOH-humate which was isolated from the garden soil. Only one small peak, maximum Fe content 9 µg, is clearly discernible and it coincides with the principal UV-absorbing peak. There is no second peak of Fe to match the minor UV-absorbing peak occurring at fraction 41. This poor definition of the Fe elution profile must be contrasted with the elution profile of Fe in the TMSi-humate extract from garden soil (Fig 1) . In this elution profile Fe appears in two fractions with a major peak, maximum Fe content 126 μg, centred around fraction 30. A clearly distinguishable but smaller second peak mirrors the UV absorption profile.

The higher recovery of Fe in the humic acid fraction isolated by <u>in situ</u> silylation indicates that <u>in situ</u> silylation is superior to the more conventional extraction procedure for demonstrating the nature of the humate-determined speciation of Fe in soils. However, when this non-aqueous permeation chromatographic procedure, using Sephadex LH20, was applied to the fractionation of ^{239}Pu in TMSi-humate extracted from salt marsh soil, the separation of ^{239}Pu into more than one fraction was not possible (Fig 2) although it was possible to resolve Fe into two peaks. As the possession of partition and adsorptive chromatographic characteristics by Sephadex LH20 might have contributed to the observed elution pattern of Pu, an alternative fractionation procedure based on molecular weight sizing was evaluated. Passage of the TMSi-humate through the silica media fractionated ^{239}Pu into two components (Fig 3) but led to the recovery of Fe primarily in fractions 35-44. However, from the shape of the elution profile in this region it would appear that Fe is perhaps present in at least 3 peaks. Confirmation of the possible presence of the binding of Fe by several materials of similar chromatographic character can be obtained by using the analytical procedures of the software *Wavescan*. These analytical features demonstrate the presence of least five sub-peaks within the main peak recorded during the collection of fractions 33-45.

As indicated above, accurate assignment of molecular weights to the eluted fractions is not possible. However, it is possible that the material collected in fraction 25 (Fig 3) has a molecular weight of approximately 50 kDa whereas fractions 40 and 44 represented material of 2.5 and 1 kDa, respectively. Some proof that fraction 25 was representative of humic acid was obtained by the recovery of silylated humic acid (Aldrich) in this fraction. Similarly, a silylated fulvate, isolated from garden soil by alkaline extraction, was recovered in fractions 37-42. In this investigation we have demonstrated the separation of ^{239}Pu in three molecular weight fractions but, as yet, it has not been possible to compare the chemical characteristics of the Pu in these fractions with those of Pu in the three fractions isolated by Livens *et al* (1987), who chromatographed alkaline extracts of a gley soil on Sephadex G150 using 0.1M NaOH as the eluent. As cationic forms Pu are spectroscopically silent, characterization of the environments in which they are bound is not possible. Perhaps some insight into the nature of the ligands binding Pu might be possible from a study of the interaction of the humate-associated Pu with low MW complexing agents.

The presence of Fe primarily in a fulvate form in the salt marsh soil cannot be readily explained at the moment but it might be a consequence of the affinity of iron (II), the eoxidation state which predominates in the soil solution in the salt marsh soil (Bulman, 1990), for carboxylates which are the predominant acidic function in fulvic acids and are conversely low in humic acids where phenolates predominate (Halbach *et al*, 1980).

To summarize: this study has shown that organic forms, possibly humates, of Pu and Fe can be extracted from the salt-marsh soil into DMF. By non-aqueous gel permeation chromatography of these extracts on silica columns the Pu-humates can be resolved into 2 widely different MW sizes. In contrast, the Fe-humates are resolved into a MW range which does not extend to a high MW. As humates might influence the uptake of Fe into plants (Linehan, 1977), further investigation of these characteristics are required as it can not be excluded that there might be significant differences in the molecular characteristics of metal ion-binding humates and that it is these differences which regulate the uptake of metal ions into plants. Improvements in the separation of the humic substances extracted by in situ trimethylsilylation might be possible by using high performance liquid chromatography.

5. ACKNOWLEDGEMENTS

Gyula Szabo's contribution to this study was made possible under the auspices of the International Atomic Energy Agency, Vienna. This work was supported in part by Commission of the European Community, under contract BI6-B-048-UK.

6. REFERENCES

Bulman R A (1983) Complexation of transuranic elements: a look at factors which may enhance their biological avail ability. In Ecological aspects of radionuclide release, Coughtrey P J (editor) Blackwell, Oxford.

Bulman R A, Johnson, T E, Reed A L (1984) An examination of new procedures for fractionation of plutonium- and ameri cium-bearing sediment. Sci Total Environ 35: 239-250.

Bulman R A, 1990, *to be published*.

Bremner J M, Lees H (1949) Studies on soil organic matter: II. The extraction of organic matter from soil by neutral reagents. J Agric Sci 32: 274-279.

Cameron R S, Thornton B K, Swift R S, Posner A M (1972) Molecular weight and shape of humic acid from sedimentation and diffusion measurements on fractional extracts. J Soil Sci 23: 394-408.

Choppin G R, Allard B (1985) Complexes of actinides with naturally occurring ligands. In Handbook on the chemistry and physics of actinides, Freeman A J, Keller C (editors) Elsevier, Amsterdam.

Coffin D E, DeLong W A (1960). Extraction and characterization of organic matter of a podzol B horizon. Trans Intern Congr Soil Sci 7th Congr Madison. II: 91-97.

Halbach P, von Borstel D, Gundermann K D (1980) The uptake of uranium by organic substances in a peat bog environment on a granite bedrock. Chem Geol 29: 117.

Kanwar J J (1954) Influence of organic matter on copper fixation in soil. J Indian Soc Soil Sci 2: 73-80.

Linehan D J (1977) Humic acid and uptake by plants. Plant Soil 50: 663-670.

Livens, F R, Baxter, M S and Allen, S E (1987) Association of plutonium with soil organic matter. Soil Sci 144: 24-28.

Martin A E, Reeve R (1957) Chemical studies on podzolic illuvial horizons: I. The extraction of organic matter by organic chelating agents. J Soil Sci 8: 268-278.

Szabo G, Wedgwood A J, Bulman R A, Comparison and development of new extraction procedures for ^{239}Pu, Ca, Fe and Cu organic complexes in soil. J Environ Radioactivity *in press*.

Talvitie, N A (1971) Radiochemical determination of plutonium and americium in environmental and biological samples by ion exchange. Anal Chem 43: 1827-1830.

DEVELOPMENT AND IMPROVEMENT OF ANALYTICAL METHODS FOR SPECIFICATION SCHEME OF AL IN THE MOBILE SOIL PHASE

H. Klöppel, W. Kördel, S. Schmid and W. Klein
Fraunhofer-Institut für Umweltchemie und Ökotoxikologie,
5948 Schmallenberg-Grafschaft, FRG

1. INTRODUCTION

The release of Al in the soil is often discussed as one cause of forest decline due to soil acidification values under 4.2 are reached, elevated concentrations of Al appear in the soil leachates. In order to assess the phytotoxic potential of aluminium present in the soil solution a differentiated determination of the Al-species is necessary. In recent years especially photometric and ion exchange procedures as well as various filtration, radiation and decomposition techniques have been applied to obtain a physical or chemical speciation of aluminium. Physical speciation is applied to differentiate with regard to the particle sizes (e.g. distinction between colloidal and non-colloidal aluminium); chemical speciation allows a classification of the Al-species with respect to their kinetic and thermodynamic properties. Campbell [Campbell et al. 1983] developed an ion exchange/UV radiation procedure which is applicable to differentiate between organic and inorganic forms of non-exchangeable aluminium. They found that in the investigated natural waters the major part of the non-exchangeable Al occurred as organic complexes.

Further speciation schemes were developed by Driscoll [Driscoll 1984] separating acid-soluble kinetically stable monomeric and kinetically labile monomeric aluminium and by Lazarte [Lazarte 1984] using dialysis procedures and photometric methods to separate highly reactive inorganic and organic Al compounds. Several authors [Turner 1969, James et al. 1983, Wilson 1984] describe speciation concepts for Al using specific reactions with chromogenic reagents mainly basing on the varying kinetic stabilities of the Al compounds.

NATO ASI Series, Vol. G 23
Metal Speciation in the Environment
Edited by J. A. C. Broekaert, Ş. Güçer, and F. Adams

The existing analysis schemes are not applicable to distinguish between the free Al^{3+}-ion, which has been proved by experiments to be toxic in hydroponic culture [Rost-Siebert 1985] and the labile inorganic Al-complexes like $(Al(OH))^{2+}$ and $(Al(OH)_2)^+$ in the soil solution. Since there is also no combined use of photometric methods and ion exchange procedures a detailed specification is not possible with these methods.

Most of the methods described in the literature have to be adjusted to definite reaction conditions for separation and detection which causes changes of the natural sample by addition of chemicals. Consequently the chemical equilibria between the kinetically labile Al-species in the natural system are changed.

The first aim of the present investigations was to establish a method, which is applicable to determine the free hydrated Al^{3+}-ion in soil solution without changing the originality of the sample. In addition a specification scheme of Al compounds in the mobile soil phase is developed, which is applicable for the determination of free Al^{3+}-ions and which allows further detailed speciation steps by combined application of photometrical methods and by the ion exchange procedure. By this way changes of the original samples by addition of chemicals can be avoided in many cases.

2. DEVELOPMENT OF AN ISOTACHOPHORETIC METHOD

For the determination of environmentally relevant concentrations of free Al^{3+}-ions in the leachate and in soil water a new method of isotachophoresis (ITP) was developed. An LKB 2187 Tachophor equipped with a high-voltage power supply (0,1-30 kV, 10-500 µA) and an integrated conductivity detector was used (Figure 1). Additionally, a precapillary tube (6 cm x 2 mm I D) from ITABA (Sweden) followed by a main capillary tube (20 cm x 0,3 mm I D) was installed to increase detection limit. To separate the Al^{3+}-ions of the samples from K, Ca, Mg as well as from mononuclear and polynuclear aluminium species, a high negative voltage and an electrolyte system consisting of 0,01 M sodium acetate (leading electrolyte) were applied (Table 1). The Al^{3+}-ions were detected on the basis of their linear and differential conductivities [Schmid et al. 1989].

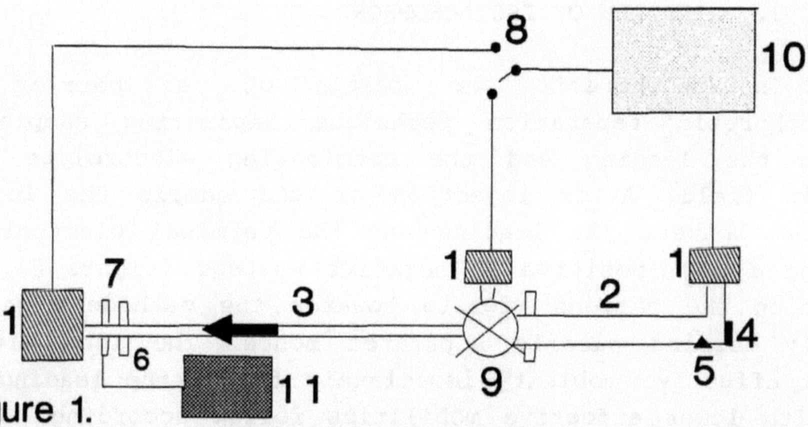

Figure 1.
SCHEME OF ISOTACHOPHORESIS

1 ELECTROLYT-CHAMBER
2 PRECAPILLARY TUBE
3 MAINCAPILLARY TUBE
4 SEPTUM
5 VALVE TO THE DRAIN
6 CONDUCTIVITY DETECTOR
7 UV DETECTOR
8 SWITCH
9 VALVE
10 HIGH VOLTAGE POWER SUPPLY
11 INTEGRATOR

	LEADING ELECTROLYTE	TERMINATING ELECTROLYTE
CATION	NA	TRIS
CONCENTRATION	0,01 N	0,01 N
pH	3,6	3,3
ADDITIVE	1% HPMC	1% HPMC
COUNTER ION	ACETATE	ACETATE
TEMPERATURE	ROOM-TEMPERATURE	
CURRENT	30 µA	
VELOCITY OF THE INTEGRATOR	600 cm/min	

Table 1. BUFFER SYSTEM FOR AL^{3+} -ION ANALYSIS BY ISOTACHOPHORESIS

3. PRINCIPLE OF ISOTACHOPHORESIS

Isotachophoresis is based on a carrier free electrophoretic separation technique separating sample ions between the leading and the terminating electrolyte in an electric field. After injection of the sample the ions are separated between the leading and the terminal electrolyte by applying a high positive or negative voltage (Figure 2). After separation the cations migrate towards the cathode with equal velocity forming clearly separated zones. The ions with the highest effective mobility immediately follow the leading ions; ions with lower effective mobilities follow according to their effective mobilities.

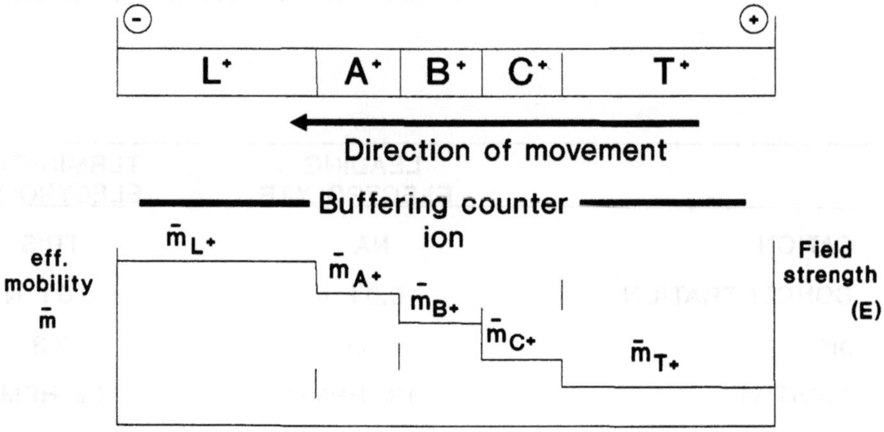

Figure 2.
PRINCIPLE OF ISOTACHOPHORETIC SEPARATION

$v_{L^+} = v_{A^+} = v_{B^+} = v_{C^+} = v_{T^+}$: velocities of cations

$\bar{m}_{L^+} \geq \bar{m}_{A^+} > \bar{m}_{B^+} > \bar{m}_{C^+} > \bar{m}_{T^+}$: eff. mobilities of cations

$v = \bar{m} \times E$ (E = field strength)

4. INVESTIGATIONS BY ISOTACHOPHORESIS (ITP)

First an adequate buffer system for separation and determination of free Al^{3+}-ions by isotachophoresis was chosen. With the buffer system shown in Table 1 the free Al^{3+}-ions could be separated from interfering ions and specifically determinated in aqueous solutions. Upon injection of 10 µl $Al(NO_3)_3$-solution a linear function was obtained for the zone length of Al^{3+}-ions and the amount of the injected Al^{3+}-ions in the concentration range from 0,1 - 2 mg Al^{3+}/l. The correlation coefficient was r = 0,999. The aluminium standard solutions were acidified to pH = 2,0 to shift the equilibrium in favour of the initial products. (1)

$$Al^{3+} + H_2O \rightleftharpoons Al(OH)^{2+} + H^+ \quad (1)$$

When $Al(NO)_3^-$, $AlCl_3^-$, $Al_2(SO_4)_3^-$ or a $KAl(SO_4)_2^-$ aqueous solutions were used for calibration, identical linear calibration functions were obtained. The relative standard deviation which was determined by repeated injection of $Al(NO)_3$ and the aluminium salt solutions $AlCl_3$, $Al_2(SO_4)_3$ and $KAL(SO_4)_2$ was in the range of 3-9 % for concentrations of 1 mg Al^{3+}/l in artificial solutions and in natural soil solutions and a DOC < 20 mg/l (DOC = dissolved organic carbon).

To test the developed methods with regard to a specified determination of free Al^{3+}, the zone length of free Al^{3+}-ions was determined in dependence of the pH value. Therefore samples of aluminium salt solutions with equal amounts of total aluminium and different pH values were injected. No influence of anionic compounds on the Al^{3+}-determination was observed. Different aluminium salt solutions [$AlCl_3$, $Al(NO_3)_3$, $Al_2(SO_4)_3$, $KAl(SO_4)_2$] adjusted to pH values from 1,7 to 6,5 give identical pH dependent curves (Figure 3).

A comparison with the calculated pH curve revealed no differences between the shapes. Therefore it was possible to calculate a pKs value of 5,05 ± 0,15 from the experimental pH curves for the first hydrolysis equilibrium (eqn. 1) which is comparable to literature values ranging from 4,60 to 5,69 [Hem, Robersen 1967, Havas, Jaworski 1986, Ares 1985].

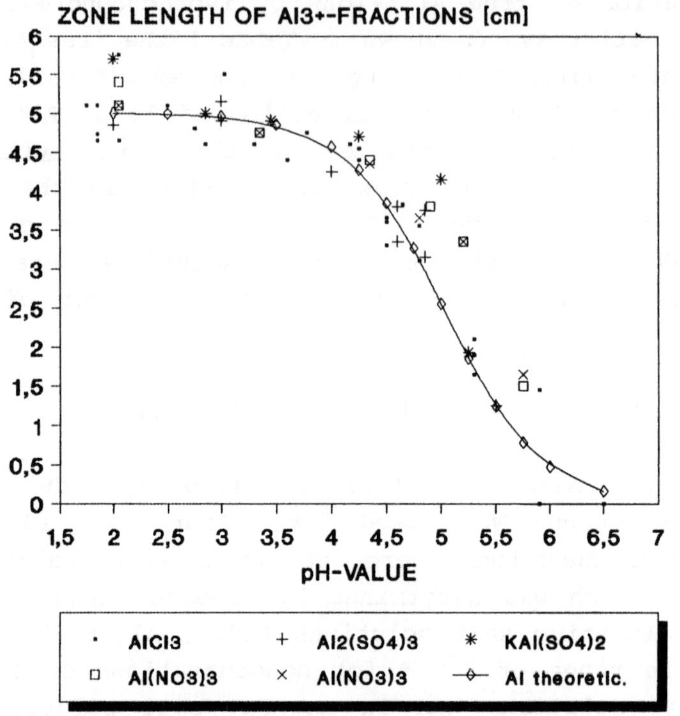

Figure 3. DIFFERENT AL-SALTS AND THE pH-DEPENDENCE

Another verification of the method is the determination of the formation constants of weak Al complexes. The tridentate ligand citrate forms different metalligand-complexes with aluminium (eqn. 2).

$$2\ Al^{3+} + 3\ Cit^{3-} \rightleftharpoons (Al^{-}Cit)^{o} + [Al(Cit)_2]^{3-} \quad (2)$$

The method was tested by changing the citrate-concentration while the Al-concentration was kept constant resulting in a molar Al : citrate ratio of the injected sample varying between 1 : 0 and 1 : 2,14. The shapes of the curves obtained for the Al^{3+}-ion zone length upon addition of different citrate-concentrations are shown in Figure 4.

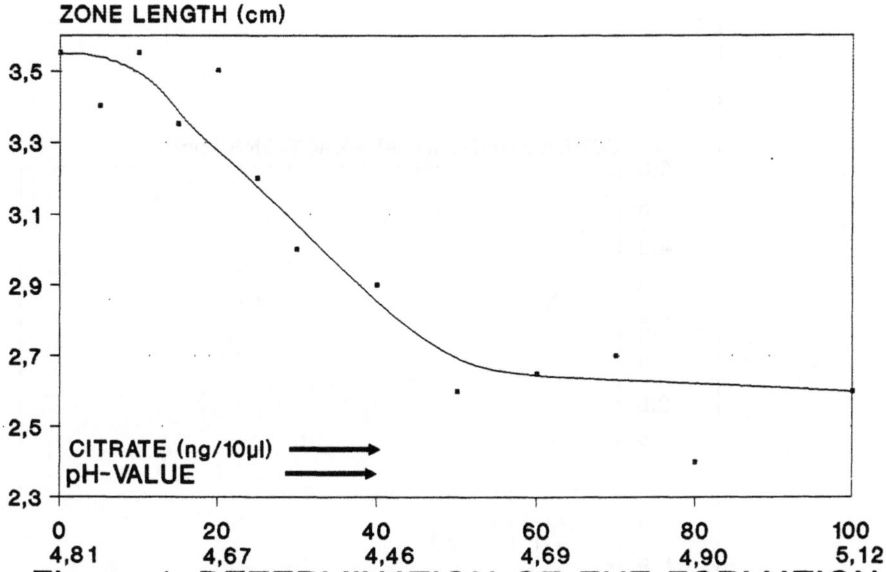

Figure 4. DETERMINATION OF THE FORMATION CONSTANT FOR THE AL-CITRATE-COMPLEX

From the inflection point of the curve the formation constant pK was calculated as pK(Al-Cit) = -7,87 ± 0,32 (literature values -7,37 Ares 1985 and -7,87 ± 0,32 Jackson 1982).

The tested method for the determination of free Al^{3+}-ions in the mobile soil phase was applied to investigate the leachates of lysimeters (25 - 70 cm depth). The shapes of the calculated curve for $Al^{3+}/Al(OH)^{2+}$ and the pH dependent curve for an $AlCl_3$ solution correspond to the shapes of the pH dependent curves obtained by leachates of lysimeters (Figure 5).

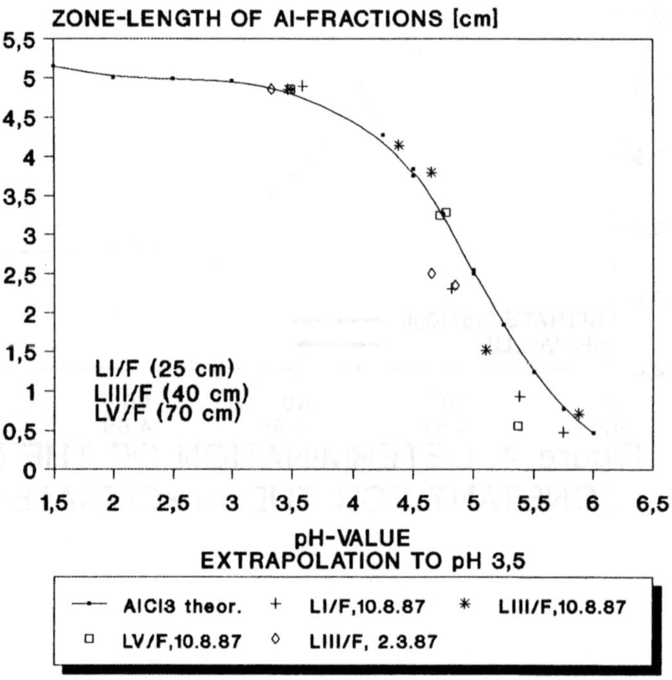

Figure 5. DETERMINATION OF AL3+ IN SOIL SOLUTION BY VARIATION OF THE pH-VALUE

5. DEVELOPMENT OF THE SPECIATION SCHEME

In addition to the isotachophoretic determination of free Al^{3+}-ions the identification of the mobile Al-species by photometric methods allows the confirmation and improvement of the speciations given in the literature [Schmid 1989]. By means of the 8-hydroxychinolinmethod monomeric Al-compounds as $[Al(OH)]^{2+}$, $[Al(OH)_2]^+$, $[AlSO_4]^+$, $[Al(OH)Cl]^+$ and the free Al^{3+}-ion can be determined [James, Clark, Riha 1983]: A buffer mixture consisting of phenanthroline-hydroxylamine hydrochloride, sodium-acetate and oxin solution is added to the sample. The Al-complex is extracted in n-butylacetate for 15 seconds and photometrically measured at 395 nm.

The catecholviolet method was applied to detect kinetically labile Al-organo-complexes [Wilson 1984]. Application of a phenanthroline-hydroxylamine solution, catecholviolet solution and hexamine-buffer to the sample leads to the formation of an Al-complex which was photometrically measured at 585 nm.

Since 8-hydroxyquinoline and catecholviolet react with Al at pH = 4,5 resp. 6,1, the application of these analytical procedures influences the pH of the natural soil solution.

A differentiation between exchangeable and non-exchangeable Al-complexes is possible by using the weakly acidic ion exchange resin Amberlite IRC-718 with iminoacetic acid groups in its Na-form in the batch process. The polymer Al-hydroxo-complexes, colloidal $Al(OH)_3$, kinetically stable Al-organo-complexes and the Al-fluorid-complexes are included in the non-exchangeable Al-complexes, which are considered to be the kinetically stable Al-compounds.

The total amount of soluble Al was determined by AAS and ICP.

By application of the described analytical methods, the following speciation scheme of Al-compounds in the mobile soil phase has been developed (Figure 6).

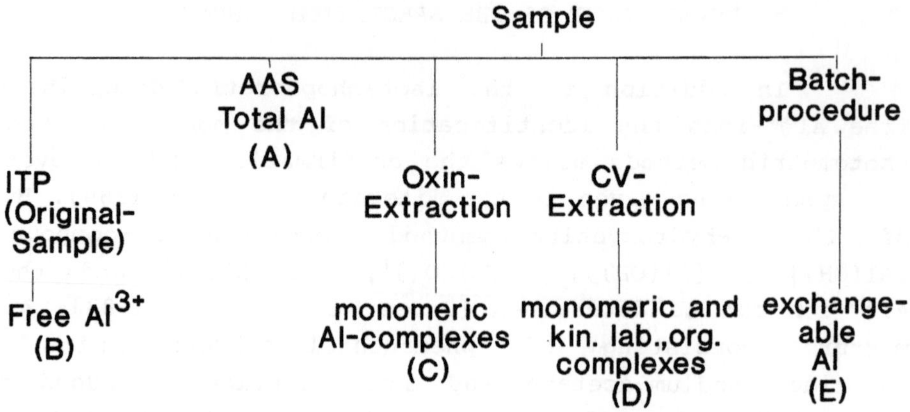

(A)-(E) = non exchangeable Al-complexes (A)-(B) = complexed Al
(D)-(C) = kinetically labile, organic complexes (E)-(D) = $Al(OH)_3$ (1)
(C)-(B) = monomeric Al-complexes: $Al(OH)^{2+}$, $[Al(OH)_2]^+$
$Al(SO_4)^+$, Al(OH)Cl, etc.

Figure 6. AL-SPECIATION CONCEPT

6. SPECIATION OF AL COMPOUNDS IN NATURAL SOIL SOLUTIONS

The feasibility and practicability of the analysis scheme was checked by means of lysimeter experiments including liming (3 t/ha coarse granulated dolomite) and acid fortified irrigation (Figures 7 - 11), as well as by application to samples from forest stands.

The results obtained for the differently treated lysimeters (Ca-fertilization, acid irrigation, untreated) show that the free Al^{3+}-ion concentrations increased with increasing soil depths. In the upper soil layers the organic carbon content (Dissolved Organic Carbon DOC < 20 mg/l), was comparatively higher; accordingly the fraction of stable and labile organically bound aluminium was relatively high.

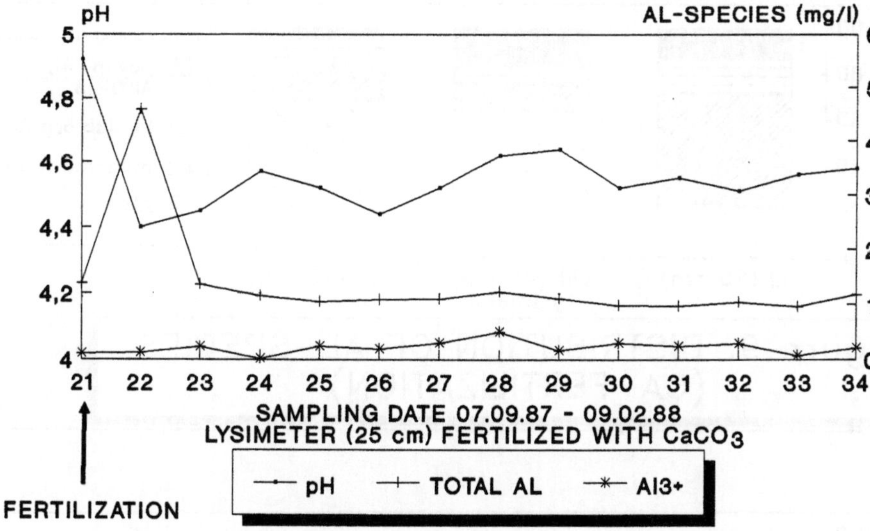

Figure 7. pH, TOTAL AL AND AL^{3+} - CONCENTRATIONS IN THE LEACHATE

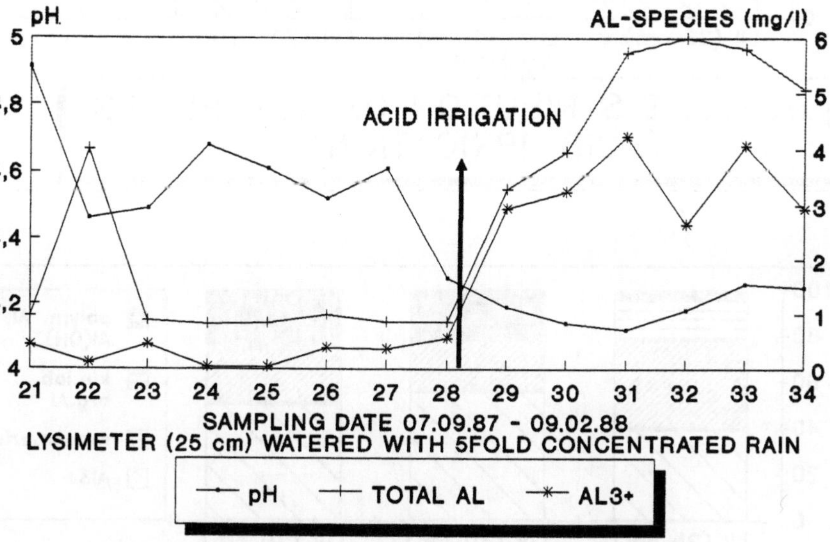

Figure 8. pH, TOTAL AL AND AL^{3+} - CONCENTRATIONS IN THE LEACHATE

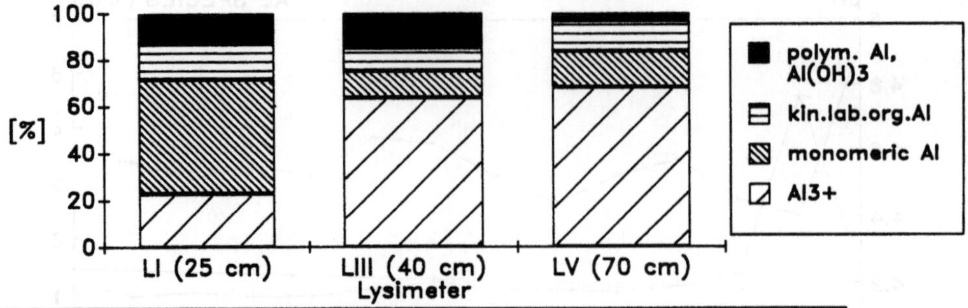

Figure 9. DISTRIBUTION OF AL-SPECIES (CA-FERTILIZATION)

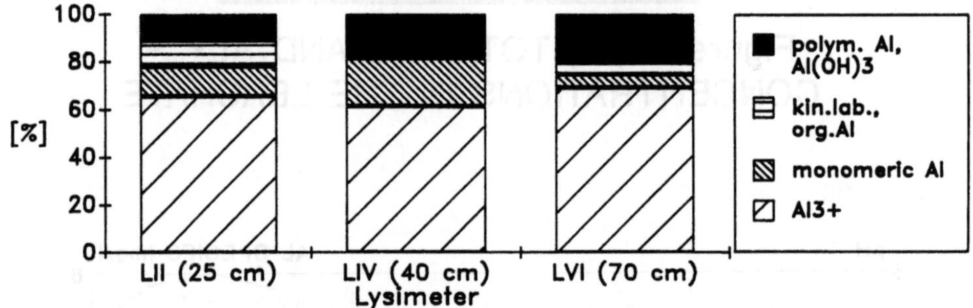

Figure 10. DISTRIBUTION OF AL-SPECIES (ACID-IRRIGATION)

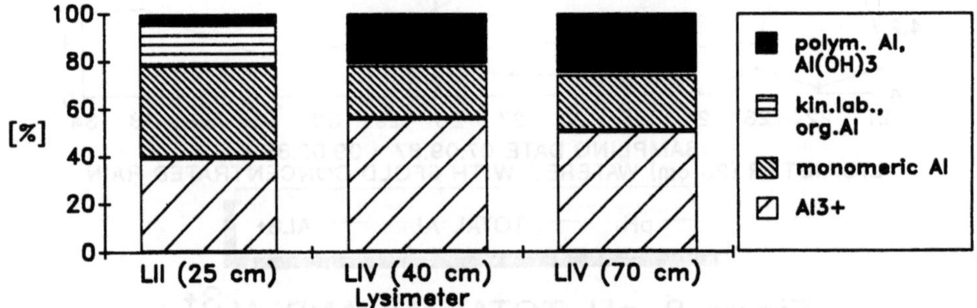

Figure 11. DISTRIBUTION OF Al-SPECIES (UNTREATED)

Due to acid irrigation the release of the total Al and of free Al^{3+}-ions increased by a factor greater than 2 (Figure 8), whereas the fraction of the monomeric and kinetically labile, organic Al-species decreased. The fraction of stable organic Al-complexes was minimal. With increasing soil depth the fraction of kinetically labile inorganic and organic Al-compounds decreased, while the fraction of the more stable Al-species increased. Although the concentration of the free Al^{3+}-ions was less elevated in 40 cm depth than in the leachates from 70 cm depth, these ions predominated in all soil layers. The variety of the Al-species was minor in the mineral soil than in the other soil layers.

As a result of liming, the Al^{3+}-ions decreased in 25 cm soil depth (Figure 9). Nearly 80 percent of the total aluminium was complexed to result in monomeric and polymeric Al-species. In 40 cm soil depth the liming first resulted in an acidification. The concentration of total Al and of the Al^{3+}-ion fraction increased while the content of kinetically labile Al-organo-complexes decreased.

7. CONCLUSION

The presented results show that ITP is a suitable method for detecting the species "$[Al(H_2O)_6]^{3+}$" in the mobile soil phase at environmentally relevant concentrations {> 0,05 mg Al^{3+}/l} without changing the originality of the sample. Since only 5 µl sample volumes are needed this micromethod allows to analyze microvariations of Al^{3+}-concentrations in the soil structure. By establishing this method the additionally developed speciation scheme of Al allows the differentiation between free Al^{3+}-ions and Al-complexes of different stabilities. Since phytotoxicity, formation kinetics and behaviour of the Al-species vary considerably, a specific assessment of the state of soil conditions as well as hazard for plants should only be performed on the basis of a species analysis. The effects of acid irrigation and liming on formation and mobilization of Al-species in soil can be elucidated.

REFERENCES

Ares J O (1985) Eigenschaften der Al-Komplexe in sauren Waldbodenlösungen. Dissertation Universität Göttingen

Campbell P G C, Bisson M, Bougie R, Tessier A, Villeneuve J P (1983) Speciation of Aluminium in Acidic Freshwaters. Anal Chem 55: 2246 - 2252

Driscoll C T (1984) A Procedure for the Fractionation of Aqueous Al in Dilute Acidic Waters. Intern J Environ Anal Chem 16: 267-283

Havas M, Jaworski J F (1986) Al in the Canadian Environment. In: Adv Chem Sor Publication NRCC 24759 of the Environmental Secretariat, Ottawa (Canada), p 51-79

Hem J D, Roberson C E (1967) Form and Stability of Aluminium Hydroxide Complexes in Dilute Solution. Geological Survey Water Supply Paper 1827-A, United States Governm Printing Off, Washington

Jackson G E (1982) Studies on the Chelation of Aluminium for Biological Application I. Citric Acid. South Afr J Chem 35: 87-92

James B R, Clark C J, Riha S J (1983) An 8- Hydroxyquinoline Method for Labile and Total Aluminium. In: Soil Extracts Soil Sci Soc Am J 47: 893 - 897

Lazerte B D (1984) Forms of Aqueous Aluminium in Acidified Catchinents of Central Ontario: A Methodological Analysis. Can J Fish Aquat Sci 41: 766 - 776

Rost-Siebert K (1985) Untersuchungen zur H- und Al-Ionen-Toxizität an Keimpflanzen von Fichte (Picea abies, karst.). Berichte des Forschungszentrums Waldökosysteme/Waldsterben 12

Schmid S (1989) Aluminium-Species im Boden und ihre Mobilität. Dissertation, Universität-Gesamthochschule Duisburg

Schmid S, Kördel W, Klöppel H, Klein W (1989) Differentiation of Al^{3+} and Al Species in Environmental Samples by Isotachophoresis. J Chromatogr 470: 289 - 297

Turner R C (1969) Three Forms of Aluminium Aqueous Systems Determined by 8-Quinolinolate Extraction Methods. Can J Chem 47: 2521 - 2527

Wilson D O (1984) Determination of Al in Plant Tissue Digests Using a Catechol Violet Colorimetric Method. Common Soil Sci Plant Anal 15: 1269 - 1279

The Analysis of Arsenic in Turkish Coals and Ashes by Photometric Methods and MECA

E.Henden, and A.Çelik
Dept. of Chemistry
Faculty of Science
University of Ege
Bornova, İzmir, Turkey

1. Introduction

It is well known that one of the main materials, responsible for environmental pollution is coal, that is used in vast amounts in most countries. Coal utilization affects terrestrial, aquatic and atmospheric environments. Effects on the terrestrial environment include disruption of land, soil and their surfaces. Effects on aquatic environment include disruption of surface and underground aquatic systems, and the thermal, particulate and chemical contamination of waters. Effects on atmospheric environment include deterioration of air quality by particulate and gaseous emissions and modification of the climate. The latter effect becomes more serious when the combustion of coals proceed in an uncontrolled way. The various effects on air, land and water may result in damages to plant and animal life, damages to food resources and modification of the landscape.

Most of the coal used is burned directly, and large volumes of sulphur and nitrogen oxides, particulates along with smaller quantities of hydrocarbons, carbon monoxide and trace elements are emitted in the effluent gases. The usual polluting elements in coal are Be, B, N, S, V, Cr, Mn, Ni, Cu, Zn, As, Se, Cd, Sb, Hg and Pb [Gluskoter, 1981]. Part of these elements are also released to the environment as a result of the combustion and disposal of waste materials. Some of these elements are even at fairly low concentrations toxic to animals and plant species. These types of toxic contaminants are not degraded in the environment, so they persist for extended periods. The effect on animals and plants is cumulative and will often appear only after years of exposure. The elements are concentrated mainly in coal ash, fly ash and particles trapped by the pollution control systems. These elements can be leached from the

NATO ASI Series, Vol. G 23
Metal Speciation in the Environment
Edited by J. A. C. Broekaert, Ş. Güçer, and F. Adams

solid waste by runoff and seepage water that comes in contact with them. Most of the trace elements in leachates are retained and fixated in soils, so that their migration to ground water is minimized. However, some of the trace ions can be taken up by the plants.

The amount of trace elements varies largely in coals. Variations in composition occur even for coal samples derived from the same seam. In this work, the arsenic content of different Turkish coals was, therefore, determined by sampling throughout different years in order to obtain information on the contribution of coal utilization on arsenic pollution of the environment.

2. Experimental

2.1. Apparatus

For the measurement of arsenic, two photometric methods, respectively a silver diethyldithiocarbamate and a molybdenum blue method were used [Turkish Standards Institute 1968, 1974, Analytical Methods Commitee, 1975].

In the determination of arsenic by means of molecular emission cavity analysis (MECA) a Pye Unicam SP 90 A flame spectrometer, with a 1.4-mm slit (bandwidth of 45 nm at 400 nm) was used to measure the emissions as described earlier [Henden, 1985]. The output was recorded on a Varian G-2500 chart recorder (response time 0.5 sec. for full scale deflection). The cavity-holding assembly and the stainless steel, flame containing cavity are shown in Figure 1 and Figure 2, respectively. The volatilization and gas chromatographic separation system used was similar to that described earlier [Belcher et al, 1977], with a glass reaction vessel (12.5 cm long, 1.5 cm i.d.)(Figure 3). The arsenic was reduced in solution with sodium tetrahydroborate(III) and arsine generated was separated from germane, stannane and stibine on a gas chromatographic column. The column was a PTFE tube (55 cm long, 3 mm i.d.) packed with 10% E-301 silicone gum rubber on Porapak Q (80-100 mesh) and operated at 15 ± 0.5°C.

2.2. Reagents

All chemicals used were of analytical-reagent grade, unless stated otherwise. Stock solution (1 mg/ml) of arsenic(III) was prepared from arsenic(III) oxide.

Sodium tetrahydroborate(III) solution (5% m/v) was prepared, as required, by

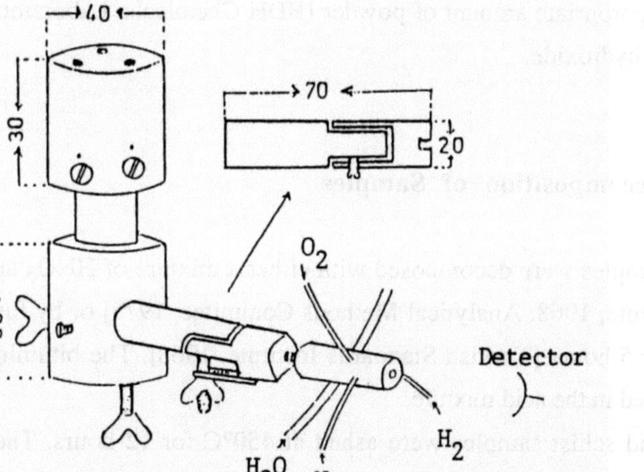

Figure 1. MECA cavity holding assembly (dimensions in mm)

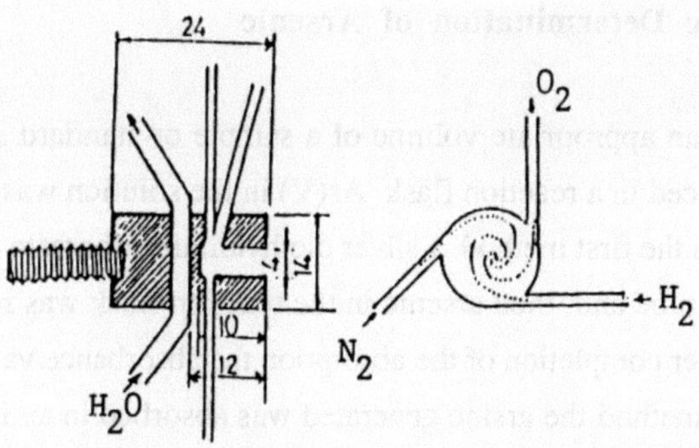

Figure 2. The cavity, which contains the flame (dimensions in mm)

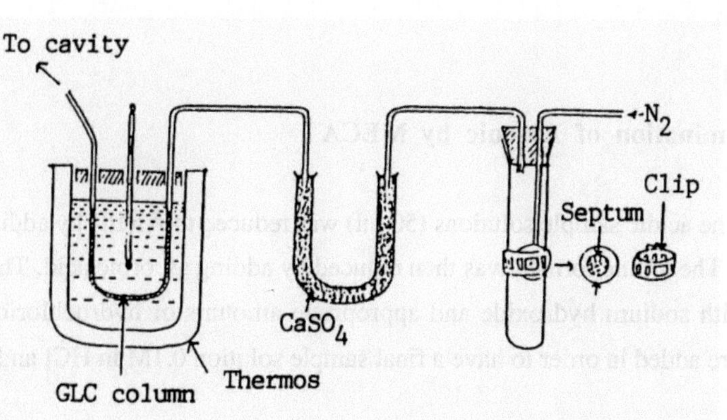

Figure 3. Volatilization and gas chromatographic separation system

dissolving an appropriate amount of powder (BDH Chemicals, Laboratory-reagent grade) in 0.01 M sodium hydroxide.

2.3. Decomposition of Samples

Coal samples were decomposed with either a mixture of HNO_3 and H_2SO_4 [Turkish Standards Institute, 1968, Analytical Methods Commitee, 1975] or by fusion with Eschka at 800 ± 25°C for 5 hours [Turkish Standards Institute, 1968]. The bituminous schist samples were decomposed in the acid mixture.

Coal and schist samples were ashed at 450°C for 12 hours. The ash samples were decomposed by fusion with sodium hydroxide in a silver crucible at 600°C for 30 min [Wilson and Wilson, 1962, Musgrave, 1962].

2.4. Photometric Determination of Arsenic

In both methods, an appropriate volume of a sample or standard solution containing 0-10μg of arsenic was placed in a reaction flask. As(V) in the solution was first reduced to As (III) with KI and $SnCl_2$. In the first method, a silver diethyldithiocarbamate solution was placed in a U-shaped absorption tube and, then arsenic in the reaction flask was reduced to arsine by addition of zinc metal. After completion of the absorption the absorbance values were measured at 540 nm. In the second method the arsine generated was absorbed in an iodine solution. The molybdenum blue colour was generated by adding ammonium molybdate and hydrazine sulphate to the absorbing solution and by warming for 10 min. in boiling water. The absorbances were measured at 680 nm. The arsenic contents of samples were calculated using calibration graphs.

2.5. Determination of Arsenic by MECA

Arsenic in the acidic sample solutions (50 ml) was reduced to As(III) by adding 2 ml of a 40% KI solution. The iodine formed was then reduced by adding ascorbic acid. The solution was neutralized with sodium hydroxide and appropriate amounts of hydrochloric acid and EDTA solution were added in order to have a final sample solution 0.1M in HCl and 0.05M in

EDTA. One ml of this solution was injected into the reaction vessel containing 0.5 ml of a 5% sodium tetrahydroborate(III) solution. AsO molecular band emission was observed as arsine was eluted from the GLC column and its peak height was measured at 490 nm. The arsenic content was calculated using the linear calibration graph in the range 0.1-5µg of arsenic.

3. Results and Discussion

The photometric techniques used are well-known and, therefore, were not further studied. However, it was the first application of MECA to the determination of arsenic in coal. The MECA technique was, therefore, further evaluated. The coefficient of variation (n=8) for the determination of 1 µg of arsenic was 4.1%, and 35 ng of arsenic could be detected [Henden, 1985]. Various metal ions are known to interfere with the formation of arsine. EDTA was, therefore, added to the sample solutions in order to mask possible metal ion interferences [Henden, 1982]. The recovery of added arsenic to the lignites, schist, and their ashes was high and varied between 96% and 104%. Moreover, arsenic was also determined by standard addition technique in samples of lignites, schist and their ashes. No significant difference was found between the results obtained using a direct calibration graph and a standard addition graph. Without EDTA addition the recovery of arsenic was only 25% and 50% with the schist and coal ash samples, respectively.

The arsenic content of the samples was determined using the techniques described. Samples were taken in 1980, 1985, 1986 and 1988. The arsenic content of some of the sample ashes was also determined in order to determine the loss of arsenic during ashing. The results are shown in Table 1. The ash contents of the samples were also included in the Table, so that the loss of arsenic during ashing could be calculated. It is apparent that two lignites samples identified as Akhisar and Şahineli, and the schist sample contained a large arsenic concentration, much above the estimated average arsenic content of word coals (14 mg/kg) [Gluskoter, 1981]. The arsenic was not homogeneously distributed in the lignites. Moreover, loss of arsenic during ashing was high which has its immediate implications with regard to environmental pollution.

4. Conclusion

Some of the lignites and bituminous schist analysed contain high amount of arsenic, most of which is lost during ashing. These types of minerals are to be controlled for arsenic content before use in order to minimize arsenic contamination of the environment. The MECA technique is fast and reliable for the determination of arsenic in such samples.

Table 1. Arsenic Content of Coal Samples, and Loss of Arsenic During Ashing

Sample name	Year of sampling	Ash %	Arsenic content (mg/kg) in original sample	Arsenic content (mg/kg) in ash	Loss of Arsenic during ashing (%)
Bituminous schist,					
Demirci	1980	82	483	382	35
Lignites,					
Akhisar	1980	34.0	342	35	97
Tunçbilek	1985	39.7	8.2		
	1986		24.7		
	1988	59.1	22.3		
Soma	1986	27.6	18.4		
	1988	26.4	26.7		
Aydın, Şahineli	1986	27.2	285		
	1988		50		
	1989	37.9	25.0	13.9	79
Yatağan	1986		19.1		
Beypazarı	1986		5.0		
Kütahya	1986		24.4		
Denizli	1988	13.0	25.0	60	69
Coals,					
Zonguldak, Centre	1986		3.3		
	1988	49.3	4.3		
Armutcuk	1988	39.9	5.2		

Acknowledgement

We would like to thank Nilgün Ulsever and Şeniz Su for their contribution to this work.

References

Analytical Methods Commitee (1975) The determination of small amounts of arsenic in organic matter. Analyst 100 : 54-62

Belcher R, Bogdanski SL, Henden E, Townshend A (1977) Molecular emission cavity analysis, part x. The separation and simultaneous determination of arsenic and antimony by hydride generation. Anal Chim Acta 92 : 33-41

Gluskoter HJ, Shimp NF, Ruch RR (1981) Coal analyses, trace elements and mineral matter. In : Elliott MA (ed) Chemistry of coal utilization, 2 nd supplementary volume. John Wiley, New York Chichester Brisbane Toronto, p 369

Henden E (1982) Attempts to eliminate interferences in the determination of arsenic, antimony, tin and germanium by molecular emission cavity analysis with the hydride generation technique. Analyst 107 : 872-878

Henden E (1985) Simultaneous determination of germanium, arsenic, tin and antimony by molecular emission cavity analysis after hydride generation and gas chromatographic separation. Anal Chim Acta 173 : 89-95

Musgrave JR (1962) Germanium. In : Kolthoff IM, Elving PJ (ed) Treatise on analytical chemistry, part 2, vol 2. Interscience, New York London, p 207

Turkish Standard Institute (1968) Determination of arsenic in coal and coke (in Turkish). Turkish standard no TS 651, Ankara

Turkish Standard Institute (1974) General method for the determination of arsenic (in Turkish). Turkish standard no TS 1463, Ankara

Wilson CL, Wilson DW (1962) Comprehensive analytical chemistry, vol 1c. Elsevier, Amsterdam London New York Princeton, p 237

THE ECOLOGICAL ROLE OF AQUATIC ORGANIC AND INORGANIC COMPONENTS, DEDUCED FROM THEIR NATURE, CIRCULATION AND INTERACTIONS

J. Buffle
Department of Inorganic, Analytical and Applied Chemistry,
University of Geneva
Sciences II, 30 quai E. Ansermet
1211 Geneva 4
Switzerland

The goal of this chapter is to give an overview of the main organic and inorganic components of natural waters, their circulation, their interactions with trace elements, and the role of the latter on preservation of life. Several of the topics outlined are discussed in more detail in other chapters of this book and in BUFFLE (1988).

I. ORGANIC AND INORGANIC WATER COMPONENTS

I.1. Classification of water components

The chemical composition of natural waters is highly complex: they contain most, if not all, of the elements of the periodic table plus a vast number of organic components. Innumerable interactions can occur in such a mixture, often yielding products that are poorly characterized. The most important organic and inorganic components are shown in Fig. 1. Most small, well defined, organic molecules (free amino-acids, sugars, small hydroxy-acids) do not play a significant role in controlling trace element concentrations because their concentrations and complexation energies are too small. About 90 % of the aquatic organic components are macromolecules or particles which may be classified into three major groups (see below): proteins and peptides, polysaccharides and fulvic-humic compounds. In most cases, however, it is still impossible to isolate and identify all of the organic components in these groups. Inorganic components are classified according to the nature of the elements of which they are composed (Figs. 1, 2). Many simple, low molecular weight ions or complexes are well-known, but, as

NATO ASI Series, Vol. G 23
Metal Speciation in the Environment
Edited by J. A. C. Broekaert, Ş. Güçer, and F. Adams

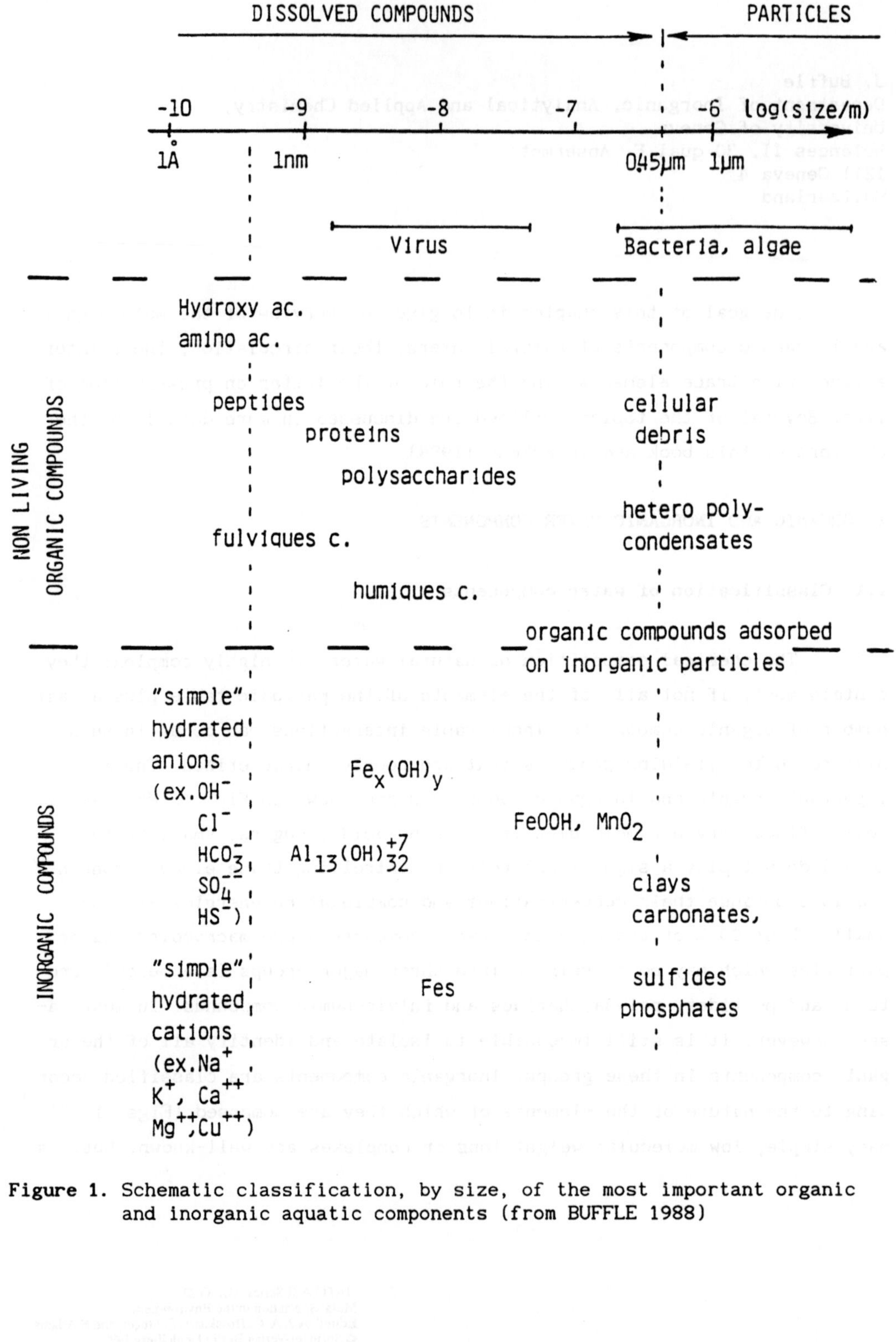

Figure 1. Schematic classification, by size, of the most important organic and inorganic aquatic components (from BUFFLE 1988)

for organic compounds, most of those which play a significant role in trace element circuculation and uptake by organisms are ill-defined polynuclear or colloidal compounds (metal hydroxides, clays, solid carbonates, inorganic solids with organic coatings). As figure 1 suggests, a physical classification of water components based on particle size does not allow to define clear boundaries between different groups, for both organic or inorganic components. In particular, the traditional use of 0.45 μm pore filters to separate dissolved and particulate components is purely a matter of convention. Furthermore, experimental results also depend on the nature of the filters and on the experimental conditions used (PERRET et al. 1989). Nevertheless, it is useful to note that the concentrations of dissolved and particulate inorganic matter (in weight per volume) are of similar orders of magnitude in fresh waters, whereas, in open oceans, the concentrations of dissolved inorganic components are about 10^5 times higher than the particulate ones. Intermediate factors apply to continental shelves and estuaries. For organic components, the concentration of dissolved matter is often found to be about 10 times greater than that of the particulate (BUFFLE 1988). However it must be borne in mind that most of organic and a majority of inorganic "dissolved" compounds occur in fact in a macromolecular or colloidal form (Fig. 1).

The nature and concentration of organic and inorganic compounds in water results from a complex set of processes amongst which are production and degradation of terrestrial and aquatic biomasses, weathering of rocks and leaching of soil by rainfall, adsorption reactions on suspended particles, particle coagulation and sedimentation in water bodies, chemical and biological activity in sediments and entrainment by aerosols (BUFFLE 1988) The composition and concentrations of these substances can, therefore, vary significantly from one water system to another, and a given system may exhibit variations in both horizontal and vertical directions. Fig. 2 compares the concentrations of the most important organic and inorganic complexants in two different water systems. The organic component concentrations were estimated on the basis of average DOC values and the average composition of organic matter for these two systems (BUFFLE 1988). The unidentified organic component concentrations are represented by those of either their functional

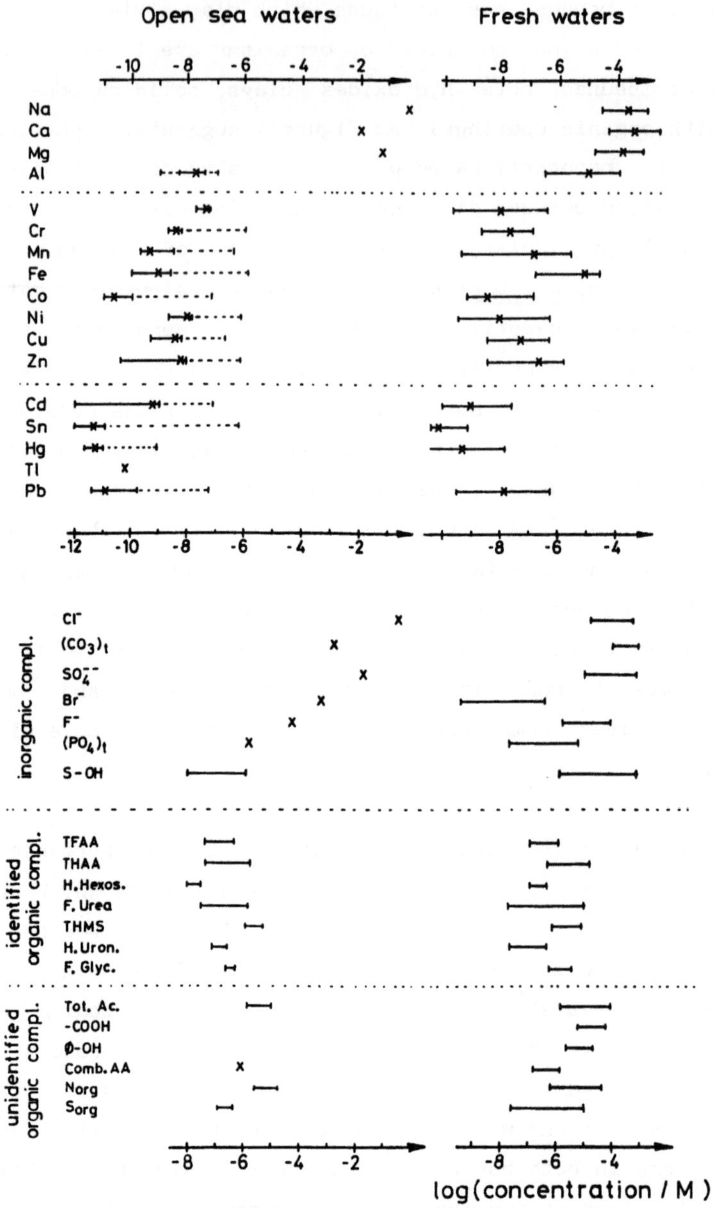

Figure 2. Range (--) and average values (x) of molar concentrations of complexants (a) and metal ions (b) in two types of waters (from BUFFLE 1988). ≡S-OH are -OH sites at the surface of inorganic suspended particles. TFAA = total free amino-acids. THAA = total combined hydrolyzable amino-acids. THMS = total combined hydrolyzable monosaccharides. -ϕ-OH = phenolic groups. Comb AA = amino-acids combined to unidentified organics. N_{org}, S_{org} = organic nitrogen and sulfur

sites or their donor atoms (N or S) expressed as mole/l. Note that, for un identified organic components, which constitute 70-80% of all organic compounds, total acidity (total concentration of acid sites) can be used as an estimate of the total concentration of oxygen containing functional sites.

I.2 Dissolved and solid inorganic components

Many inorganic elements may exist under various oxidation states. The thermodynamically stable oxidation state of any given element in a given aquatic medium can be predicted from the Nernst equation, the standard redox potential of this element, and the degrees of complexation, $\alpha_{Ox} = |Ox|_t/|Ox|$ and $\alpha_R = |Red|_t/|Red|$, of the corresponding oxidized and reduced forms (STUMM and MORGAN 1981, TURNER et al. 1981) ($|X|_t$ = total concentration of X; $|X|$ = concentration of non complexed X). On this basis, it may be shown that, in air-saturated ocean or fresh waters, all trace elements should be in their maximum oxidation states, because the redox potential of the O_2/H_2O couple is higher than that of any other couple (the only exception being Au(III)/Au(I) in sea water for which the lower oxidation state, Au(I), is stabilized by its strong complexation by Cl^-). The relative position of the various couples on the redox potential axis, however, may vary much from water to water, because of the varying complexation effects due to differences in pH and in the composition of complexants.

Suspended solid surfaces (particles or colloids) in waters play a prominent role in controlling the concentration of dissolved trace elements. Most of these elements are eliminated by sedimentation after incorporation onto or into particles, generally by complexation with the surface sites (see sec. II and III). The nature of the inorganic particles and colloids, most important for these processes, is given in Table 1. The organic particles of water bodies are mostly debris of plankton and bacteria, and macromolecules (humic, polysaccharides or proteins) adsorbed on inorganic particles. Particles and colloids in a water body may be classified as a function of their origin:

- allogenic compounds (mainly pedogenic, i.e. originating from soils), are mostly **clays** and other **silicates**, **Fe and Mn oxyhydroxydes**, and **humics, fulvics and polysaccharides**, leached out by rain water;
- endogenic compounds (or aquagenic, i.e. formed in the water column)

Mineral	Type of source		
	Allogenic[a]	Endogenic[b]	Authigenic[c]
Non-Clay Silicates			
QUARTZ – SiO_2	X		
POTASH FELDSPAR – $KAlSi_3O_8$	X		
PLAGIOCLASE – $(Na,Ca)(Al,Si)Si_2O_8$	X		
Mica – $K(Mg,Fe,Al)_3AlSi_3O_{10}(OH)_2$	X		
Amphibole – $(Ca,Mg,Fe,Al)_{3.5}Si_4O_{11}(OH)$	X		
pyroxene – $(Ca,Mg,Fe)_2Si_2O_6$	X		
OPALINE SILICA (diatoms)		X	
Clays			
ILLITE – $K_{0.6}Mg_{0.25}Al_{2.26}Si_{3.43}O_{10}(OH)_2$	X		
SMECTITE – $X_{0.3}Mg_{0.2}Al_{1.9}Si_{3.9}O_{10}(OH)_2$	X		
Chlorite – $Mg_5Al_2Si_3O_{10}(OH)_8$	X		
Kaolinite – $Al_2Si_2O_5(OH)_4$	X		
mixed-layer clays, vermiculite–intermediate	X		?
palygorskite – $(Ca,Mg,Al)_{2.5}Si_4O_{10}(OH) \cdot 4H_2O$	X		?
nontronite – $X_{0.5}Fe_2Al_{0.5}Si_{3.5}O_{10}(OH)_2$	X		X
Carbonates			
CALCITE – $CaCO_3$	X	X	x
DOLOMITE – $CaMg(CO_3)_2$	X		?
Aragonite – $CaCO_3$	x	X	
Mg-Calcite – intermediate		x	X
rhodochrosite – $MnCO_3$			X
monohydrocalcite – $CaCO_3 \cdot H_2O$		X	?
siderite – $FeCO_3$	?		?
Fe-Mn Oxides			
GOETHITE, Lepidocrocite – FeOOH	X	x	X
MAGNETITE – Fe_3O_4	X		
Hematite, maghemite – Fe_2O_3	X		?
birnessite – $(Na,Ca)Mn_7O_{14} \cdot 3H_2O$	?		X
todorokite – $(Na,Ca,K,Ba,Mn)_2Mn_5O_{12} \cdot 3H_2O$	?		X
psilomelane – $(Ba,K)(MnO_2)_{2.5} \cdot H_2O$			X
ilmenite – $FeTiO_3$	X		
Phosphates			
APATITE – $Ca_5(PO_4)_3(OH,F)$	X		x
Vivianite – $Fe_3(PO_4)_2 \cdot 8H_2O$			X
ludlamite – $(Fe,Mn,Mg)_3(PO_4)_2 \cdot 4H_2O$			X
(?) lipscombite – $Fe_3(PO_4)_2(OH)_2$			X
(?) phosphoferrite – $(Mn,Fe)_3(PO_4)_2 \cdot 3H_2O$			X
(?) anapaite – $Ca_2Fe(PO_4)_2 \cdot 4H_2O$			X
Sulphides			
MACKINAWITE – $FeS_{0.9}$		x	X
pyrite – FeS_2	X		x
griegite – Fe_3S_4			X
sphalerite – ZnS		x	
Fluoride			
fluorite – CaF_2			X

Table 1. Most important minerals reported for freshwater lake sediments.
(a) Sources external to the lake (rivers, leaching of soils..).
(b) Compounds formed within the water column. (c) Compounds formed within the sediments. X in the formula refers to univalent cation exchange. In "type of source" size of X is varied to illustrate the frequency of occurrence. (from JONES and BOWSER 1978)

may be subdivided into i) inorganic compounds formed by chemical precipitation (mostly **Fe and Mn oxyhydroxydes; Ca and Mg carbonates; Fe sulfides** in anoxic waters), ii) inorganic compounds (**SiO_2, $CaCO_3$**) that are the backbones of certain microorganisms and are released after their death, and iii) organic debris and compounds resulting from the degradation of microorganisms (**aquagenic fulvic and humics, polysaccharides, cell wall debris**).

I.3. Natural Organic Components

The nature of the major classes of organic matter found in water bodies is described below (For details see BUFFLE 1988, THURMAN 1985). Although the global oceanic biomass (2.10^{15} g C) is much smaller than the terrestrial one (827.10^{15} g C), the corresponding masses of detrital organic matter are of similar magnitude (1650.10^{15} and 1080.10^{15} g C respectively). Their residence times (average values are 66 and 22.4 years respectively, but the actual range of values is between days and millenaries) indicate that in both cases most of these compounds are rather refractory. Their nature, however, is quite different because of the difference in their origin. It is therefore important to distinguish the following classes of natural organic matter (NOM) (Fig. 3):

Pedogenic NOM is due to rain water leaching of soil NOM, which in turn results from the decomposition of higher plants by bacteria and fungi

Aquagenic NOM is that formed in the water column itself, essentially by excretion by and decomposition of plankton and aquatic bacteria

Sedimentary NOM results from accumulation and transformation in sediments, of organisms debris and aquatic NOM (predominantly the aquagenic part; see sec. III).

The **biochemical** classification (Fig.3) may also be applied and would be most appropriate. However, the three main biochemical classes (polysaccharides, proteins, lipids) represent only about 20% of the NOM, both in soils and water bodies, the remaining being still ill characterized pedogenic refractory organic matter (PROM) in soils, and mixture of varying proportions of PROM + aquagenic organic matter (AROM) in waters.

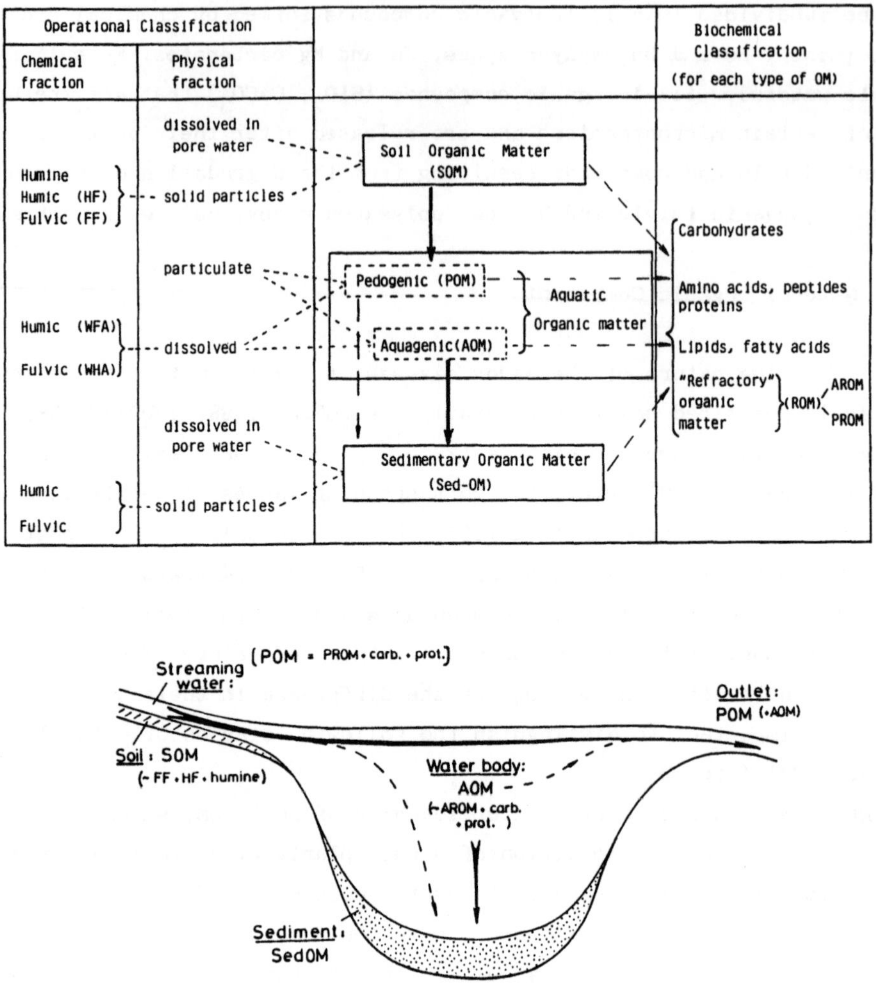

Figure 3. Classification of various types of natural organic matter, and schematic picture of the circulation of the major groups of organic compounds in waters (from BUFFLE,1988). Carb.=carbohydrates; Prot.=proteins HA,FA = humic and fulvic fractions; WFA,WHA = water fulvic and humic fractions, subsequently denoted by AROM

Classifications based on **Physical** and **Chemical fractionation procedures** are purely operational, but sometimes useful for studying PROM and AROM in the absence of more theoretically sound discrimination mode. In particular, the "fulvic" and "humic" fractions of soil or water NOM are distinguished basically on their differences in solubilities, humic compounds being only soluble in alkaline solutions, and fulvic above pH ~ 2.

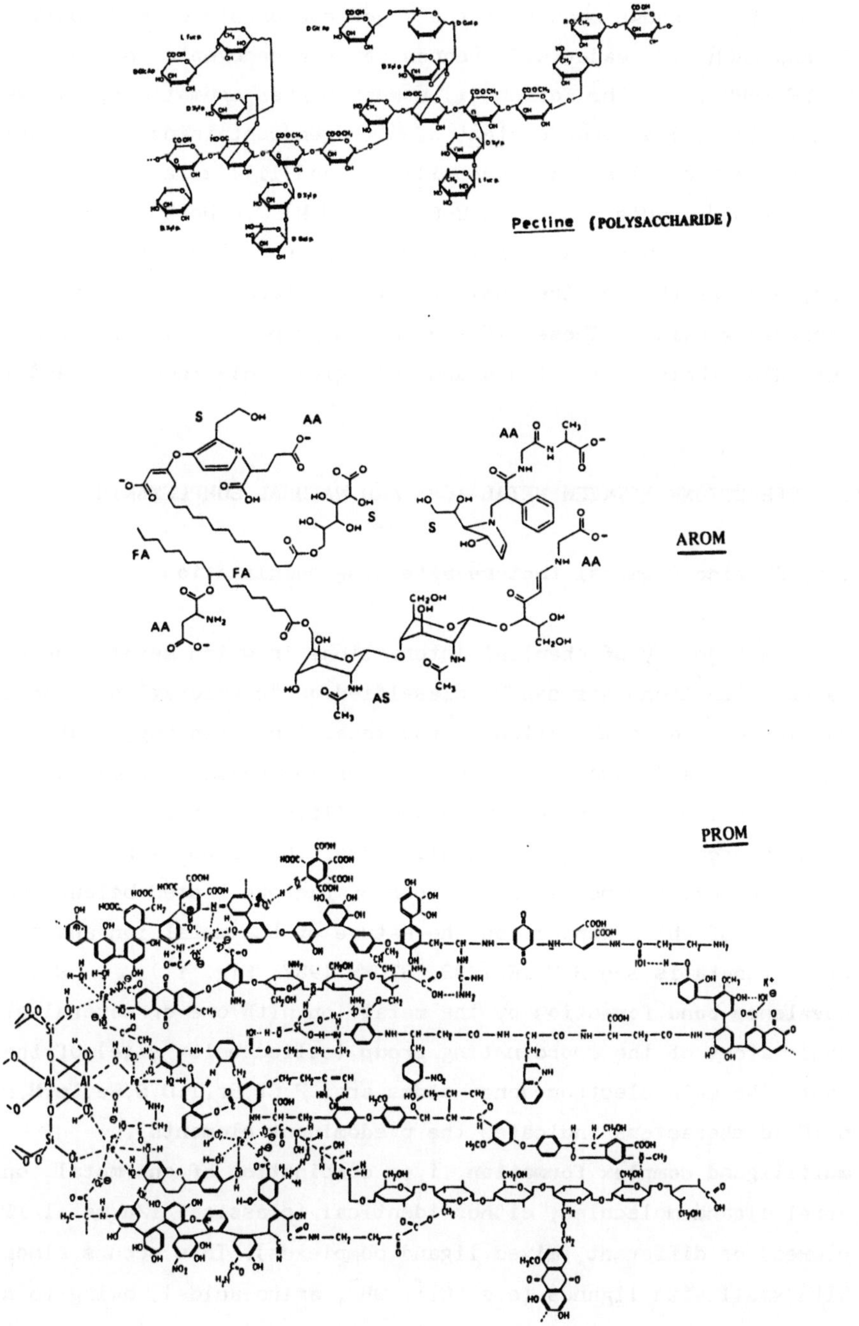

Figure 4. Example of polysaccharide and models of pedogenic ROM (PROM) and aquagenic ROM (AROM). AROM are recombinations of basic biochemical units (AA= animo-acid, S= sugar, FA= fatty acid, AS =amino-sugar).(From BUFFLE 1988)

Fig. 4 shows models of the major components of NOM: polysaccharides, PROM and AROM. The major differences between pedogenic and aquagenic ROM are (BUFFLE 1988), for the former: i) a much higher aromaticity (41 vs 9 % of aromatic C), ii) a lower content of nitrogen containing functional groups (1-2 vs 6-7 % of elemental composition), and iii) a larger total acidity (mostly -COOH groups; 10-17 vs 2-6 meq/g NOM). In both cases the molecular weight of the fulvic fraction lies between several hundreds and a few thousands, whereas that of the humic fraction extends from thousands to hundreds of thousands Dalton. These differences in properties are important for understanding their interactions and ecological role (sec. III and IV).

II. INTERACTIONS BETWEEN METAL IONS AND NATURAL COMPLEXANTS

II. 1. Physico-chemical factors affecting complexation

A majority of chemical interactions in which metal ions participate in aquatic environments can be classified as "complexation" reactions in the broad sense: i.e. combination of the metal ions with any aquatic component (e.g. small ligand such as CO_3^{--}, organic or inorganic macromolecule such as proteins or iron hydroxide, suspended colloid or particle such as clay or debris of organism; Fig. 1). In all cases, the proportion of metal retained by the complexant depends on a number of factors listed below, the relative importance of which depends on the nature of the metal ion and the complexant (For details see BUFFLE 1988, STUMM 1987; Fig. 1):

- **covalence bond formation** by the metal ion with one or several electron-donor atoms of the *coordinating group* (called below *site*) of the complexant. The main electron-donor atoms are: **F**, **Cl**, Br, I, **O**, **S**, Se, Te, **N**, P, As, and **C** (Bold characters indicates the predominant elements).
- **multiligand complex formation**, i.e. combination of one metal ion with several ligand molecules, either identical (classical 1/n metal/ligand complexes) or different (mixed ligand complexes). This occurs almost only with small size ligands (e.g. Cl^-, OH^-, amino-acids), owing to steric hindrance occurring with large complexants.
- **polyfunctionality** is important for medium to large size complexants (PROM, AROM, humics, proteins) which may possess a large number of different coordinating groups. The stronger complexing sites being occupied first by metals, the overall retention energy of these complexants depend on the

total metal to total complexant concentration ratio (BUFFLE et al. 1987).

- **polyelectrolyte properties** also play a role in the case of medium to large size complexants (fulvics, humics, polysaccharides, metal oxyhydroxides, clays). By definition, coordinating sites may retain or release cations, including protons. Most of these sites are therefore ionised (e.g. $-COO^-$, $\equiv S-O^-$). The net result is that highly polyfunctional macromolecules are often highly charged. In natural waters, complexants are generally negatively charged, and hence many cations are retained not only by covalent bonding, but also by electrostatic attractions.
- **conformational factors** play a role with large macromolecules and suspended particles (humics,polysaccharides,cell wall debris,clays,metal oxyhydroxides). Degrees of hydration and ionization of macromolecules change with environmental conditions (e.g. pH, or ionic strength). This may induce changes in the conformation of the macromolecules, which in turn may affect the local environment and then the retention energy of specific coordinating groups. Retention of metals at the surface or inside particles are influenced by similar effects.

II.2. Classification of metals and ligands based on their hard and soft character

Although classical mathematical models have been widely used (NORDSTROM and BALL 1984) to estimate the degree of metal-ligand interactions, they are generally unable to make correct predictions of metal speciation in natural waters (CABANISS et al.1984). This is due to the fact that the numerous and complicated factors mentioned in section II.1 cannot be incorporated easily into numerical models. Indeed, at present, there is very little quantitative information on these factors, our detailed knowledge regarding complexation processes being mainly limited to reactions with small ligands in solutions containing only one or few of them, i.e. for systems much simpler than natural ones.

The overall interaction between the metals and ligands in waters can however be estimated by taking into account i) the hydrolysis properties of the elements and ii) the concept of "hardness" and "softness" of metals and electron-donor atoms:

- Inorganic elements of the whole periodic table can be divided into three categories (TURNER et al, 1981):

a) elements which form non dissociated oxo-complexes (e.g. SO_4^{--}) and oxy-acids (e.g. $As(OH)_3$),

b) elements that are highly hydrolysed, even in acidic conditions, but which may also exist as hydrated cations (e.g. Fe(III)), and

c) elements whose hydroxo-complexes are not very stable, even at relatively high pH (e.g. Ca^{++}, Zn^{++}).

	I	II	III	
z^2/r \ $\Delta\beta$	> 2	2 - 0	0 - (-2)	<-2
< 2.5	Na^I Cs^I Li^I K^I Rb^I (<0.1 FW, SW)		Cu^I Tl^I (<0.1 FW) (0.3-5.2 SW)	Ag^I Au^I (0.2-6.1 FW) (5.3-12.9 SW)
2.5 - 7.0	Ca^{II} Mg^{II} Ba^{II} Sr^{II} (<0.2 FW, SW)	Co^{II} Cu^{II} Fe^{II} Mn^{II} Ni^{II} Zn^{II} Pb^{II} (0.1 - 2.7 FW) (0.2 - 1.5 SW)	Cd^{II} (<0.1-0.3 FW) (1.6 SW)	Hg^{II} (6.9-11.8 SW) (14.2 SW)
7.0 - 11.0	Ln^{III} Y^{III} (0.1 - 4.2 FW) (0.4 - 1.3 SW)			
> 11.0	Hydroxo or oxo complexes (in particular Al(III), Be(II), Cr(III), Fe(III), Bi(III), U(IV), In(III), Tl(III).			

Table 2. Reactive metals (type b) and c), see text), classified as a function of their softness and hardness. Degrees of complexation of M by "simple" inorganic anions, $\alpha_M = |M|_t/|M|$, are given in parentheses for sea-water (SW, pH=8.2) and freshwater (FW, pH=6-9) ($|M|_t$=total M conc.; $|M|$ = free M conc.).(BUFFLE 1988, adapted from TURNER et al 1981)

OH^- is the major ligand of water, and therefore clearly, ligands other than OH^- will readily combine only with group c) elements and, to a lesser degree, group b). A list of these elements is given in Table 2.

- The elements can also be subdivided according to their hard and soft character (PHILLIPS et al. 1965). Hard cations participate preferentially in electrostatic interactions, while soft ones form more covalent bonds. The degree of hard character can be estimated semi-quantitatively from the term z^2/r where z and r are the cation charge number, and radius respectively. The following order has been observed for the affinity of donor atoms for hard metals: F > O > N ~ Cl > Br > I > S whereas this

	Ligands preferred by hard cations from Group (I)	Intermediate ligands	Ligands preferred by soft cations from Group (III)
Inorganic ligands	F, O^{2-}, OH^-, H_2O	Cl^-, Br^-	H^-, I^-
	CO_3^{2-}, HCO_3^-	N_3^-, NH_3	CN^-, CO
	SO_4^{2-}, NO_3^-	NO_2^-, SO_3^-	S^{2-}, SH^-
	$H_2PO_4^-$, HPO_4^{2-}, $[-O-P(=O)(O^-)-O-]_n$	O_2, O_2^-, O_2^{2-}	
Organic ligands	$R-C(=O)O^-$, $R-C(=O)-O-R'$	$R-NH_2$	R^-
	$R-OH$, $R-O-R'$	$RR'NH$, $RR'R''N$	$R-SH$, $R-S^-$
	$R-C(=O)-R'$,	$^*R-NH-C(=O)-R'$	$^*R-S-S-R'$, $^*R-S-R'$
	$^*R-O-P(=O)(O^-)-O^-$, $^*R-O-P(=O)(O^-)-O-R'$		*N*-heterocycles[b]
	$R-O-SO_3^-$		
	$\phi-COO^-$, $\phi-OH$		

Table 3. Ligands or coordinating sites encountered in biological systems and waters. Groups I and III refers to Table 2 (From BUFFLE 1988)

order is reversed for soft metals. Therefore, it is also possible to estimate the degree of soft character of a metal, M, by means of the parameter: $\Delta\beta = \log\beta^{o}_{MF} - \log\beta^{o}_{MCl}$ where β^{o}_{MF} and β^{o}_{MCl} are the stability constants at infinite dilution for fluoride and chloride complexes of M. The soft character increases with increasing negative values of $\Delta\beta$.

Table 2 gives a classification of type b) and c) metals as function of $\Delta\beta$ and z^2/r. The elements grouped together possess similar reactivities. The indicated values of α_M allow to estimate their reactivities. The larger is α_M, the more it is complexed. It can be seen that alkali-metal and alkaline-earth metal cations are not very reactive in either fresh water or sea water. In contrast, soft metals can form very strong complexes, the stability of which may vary greatly from one aquatic system to another.

Table 3 lists the principal complexing sites present in both aqueous and biological systems, classified as a function of their affinity for hard (group I in Table 2) and soft (group III in Table 2) metals.

II.2. Main interactions between metal and complexant groups

On the basis of the above considerations, the following overall picture, for metal-complexant interactions in aquatic systems, can be drawn. In this picture, discrimination is made between metal ion groups I, II and III (from hard to soft, Table 2) and between the following groups of complexants

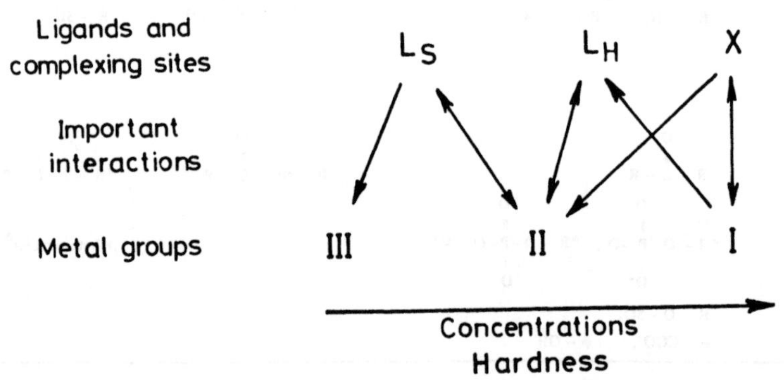

Figure 5. The most important interactions between the different metal and ligand groups. The head of the arrow indicates which group (ligand or metal) is the most affected by the interaction (see text)

(see also Fig.2): i) the "simple" inorganic ligands, denoted by X (these are the major anions; except for Cl^-, their donor atom is O and they prefer hard metals), ii) the "hard" sites of NOM, denoted by L_H (they are mostly caboxylic and phenolic sites), and iii) the "soft" sites of NOM, denoted by L_S (in particular N and S containing sites).

On the basis of total concentrations of these ligands and metals given in Fig.2, the following scheme for metal-complexant interactions can be written (Fig. 5):

GROUP I (mainly alkali-metal and alkaline earth metal cations)

$$(|M|_t \leq |X|_t; \quad |M|_t >> |L_H|_t; \quad |M|_t >> |L_S|_t)$$

These hard cations prefer hard ligands, but form rather weak complexes with them. Consequently, complex formation is possible only under conditions of high metal or ligand concentrations: most of these complexes therefore are formed with the major inorganic anions X. Similarly, only those cations having the highest concentrations (Ca^{++}, Mg^{++} and Na^+) can combine with L_H sites. In this case, since $|M|_t >> |L_H|_t$, only a small fraction of M is bound to L_H. Such interactions have thereforee little importance for group I cations, but they can have an important indirect competitive effect on the complexation of group II and III metal ions. Because of their high concentrations, group I cations also act as counterions for the polyelectrolytic complexing agents. They therefore influence the electric field in the neighbourhood of the complexing sites, which may in turn play a predominant role in macromolecule conformation, colloid coagulation, and complex stability for cations of groups II and III.

GROUP II (especially bivalent transition metals)

$$(|M|_t << |X|_t; \quad |M|_t \leq |L_H|_t; \quad |M|_t \geq |L_S|_t)$$

These metals possess appreciable affinity for both hard and soft sites. They can therefore react with all three groups of ligands. Since, in addition, some of them (Cu, Fe, Mn) can present several stable oxidation states, it is often complicated to describe their reactivity. In their bivalent form however, it can be predicted that they will compete i) for L_H sites with metals from group I, which are less strongly bound but are present at higher concentrations, and ii) for L_S sites with metals from group III which are present at lower concentration but which are more strongly bound.

GROUP III (typical soft metal)

$$(|M|_t << |X|_t ; \quad |M|_t << |L_H|_t ; \quad |M|_t << |L_S|_t)$$

These metals have more affinity for soft sites, L_S, than for the hard sites, L_H, or the ligands X (with exception of Cl^- which plays an important role in sea water). Since $|M|_t << |L_S|_t$ complexation of group III cations can be effectively controlled by L_S sites.

Fe(III), Al(III), Mn(IV)

These metals, which are strongly hydrolysed under natural conditions, can form mononuclear complexes of the classical type only with some very strong specific ligands (e.g.. for Fe(III), the siderochromes excreted by organisms). In most cases, however, these metals are found in the form of hydroxide or hydrous oxide particles having very reactive surfaces. These particles may adsorb on their surfaces other metals or ligands that are too weak to cause dissociation of the whole particle.

III. CIRCULATION OF METALS AND ORGANIC MATTER

III.1. Circulation of inorganic compounds

It may be shown (BREWER, 1975) that, among the factors controlling the concentration of inorganic elements in water, geochemical processes dominate on the global scale of the earth. As will be discussed below, this is not true at local and short time scales, where microbiological processes often play a dominant role. Globally, and on the long term range, however, the aquatic cycles of most of the elements depend mainly on the dynamic action of i) erosion of the terrestrial crust and leaching by rainfall, and ii) sedimentation of dissolved inorganic ions in oceans, after formation of insoluble compounds or adsorption on particles.

The most important inputs and outputs of compounds in aquatic reservoirs are, i) for inputs: inflowing rivers and soil leachates, and atmospheric deposition, and ii) for outputs: outflowing rivers (in lakes) and particle sedimentation (in some cases diffusive transport to and from the sediments may also be important). Discussion on the circulation of element E in reservoir X is most often based on the concept of its residence time, τ_X^E,

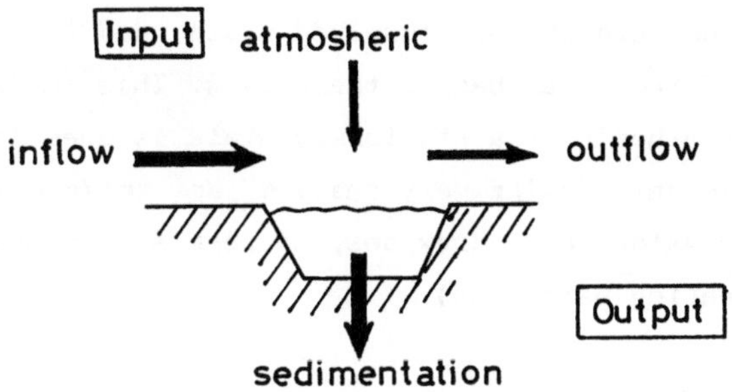

Figure 6. Inputs-outputs of compounds in an aquatic reservoir

defined as:

$$\tau_X^E = \frac{m_X^E}{dm_X^E/dt} \qquad (1)$$

where m_X^E is the mass of element E in reservoir X, and dm_X^E/dt is the flow of E through it. Simple discussion is possible only when τ_X^E is constant, which is only valid when a stationary state is established for E and X, i.e. (Fig.6) when the sum of input flows equals that of output flows (each sum being equal to dm_X^E/dt). In the case of the global circulation of elements between the terrestrial crust reservoir (X = S) and the ocean reservoir (X = O) (i.e. a cycle based on erosion + sedimentation followed by upwelling), the hypothesis of a stationary state is reasonable, considering the very long period (millions of years) during which water and sediments have been recycled (WHITFIELD and TURNER, 1979). Then $dm_S^E/dt = dm_O^E/dt$. Furthermore, considering the erosion rate of E to be a first order reaction, i.e. $dm_S^E/dt = k_S^E.m_S^E$, the residence time of E into the ocean, τ_O^E, becomes (eq.1):

$$\tau_O^E = \frac{m_O^E}{k_S^E.m_S^E} = \frac{m_O.Y_O^E}{k_S^E.m_S.Y_S^E} \qquad (2)$$

where m_S and m_O are the total mass of solid in the crust and of dissolved + suspended compounds in the ocean, and $Y_O^E = m_O^E/m_O$ and $Y_S^E = m_S^E/m_S$. Eq. 2 shows that for any element, τ_O^E is proportional to Y_O^E/Y_S^E, since $m_O/k_S^E.m_S$ is a constant. A correlation has indeed been found between these two terms by WHITFIELD et al (1979). The elements possessing the longest residence times (or the largest Y_O^E/Y_S^E ratio; Cl, Br, Na : see Fig. 7) are those which have the weakest reactivities and which are therefore less subject to sedimentation.

They are likely to accumulate in the dissolved state, and their τ_0^E values are close to the age of the ocean basins themselves. This implies that, for these elements, the hypothesis of a stationary state is questionable. On the other hand, the elements having very small τ_0^E are those which readily precipitate as hydrous oxides or hydroxides, or form strong complexes (in particular with sedimenting particles).

WHITFIELD et al.(1979) have also shown that the residence time of elements is directly linked to their chemical reactivity. As it has been mentionned previously, many of these reactions occur with particles via complexation at surface sites. The structural matrices of most suspended inorganic solids (clays, oxides and hydroxides, carbonates) and the main complexing groups of organic particles are oxygen-based (Fig.2, Table 1). Therefore the following reaction may be considered as representing most adsorbtion reactions (at least for groups I and II metals):

$$(\equiv|-\mathrm{OH})_n + \mathrm{M} \rightleftharpoons n.\mathrm{H} + \mathrm{M}\text{-}(\text{-O-}|\equiv)_n \qquad (3)$$

Furthermore, the M-O bond energy can be related to $\Delta^2 = (X_M - X_O)^2$, where X_M and X_O are the electronegativities of M and O respectively. Fig.7 shows that Y_0^E/Y_s^E as well as τ_0^E decrease as Δ^2 increases, indicating clearly the important role of chemical reactivity in the elimination process of metals. Figure 7 permits the distinction of four metal-ion groups having different reactivities: 1) alkali metal ions, only slightly or non reactive, 2) alkaline-earth metal ions, slightly reactive 3) the very reactive cations (Fe(III), Al(III)), eliminated by insoluble hydrous oxide formation, and 4) all other metal ions which follow the main correlation. For the latter, the dispersion of the points is due to "secondary effects" specific to each cations (e.g. influence of N or S sites for soft metals, stereochemical factors which depend on cation size..). It must be noted that the existence of a relationship between chemical reactivity of metals and their circulation, has been demonstrated experimentally both in lake and ocean water (STUMM 1983): in both cases the residence time of metal M was found to decrease with its distribution coefficient defined as $K_D = |M|_p/|M|_d$ where $|M|_p$ and $|M|_d$ are the concentration of M measured in the particulate and dissolved state respectively.

As it has been mentioned above, special effects may affect, locally, the distribution of trace metals. Two important ones are illustrated

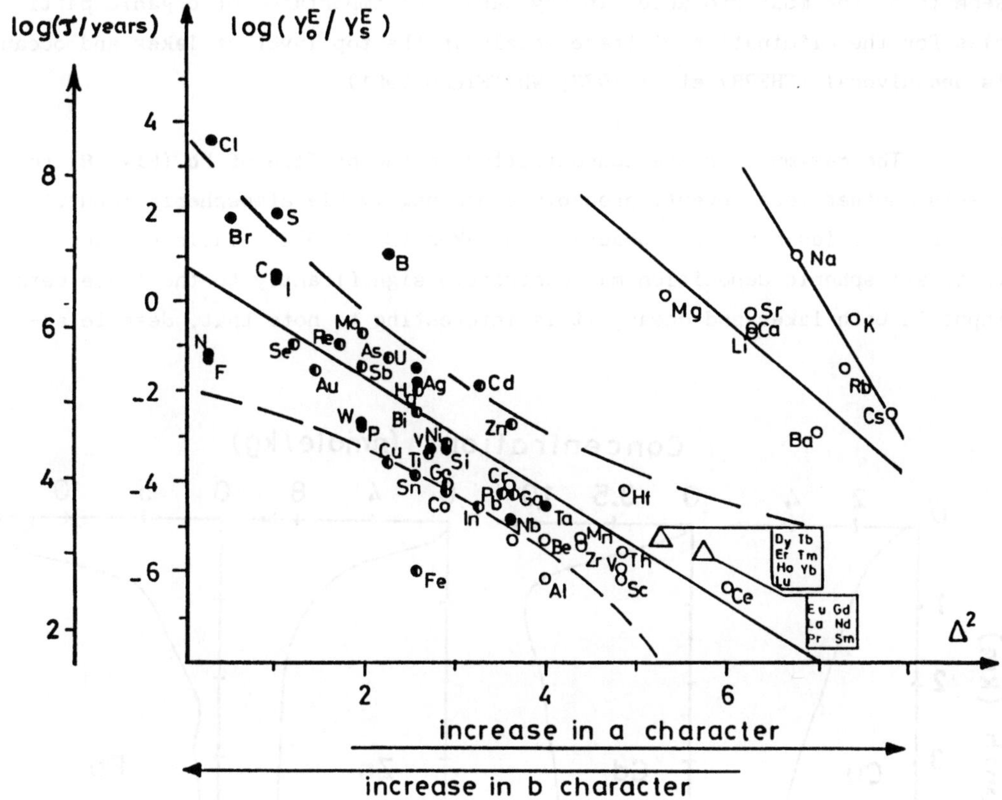

Figure 7. Relationship between the residence time of element E (or its partition between terrestrial crust and ocean water), and the difference between its electronegativity and that of oxygen. (From WHITFIELD and TURNER 1979)

in Fig. 8, which shows concentration profiles of dissolved trace metals in Northern Pacific Ocean. Often, these profiles show a minimum at ocean surface (e.g. for Cu, Zn and Cd). This is related to the biological activity, but the exact reason is not fully known (MOREL and HUDSON 1984; SIGG 1984). The three most probable alternatives are:

- generation of downward particle flux of organism debris which carry down, by coagulation, inorganic particles to which trace metals are bound
- coating of inorganic particles by organics released by organisms: this coating may bind trace metals, more efficiently than inorganic particles
- uptake of trace elements by living matter (many of these elements being as necessary for life as the well-known N and P nutrients) and resulting in their incorporation into the organic particle flux.

Although the exact mechanism is still unknown, the last two possibilities seem to be the most probable. In any case, the importance of organic parti cles for the elimination of trace metals in the top layer of lakes and ocean is unequivocal (CHERRY et al 1978; WHITFIELD 1981).

The maximum surface concentration in the profile of Pb (Fig. 8) reflects another local event, probably a non negligible atmospheric input. Similar profiles have been observed in lakes (SIGG 1984). Table 4 shows that, atmospheric deposition may contribute significantly to the trace metal input in both lakes and ocean. It is interesting to note that, despite at-

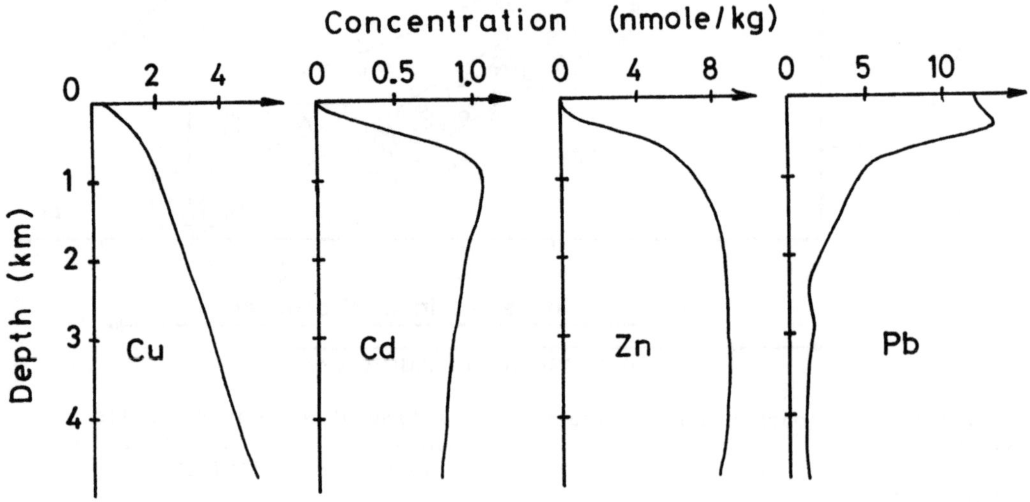

Figure 8. Dissolved concentration profiles of a few trace metals in North Pacific Ocean. Cu, Cd, Zn, taken from BRULAND 1980; Pb taken from SCHAULE and PATTERSON, 1981

mospheric pollution in trace metals is 10-100 times greater in the lake than in the ocean, the order of magnitudes of their metal concentrations are similar. It just reflects that elimination mechanism by particle sedimentation is much faster in lakes.

NORTH ATLANTIC	Cu	Pb	Cd	Zn
Atmospheric deposition ($ng.cm^{-2}.yr^{-1}$)	25	310	-	130
Sinking suspended particles (")	234	330		1040
Concentration in water ($ng.l^{-1}$)	30-300	1-15	1-120	10-600
LAKE CONSTANCE				
Atmospheric deposition ($ng.cm^{-2}.yr^{-1}$)	714	11000	20	8400
Sinking suspended particles (")	6500	9500	100	36000
Concentration in water ($ng.l^{-1}$)	300-800	50-100	6-20	1000-4000

Table 4. Atmospheric deposition and sedimentation rates of heavy metals in the North Atlantic and in lake Constance (From STUMM, 1983)

III.2. Circulation of organic compounds

Natural organic matter comprises very important complexants, in particular for soft metals since they are the main sources of "soft" coordinating groups (except for Cl^- in sea water and S^{-II} in anoxic systems). In that respect, it is important to differentiate between aquagenic and pedogenic ROM (AROM and PROM respectively) because the former contain much more soft sites than the latter. It is therefore important to understand their respective circulation process separately, despite the experimental difficulties to determine them selectively.

For that purpose, separation methods, like chromatographic, or operational physical or chemical fractionation techniques are most often inadequates, either because they introduce too many artefacts or because they are not selective enough. However the structure characteristics of PROM and AROM can be used to determine their relative proportions: compared to the latter, the former has a much higher content of phenolic and benzene-carboxylic groups which are known to absorb in the near UV at about 280 nm. Indeed it has been shown (BUFFLE 1988, ZUMSTEIN and BUFFLE 1989) that absorbance per gram of organic carbon is about 10 times greater for PROM than for AROM. The respective values are 18 l/g and 2 l/g. They are analogues of molecular extinction coefficients and may be obtained by measuring the ratio absorbance/DOC (DOC=dissolved organic carbon) of "pure" PROM (found in non productive ponds) and "pure" AROM (found in open sea water, where there is no terrestrial input). Thus, the simultaneous measurements of DOC and absorbance at 280 nm, in a water sample, allows to compute selectively the concentrations of its PROM and AROM, and even allows to estimate the concentration of the group (polysaccharides + proteins). It has been shown (BUFFLE et al, 1980, ZUMSTEIN et al, 1989) that similar results can be obtained by fluorescence determination instead of UV absorbance. By these methods, it is possible to monitor separately, without using any separation technique, the circulation of the two major (and qualitatively very different) groups of organic matter :

- PROM which contains mostly "hard" oxygenated sites,and which is largely hydrophilic because of its large number of dissociable sites (sec. I.3);
- AROM + proteins + polysaccharides, which contains most of the organic "soft" sites (N and S containing sites), and which may be expected to be much more hydrophobic than PROM because of their much lower content of dissociable groups (sec. I.3).

Figure 9 shows two profiles of absorbance, A, and DOC in the water column of an eutrophic lake, during the productive (b1) and non productive (b2) periods respectively, and the seasonal evolution (a) of PROM, AROM, and (polysaccharides + proteins) at the lake surface. It is seen that, during the non productive period (b2) or in the non productive part of the lake (deepest part of b1), the profiles of A and DOC coincide. In contrast, during summer, DOC increases at the lake surface, while A remains relatively constant. These observations reflect the fact that A measures, predominantly PROM, whose concentration remains approximately constant in the whole water column all through the year (a). This is because its inflow, due to soil leaching, does not vary much with season (specific events, like storms, produce peak concentrations (ZUMSTEIN and BUFFLE 1989), but their impact on the annual evolution of concentration is negligible). As it has been discussed

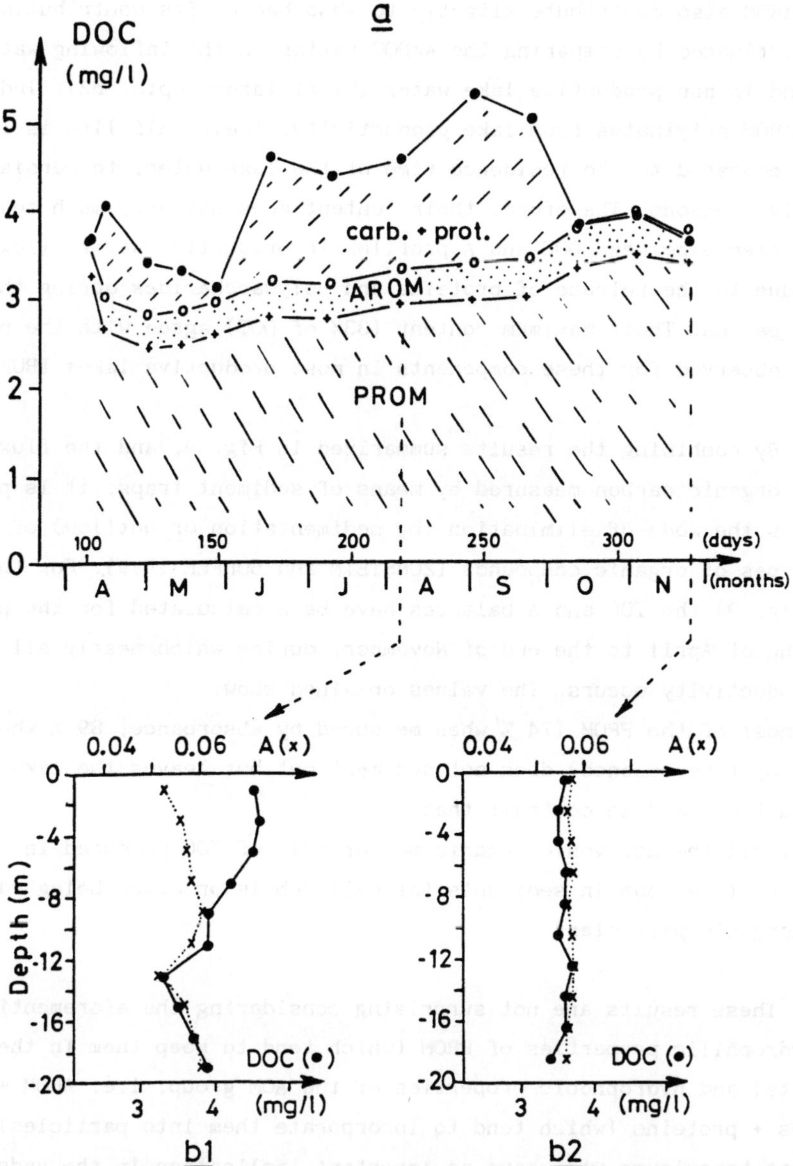

Figure 9. Variation of DOC and UV absorbance, A, at 285 nm for the eutrophic lake Bret (Switzerland) a) annual variation at the surface (1m) of total DOC and of its PROM, AROM and (polysaccharides + proteins) components (computed as explained in the text). b) Examples of variations as function of depth during a period of productivity (b1) and non productivity (b2). A and DOC scales are positionned in such a manner that A/DOC = 16 l/g C, which is the common value observed in all non productive water of the lake

above, AROM also contribute slightly to absorbance. Its contribution to DOC can be estimated by comparing the A/DOC ratios in the inflowing water (no AROM) and in non productive lake water (b1 at large depth, b2). Indeed, although AROM originates from lake productivity, their half-life is long enough, compared to the residence time of the lake water, to persist in non productive seasons. Therefore, their content does not vary much seasonally. The difference between DOC and A profiles in productive waters (low depths in b1) is due to the release of proteins and polysaccharides during the corresponding period. Their maximum content (33% of DOC) agree with the proportions (3-27%) observed for these components in most productive lakes (BUFFLE 1988).

By combining the results summarized in Fig. 9, and the flux of sedimenting organic carbon measured by means of sediment traps, it is possible to assess the mode of elimination (by sedimentation or outflow) of the various types of organic compounds (ZUMSTEIN and BUFFLE 1989). For the lake BRET (Fig. 9) the TOC and A balances have been calculated for the period: beginning of April to the end of November, during which nearly all the annual productivity occurs. The values obtained show:

- that most of the PROM (74 % when measured by absorbance, 89 % when measured by fluorescence) does not sediment out but leaves the lake with the outflow, and in contrast that
- nearly all the aquagenic organic matter (91% of TOC produced in the lake) settles down in sediments (as cell debris or after being adsorbed on inorganic particles).

These results are not surprising considering the aforementioned rather hydrophilic properties of PROM (which tend to keep them in the dissolved state) and hydrophobic properties of the AOM group, i.e. AROM + polysaccharides + proteins (which tend to incorporate them into particles) These different behaviours will have an important implication in the understanding of the overall circulation of trace metals (sec. IV) owing to the differing complexing properties of these two groups of organic compounds. Figure 10 summarizes the different behaviours of PROM and AOM in the whole hydrosphere: schematically, PROM follow a "horizontal" pathway, from soil to ocean, through lakes and rivers, whereas AOM follow mainly "vertical" pathways.

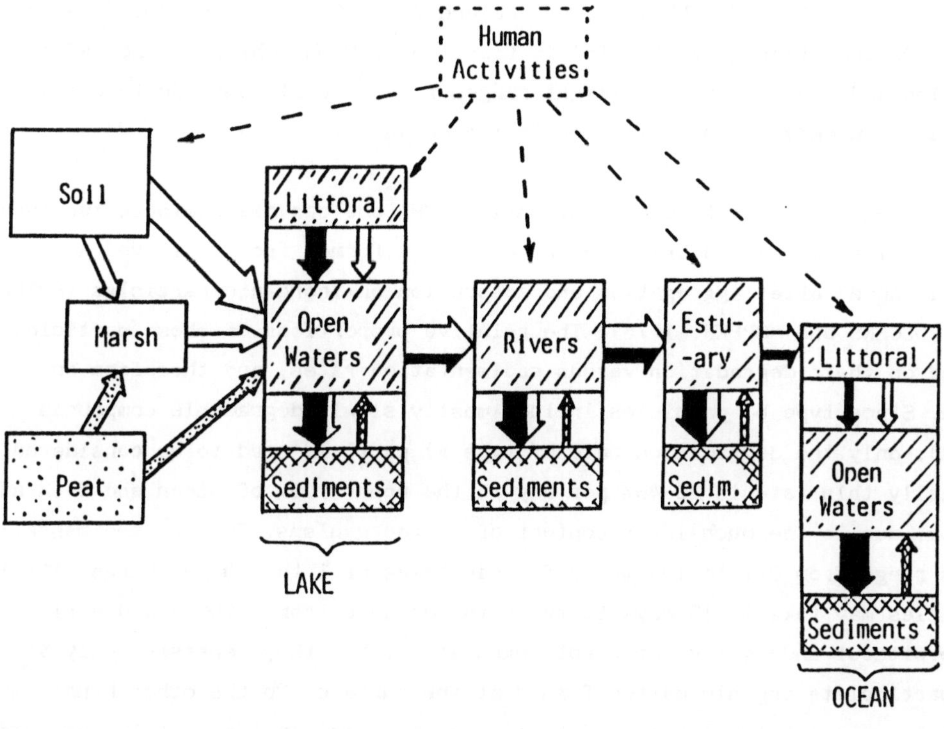

Figure 10. Schematic representation of circulation of aquagenic (AOM) and pedogenic (PROM) organic matter in the hydrosphere. White and black colors represent "pure" pedogenic and aquagenic organic matter respectively. In water bodies: the darker the color, the larger the content of aquagenic components. The white and black parts of the arrows suggest the respective proportions of pedogenic and aquagenic organic matter transferred from one compartment to the other. [hatched box] :mixture of PROM and AOM in water bodies; [cross-hatched box] :sediment organic matter

Before discussing the ecological role of aquatic components, the separate fate of AOM and PROM should be described briefly. Since PROM does not sediment appreciably in freshwaters, it may either be transported up to and diluted into the ocean, or sediment in estuaries or coastal ocean, after coagulation with inorganic colloids under the influence of high salt concentrations. The study of different types of organic matter along estuarine gradients show that coagulation occurs mostly with the humic fraction of PROM, i.e. the high molecular weight fraction (MW > 100,000), and only for salinity greater than 0.1 % (BUFFLE 1988; BUFFLE and DELADOEY 1982; SHOLKOVITZ et al. 1978). On the other hand, the fulvic fraction of PROM (which is the major fraction and whose molecular weight ranges from several hundreds to a few thousands) undergoes just dilution in the ocean.

Regarding the fate of sedimenting AOM, one can discriminate two types of particles: a) cell debris and b) particles formed from dissolved AOM (mostly AROM) after aggregation and adsorption on inorganic particles (BUFFLE 1988; DEGENS and MOPPER, 1976). The relative proportions of these particles depend on their degradation versus sedimentation rates, and therefore on depth. Since type b) particles include mostly slowly degradable compounds (AROM), only the degradation rate of type a) particle need to be considered. Typically this rate is 20-40% per day in the upper 100m of ocean and 0-1% at bottom, due to the much lower content of microorganisms. Typical sedimentation rates range from 0.1 to 100 m/day for particles of 1 to 100 μm. Large settling particles may take 10-15 days to reach the ocean bottom (4000m) and arrive there without undergoing important degradation, but they represent only 4% of the particulate organic matter formed at the surface. On the other hand, small particles are totally decomposed in the upper layer. In fact, it is observed that ~ 90% of plankton debris is decomposed in the upper 10m or so.

As a result of these processes, it is observed that type a) particles, containing a high proportion of proteins and polysaccharides with carboxylic sites of uronic acids, dominate in the upper layer of the ocean pelagic zones and in lakes. As depth increases, the proportion of type b) particles increases, relative to that of type a), due to the increased degradation of these latter.

From the above considerations, it may be expected that sediment organic matter will contain little PROM, but mostly AROM, polysaccharides and proteins. This is indeed observed (KEMP et al 1979). It might also be expected that the proportions of the latter two relative to AROM will decrease when the depth of the water column increases, due to their degradation in the water body before reaching the sediments. An opposite relationship has, however, been observed in "shallow" waters, these proportions decreasing from 25 to 14 %, from 250 to 10 m (KEMP et al. 1979). This probably results from an accrued microbiological activity in surface sediments of shallow waters, owing to the increase in light and temperature and to adequate oxygenation. By combining the above two effects, (degradation in the water column and in sediments) one can therefore expect the proportions of polysaccharides and proteins in sediments to be minimum in either shallow waters (~ 10 m) or very deep waters (> 400 m). For interpretation of metal complexation, it is important to note that, like AROM, sediment organic matter contains a larger proportion of N containing sites than PROM, and a much larger (about 10 times) content of S containing sites.

IV. ECOLOGICAL ROLE OF THE VARIOUS GROUPS OF METALS AND COMPLEXANTS, DEDUCED FROM THEIR PHYSICO-CHEMICAL PROPERTIES

IV. 1. Vital metal supply of organisms and cleaning of the medium from toxic metals

Before discussing the ecological role of metals and complexants, it is interesting to note the similarity in behaviour of metals of groups I, II and III, in aquatic and biological systems. The role of metals **inside biological organisms** can be schematically represented as follows (NIEBOER and RICHARDSON 1980; WILLIAMS 1967; see also sec. II.2):

Na and K are highly mobile because of their weak complexing ability. They facilitate electrical flow through membranes and regulate osmotic pressure.
Ca and Mg have their principal function in stabilizing or modifying the conformation of biological structures (the cellular "skeleton") by means of weak electrostatic interactions with dissociated oxygen-donor sites on macromolecules and membranes.
Metals from group II provide the majority of metal co-enzymes, frequently as a result of their redox properties. They are then bound to L_S sites. They can also play a role in macromolecular structures by forming metal bridges.
Metals from group III are very toxic because their complexes with L_S sites are often more stable than those formed by group II metals. They can therefore replace group II metals and thereby block catalytic action or modify vital structures.

By analogy with the internal biological media, the aquatic systems external to the organisms can be related to the functioning of the globe and its biosphere considered as a giant organism (Gaia in LOVELOCK 1979). A natural aquatic medium is then viewed as one of the regulating systems essential for the maintenance of life on Earth. To permit this regulating action, the aquatic medium should fulfill two roles: a) to ensure a sufficient supply of vital elements (group I and II) and b) to prevent excessive increases in concentration of the toxic elements from group III. The various groups of natural complexing agents described in sec I-III, either particulate or dissolved, seem to play a crucial role in this regulation process, thanks to their complementary physico-chemical properties and circulation pathways in aquatic systems (Figure 11; see also BUFFLE 1988 for more details and references):

Cations in group I, which occur in great quantity in water bodies owing to rock dissolution, and form only a few insoluble compounds, do not require a particular complexing reaction, as do group II and III metals, to be available to organisms in sufficient concentration. They can be directly assimilated in their dissolved ionic form. As in biological systems, they play an important ecological role in regulating the formation and structure of vital aggregates (see below).

Vital cations in group II occur in water systems, many in the form of complexes with oxygen-donor sites of suspended particles. In order to be assimilated by unicellular organisms, they must dissociate from their inorganic "support" and diffuse towards the cell (Process B). These two rather slow processes are accelerated when aggregates are formed between organisms and inorganic particles (Process C). Such aggregates are also well-known in soil. Cohesion is facilitated, in particular, by polysaccharide fibrillary macromolecules present in the bulk water phase and in the immediate neighbourhood of unicellular algae and bacteria. As mentioned above, group I cations acting as electrolytes play an important role by regulating the extent of unfolding of macromolecules (sec. II.2), and therefore, aggregate cohesion.

Organism-inorganic particle association seems to play a primary role in soils, vis-a-vis microelement assimilation and it is probable that it is of even greater importance in water masses where the availability of

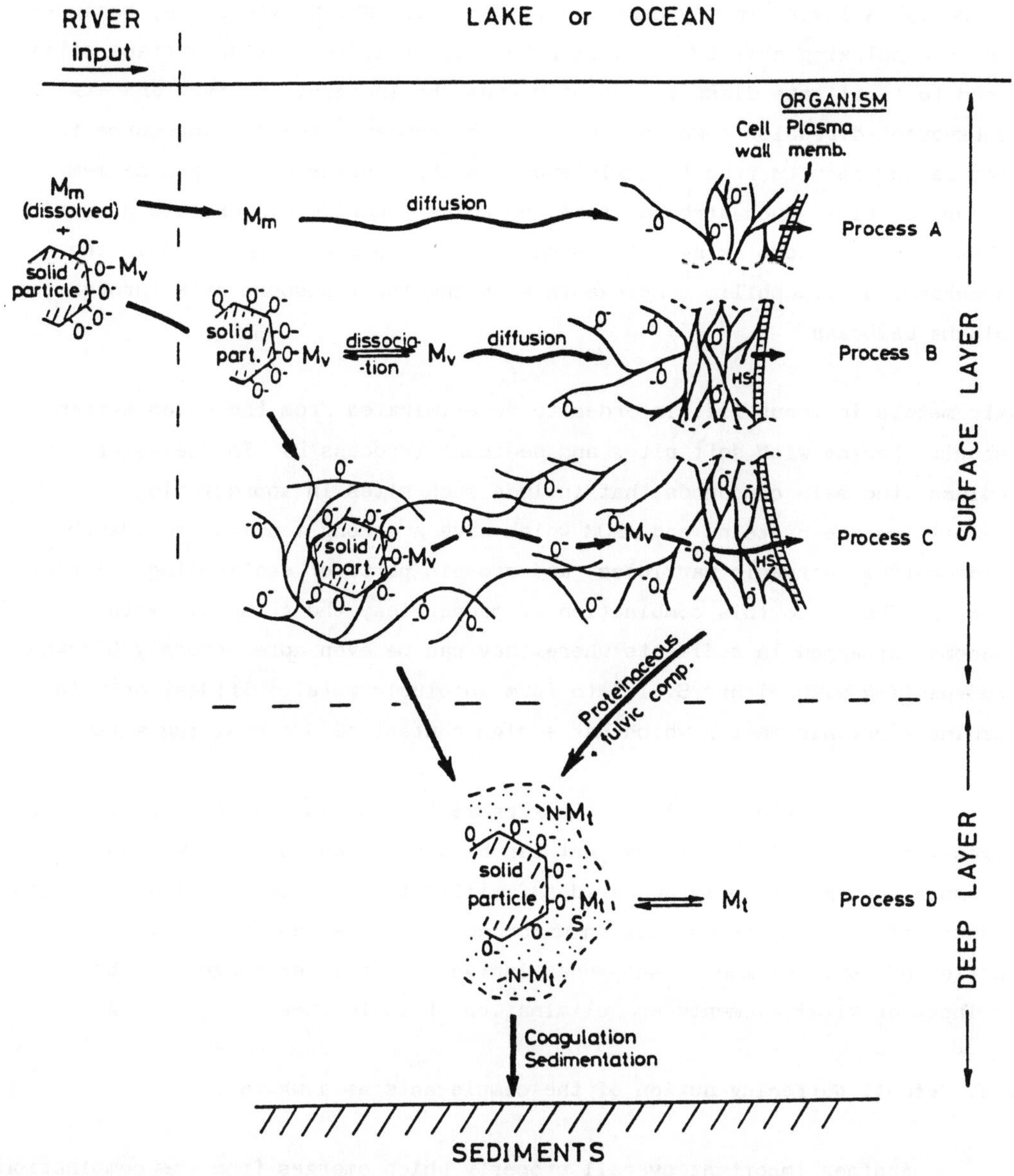

Figure 11. Schematic representation of the role of suspended inorganic particles, of polysaccharides and protein and AROM components on the cycle of major metals (M_m), vital metals (M_v) and toxic metals (M_t). $-O^-$ = oxygenated site on inorganic solid particle, or on polysaccharides in solution or in the cell wall of organism. -N, -S, = nitrogen or sulfur sites at the plasma membrane of organism, or from protein or AROM adsorbed on inorganic particles. (From BUFFLE 1988)

elements is lower. In any case, polysaccharide carboxylate sites, because their complexing ability is similar to that of hydrous oxide surface sites, tend to facilitate dissociation of M from the inorganic surface and its subsequent diffusion towards the cellular membrane. The driving force for the latter results from the existence, on the surface of the plasma membrane, of sites much stronger than the polysaccharide carboxylate groups: these are soft complexing sites such as -SH of proteins included in the membrane, or lipophilic complexants allowing the transport of M through the plasma membrane.

Toxic metals in group III, in order to be eliminated from the water system, should combine with soft sites and sediment (Process D). In the water column, the main compounds that include such sites in appreciable quantities are proteins and AROM which have a strong tendency to adsorb on sedimenting inorganic particles, and proteic parts of sedimenting cellular debris. Thanks to this combination of properties, the group III metals become entrapped in sediments where they can be even more strongly blocked by reacting with either S(-II) to form insoluble metal sulfides, or with sediment organic matter which has a high content of S containing sites.

This obviously simplistic representation has, as its only goal, the suggestion of certain key roles that can be attributed to the different groups of complexing agents: polysaccharides facilitate the uptake of vital elements; products of organism decomposition which contain L_S sites facilitate the elimination of toxic elements; suspended inorganic particles serve as substrate for input of vital elements and elimination of toxic ones.

IV.2. Overall buffering action of the complexants as a whole

Another important overall property which emerges from the combination of all the individual complexation properties of the aquatic complexants is illustrated in Fig. 12. It shows a distribution of the major complexants in terms of the total concentrations, $|L|_t$, and the equilibrium constants, K, (for complexation of Cu(II)), of their sites. Such a representation can only be semiquantitative because, for the same type of complexant, K and $|L|_t$ vary from one aquatic system to another. Nevertheless, the various complexants groups can be viewed as reservoirs for M accumulation, and the overall response of the aquatic system to any change (e.g. metal pollution due to anthropogenic impact) can be deduced: as $|M|_t$ increases, so does the degree of occupancy of sites, with those sites having the lowest free energy level (maximum K) being occupied first. Therefore, saturation progresses vertically, from bottom to top of Fig. 12, regardless of the nature of the sites: the strongly complexing FA sites are the first to be saturated with Cu(II), followed by the weaker FA sites, then those of proteins etc..Fig. 12 shows that, as a result

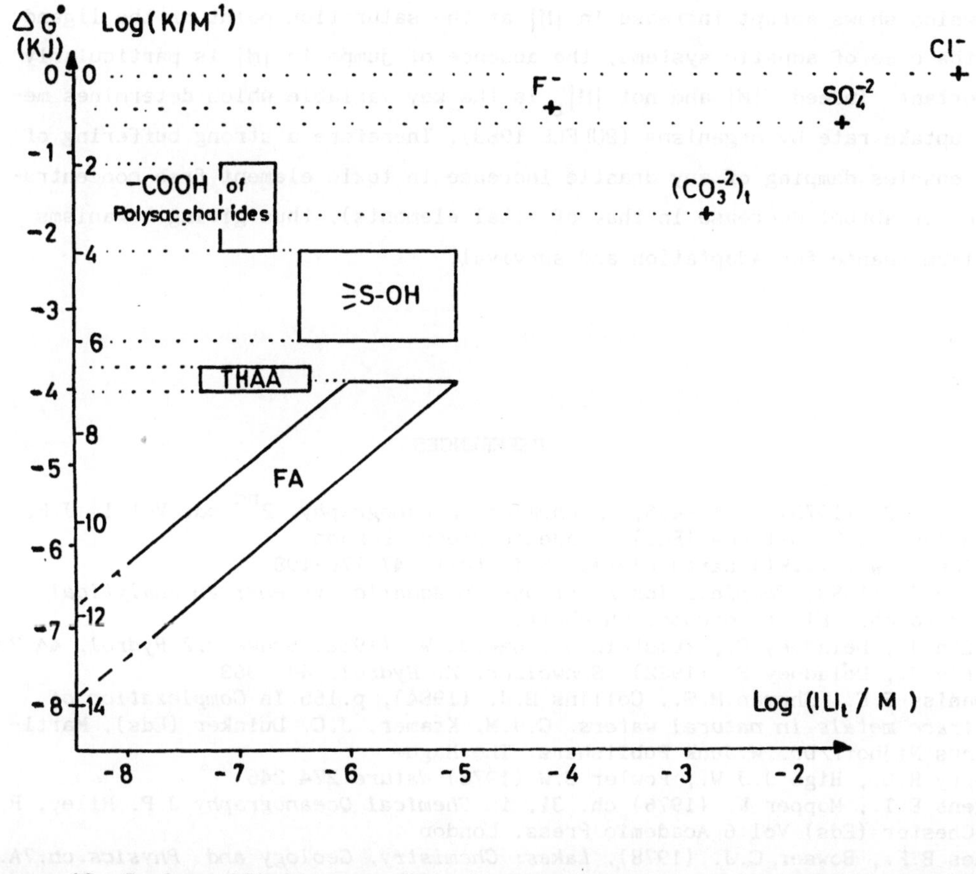

Figure 12. Semiquantitative complexant classification in sea water, according to complexing ability (K, ΔG°) vis a vis Cu(II), and total site concentrations ($|L|_t$).(from BUFFLE 1988). For details on K and $|L|_t$ values, see the above reference. THAA = proteins; FA = fulvic compounds. Variability in K values in each complexant group depend on polyfunctional and polyelectrolytic character of each group; variability in $|L|_t$ is the same as that in Fig. 2

of the great diversity of complexants, there exists no significant range for K (or $\Delta G°$) that is not represented by some complexant group, and a general decrease in K with increasing $|L|_t$ is observed. It may be shown (BUFFLE 1988) that thanks to the continuous quality of this $\log|L|_t$-logK relationship, the whole system yields a continuous buffer-type response for M complexation over at least seven decades of variation of $|M|_t$. This means that the increase in free M concentration, $|M|$, is gradual over all this range of $|M|_t$. This is at complete variance with the results of classical simple ligand titration by M, which shows abrupt increase in $|M|$ at the saturation point of the ligand. In the case of aquatic systems, the absence of jumps in $|M|$ is particularly important. Indeed, $|M|$ and not $|M|_t$ is the key variable which determines metal uptake rate by organisms (BUFFLE 1988). Therefore a strong buffering of $|M|$ enables damping of any drastic increase in toxic element free concentration (or abrupt decrease in that of vital elements), thus giving organisms maximum chance for adaptation and survival.

REFERENCES

Brewer P.G. (1975) P.415-496, *in Chemical Oceanography,* 2nd ed. Vol.1, J.P. Riley and G. Skirrow (Eds), Academic Press, London

Bruland K.W., (1981) *Earth Planet. Sci. Lett.* **47** 176-198

Buffle J. (1988) *Complexation reactions in aquatic systems; an analytical approach.* Ellis Horwood, Chichester

Buffle J., Deladoey P., Zumstein J., Haerdi W. (1982) *Schweiz.Z.Hydrol.* **44** 325

Buffle J., Deladoey P. (1982) *Schweizer. Z. Hydrol.* **44** 363

Cabaniss S.E., Shuman M.S., Collins B.J. (1984), p.165 In *Complexation of trace metals in natural waters.* C.J.M. Kramer, J.C. Duinker (Eds), Martinus Nijhoff/Dr. W.Junk Publishers. The Hague

Cherry R.D., Higo J.J.W., Fowler S.W (1978) *Nature* **274** 246

Degens E.T., Mopper K. (1976) ch. 31, in *Chemical Oceanography* J.P. Riley, R. Chester (Eds) Vol.6 Academic Press, London

Jones B.F., Bowser C.J. (1978), *Lakes: Chemistry, Geology and Physics.*ch.7A. Lerman (Ed), Springer-Verlag, Berlin

Kemp A.L.W., Johnston L.M. (1979) *J. Great Lake Res.* **5** 1

Lovelock J.E. (1979) *Gaia*, Oxford University Press, Oxford

Morel F.M.M., Hudson R.J.M. (1984) p.251-281 in *Chemical Processes in Lakes,* W. Stumm (Ed), J. Wiley, N.Y.

Nieboer E., Richardson D.H.S. (1980) *Environ. Pollut.* **B1**, 3

Nordstrom D.K., Ball J.W. (1984), p.149, In *Complexation of trace metals in natural waters.* C.J.M. Kramer, J.C. Duinker (Eds), Martinus Nijhoff/Dr. W. Junk Publishers, The Hague

Perret D., De Vitre R.R., Leppard G.G., Buffle J. (1990) in *Ecological Structure and Function in Large Lakes.* M.M. Tilzer, C. Serruya (Eds), Lewis Pub, Mi

Phillips C.S.G., Williams R.J.P. (1965) *Inorganic Chemistry (2 vols.)*, Clarendon Press, Oxford
Schaule B.K., Patterson C.C. (1981) *Earth Planet. Sci. Lett.* **54** 97
Sholkovitz E.R., Boyle E.A., Price N.B. (1978) *Earth Planet. Sci.Lett.* **40**, 130
Sigg L. (1984) P.283- 310 in *Chemical Processes in Lakes,* W. Stumm (Ed) J. Wiley N.Y.
Stumm W. (1983) p.1-8 in *Heavy Metals in the Environment*, Vol.1, Proceedings of an intern. conference held in Heidelberg, Sept. 1983
Stumm W (Ed) (1987) *Aquatic Surface Chemistry*. J. Wiley, N.Y.
Stumm W., Morgan J.J. (1981) *Aquatic Chemistry* J. Wiley, N.Y.
Thurman E.M. (1985) *Organic Geochemistry of Natural Waters*, Martinus Nijhoff/ Dr. W. Junk Pub., Dordrecht
Turner D., Whitfield M., Dickson A.G. (1981) *Geochim. Cosmochim. Acta* **45**, 855
Whitfield M, Turner D.R. (1979) *Nature* **278** 132
Whitfield M. (1981) *Interdiscipl. Sci. Rev.* **6** 12
Williams R.J.P. (1967) *Endeavour* **26** 96
Zumstein J, Buffle J (1989) *Water Research* **23** 229

EXCHANGE OF HEAVY METALS BETWEEN SEDIMENT COMPONENTS AND WATER

Wolfgang Calmano, Wolfgang Ahlf, and Ulrich Förstner
Technical University of Hamburg-Harburg, Eissendorferstr. 40, D-2100 Hamburg 90, F.R. Germany

I. INTRODUCTION

Sediments are both carriers and potential sources of contaminants in aquatic systems, and these materials may also affect groundwater quality and agricultural products when disposed on land. Such problems have initially been recognized for inorganic chemicals in the early and middle sixties from the studies on artificial radionuclides in the Columbia and Clinch Rivers by Sayre et al. (1963) and on heavy metals in the Rhine River system by De Groot (1966). In the early seventies, following the catastrophic events of cadmium- and mercury-poisoning in Japan, sediment-associated metal contaminants received public attention, for example, with severe effects on aquatic ecosystems in the Wabigoon River, Laurentian Great Lakes, Swedish Lakes, and in many past and present mining areas all over the world (Förstner & Wittmann, 1979).

During the last decade the major objectives of research on metal-polluted waters have changed from the initial surveys of sources and pathways to more detailed investigations of the mechanisms controlling the mobility and bioavailability of different metal species. The general experience that the environmental behavior and toxicity of an element can only be understood in terms of its actual molecular form led to the introduction of the term "speciation", which is used in a vague manner both for the operational procedure for determining typical metal species in environmental samples and for describing the distribution and transformation of such species in various media (Leppard, 1983; Bernhard et al., 1986; Landner, 1987; Patterson & Passino, 1987; Batley, 1989, Allen et al., 1989).

NATO ASI Series, Vol. G 23
Metal Speciation in the Environment
Edited by J. A. C. Broekaert, Ş. Güçer, and F. Adams

Problems of "speciation" are particularly complex in heterogenous systems, such as in soils, aerosol particles and sediments; thermodynamic models may give suggestions as to the possible species to expect, but due to the important role of kinetically controlled processes in biogeochemistry, the actual speciation is often different from what can be expected (Andreae et al., 1984).

Conceptually, a sediment can be considered as a heterogeneous mixture of dissimilar particles. Solid phases interacting with dissolved constituents in natural waters consist of a variety of components including clay minerals, carbonates, quartz, feldspar and organic solids. The "matrix vehicle" or residual fraction (Jenne, 1977) is associated with more labile and thermodynamically unstable components such as carbonates, amorphous aluminosilicates, and organic matter (Martin et al., 1987). These fractions are usually coated with Fe- and Mn-oxides and living or non-living organic material. A primary medium for sorption of inorganic components by sediments are metastable iron and manganese oxides and oxyhydrates, which have a high degree of isomorphic substitution (Jenne, 1977).

Organic surfaces for metal sorption could form in three possible ways (Hart, 1982):

(i) from organisms such as bacteria and algae;
(ii) by the breakdown of plant and animal material and by the aggregation of lower-molecular weight organics; and
(iii) by organic matter of lower-molecular weight sorbed onto clay or metal oxide substrates (Davis & Gloor, 1981).

Although the difference between these three surface types is not well understood with respect to metal uptake, there is a general agreement that at least one major binding mechanism involves salicylic entities. Other strong binding entities (such as peptides) may also be present in some systems. At least part of the organic matter adsorbed onto the particulate matter in natural waters has carboxylic and phenolic functional groups available for binding with trace metals. The trace metal adsorption capacity of organic matter is generally between that for metal oxides and clays (Tipping, 1981).

Mineral phases as substrates of pollutants originate from two major sources: Endogenic fractions of particulate matter include minerals that result from processes occurring within the water column (Jones & Bowser, 1978). Enrichment of minerals generated by endogenic processes may be influenced by settling of particulates, filtering organisms, and flocculation. Endogenic processes exhibit a distinct temporal character, often as a result of the variation of the organic productivity. In lakes, the total particulate concentration of trace metals is generally lowest in the hypolimnion due to the decomposition of organic matter. Consequently, net biogenic flux, for example, of metals depends on the lake's capacity to produce organic particulate matter and to decompose it before it is buried definitely in the sediment (Salomons & Baccini, 1986).

Authigenic (or diagenetic) fractions include minerals that result from processes within deposited sediments. Decomposition of organic matter, which is mediated by microorganisms, generally follows a finite succession in sediments depending upon the nature of the oxidizing agent (see Berner, 1981); the successive events are oxygen consumption (respiration), nitrate reduction, sulfate reduction, and methane formation.

The composition of interstitial waters in sediments is perhaps the most sensitive indicator of the types and the extent of reactions that take place between pollutant-loaded sediment particles and the aqueous phase that contacts them. The large surface area of fine-grained sediment in relation to the small volume of its trapped interstitial water ensures that minor reactions with the solid phases will be shown by major changes in the composition of the aqueous phase.

II. MOBILITY, ACID POTENTIAL AND BUFFER CAPACITY

Sediments are an important storage compartment for the metals released into surface waters. Furthermore, because of their ability to accumulate metals, sediments can reflect water quality and record the effects of anthropogenic emissions. Regarding the potential release of contaminants from sediments, changes of pH and redox conditions are of prime importance. In practice, therefore, characterization of sediment substrates with respect to their buffer capacity is a first step for the prognosis of middle- and long-term processes of mobilization, in particular, of toxic chemicals in a certain milieu.

By dredging activities anoxic sediments are suspended, removed from the river bottom, deposited on land or redeposited into the oxic water body. Evaluation of the pH- and redox-changes resulting from the oxidation of anoxic sediment constituents can be performed in the laboratory by ventilation of sediment suspensions with air or oxygen. Another approach for evaluating pH-effects is titration with acids. For quantifying pH-properties and for better comparison of sediment samples it is proposed to use the term "pH_{diff}", which is characterized by the difference of pH-values of 10-percent sludge suspensions in distilled water (pH_o) and in 0.1 N acid (pH_x) after 1 h shaking time (Calmano et al. 1986):

$$pH_{diff} = pH_o - pH_x$$

Three categories of pH_{diff}-values can be established ranging from $pH_{diff} < 2$ (strongly buffered), pH_{diff} 2-4 (intermediate) to $pH_{diff} > 4$ (poorly buffered).

Particularly a short time mobilization of heavy metals takes place, if the buffer capacity of a sediment is low and the acid potential is high. Under "acid potential" we understand the release of hydrogen ions as it can be observed at the oxidation of anoxic sediments e.g. by the following reactions:

$$NH_4^+ + 3/2\ O_2 \longrightarrow NO_2^- + 2\ H^+ + H_2O$$

$$2\ FeS + 7/2\ O_2 + H_2O \longrightarrow 2\ Fe^{2+} + 2\ SO_4^{2-} + 2\ H^+$$

$$Fe^{2+} + 1/4\ O_2 + H^+ \longrightarrow Fe^{3+} + 1/2\ H_2O$$

$$Fe^{3+} + 3\ H_2O \longrightarrow Fe(OH)_3 + 3\ H^+$$

For a classification of sludges regarding their acid potential which can be produced by oxidation of sulfidic components one can preferentially use the data of calcium and sulfur from the sequential extraction scheme as proposed, for example, by Tessier et al. (1979). In anoxic, sulfide-containing sediments the two elements are selectively released during anaerobic experimental procedures (argon or nitrogen atmosphere in glove box) by the Na-acetate step (Ca from carbonates) and peroxide step (S from oxidizable sulfides, mainly iron sulfide).

Reaction of oxygen with iron sulfide will produce H^+-ions; by reaction with carbonate H^+-ions are buffered. For an initial estimation, one may compare total calcium and sulfur concentrations in the sediment sample to get an index of acidity:

$$I_{Ac} = \frac{[S]}{[Ca]}$$

Experimental approaches for prognosis of acid producing potential of sulfidic mining residues have been summarized by Ferguson and Erickson (1988). A test described by Sobek et al. (1978) involves analysis of total or pyritic sulfur. Neutralization potential is obtained by adding a known amount of HCl and titrating with standardized NaOH to pH 7. Potential acidity is subtracted from neutralization potenital. A value below 5 t $CaCO_3$/1000 t of rock indicates a potential acid producer. Bruynesteyn & Hackl (1984) calculated acid-producing potential from total sulfur analysis. Acid-consuming ability is obtained by titration with standardized sulfuric acid to pH 3.5

(Bruynestein & Duncan,1979). Acid-producing potential is subtracted from acid-consuming ability; a negative value indicates a potential acid producer.

Results from titration experiments using 1 M nitric acid on sediment suspensions of 100 g/l are presented in Figure 1.

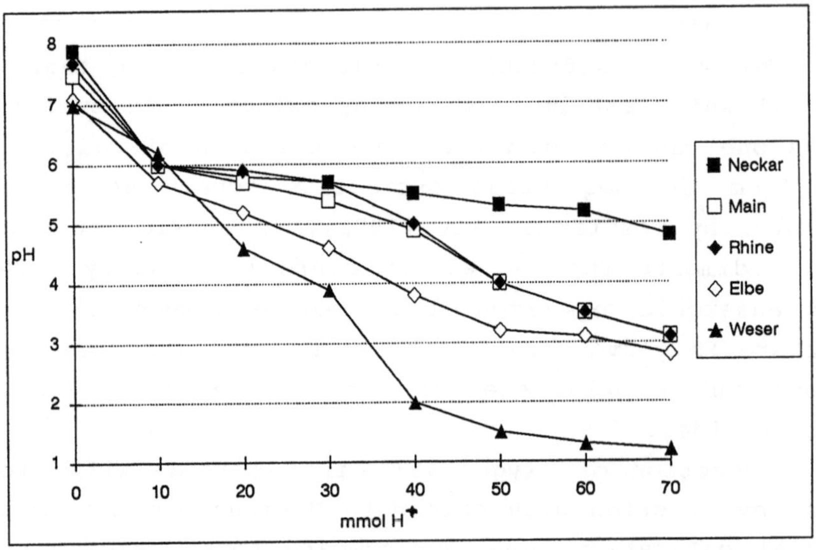

Figure 1. Acid titration curves for 5 German river sediments

The titration curve of the Neckar River sediment exhibits a large plateau in the pH-range of 5 and 6, probable due to the fraction of carbonate which is consumed by the addition of acid. In contrast to this observation the titration curves of both Elbe River and Weser River sediments are continously decreasing because of their small contents of carbonate. The sediments from the Rhine and Main River show small plateaux in the pH-region 5.5-6 which indicate a medium content of buffering substances.

An originally sulfide-rich material from an inland harbor basin of Harburg which had only been stored for 1 year in a closed bottle reached a pH value of 4.3. This was probably due to the consumption of the small residual buffer capacity by oxidation of parts of the sulfide fraction. Similarly, a reduction of the pH has been found at upland disposal sites of dredged sediments from Hamburg harbor (Tent, 1982). Due to the small carbonate content which is consumed during several months or years and subsequent lowering of the pH, metals are easily transferred to agricultural crops, and permissible limits of cadmium have been surpassed in as much as 50% of the wheat crops grown on these materials (Herms & Tent, 1982).

High concentrations of metals have been measured in pore waters from sedimentation polders in the Hamburg harbour area, in the older, oxidized deposits (Maaß et al., 1985; Table 1). Similar effects occured in the aquatic system, particularly on tidal areas affected by periodical drying and wetting (Kersten 1989), and at other sites exhibiting strong resuspension activities. The situation in the Elbe River estuary is particularly critical since small buffering capacities of the sediments coincide with a relative long residence time of suspended particles (Tent, 1987).

Table 1. Composition of anaerobic and oxidized pore waters from sedimentation polders (Maaß et al., 1985)

	"Anaerobic pore water"	"Oxidized pore water"
Nitrate	< 3 mg/l	120 mg/l
Ammonium	125 mg/l	< 3 mg/l
Iron	79 mg/l	< 3 mg/l
Zinc	< 30 µg/l	> 5000 µg/l
Cadmium	< 0.1 µg/l	80 µg/l
Arsenic	150 µg/l	15 µg/l

III. METAL TRANSFER

For a better understanding of the behavior of heavy metals in sediments and pore waters and their potential transfer to biota more experimental speciation studies and identification of solid phases are urgently needed (Hart & Davies, 1977; Batley & Giles, 1980; Elderfield, 1981).

With respect to the modeling of metal partitioning between dissolved and particulate phases in a natural system such as estuarine sediments, the following reqirements have been listed by Luoma & Davis (1983):

- the determination of binding intensities and capacities for important sediment components,
- the determination of relative abundance of these components,
- the assessment of the effect of particle coatings and of multi-component aggregation on binding capacity of each substrate,
- the consideration of the effect of major competitors (Ca^{2+}, Mg^{2+}, Na^{+}, Cl^{-}),
- the evaluation of kinetics of metal redistribution among sediment components.

It seems that models are still restricted because of various reasons: (i) adsorption characteristics are related not only to the system conditions (i.e., solid types, concentrations, and adsorbing species) but also to changes in the net system surface properties resulting from particle/particle interactions such as coagulation; (ii) influences of organic ligands in the aqueous phase can as yet rarely be predicted; (iii) effects of competition between various sorption sites, and (iv) reaction kinetics of the individual constituents cannot be evaluated in a mixture of sedimentary components. These restrictions have been recently discussed in detail by Honeyman & Santschi (1988), who stated that even for aquatic environments with small concentrations of particles the non-deterministic and interactive effects described above generally influence the estimation of an apparent partitioning coefficient by 1 to 3 orders of magnitude in either direction. With respect to environments of moderate to high particle concentration such as in soils and sediments those authors concluded that theoretical approaches have failed, thus far, to provide a sound basis for the prediction of trace-element behaviour.

At present experimental studies on the solid/solution interactions in such complex systems seem to be strongly needed. One approach is the use of a multi-chamber device, which still permits phase interactions via solute transport of the elements (Calmano et al., 1988). In this way, exchange reactions and biological uptake can be studied for individual phases under the influence of e.g. pH, redox, ionic strength, or solid and solute concentration.

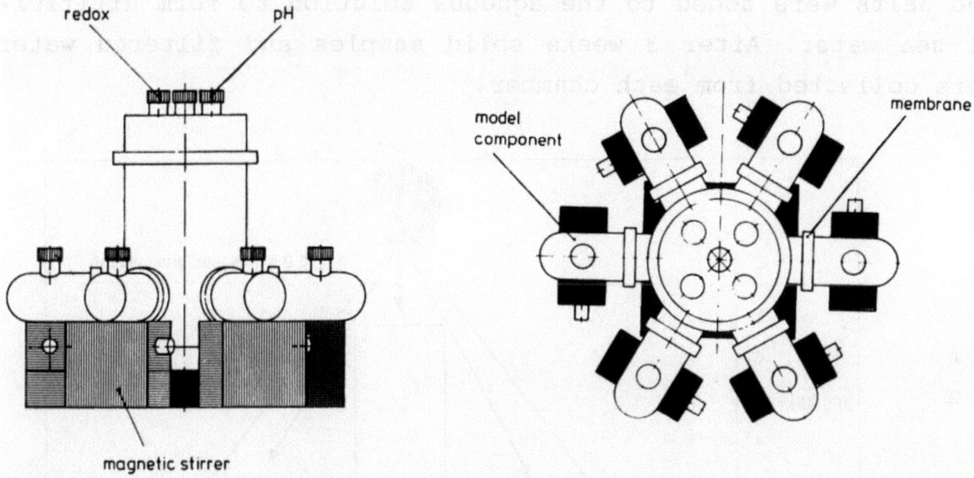

Figure 2. Schematic view of the multi-chamber device

The laboratory system used in these studies was developed from the experience on sediment/algae interactions with a modified two-chambered device (Ahlf et al., 1986). The system is made of a central chamber connected with 6 external chambers and separated by membranes of 0.45 um pore diameter (Figure 2).

The volume of the central chamber is 6 litres and each of the external chambers contains 250 ml. Either solution or suspension can be placed into the central chamber. The external chambers are filled with suspensions of model sediment components like clay, iron- and aluminium oxide, quartz, and algal cell walls. The solid components in each chamber are kept in suspension by magnetic stirring. Redox potential, pH values, and other parameters may be controlled and adjusted in each chamber.

In an experimental series on the effect of salinity on the metal transfer of anoxic dredged mud which has to be disposed into sea water, quantities of model components were chosen in analogy to an average sediment composition. In the central chamber a mud suspension from Hamburg harbour was placed and salts were added to the aqueous solution to form artificial sea water. After 3 weeks solid samples and filtered water were collected from each chamber.

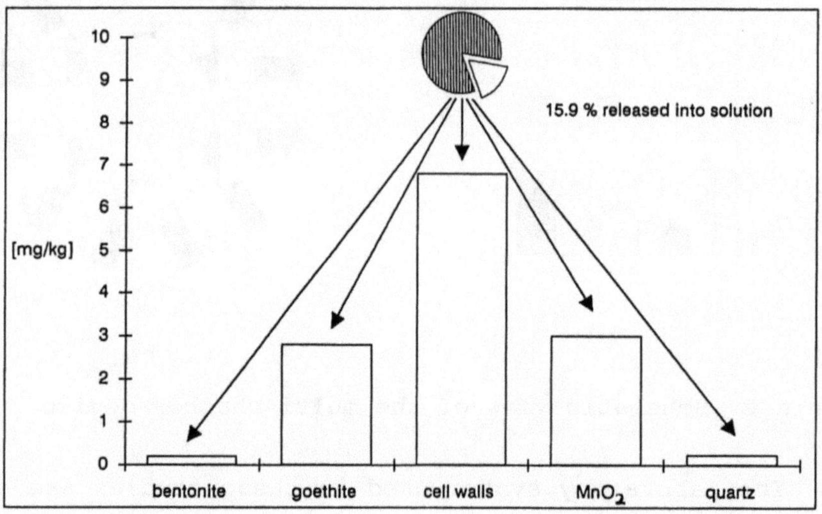

Figure 3. Mobilization of Cd from dredged mud and readsorption on different model sediment components in artificial seawater

The effect of salinity on metal remobilization from contaminated sediments was different for each of the elements studied. While approximately 16% of cadmium in the dredged mud from Hamburg harbour was released (see fig. 3), for metals such as copper the effect of salinity increase seems to have been less important in the transfer both among sediment substrates and to aquatic biota. This is, however, not true as can be demonstrated with a mass balance for the element copper in Figure 4.

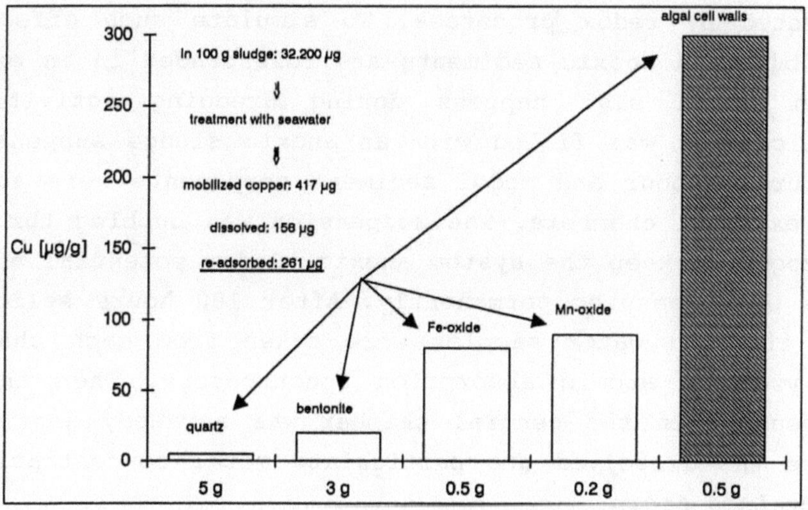

Figure 4. Mass balance for Cu: Distribution between model components and solution

It was found that only 1.3 % of the copper in the sludge sample was released when treated with seawater. Only one third remained in solution, which was equivalent to approximately 40 µg/l, and there was no significant difference to the conditions before salt addition. Two thirds of the released copper was readsorbed but with different affinities to the various model substrates. Copper concentrations in quartz and bentonite clay were not significantly different from their natural contents. Slight enrichment of copper occurred in the hydrous iron oxide and manganese oxide, whereas the cell walls - a minor component in the model sediment - accumulated nearly 300 mg/kg of copper.

The dominant role of organic substrates in the binding of metals such as Cd and Cu is of particular relevance to the transfer of these elements into biological systems. It can be expected that even at relatively small percentages of organic substrates these materials are primarily involved in metabolic processes and thus may constitute the major carriers by which metals are transferred within the food chain.

Metal mobility and transfer from solid matter is strongly affected by redox processes. To simulate such effects, when low buffered anoxic sediments are resuspended in an oxic river water which e.g. happens during dredging activities, the multichamber was filled with an anoxic sludge suspension from Hamburg harbour and model sediment components were added into the external chambers. The suspension was bubbled through with nitrogen to keep the system anoxic. Redox potential and pH values were measured permanently. After 100 hours solid samples and filtered water samples were taken from each chamber and analyzed by atomic absorption spectroscopy. Then the sludge suspension in the central chamber was aerated, and after 240 hours the dissolved and particulate metal concentrations were determined again.

Figure 5 shows the pH and redox development during the experiment. There exists a clear interdependence between these two parameters. During the anoxic run the pH value in the sludge suspension remains at about 8 and redox potential at about -200 mV. After aeration the redox potential increases from -230 mV to about +600 mV and the pH decreases from 8 to 4.2.

Metal concentrations in solution and in solids of the multichamber device are given in figure 6. The bars represent the concentrations of copper, zinc, and cadmium in the anoxic and the oxic system. Under anoxic conditions metal concentrations are generally low. In solution only copper is found in significant amounts, and the transfer to the single model

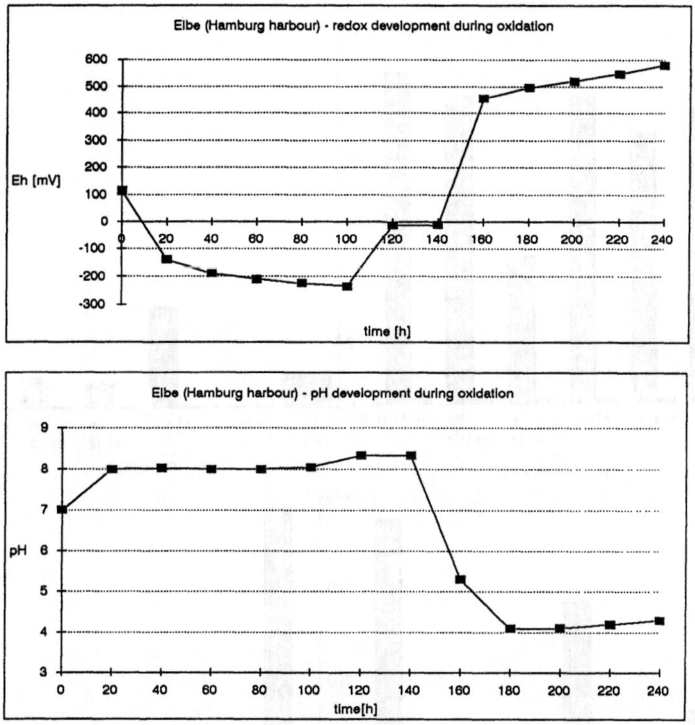

Figure 5. Redox and pH development during the experiment

sediment components is limited. After oxidation of the sediment suspension all dissolved metal concentrations are distinctly higher, and the same can be observed for the solids. The distribution of copper is dominated by the transfer and sorption on the the algal cell walls.

Cadmium and zinc are mainly sorbed on bentonite, but the highest contents for these metals are found in the control chamber, which is due to the precipitation of iron hydroxide after oxidation and sorption on these highly active surfaces. It can be concluded that at least part of the metal concentrations on the other substrates are affected as well by this mechanism. The binding intensities of the metals, in particular on bentonite and iron oxihydrates and additionally for copper on algal cell walls must be relatively strong because of the distinct sorption at these low pH values.

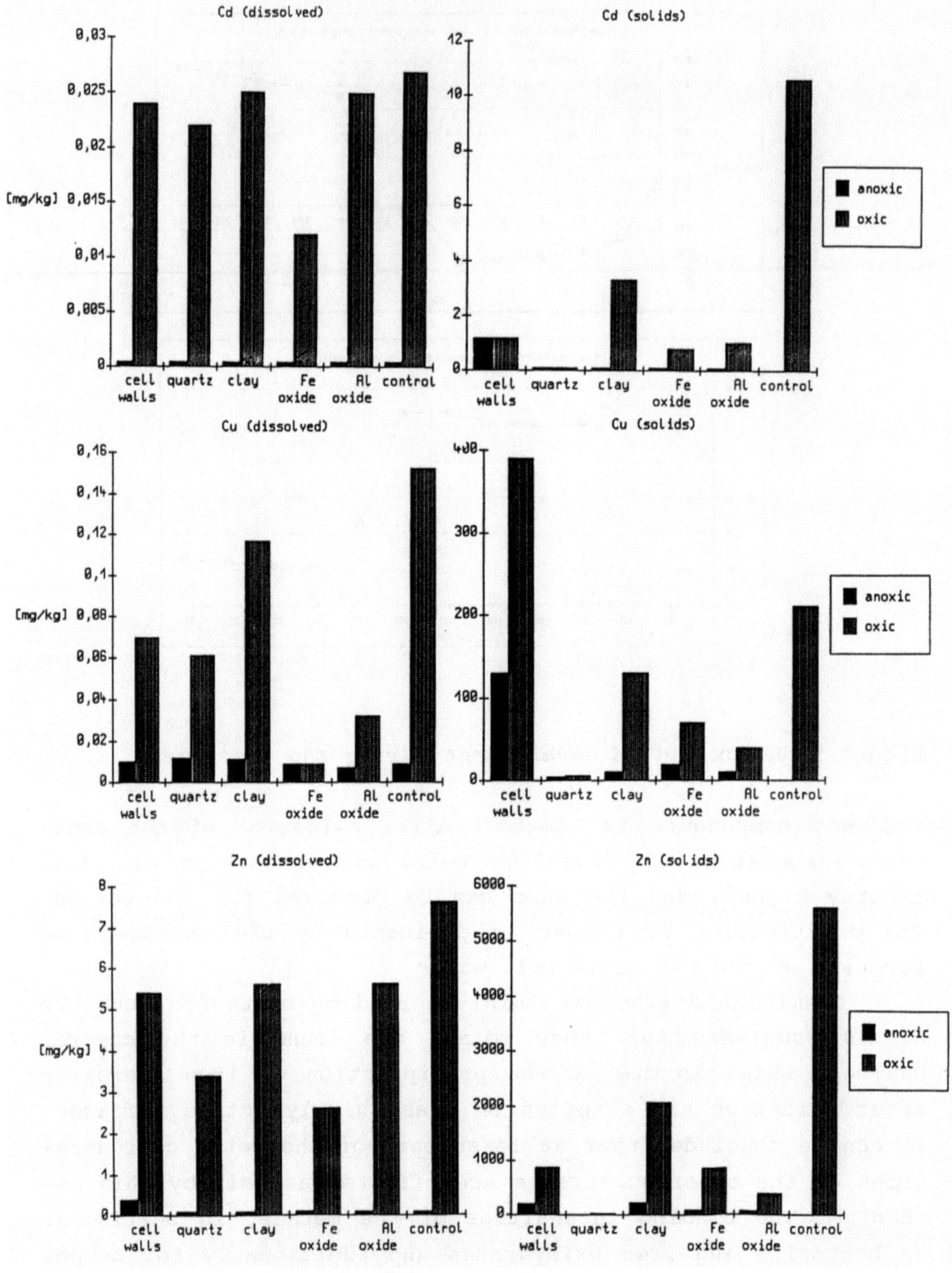

Figure 6. Dissolved and solid concentration of Cd, Cu, and Zn under anoxic and oxic conditions

To study the separate influence of redox potential on metal transfer and redistribution among sediment components a further experiment was carried out, in which pH values were kept constant at 7.5. Apart from this the experimental conditions were the same. Results for copper, cadmium, and zinc showed that metal concentrations in solution, under both anoxic and oxic conditions, were low and no significant transfer to the model sediment components could be observed, with the exception of copper and cadmium for algal cell walls, where an increased content was found under oxic conditions.

Chemical extraction procedures provide a convenient means to determine the major accumulative phases for metals in sediments and mechanisms of their diagenetic transformation. We applied a sequential method (Tessier et al., 1979) to 5 river sediments before and after oxidation. For anoxic sediments the maintenance of oxygene-free conditions during sampling and extractions is of critical importance.

Figure 7 shows the results by example of an Elbe River sediment. Cadmium, e.g., is under anoxic conditions nearly quantitatively found in the organic/sulfidic (H_2O_2-oxidizable) fraction, supporting the hypothesis that it is predominantly bound to sulfides in anoxic sediments. Exposure of the sample to oxic conditions caused the cadmium to be shifted in the extraction scheme to easily reducible and carbonatic fractions, respectively. From these fractions cadmium may be easily mobilized by acid conditions.

The element cadmium showed a similar behaviour in all studied sediment samples (see fig. 8). In anoxic sediments only very small portions were found in the first two steps of the sequential extraction procedure. After oxidation of the sediment samples the portions in both the ammonium- and sodiumacetate fractions were distinctly higher. The somewhat lower contents in these fractions of the Elbe and Weser sediments are related to their low buffer capacities which resulted in pH decrease of the suspension by oxidation and higher cadmium concentrations in solution.

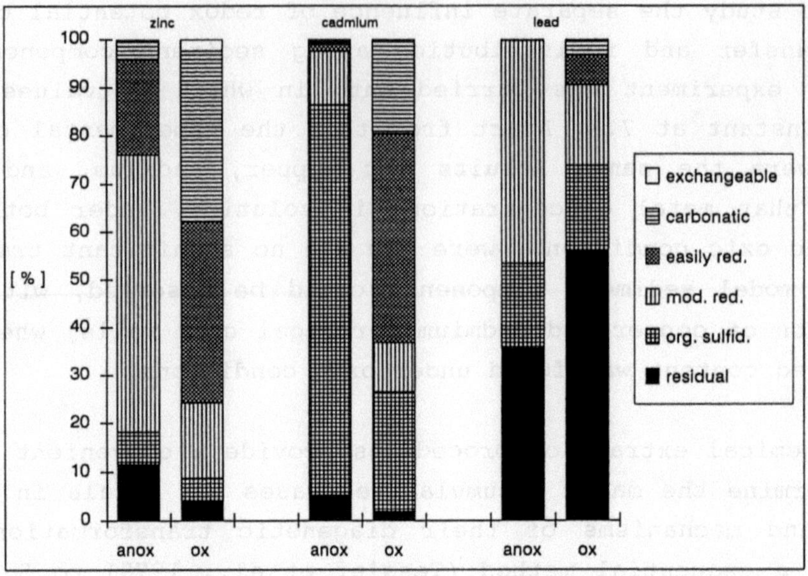

Figure 7. Sequential chemical extraction of an Elbe sediment under anoxic and oxic conditions

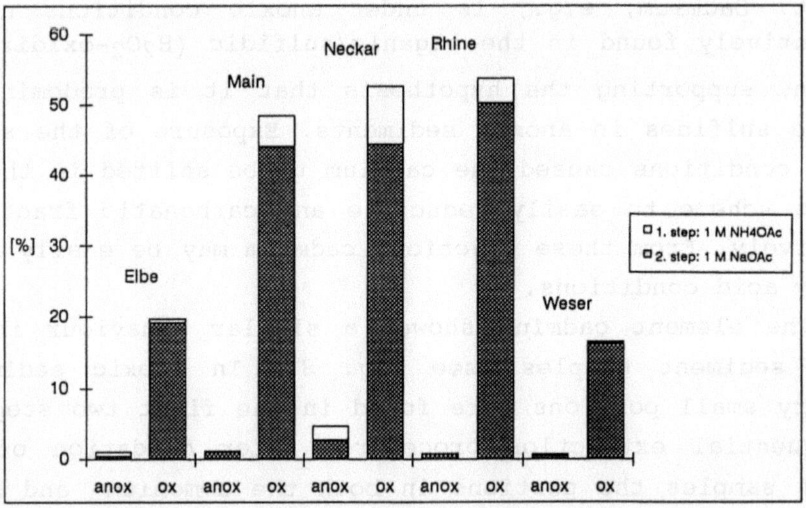

Figure 8. Extractable Cd-portions in the first two steps of the sequential extraction procedure under anoxic and oxic conditions

IV. CONCLUSIONS

The multichamber device allows the investigation of heavy metal sorption/desorption processes on typical sediment components with different surface binding intensities, which are in competition for the metals. It is possible to study the influence of single components on the metal transfer from contaminated material via aqueous phase to solids under changing environmental conditions. The investigations reported here related to the effects of salinity, redox, and pH changes in a low buffered sediment. By oxidation of anoxic sediments hydrogen ions are formed, as a result of e.g. sulfide and ammonium oxidation. The oxidation is catalyzed by microorganisms and the pH value decreases. In systems with higher buffer capacities these pH changes may not be measureable, but there is the possibility of the existence of low pH microenvironments. Metals, e.g. bound as sulfides under anoxic conditions, are mobilized, dissolved within a short time as ions or complexed by dissolved organic acids, and then transported and re-adsorbed on more reactive solid surface sites, like freshly precipitated iron hydroxides or biotic material.

These processes lead to changes of heavy metal binding forms on solids and to a translocation of distinct metals into more reactive sedimentary phases. The present state of knowledge of solid matter speciation of heavy metals is still somewhat unsatisfactory because the appropriate techniques are only operational tools with associated conceptual and practical problems (Kersten & Förstner, 1989). Most instrumental techniques available to date are too sophisticated to be routinely included in trace element speciation studies. The method of sequential chemical extraction is the least sophisticated but most convenient technique available for a speciation assessment. The usefulness of a differentiated approach to the interactive processes between water and operationally defined solid phases has been clearly evidenced. Possible environmental implications, e.g. during dredging operations, after land disposal of sediments, and other processes which lead to changes in the physico-chemical conditions of the interacting environmental milieu, can be qualitatively estimated. A combination of the different methods reported here should therefore help to explain mechanims which determine the interactions of metals between aquatic solids and solution.

V. REFERENCES

Ahlf W, Calmano W & Förstner U (1986) The effects of sediment-bound heavy metals on algae and importance of salinity. In: Sly PG (ed) Sediments and Water Interactions, pp. 319-324, Springer-Verlag New York

Allen HE et al. (eds) (1989) Metal Speciation and Transport in Groundwaters. Workshop organized by U.S.EPA and American Chemical Society, May 24-26, 1989, Jekyll Island/Georgia.

Andreae MO et al. (1984) Changing biogeochemical cycles - group report. In: Nriagu JO (ed) Changing Metal Cycles and Human Health. Dahlem-Konferenzen, Life Sciences Res. Rept. 28, pp. 359-374. Springer-Verlag Berlin

Batley GE , Giles MS (1980) A solvent displacement technique for the separation of sediment interstitial waters. In: Baker RA (ed) Contaminants and Sediments. Vol. 2, pp. 101-117. Ann Arbor Sci Publ.

Batley GE (1989) Trace Element Speciation - Analytical Methods and Problems. Boca Raton/Fla., CRC Press

Berner RA (1981) A new geochemical classification of sedimentary environments. J Sediment Petrol 51:359-365.

Bernhard M, Brinckman FE & Sadler PS (eds) (1986) The Importance of Chemical "Speciation" in Environmental Processes. Dahlem-Konferenzen, Life Sciences Research Report 33, 763 p. Springer-Verlag Berlin

Bruynesteyn A, Duncan DW (1979) Determination of acid production potential of waste materials. Met. Soc. AIME, Paper A-79-29, 10 p.

Bruynesteyn A, Hackl RP (1984) Evaluation of acid production potential of mining waste materials. Miner Environ 4:5-8

Calmano W, Förstner U, Kersten M & Krause D (1986) Behaviour of dredged mud after stabilization with different additives. In: Assink JW, Van Den Brink WJ (eds) Contaminated Soil. pp 737-746. Dordrecht/The Netherlands, Martinus Nijhoff Publ.

Calmano W, Ahlf W & Förstner U (1988) Study of metal sorption/desorption processes on competing sediment components with a multi-chamber device. Environ Geol Water Sci 11:77-84.

Davis JA, Gloor R (1981) Adsorption of dissolved organics in lake water environments by aluminium oxide: Effect of molecular weight. Environ Sci Technol 15:1223-1227.

De Groot AJ (1966) Mobility of trace metals in deltas. In: Meeting Int. Comm. Soil Sciences, Aberdeen, Jacks GV (ed) Trans. Comm. II & IV, pp. 267-297

Elderfield H (1981) Metal-organic associations in interstitial waters of Narragansett Bay sediments. Am J Sci 281:1184-1196

Ferguson KD, Erickson PM (1988) Pre-mine prediction of acid mine drainage. In: Salomons W, Förstner U (eds) Environmental Management of Solid Waste - Dredged Material and Mine Tailings pp. 24-43, Springer-Verlag Berlin

Förstner U, Wittmann G (1979) Metal Pollution in the Aquatic Environment. Springer-Verlag Berlin

Hart BT (1982) Uptake of trace metals by sediments and suspended particulates. In: Sly PG (ed) Sediment/Freshwater Interactions, pp. 299-313. Dr. W. Junk Publ. The Hague

Hart BT, Davies SHR (1977) A new dialysis-ion exchange technique for determining the forms of trace metals in water. Aust J Mar Freshwater Res 28:105-112

Herms U, Tent L (1982) Schwermetallgehalte im Hafenschlick sowie in landwirtschaftlich genutzten Hafenschlickspülfeldern im Raum Hamburg, Geol Jb F12:3-11

Honeyman BD, Santschi PH (1988) Metals in aquatic systems - predicting their scavenging residence times from laboratory data remains a challenge. Environ Sci Technol 22: 862-871.

Jenne EA (1977) Trace element sorption by sediments and soils - sites and processes. In: Chappell W, Petersen K (eds) Symposium on Molybdenum, Vol. 2, pp. 425-553, Marcel Dekker New York

Jones BF, Bowser CJ (1978) The mineralogy and related chemistry of lake sediments. In: Lerman A (ed) Lakes - Chemistry, Geology, Physics, pp. 179-235, Springer-Verlag New York

Kersten M (1989) Mechanismus und Bilanz der Schwermetallfreisetzung aus einem Süßwasserwatt der Elbe. Dissertation Technische Universität Hamburg-Harburg

Kersten M, Förstner U (1989) Speciation of trace elements in sediments. In: Batley GE (ed) Trace Element Speciation: Analytical Methods and Problems, CRC Press, Inc., pp. 245-317

Landner L (ed) (1987) Speciation of Metals in Water, Sediment and Soil Systems. Lecture Notes in Earth Sciences No. 11, 190 p. Springer-Verlag Berlin

Leppard GG (ed) (1983) Trace Element Speciation in Surface Waters and its Ecological Implications. Proc. NATO Advanced Research Workshop, Nervi/Italy, Nov. 2-4, 1981, 320 p. Plenum Press New York

Luoma SN, Davis JA (1983) Requirements for modeling trace metal partioning in oxidized estuarine sediments. Mar Chem 12:159-181

Maaß B, Miehlich G & Gröngröft A (1985) Untersuchungen zur Grundwassergefährdung durch Hafenschlick-Spülfelder. II. Inhaltsstoffe in Spülfeldsedimenten und Porenwässern. Mitt Dtsch Bodenkundl Ges 43/I:253-258

Martin JM, Nirel P & Thomas AJ (1987) Sequential extraction techniques: Promises and problems. Mar Chem 22:313-342

Patterson JW, Passino R (eds) (1987) Metals Speciation, Separation, and Recovery, 779 p., Chelsea/MI, Lewis Publ. (Second International Symposium May 14-19, 1989, Rome)

Salomons W, Baccini P (1986) Chemical species and metal transport in lakes. In: Bernhard M et al. (eds) The Importance of Chemical "Speciation" in Environmental Processes, pp. 193-216, Springer-Verlag Berlin

Sayre WW, Guy HP & Chamberlain AR (1963) Uptake and transport of radionuclides by stream sediments. U.S. Geol. Surv. Prof. Paper 433-A, 23 p.

Sobek AA, Schuller WA, Freeman JR & Smith RM (1978) Field and Laboratory Methods Applicable to Overburden and Mine Spoils. U.S. Environmental Protection Agency Report EPA-600/2-78-054. Washington D.C.
Tent L (1982) Auswirkungen der Schwermetallbelastung von Tidegewässern am Beispiel der Elbe. Wasserwirtschaft 72:60-62
Tent L (1987) Contaminated sediments in the Elbe estuary: Ecological and economic problems for the Port of Hamburg. In: Thomas RL et al. (eds) Ecological Effects of In-Situ Sediment Contaminants, Hydrobiologia 149:189-199
Tessier A, Campbell PGC & Bisson M (1979) Sequential extraction procedure for the speciation of particulate trace metals, Anal Chem 51:844-851.
Tipping E (1981) The adsorption of aquatic humic substances by iron hydroxides, Geochim Cosmochim Acta 45:191-199.

BIOGEOCHEMICAL CYCLING OF ATMOSPHERIC SELENIUM

Howard B Ross
Department of Meteorology
University of Stockholm
S-106 91 Stockholm SWEDEN

1. Introduction

Selenium was first discovered in 1818 by Jöns Jacob Berzelius in deposits from the lead chambers of a sulfuric acid plant at Gripsholm, Sweden. Berzelius realized that the new element was related to tellurium which had been discovered in the late 1700's. For the first 140 years after its discovery Se was thought to be deleterious to humans and animals. In the 1930's it was shown that certain plants could accumulate Se to levels which are toxic to livestock. The metabolic disorders "alkali disease" and "blind staggers" were shown to be caused by excess dietary intake of Se. In was only in the late 50's that Se was shown to be an essential nutrient to animals. The discovery of the "Factor-3 effect" (the anti-oxidant capacity of Se, vitamin D and coenzyme Q) by Schwarz and Foltz (1957) brought about this abrupt change. Further studies then indicated that Se interacted with vitamin D on a variety of deficiency disorders in livestock and poultry. In 1984 it was shown that Se was able to prevent cancer in humans (Salonen, 1984).

The biogeochemical cycling of Se and S are very similar. Indeed when Se data have been lacking, information on sulfur fluxes has been used to infer the respective Se fluxes (Ross, 1985; Mosher and Duce, 1987b). Perhaps the major difference in Se and S biogeochemistry is that Se rapidly changes oxidation state with respect to pH. Thus in environmental systems Se(0); Se(IV), selenite; and Se(VI), selenate; are observed in varying quantities. It has been postulated that the Se(IV)/Se(VI) ratio in precipitation can be used to infer the oxidation potential of an air-mass (Cutter and Church, 1986).

Budget calculations indicate that the atmospheric transport of Se is the major mechanism of transport in the biosphere (Lantzy and Mackenzie, 1979). Anthropogenic sources are very similar to those of S (fossil fuel combustion and sulfide ore roasting). The major natural sources are thought to be biogenic emissions of reduced organo-selenides from the oceans and continents (Ross, 1985; Mosher and Duce, 1987a,b). While the chemical speciation of biogenic emission is known ($(CH_3)_2Se$, $(CH_3)_2Se_2$ and $(CH_3)_2SeO_2$), there is little information on their magnitude, temporal and spatial variations.

NATO ASI Series, Vol. G 23
Metal Speciation in the Environment
Edited by J. A. C. Broekaert, Ş. Güçer, and F. Adams

Budget calculations also indicate that on a global scale anthropogenic and natural fluxes are of comparable magnitude. In Europe however, it is thought that anthropogenic emissions dominate the Se cycle (Ross, 1985). For example, the atmospheric input of Se has increased the Se content of Danish field crops (Gissel-Nielsen, 1980). However, the biological availability of Se in soils is decreasing in acidified areas, because selenite (and even sulfate) adsorbs onto ferrous oxides (like goethite) and hydrous oxides of aluminium at low pH (Balistrieri and Chao, 1987; Reuss and Johnson, 1986). It has also been postulated that the toxic levels of Hg in Swedish lakes and fish, due to anthropogenic emissions, can be reduced by adding Se to the water. Additions of Se would cause the formation of HgSe, a highly insoluble compound ($K_{sp} \approx 10^{-59}$; Fischer and Peters, 1970) and thus would remove the Hg from the food chain.

2. Atmospheric Burden - particulate phase

To analyze the importance of various sources to an element's aerosol burden an enrichment factor can be calculated where:

$$EF = (X/R)_{aero} / (X/R)_{ref} . \tag{1}$$

X is the element of interest and R is a reference element whose atmospheric burden is dominated by the source. Usually Al is used as the reference element for wind blown soil dust and Na for sea salt formation. Se enrichments with respect to crustal material and sea salt are on the order of 10^3 to 10^4 (Figure 1). This indicates that atmospheric Se has other sources than wind blown soil dust and sea salt formation.

Figure 2 shows schematically the concentration of Se, $exSO_4^=$ (i.e. that amount of $SO_4^=$ remaining after the sea salt component is removed) and Pb in atmospheric sub-micron particulate matter (Wiersma and Davidson, 1986). The data have been divided into four regions; rural sites from industrialized countries in North America and Europe, the north Atlantic, the remote Pacific and Antarctica.

For Pb, large variations in the atmospheric burden are observed; concentrations in Antarctica are about 1000 times lower than in rural regions. This is consistent with observations and processes which govern the cycling of Pb in the atmosphere (Nriagu, 1978):

1. Anthropogenic emissions (mostly in the northern hemisphere) are the dominant source of airborne Pb.
2. Nearly all Pb is found on sub-micron particles.
3. The residence time of atmospheric particles is much shorter than the characteristic time for interhemispheric mixing.

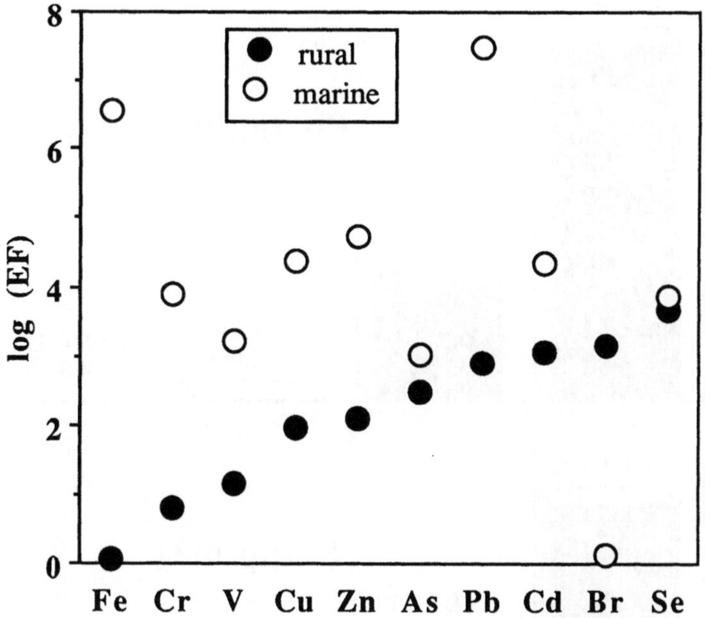

Figure 1. *Average enrichment factors (EF) for a variety of elements in atmospheric particulate matter from rural and marine regions. The EFs for rural data are relative to crustal matter (Al as a reference material) while EFs for marine data are relative to sea salt (Na as a reference material). The composition of crustal material is taken from Mason (1966). Average sea salt composition is from Broecker and Peng (1982). The particle concentrations are from the compilation of Wiersma and Davidson (1986).*

For S and Se variations in their atmospheric burden are much less pronounced. The difference between maximum and minimum concentrations are of the order of 15-65. This indicates that there are other processes than transport and removal of pollution particles that determine their concentrations in the atmosphere. For S extensive research has shown that the oceans release large amounts of particulate S (as $SO_4^=$) in the formation of sea spray and gas phase reduced organo-sulfur compounds from biogenic marine sources (Ryaboshapko, 1983). The predominate biogenic compound released is dimethyl sulfide: $(CH_3)_2S$ (DMS) followed by lesser amounts of methyl mercapatin (CH_3SH), hydrogen sulfide, carbonyl sulfide and carbon disulfide (Andreae, 1986).

The amount of Se in the marine atmosphere which arises from sea spray can be calculated by using the ratio of Se to Na in bulk sea water (Broecker and Peng, 1982), Na aerosols concentrations and assuming that all Na in the marine atmosphere is derived from the ocean:

$$Se_{sea\ spray} = Na_{aero}\left(\frac{1.7 \times 10^{-9} \text{ mole Se kg}^{-1} \text{ sea water}}{4.7 \times 10^{-1} \text{ mole Na kg}^{-1} \text{ sea water}}\right). \quad (2)$$

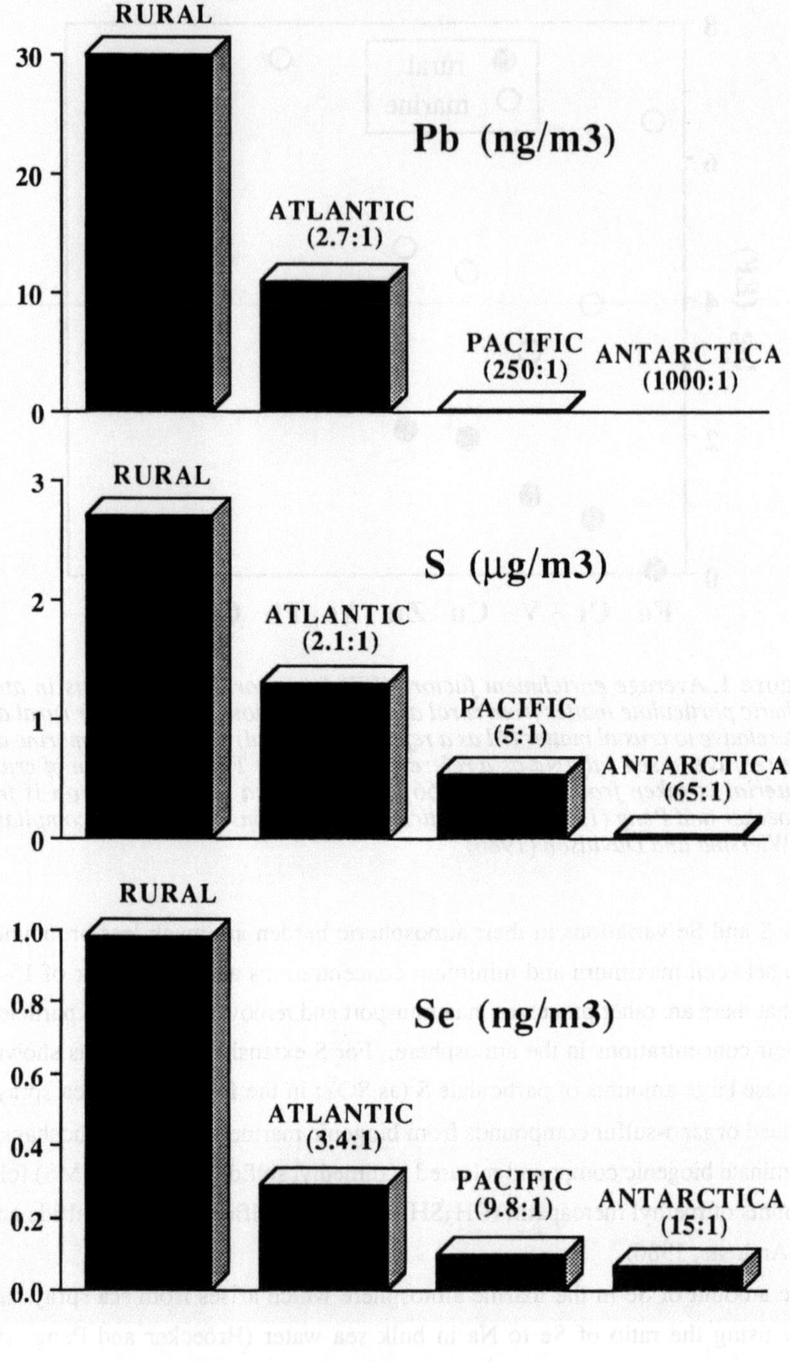

Figure 2. *Average particle concentrations of Pb, exSO$_4^=$ (i.e that SO$_4^=$ remaining after the sea salt component is removed) and Se at rural locations in N. America, Europe, N. Atlantic, the Pacific and Antarctica. The numbers in parenthesis are the ratio of the rural data with respect to the regions. The data are from the compilation of Wiersma and Davidson (1986).*

Typical concentrations of Na over the world's oceans are around 5 μg m^{-3} (Wiersma and Davidson, 1986). This would imply Se levels of about 2×10^{-14} g m^{-3}. Even if one allows for an enrichment of Se by two order of magnitude during the formation of sea salt (Buat-Ménard, 1983), levels would not approach measured concentrations (Figure 2). Hence to explain the enrichment of Se on the marine aerosol, Mosher and Duce (1983) have postulated that emissions of vapor phase Se from the oceans are an important source of atmospheric Se.

It is interesting to note that Se relative to S is depleted in the oceans compared to other geologic and biological reservoirs (Table 1). The depletion is probably due to differences in removal processes. In the ocean the predominant forms of Se are selenite and selenate (Measures et al. 1980). These ions, but not $SO_4^{=}$, are readily adsorbed on metal hydroxide precipitates of Fe and Mn, organic matter and ferric sulfides (Goldschmidt, 1954).

Further evidence for a biogenic source of Se from the oceans is given in Figure 3. Higher concentrations of particulate Se are found towards the equator, father away from anthropogenic sources and in regions of nutrient rich waters where there is increased primary production. In addition, Se levels are very well correlated to simultaneous measurements of $exSO_4^{=}$ (Mosher and Duce, 1989) a substance which is associated with biogenic emissions from the oceans (Charlson et al., 1987).

Table 1. *Concentration of S and Se in various biogeochemical reservoirs.*

	Se (ppm)	S (ppm)	Se/S
sea water[1]	1.7×10^{-6}	28000	6.1×10^{-11}
earth's crust[2]	0.05	260	1.9×10^{-4}
coal[3]	3	20000	1.5×10^{-4}
oil[3]	0.17	3400	5.0×10^{-5}
atmospheric particles	(ng m^3)[4]	(μg m^3)	
rural	1.2	27[5]	4.4×10^{-5}
remote continental	0.170	0.10[6]	1.7×10^{-3}
marine	0.290	0.61[6]	4.8×10^{-3}
plants[7]	0.5	260	1.9×10^{-3}
animal tissue[7]	2.2	7500	2.9×10^{-4}

[1]Broecker and Peng (1982)
[2]Mason (1966)
[3]Bertine and Goldberg (1971)
[4]Ross (1985)
[5]Ryaboshapko (1983)
[6]Warneck (1988)
[7]Borek (1980)

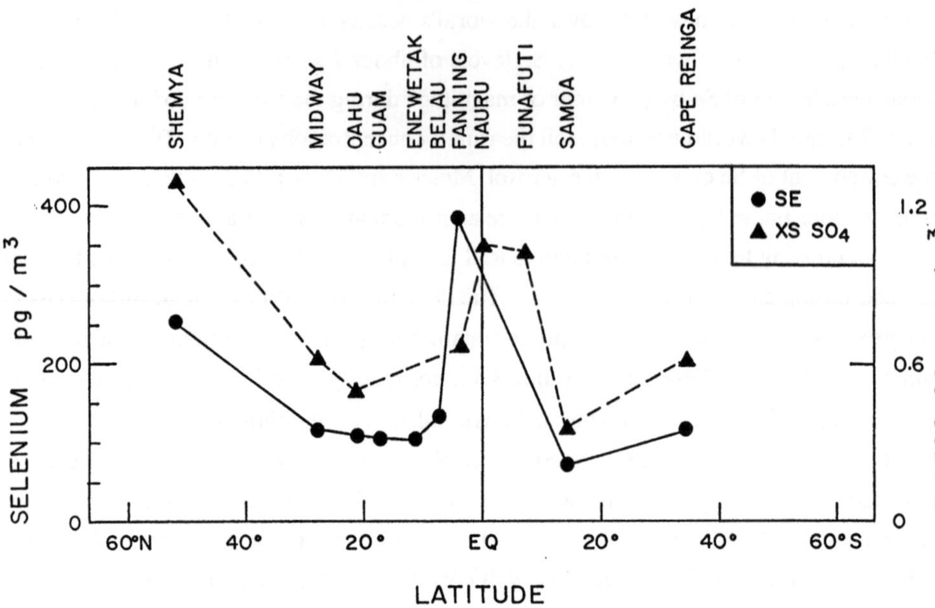

Figure 3. *Se and exSO$_4^{=}$ concentrations in atmospheric particles as a function of latitude in the north and south Pacific Ocean (source: Mosher and Duce, 1987a). Samples were collected on a weekly basis using high volume collectors (700 samples) from nine different islands in the Pacific.*

2.2 particle size distribution

How the Se mass is distributed with respect to particle size can be expected to yield information on the processes which emit Se into the atmosphere. The particle size distributions of Se mass for a variety of rural and marine locations are presented in Figure 4. In general, the Se concentration is highest on the smallest particles. The exception is the very remote marine atmosphere which will be discussed separately.

Numerous authors have postulated that the enrichment of Se on sub-micron pollution particles is due to volatilization during combustion and metal manufacturing (Bertine and Goldberg, 1971; Rahn, 1971; Natusch et al., 1974; Andren et al., 1975; Kaakinen et al., 1975; Klein et al., 1975). During high temperature industrial processes, Se and other elements with low boiling points (<1200°C) will be volatilized. When the emissions cool the elements will condense on existing particles and be preferentially enriched on sub-micron particles, since the surface area of particles is inversely related to particle size and the number of particles increases with decreasing radii.

Further evidence for this hypothesis is found from mass balance studies in coal fired power plants. Se was greatly enriched in the fly ash compared to the bottom ash and most of the Se was found to escape particle abatement devices (Natusch et al., 1974; Kaakinen et al., 1975; Klein et al., 1975;). The most detailed study was reported by Andren et al. (1975) on a coal fired power plant having at that time the latest in pollution abatement devices. The authors found that 68% of the Se occurred in the fly ash, negligible amounts in the bottom ash, and that 32% was in the vapor phase. After analyzing the precipitators efficiency it was found that 93% of the Se in the coal was released to the atmosphere in the gas phase.

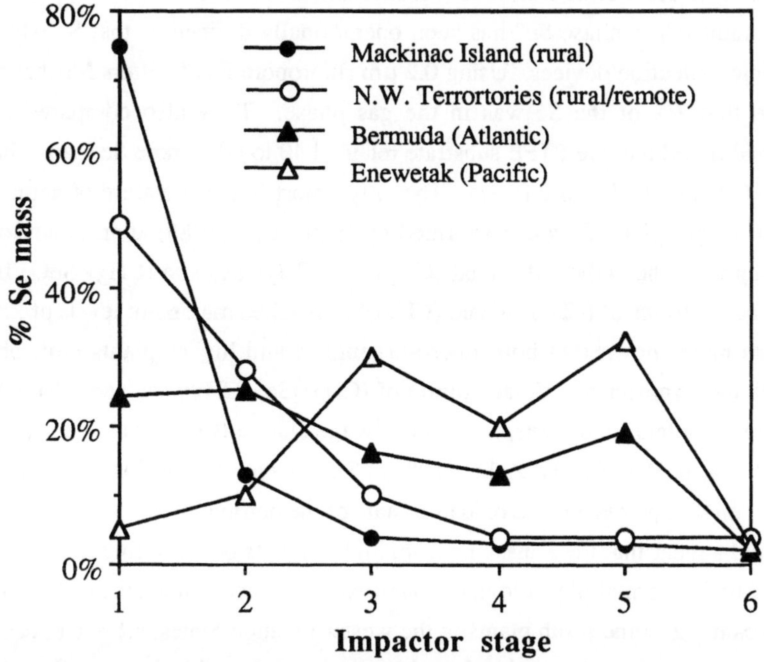

Figure 4. *Distribution of Se mass with particle size. The impactor stage 50% cut-off diameters for the Mackinac Island and N.W. Territories are 1: <1.0 μm, 2: 1.0 μm, 3: 2.0 μm, 4: 3.3 μm, 5: 5.5 μm and 6: 9.2 μm. The data is from Rahn (1971). The cut-off diameters for the marine data are: 1: <0.5 μm, 2: 0.5 μm, 3: 1.0 μm, 4: 1.5 μm, 5: 3.0 μm and 6: 7.2 μm. The data from Bermuda is from Duce et al. (1976) and the Enewetak data is from Duce et al. (1983).*

The particle size distributions from rural areas and the north Atlantic Ocean indicate that particulate Se has a gaseous precursor. From budget calculations it is thought that a large majority of the gaseous emissions derives from anthropogenic sources (Ross, 1985). In the remote marine atmosphere the particle size distribution is different (Figure 4). The question arises: how can this be when it is postulated that the oceans release vast amounts of gaseous Se? Mosher and Duce (1987a) have attempted to resolve this apparent contradiction by first noting that Se distributions in the remote atmosphere follow the distribution of sea salt particle surface area. Hence they argue that reaction of Se with sea salt particles is a sink for gas phase Se while acidic sub-micron particles (which are mostly H_2SO_4) in some way volatilize Se.

2.3 gas phase selenium

The partitioning of atmospheric Se between the gas and particulate phase is not well known. Estimates range from about 50% in the gas phase at an urban location (Pillay et al., 1971) to about 25% for marine locations (Mosher and Duce, 1987a). It is important to note that in most studies "gas phase Se" has been operationally defined as that Se which passes through particle collection devices. Using 0.2 μm fluoropore PTFE filters Mosher and Duce (1989) found that 7% of the Se was in the gas phase. They also compared collection efficiencies and found that the PTFE substrate retained 40 to 60% more Se than Whatman 41 cellulose and 0.4 μm Nucleopore filters. The only report in the literature of actual ambient measurements of gas phase Se was performed by Jiang et al. (1985) at an urban location in Antwerp, Belgium. The authors detected $(CH_3)_2Se$, $(CH_3)_2Se_2$ and $(CH_3)_2SeO_2$ in various amounts. Concentrations of $(CH_3)_2Se$ and $(CH_3)_2Se_2$ reached maximum levels of 2.5 ng m^{-3}.

Studies have shown that both micro-organisms and higher plants emit organic Se compounds to the atmosphere. Volatilization of $(CH_3)_2Se$, $(CH_3)_2Se_2$ and $(CH_3)_2SeO_2$ has been observed from lake sediments, soils and sewage sludge (Chau et al., 1976; Doran and Alexander, 1977a, b; Reamer and Zoller, 1980). The extent of emissions is highly variable and dependent on the species and the oxidation state of the nutrient Se.

It would appear that the emissions from higher plants occur during normal biological processes and under normal physiological conditions. Species which accumulate Se in their tissue, as for example some scrub plants in the western United States, release $(CH_3)_2Se_2$. Se levels in these plants can reach ppm levels and have been responsible for acute Se poisoning of livestock (Challenger, 1935). Nonaccumulating species have been shown to release $(CH_3)_2Se$ (Lewis, 1976). The mechanism of release is related to that of S biochemistry in the plant. Losses of Se from the leaves in a variety of crucerferous plants have been studied and found to be 0.05 to 0.30% $hour^{-1}$. Volatilization rates were dependent on the species and leaf temperature. Interestingly, while the uptake of Se(IV) and Se(VI) was the same, emissions of Se(IV) were 10 times higher than Se(VI) emissions (Lewis et al., 1974).

In conclusion, it is known that plants and micro-organisms in terrestrial and marine ecosystems can assimilate inorganic Se and then release reduced organo-selenide compounds to the atmosphere. The magnitude of these emissions are unknown. The partitioning of atmospheric Se is also not known because of uncertainties in the experimental methods for determining gas phase Se.

3. Sources of Atmospheric Selenium

The most detailed budgets for the sources of atmospheric Se are those published by Ross (1985) for the region 30°N to 90°N and Mosher and Duce (1987b) for the globe. The latter is reproduced in Table 2. In general it appears that anthropogenic and natural emissions of Se are more or less the same. By far the largest man-made source of Se is coal combustion, accounting for about half of the total anthropogenic sources. Since nearly all of the anthropogenic emissions are situated in the northern hemisphere and budget calculations estimate that the residence time of Se is around 1 week, the greatest alterations of the atmospheric Se budget will be on a regional scale in and around industrial areas of N. America and Europe.

Table 2. *Global estimates of Se emissions to the atmosphere (adapted from Mosher and Duce, 1987b).*

ANTHROPOGENIC SOURCES		(10^9 g Se a^{-1})
Combustion		
coal		3.0
oil		0.5
fuel wood, biomass, incineration		0.6
Nonferrous Metal Production		1.7
Manufacturing		0.3
	TOTAL	**6**
NATURAL SOURCES		
Crustal Weathering		0.003 - 0.025
Sea Salt		0.04 - 0.4
Volcanoes		0.4 - 1.2
Gas Phase Emissions		
marine biosphere		5.0 - 8.0
continental biosphere		0.7 - 3.6
	TOTAL	**6 - 13**

With regard to natural emissions it is postulated that vapor phase emissions from the biosphere are the most important. Evidence for this hypothesis has been previously presented:

1. EFs indicate that wind blown soil dust and sea salt formation inject little Se into the atmosphere.
2. The concentration of particulate Se in the atmosphere is relatively evenly distributed throughout the world.

3. Levels of particulate Se in remote parts of the Pacific are highest in the regions of high primary production.
4. Both plants and micro-organisms, under normal biological conditions, can release gas phase organo-selenides into the atmosphere.

The anthropogenic sources of S and Se are similar but there are differences. As a rule of thumb S fluxes are 10^3 to 10^4 times larger than Se fluxes. The major differences in their cycling appear to be the natural fluxes. Sea salt is a minor source of Se to the atmosphere since Se is depleted relative to S in the oceans. It is also postulated that terrestrial plants are an important source of organic Se compounds to the atmosphere while in most S budgets they are only considered a minor source.

4. Chemical Speciation of atmospheric selenium

While the chemical speciation of biogenic and anthropogenic emissions of Se has been studied, little is known about the subsequent physical and chemical transformations of the element in the ambient atmosphere. This information is essential if one wishes to understand how atmospheric Se is transported and cycled. Atmospheric Se will undergo chemical reactions in the gas, liquid and solid phases. However it is important to note that many chemical reactions in the atmosphere are multiphase processes. To understand the behaviour of atmospheric particles and the chemical reactions which can occur, they must be considered as a multiphase system comprising an electrolytic solution with dissolved gases and insoluble materials. Hence in attempting to consider the fate of Se one will need to consider both chemical and physical processes. As a first attempt to model the chemistry of atmospheric Se information gained in the laboratory can be used.

Like other group VIA elements Se has allotropic forms:

1. *Grey Se*: having a hexagonal crystalline structure.
2. *Red Se*: $(Se_8)_n$ is monoclinic and is formed by the slow cooling of molten Se. The structure is similar to that of S.
3. *Amorphous Se*: Se_x^o (x=1 to 8) exists in three forms: black, red or colloidal Se.

All of the allotropic forms of Se are insoluble in water, though at higher temperatures (>50°C) Se_x reacts with water to form H_2SeO_3 and H_2. With respect to chemical speciation and cycling through the atmosphere Se_x is probably the most important allotrope.

Hydrogen selenide (H_2Se) is a colorless and highly toxic gas. It's chemical properties are similar to those of H_2S though it is somewhat more soluble in water and is a stronger acid. H_2S is known to be formed by biological systems in anaerobic conditions as found in marshes and swamps (Andreae, 1986). Considering the similarity of S and Se biogeochemistry the possible release of H_2Se by a similar mechanism should not be excluded.

The oxides of Se; SeO_2 and SeO_3, and their corresponding oxo-acids; H_2SeO_3 and H_2SeO_4, have similar chemical properties to their S analogs but are less stable. SeO_2 and SeO_3 are very hydroscopic, the latter reacts vigorously in moist air to form H_2SeO_4. SeO_2 can be formed by the combustion of Se, but is readily reduced back to Se°. This is exemplified by the widespread use of SeO_2 as an oxidizing agent in organic chemistry. Compared to oxides of other metalloids, SeO_2 has a relatively high vapor pressure and in the gas phase its reactivity is greatly increased. H_2SeO_4 is as strong an acid as H_2SO_4 but is a better oxidizing agent. H_2SeO_3 is a weak acid and can easily be reduced to Se° in an acidic medium.

The most salient feature of aqueous Se chemistry is the ease by which Se changes oxidation state (Figure 5). The redox potential of Se is highly dependent on the pH of the solution. Also Se compounds undergo redox reactions more readily than their S analogs. The thermodynamic sink of Se in highly basic solutions is selenate (+6 oxidation state), in mildly basic or acid solutions selenite (+4) and in highly acidic solutions Se(0) or selenide (-2).

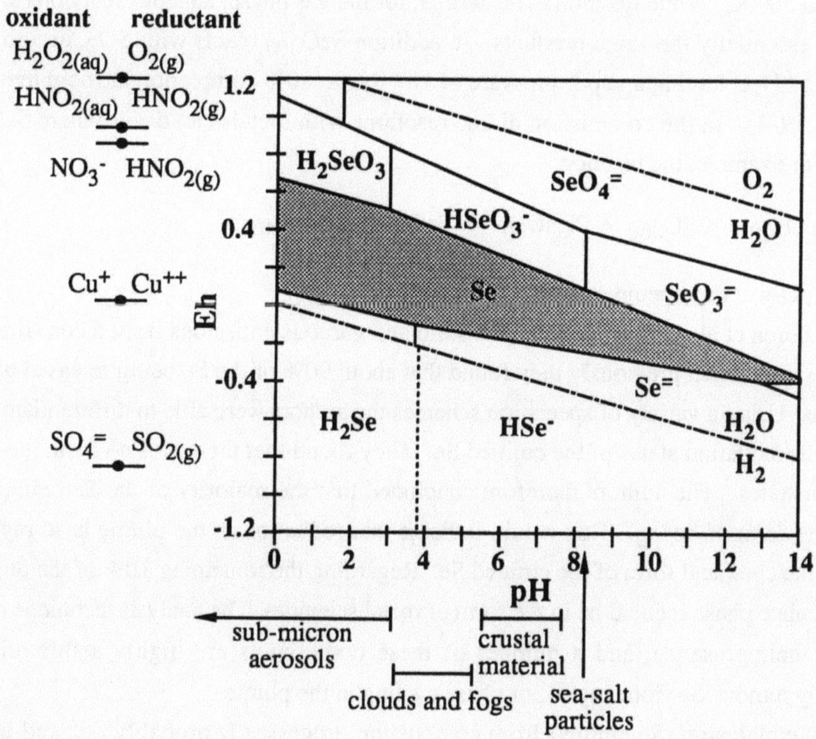

Figure 5. *Eh vs. pH diagram for Se. Under the X-axis the pH ranges for various atmospheric particulate matter are indicated. Along the Y-axis the oxidation potentials for some important atmospheric redox reactions are shown.*

With respect to metals, Se has a propensity to form highly insoluble precipitates. Metal selenides can form when the metal compounds are in the free ionic state or bound to ligands. The importance of these compounds is that they are so insoluble that they essentially remove Se from any further reactions in the atmosphere and prohibit the uptake of Se by plants and micro-organisms after it has been deposited.

During combustion processes one would expect Se to be released as SeO_2. However, there are a wide variety of reactions which can occur in the combustion plume, which may even determine the speciation of Se. While the reaction:

$$SeO_{2(g)} + 2SO_{2(g)} \rightarrow Se°_{(g)} + 2SO_{3(g)} \ , \tag{3}$$

proceeds slowly, the presence of water greatly increases the reaction rate:

$$SeO_{2(g)} + 2SO_{2(g)} + 2H_2O_{(g)} \rightarrow Se°_{(g)} + 2H_2SO_{4(g \text{ or } l)} \ . \tag{4}$$

This is exemplified by the ΔG_f° (T=295 K) for reactions (3) and (4): -2.03 and -26.17 kcal mole^{-1} (-45.73 for the formation of aqueous sulfuric acid). Theoretically reaction (4) can proceed up to 500 K. While reaction (4) is written for the gas phase, aqueous reactions also occur and give essentially the same products. In addition $SeO_{2(s)}$ reacts with SO_2, in 5 to 6 molar excess, and H_2O having a vapor pressure of 187 torr to 80% completion at room temperature (Gmelin, 1981). In the combustion plume reactions with metals can occur where SeO_2 is reduced. For example, the reaction:

$$SeO_{2(g)} + 2SO_{2(g)} + 2ZnO_{(s)} \rightarrow Se°_{(s)} + 2ZnSO_{4(s)} \ , \tag{5}$$

occurs rapidly at room temperature.

Andren et al. (1975) analyzed the solid and gaseous emissions from a coal-fired power plant. As mentioned previously they found that about 90% of the Se being released was in the gas phase. Using a variety of speciation schemes the authors were able to differentiate between the various oxidation states of the emitted Se. They found that there was no Se in the +4 or +6 oxidation states. The authors therefore concluded that the majority of the Se being released was in the form of $Se°_{(g)}$. This would indicate that reduction in the plume is so rapid that it controls the chemical form of the emitted Se. Regarding the remaining 10% of the Se found in the particulate phase it could be in the form of metal selenides. The analysis technique could not exclude their presence, and a number of these compounds are highly stable and would essentially remove Se from any vapor phase cycling in the plume.

In conclusion, Se emitted from combustion processes is probably Se° and in the gas phase. The exact partitioning between the various oxidation states and phases depends on a variety of factors including humidity, combustion temperature, acidity and even SO_2 concentrations.

The vapor pressure of Se° and SeO_2 can be estimated from classical thermodynamics using the Clausius-Clapeyron equation if the enthalpy ($\Delta H°$) and entropy ($\Delta S°$) of vaporization are known. Assuming that $\Delta H°$ and $\Delta S°$ are independent of temperature, then:

$$\log(P) = -\frac{A}{T} + B \; ; \quad \text{where } B = \frac{\Delta H°}{2.303R}, \quad A = \frac{\Delta S°}{2.303R}, \tag{6}$$

and R is the gas constant. Substituting the ideal gas law into equation (6) yields:

$$\log\rho = -\log T - \frac{A}{T} + C, \quad \text{where } C = B + \log M + 13.08, \tag{7}$$

and M the gram molecular weight of the substance. The vapor pressure of of SeO_2 and Se° in the temperature range -15° to 30°C is presented in Figure 6. While SeO_2 is much more volatile than Se°, both of these substances' gas phase concentrations are comparable to or even higher than particulate Se concentrations (Figure 2). Hence there will be a thermodynamic driving force to form gas phase Se and SeO_2. However it is important to note again the physical and chemical properties of SeO_2. SeO_2 is highly hygroscopic and hence any gas phase SeO_2 will be scavenged onto particles where it will be hydrolyzed to $HSeO_3^-$ and $SeO_3^{=}$. The exact chemical form would depend on the relative humidity and the acidity of the aerosol.

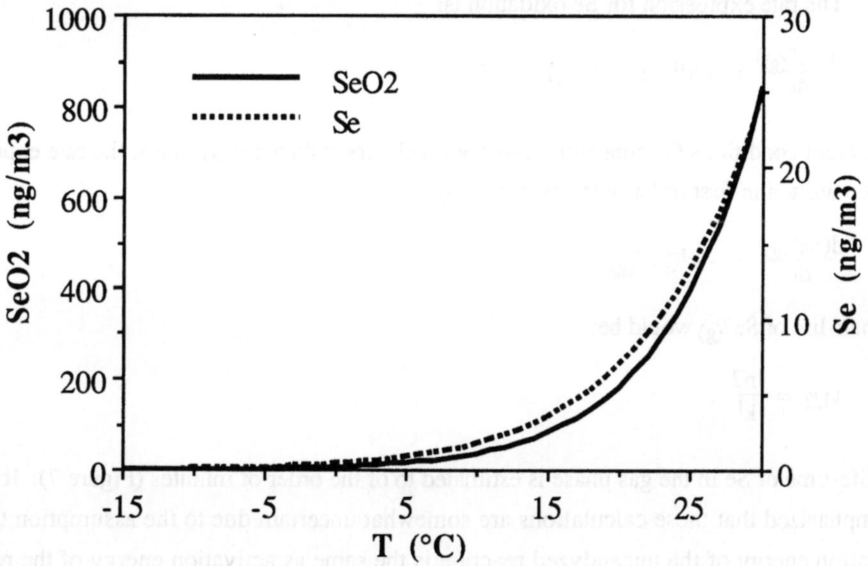

Figure 6. *Gas phase concentrations (ng m^{-3}) of Se° and SeO_2 from -15° to 30°C. The curves were calculated using the equation:*

$$\log\rho = -\log T - \frac{A}{T} + C$$

A = 6045 and C = 25.358 for SeO_2 (Pupp et al., 1974; 1975). For Se the values are A = 5061 and C = 20.582 (Gmelin, 1981). See text for the derivation of the equation.

In the ambient atmosphere Se° will react with O_2. The rate of reaction has not been evaluated at ambient temperatures or concentration. However, it has been assessed in the context of metallurgical processes (T = 500° to 600°C) where $Se°_{(g)}$ was in equal concentrations with O_2 (Dutton et al., 1971). At T = 550°C, the rate constant (k^{II}) was found to be 1.67×10^2 $l\ mole^{-1}\ s^{-1}$. Temperature dependence studies (T = 450° to 600°C) were performed when the reaction chamber was filled with spent industrial catalysts of unknown composition. The rate constant for this reaction (k_c^{II}) followed Arrhenius behaviour and fitted the following equation:

$$\ln (k_c^{II}) = -\frac{E_{ac}}{RT} + \ln(A_c) = -\frac{2.736 \times 10^3}{T} + 11.04 \ , \tag{8}$$

where E_{ac} is the activation energy of the reaction, T is in Kelvin, and A is the "Arrhenius coefficient". At T = 550°C, k_c^{II} is equal to $13k^{II}$. If one assumes that the reaction without industrial catalysts has the same activation energy, and that A is proportional to the rate constants k_c^{II} and k^{II}, the temperature dependence of k^{II} would be:

$$\ln (k^{II}) = -\frac{E_a}{RT} - \ln(\frac{A_c}{13}) = -\frac{2.736 \times 10^3}{T} + 8.47 \ . \tag{9}$$

Figure 6 presents the rate constant as a function of temperature (T = -15° to 30°C).

The rate expression for Se oxidation is:

$$\frac{d(Se°_{(g)})}{dt} = -k^{II}(Se°_{(g)})(O_2) \ . \tag{10}$$

At ambient conditions O_2 concentrations are much larger than $Se°_{(g)}$, hence the rate expression may be written in first order with respect to O_2:

$$\frac{d(Se°_{(g)})}{dt} = -k^{I}(Se°_{(g)}) \ . \tag{11}$$

The half-life of $Se°_{(g)}$ would be:

$$\tau_{1/2} = \frac{\ln 2}{k^I} \ . \tag{12}$$

The life-time of Se in the gas phase is estimated to of the order of minutes (Figure 7). It should be emphasized that these calculations are somewhat uncertain due to the assumption that the activation energy of the uncatalyzed reaction is the same as activation energy of the reaction with used industrial catalysts. Thus these calculations should be looked upon as the lower limit of Se gas phase lifetimes.

The fate of organic compounds in the atmosphere has not been studied. As a first approximation one can examine the general trends of organo-selenium chemistry in the lab-

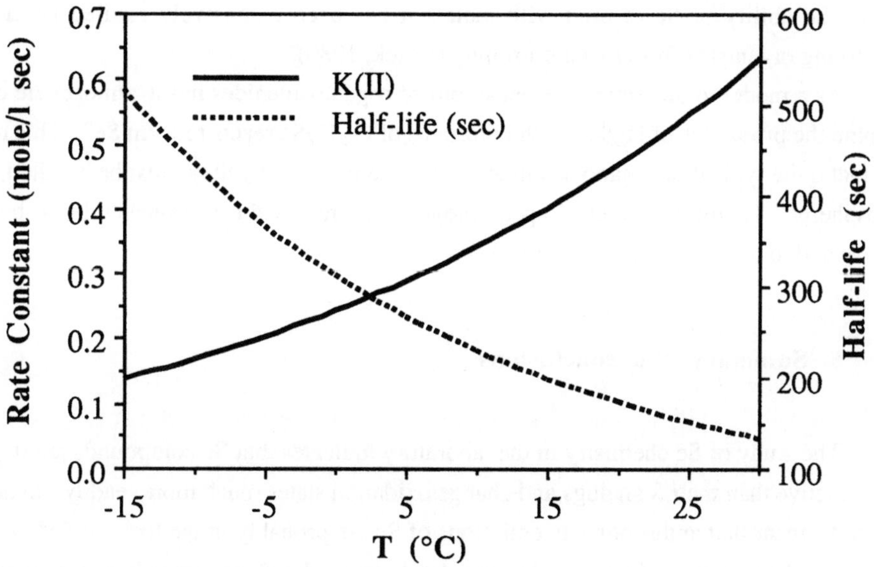

Figure 7. *Calculated second order rate constant of the reaction:*

$$Se^{\circ} + O_2 \xrightarrow{k^{II}} SeO_{2(g)} \ ,$$

where k^{II} is extrapolated from Arrhenius plots (Dutton, el al. 1971). Half-lifes of Se were determined by assuming that the reaction is first order with respect to O_2.

oratory and correlate this to biologically emitted species. As expected from the position of Se in the periodic system, C-Se and H-Se bonds are weaker then their S analogs.

In the presence of O_2, organo-selenides in the gas phase undergo the following type of reaction:

$$(CH_3)_2Se + \tfrac{1}{2}O_2 \rightarrow CH_3SeH + H_2CO \ , \qquad (13)$$

where one or more methyl groups are removed. The Se-Se bond can be cleaved by mild oxidizing agents:

$$(CH_3)_2Se_2 + ox^- \rightarrow CH_3SeH + RSe \ . \qquad (14)$$

Diorganyl selenones are prepared by oxidation of dialkyl-selenides:

$$(CH_3)_2Se + 2H_2O_2 \rightarrow (CH_3)_2SeO_2 + 2H_2O \ . \qquad (15)$$

Further work is needed to ensure that $(CH_3)_2SeO_2$ is being emitted by biogenic processes or if it is being formed by oxidative attack of $(CH_3)_2Se$. In addition, metal compounds effectively bind Se:

$$(CH_3)_2Se + M \rightarrow MSe_{(s)} + C_2H_6 \ . \qquad (16)$$

Due to the ability of Se to bind with metals it has been extensively studied as means of detoxifying organisms from metal poisoning (Borek, 1980).

As a model to determine the behaviour of organo-selenides in the atmosphere one can examine the properties of H_2Se. In the presence of O_2 H_2Se reacts to form Se°. This reaction is in fact quite typical of organo-selenides and is the reason why they must be handled in inert atmospheres. Therefore one can expect biogenically released organo-selenides to be highly reactive and to have Se° as their chemical sink.

5. Summary and conclusions

The study of Se chemistry in the laboratory indicates that Se compounds are in general more reactive than their S analogs and change oxidation states much more readily. In addition, it would appear that anthropogenic emissions of Se are probably in the form of Se°, not as the oxide as in the case of S. Organo-selenides should be oxidized very quickly and their chemical sinks should be similar to those of anthropogenic emissions. One could also postulate that the residence time of organ-selenides is very short, since it is thought that the residence time of DMS is on the order of hours (Andreae, 1986).

Figure 8 presents the summary of the proposed chemical speciation and reactions of atmospheric Se. The analysis indicates that Se will be cycled in the atmosphere between the particulate and gas phases. While SeO_2 has a high vapor pressure it is highly hygroscopic, hence it will rapidly condense onto aerosols. The chemical fate of SeO_2 will depend on the acidity of the particles (Figure 5). On highly acidic particles, SeO_2 will be reduced to Se° and be volatilized. On aerosols which are less acidic (such as sea salt particles) SeO_2 will hydrolyse and perhaps be oxidized to selenate. This internal cycling of atmospheric Se explains the findings of Mosher and Duce (1987b) that Se was depleted on sub-micron particles in the remote Pacific and was enriched on sea salt particles.

During the formation of clouds and precipitation the acidity of the hydrometer changes rapidly and a variety of aqueous phase oxidation-reduction reactions can take place. If one assumes that the redox potential of the cloud and rainwater is determined by the H_2O_2-O_2 redox couple, then Se can exist both in the Se(IV) and (VI) oxidation states (Figure 5). Evidence for the highly variable oxidation state of Se in hydrometers is presented by Cutter and Church (1986) who studied the oxidation state of Se in rainwater from a site on the Atlantic coast of USA (Lewes, Delaware). The authors found that Se could be observed in the (0), (IV) and (VI) oxidation state. The exact partitioning depended on the history of the in-coming storm.

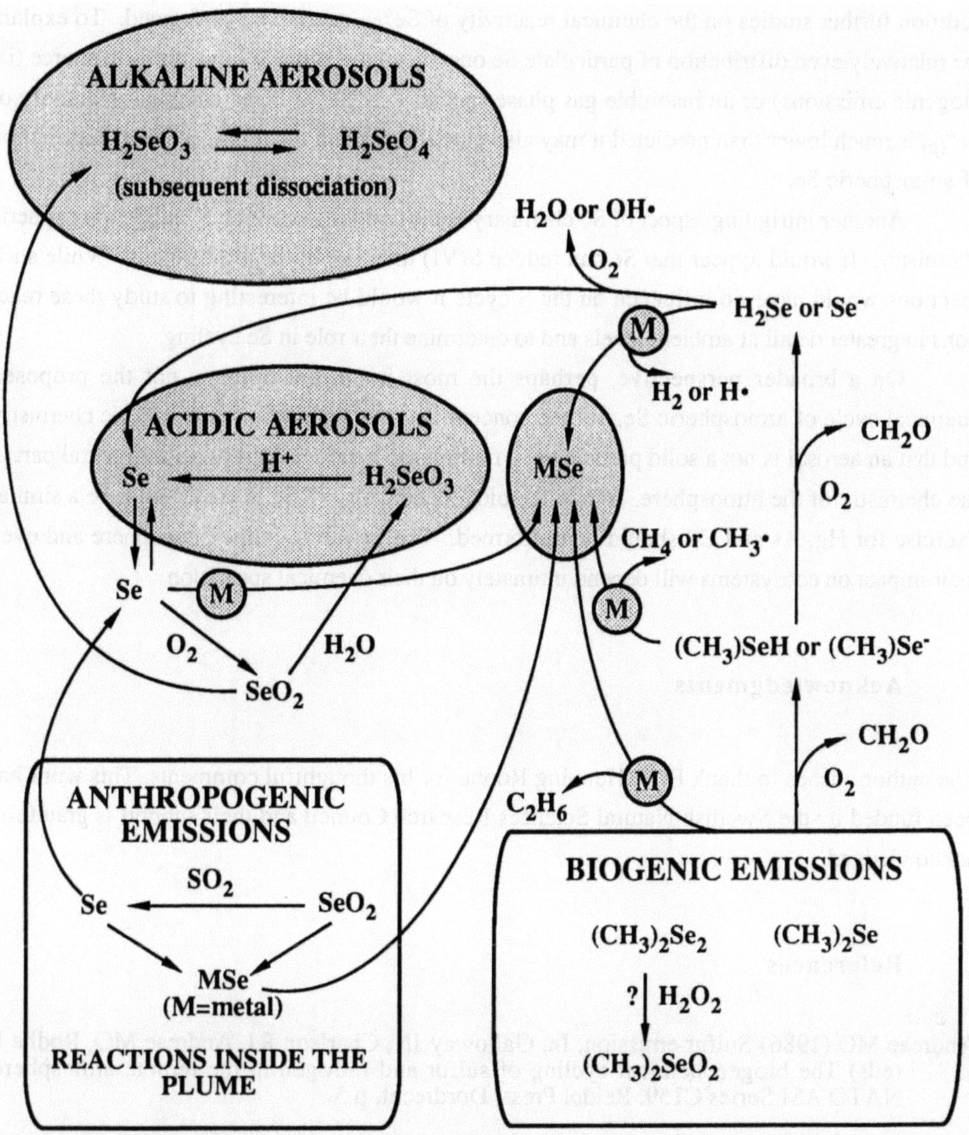

Figure 8. *Summary of the proposed chemical reactions Se will undergo in the ambient atmosphere. The shaded areas indicate reactions which occur on particles. The boxes indicate reactions which are associated with the release of Se to the atmosphere.*

To prove the hypothesis of the internal cycling of Se, it is proposed that vertical profiles of gaseous and particulate Se over the oceans be taken. Since the scale height of sea salt particles is lower than that of sub-micron particles a sink for gas phase Se will be removed. Therefore there should be more $Se^{\circ}_{(g)}$ in the free troposphere than in the boundary layer. In addition further studies on the chemical reactivity of $Se°_{(g)}$ need to be performed. To explain the relatively even distribution of particulate Se one must have either a large uniform source (i.e biogenic emissions) or an insoluble gas phase species (i.e. Se°). If the chemical reactivity of $Se°_{(g)}$ is much lower than predicted it may also partially explain the relatively even distribution of atmospheric Se.

Another intriguing aspect of Se chemistry is the coupling between S and Se atmospheric chemistry. It would appear that Se can reduce S(VI) to S(IV) in the atmosphere! While such reactions would have no influence on the S cycle it would be interesting to study these reactions in greater detail at ambient levels and to determine their role in Se cycling.

On a broader perspective, perhaps the most important point is not the proposed chemical cycle of atmospheric Se, but the concept that metals have an atmospheric chemistry and that an aerosol is not a solid particle but a multiphase system which plays an integral part in the chemistry of the atmosphere. While the biogeochemistry of Se is presented here a similar exercise for Hg, As and Cd should be performed. Their cycling in the atmosphere and even their impact on ecosystems will depend ultimately on their chemical speciation.

Acknowledgments

The author wishes to thank Prof. Henning Rodhe for his thoughtful comments. This work has been funded by the Swedish Natural Sciences Research Council and their support is gratefully acknowledged.

References

Andreae MO (1986) Sulfur emission. In: Galloway JN, Charlson RJ, Andreae MO, Rodhe H (eds) The biogeochemical cycling of sulfur and nitrogen in the remote atmosphere, NATO ASI Series C159. Reidel Press, Dordrecht, p 5

Andren AW, Klein DH, Talmi Y (1975) Selenium in coal-fired plant emissions. Environ Sci Technol 9:856-858

Balistrieri LS, Chao TT (1987) Selenium adsorption by goethite. Soil Sci Soc Amer J 51:1145-1151

Buat-Ménard P (1983) Particle geochemistry in the atmosphere and oceans. In: Liss PS, Slinn WGN (eds) Air-sea exchange of gases and particles, NATO ASI Series C108. Reidel Press, Dordrecht

Bertine KK, Goldberg ED (1971) Fossil fuel combustion and the major sedimentary cycle. Science 173:233-235, p 455

Borek HV (1980) Selenium, 2nd edn, USIP Rep 79-16. U of Stockholm, Dept of Physics, Stockholm, Sweden

Broecker WS, Peng TH (1982) Tracers in the sea. Lamont-Doherty Geological Observatory, Columbia University, New York

Challenger F (1935) The biological methylation of compounds of arsenic and selenium. Chem Ind 54:657

Charlson RJ, Lovelock JE, Andreae MO, Warren SG (1987) Oceanic phytoplankton, atmospheric sulphur, cloud albedo and climate. Nature 295:683-685

Chau YK, Wong PTS, Silverberg BA, Luxon PL, Bengert GA (1976) Methylation of selenium in the aquatic environment. Science 192:1130-1131

Cutter GA, Church TM (1986) Selenium in western Atlantic precipitation. Nature 322:720-722

Doran JW, Alexander M (1977a) Microbial transformations of selenium. Appl Environ Microb 33:31-37

Doran JW, Alexander M (1977b) Microbial formation of volatile selenium compounds in soil. Soil Sci Soc Am J 41:70-73

Duce RA, Ray BJ, Hoffman GL, Walsh PR (1976) Trace metal concentrations as a function of particle size in marine aerosols from Bermuda. Geophys Res Lett 3:339-342

Duce RA, Arimoto R, Ray BJ, Unni CK, Harder PJ (1983) Atmospheric trace elements at Enewetak Atoll: 1, concentrations, sources and temporal variability. J Geophys Res C88:5321-5342

Dutton WA, Janjua MBI, Van den Steen AJ, Watkinson AP (1971) Vapour phase oxidation of selenium. Can Metallurg Quart 10:97-104

Fischer RB, Peters DG (1970) Chemical Equilibrium. WB Saunders, Philadelphia Eastbourne Toronto

Gissel-Nielson G (1980) Selen i mark och växter. Svensk Veterinärtidning 32:50-54

Gmelin Handbuch (1981) Se ergänzungsband, A3 element. Springer-Verlag, Berlin

Goldschmidt VM (1954) Geochemistry. Oxford Univ Press, London

Jiang S, Robberecht H, Adams F, (1985) Identification and determination of alkylselenide compounds in environmental air. Atmos Environ 17:111-114

Kaakinen JW, Jorden RM, Lawasani MH, West RE (1975) Trace element behavior in coal-fired power plant. Environ Sci Technol 9:862-869

Klein DH, Andren AW, Carter JA, Emery JF, Feldman C, Fulkerson W, Lyon WS, Ogle JC, Talmi Y, Van Hook RI, Bolton N (1975) Pathways of thirty-seven trace elements through coal-fired power plants. Environ Sci Technol 9:973-979

Lantzy RJ, Mackenzie FT (1979) Atmospheric trace metals: global cycles and assessment of man´s impact. Geochim Cosmochim Acta 43:511-525

Lewis BG (1976) Selenium in biological systems, and pathways for its volatilization in higher plants. In: JO Nriagu (ed) Environmental biogeochemistry, vol I. Ann Arbor Science, Ann Arbor, p 389

Lewis BG, Johnson CM, Broyer TC (1974) Volatile selenium in higher plants: the production of dimethyl selenide in cabbage leaves by enzymatic cleavage of Se-methyl seleneomethionine selenium salt. Plant Soil 40:101-118

Mason B (1966) Principles of geochemistry, 3rd edn. Wiley & Sons, New York

Measures CI, McDuff RE, Edmund JM (1980) Selenium redox chemistry at GEOSECS 1 reoccupation. Earth Planet Sci Letters 49:102-108

Mosher BW, Duce RA (1983) Vapor phase selenium in the marine atmosphere. J Geophys Res 88:6761-6768

Mosher BW, Duce RA (1987a) Atmospheric selenium: geographical distribution and ocean to atmosphere flux in the Pacific. J Geophys Res 92D:13277-13287

Mosher BW, Duce RA (1987b) A global atmospheric selenium budget. J Geophys Res 92D:13289-13298

Mosher BW, Duce RA (1989) Selenium in the atmosphere. In: Ihnat M (ed) Occurrence and distribution of selenium. CRC Press, Boca Raton, Florida, p 296

Natusch DFS, Wallace JR, Evans CA (1974) Toxic trace elements: preferential concentrations in respirable particles. Science 183:202-204

Nriagu JO (1978) Lead in the atmosphere. In: Nriagu JO (ed) The biogeochemistry of lead in the environment. Elsevier Press, New York

Pillay KKS, Thomas CC, Sondel JA (1971) Activation analysis of airborne selenium as a possible indicator of atmospheric sulfur pollutants. Environ Sci Technol 5:74-77

Pupp C, Lao RC, Murray JJ, Pottie RF (1974) Equilibrium vapour concentrations of polycyclic aromatic hydrocarbons, As_4O_6 and SeO_2 and the collection efficiencies of these pollutants. Atmos Environ 8:915-925

Pupp C, Lao RC, Murray JJ, Pottie RF (1975) Equilibrium vapour concentrations of some polycyclic aromatic hydrocarbons, As_4O_6 and SeO_2. Atmos Environ 9:367

Rahn KA (1971) Sources of trace elements in aerosols - an approach to clean air, Tech Rep. College of Engineering, Dept of Oceanography, U of Michigan, 089030-9-T

Reamer DC, Zoller WH (1980) Selenium biomethylation products from soil and sewage sludge. Science 208:500-502

Reuss JO, Johnson DW (1986) Acid deposition and the acidification of soils and waters, Ecological Studies 59. Springer-Verlag, Berlin

Ross HB (1985) An atmospheric selenium budget for the region 30°N to 90°N. Tellus 37B:78-90

Ryaboshapko AG (1983) The atmospheric sulfur cycle. In: Ivanov MV, Frenzy JR (eds) The global biogeochemical sulfur cycle, SCOPE Report 19. Wiley Press, New York

Salonen JT (1984) Selenium in cardiovascular diseases and cancer-epidemiologic findings from Finland. In: Boström H, Ljungstedt N (eds) Trace elements in health and disease, Skandis Int Sym. Almqvist & Wiksell, Stockholm, p 172

Schwarz K, Foltz CM (1957) Selenium as an integral part of Factor 3 against dietary necrotic liver digestion. J Am Chem Soc 79:3292-3293

Warneck P (1988) Chemistry of the natural atmosphere. Academic Press, San Diego

Wiersma GB, Davidson CI (1986) Trace metals in the atmosphere of rural and remote areas. In: Nriagu JO, Davidson CI (eds) Toxic metals in the atmosphere. John Wiley & Sons, New York

Partitioning of Trace Metals in Sediments and its Relationship to their Accumulation in Benthic Organisms

A. Tessier and P.G.C. Campbell
INRS-Eau, Université du Québec
C.P. 7500, Sainte-Foy, Québec
CANADA G1V 4C7

1. Introduction

Increased industrialization has led in the last decades to important point-source discharges of trace metals from mining operations and from metal refining and processing industries. In certain parts of the world (e.g., Sweden, Eastern Canada; Dickson 1980), atmospheric deposition also represents an important direct contribution of trace metals to surface waters; the concomitant acidic deposition may also lead to increased mobilization of trace metals from terrestrial sources to the aquatic system. Once introduced into the aquatic system, trace metals undergo biological, chemical and physical reactions such as uptake by planktonic organisms followed by sedimentation of these organisms, sorption at solid-water interfaces, or diffusive fluxes across the sediment-water interface; as a result of these reactions, a large fraction of the trace metals introduced into the aquatic environment is normally found associated with the bottom sediments. Typically, sedimentary metal concentrations are three to four orders of magnitude higher than those in the dissolved phase. These high concentrations of certain trace metals (Cd, Cu, Hg, Pb, Zn) may have deleterious effects on benthic organisms themselves or on their predators.

Management of in-place contaminants is one of the major problems confronted by governmental agencies responsible for the protection of the environment. In principle, remedial actions such as dredging, capping, waste reduction, and *in situ* chemical treatment can be undertaken; these actions are however costly. It is clear that the development of rational, effective and economical strategies to solve the problems posed by sediment-bound toxic metals present in sediments must be based on an understanding of the processes governing metal accumulation by benthic organisms, and upon an objective evaluation of the effects of these metals on benthic communities.

Relating the concentration of sediment-bound trace metals to their concentrations in benthic organisms is not straightforward. These organisms present a wide spectrum of nutritional strategies and burrowing behaviors. Knowledge of these ecological aspects is necessary for the judicious choice of environmental variables to be included in models designed to predict organism metal burdens. Many of these organisms are in contact with both dissolved and particulate trace metals and can in principle accumulate the metals either directly from the water or from the solid phases. Dissolved metal concentrations in the interstitial and overlying waters are related to those present in the adjacent sediments (Tessier et al., 1989a,b). Uptake from either source will be influenced by physico-chemical factors in the aqueous (e.g., pH, ligand type and concentrations) and particulate phases (e.g., association of the trace metal with specific sedimentary phases); since the composition of water and sediments will vary from one site to another, these factors must be taken into account if models of general applicability are to be developed. In addition to these physico-chemical factors, biological factors (e.g., age, size, reproductive cycle) may also influence metal levels in benthic organisms and must be considered

The purpose of this paper is to describe recent progress in the development of models that can relate trace metal concentrations in benthic organisms to those prevailing in their environment.

2. Importance of trace metal speciation for predicting bioavailability - laboratory studies

Benthic invertebrate communities contain a wide variety of species representing a diversity of life habits and feeding mechanisms (Adams, 1987). For example, some are filter-feeders (e.g., the freshwater bivalves *Anodonta grandis* and *Elliptio complanata*) and live at the sediment surface, in oxic environments. Others live in galeries (e.g., *Hexagenia* spp., *Ephemeroptera*) or tubes (e.g., *Chironomus* spp.) which they irrigate with oxygenated water. Certain animals ingest sediments with some selectivity based on particle size or density (e.g., *Hexagenia* spp.), whereas others are indiscriminate sediment ingesters (e.g., *Oligochaeta* spp.). Benthic organisms may thus obtain trace metals from the water (overlying or interstitial) or from the particulate matter (algae, bacteria, surficial or deeper sediments) that they ingest intentionally or by accident or with which they are simply in contact; none of these routes of uptake can be rejected *a priori*. It is thus important to determine the physico-chemical factors that will influence the biological uptake of the trace metals from both sources (dissolved; particulate). This information is best obtained from bioassays performed in the laboratory, under controlled conditions.

Many bioassays conducted with unicellular organisms (algae, bacteria) have demonstrated that the free aquo-ion concentration (or activity), $[M^{z+}]$, is the key factor influencing dissolved metal uptake, toxicity or nutrition. These observations have led to the development of the "free-ion activity" model (Fig. 1). Only few exceptions, usually involving lipo-soluble molecules or complexes (e.g., methylated forms of mercury are taken up more easily than inorganic forms; *Smith et al.*, 1975), are known to this general observation. Very few bioassays have been conducted under carefully controlled conditions with benthic invertebrates; one involved the accumulation of dissolved copper in the American oyster *Crassostrea virginica* (Zamuda and Sunda, 1982), and another one the toxicity of dissolved cadmium to the grass shrimp

Palaemonetes purgio (Sunda *et al.*, 1978). Both these experiments confirmed that the "free-ion activity" model applies to other than unicellular organisms.

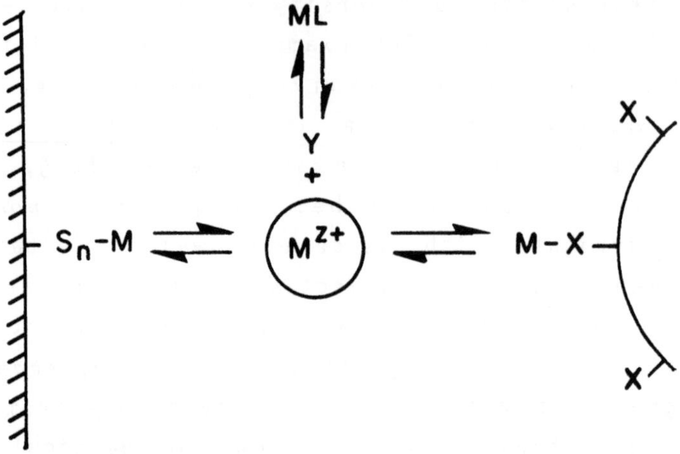

Figure 1: *Schematic representation of the "free-ion activity" model. X represents transport sites on the cell membrane; L and S are a dissolved ligand and a surface site that compete for M. Note that the central role attributed to M^{Z+} does not necessarily imply that this is the only species to react with the transport sites.*

According to this model, metal uptake or toxicity is proportional to the free-ion activity rather than to the total dissolved metal concentration, $[M]_T$, or to the complexed metal concentration, [ML]. The model assumes that the rate of reaction of the metal with transport sites at the biological surface (X-membrane) is fast, compared with the rate of transport across the membrane, and thus that there is a quasi-equilibrium state between the dissolved metal and that bound to the membrane sites. The rate of transport across the membrane would then be proportional to [M-X-membrane], and thus to $[M^{Z+}]$. The presence of dissolved ligands (L) or surface sites (S) would compete with the membrane for the metal and thus decrease the rate of uptake by the organism (Morel, 1984).

The investigation in the laboratory of factors that influence biological availability is more difficult for sediment-bound than for dissolved metals, due to the tendency of metals added in particulate forms to establish solute-solid equilibrium. Hence, trace metals added in particulate form will partly pass into solution and biological uptake from the solution will occur concurrently with that from the ingestion of particulate material. To isolate the contribution of sediment-bound metal to the total biological uptake of trace metals, it is thus necessary in bioassays to separate the two vectors of trace metal transport to the organisms. Such a condition has been achieved only in a limited number of experiments where dialysis membranes were used to separate specimens of *Macoma balthica* (an estuarine deposit-feeder) in contact with both particulate and dissolved metal from others in contact with dissolved metal only (Luoma and Jenne, 1976; 1977; Harvey and Luoma, 1985a,b); biological uptake from sediment-bound metals was obtained by subtracting the uptake of dissolved metal from the total uptake. In these experiments, the animals were fed with natural sediments or with individual components of sediments (Fe oxyhydroxides, Mn oxides, organic material, carbonates) doped with trace metals. These experiments demonstrated that for certain metals, uptake from ingestion of sediment-bound metals may be important, compared with uptake from solution, even if the transfer of trace metals to animals is more efficient from solution than from sediments; it thus seems that the much higher trace metal concentrations in the sediments compared to those in the water compensate for the differences in transfer efficiencies. Another important observation obtained from these experiments is that uptake through ingestion is highly dependent on the metal-solid association in the sediment; for example, uptake of ingested ^{110m}Ag by *M. balthica* was found to be more than 1000-fold higher if the metal was associated with calcite than if it was bound to organics (Luoma and Jenne, 1977). Differences in metal uptake were related to the strength of metal binding by the sediment components (Luoma, 1983).

An important conclusion from the laboratory bioassay studies is that the physico-chemical forms of the trace metals in the water as well as in the sediments influence metal uptake by animals. For dissolved metals, factors other than $[M]_T$ that influence $[M^{z+}]$ include the nature and concentration of dissolved ligands, concentrations of competing cations, redox potential, pH, temperature and ionic strength. For sediment-bound metals, partitioning of trace metals among the various sediment components is expected to vary with environmental variables such as pH, sediment composition, and redox conditions. These factors will vary from one site to another in the natural environment, and it is thus likely that total dissolved ($[M]_T$) or total particulate ($\{M\}_T$) trace metal concentrations will not be good predictors of metal bioavailability. Laboratory bioassays carried out under rigorously controlled conditions provide useful information concerning the factors that influence biological uptake or toxicity of trace metals, and contribute to the interpretation of field observations; however, the results obtained from such experiments cannot yet be extrapolated from the laboratory to the field to predict uptake or toxicity under natural conditions.

3. Approaches used to relate sedimentary trace metals to bioaccumulation in the field

Many researchers have tried to predict concentrations of trace metals in benthic organisms ($[M(org)]$), or the physiological or toxicological responses of target organisms exposed to the sediments, from total metal concentrations present in the sediment matrix ($\{M\}_T$). However, the results of this empirical approach have proven disappointing, due to the generally low predictive power of $\{M\}_T$ for $[M(org)]$, and to the difficulty of transposing the results obtained for a given aquatic system to another one. In more recent work, two approaches have been explored: one involves the search for reagents capable of extracting from sediments the "bioavailable fraction" of trace metals; the second is based on the free-ion activity model. The merits and limitations of these two approaches are discussed below.

3.1 Chemical extractants

Trace metals can be associated with various sediment phases; for example, they can be adsorbed at mineral surfaces, precipitated as discrete phases, coprecipitated, occluded in sediment components, bound up with organic matter in either living or detrital form, or bound in lattice positions in aluminosilicates. Laboratory bioassays have suggested, as discussed above, that only certain of these trace metal forms should be accessible to benthic organisms; identification of trace metal forms in a sediment matrix is however not easy.

Direct determination of sediment-trace metal associations is difficult, if not impossible, given the great variety of solid phases that can bind the trace metals, their amorphous character, and the low trace metal concentrations involved. As an alternative to direct measurements, reagents have been suggested for partially extracting sediments (Langston, 1980; 1982; Luoma and Bryan, 1978) or for fractionating the sediments chemically (Engler *et al.*, 1977; Tessier *et al.*, 1979; Förstner, 1982; Robbins *et al.*, 1984). However, it must be realized that an extract obtained with a partial extraction reagent cannot be ascribed with certainty to a particular geochemical phase, but will retain a certain operational character. Indeed, any partial extractant or extraction procedure will suffer from unavoidable lack of selectivity; it would be unrealistic to think that a given reagent could extract metals completely from a target sediment component while leaving untouched the same metal bound to other components. In addition, if a metal is liberated with a given reagent, it may readsorb onto the remaining phases (Rendell *et al.*, 1980; Tipping *et al.*, 1985); recent results obtained for natural sediments indicate, however, that the importance of this post-extraction readsorption problem may be minimal with certain reagents (Belzile *et al.*, 1989a). Thus, the more serious problem with partial extractants is probably their inherent lack of selectivity.

In a few field studies, sediments collected along a trace metal gradient have been subjected to partial extractants, and the resulting metal concentrations compared to those observed in the indigenous benthic animals (Luoma and Bryan, 1978; Langston, 1980; 1982, Tessier *et al.*, 1983; 1984). In general, extracted metal concentrations were found to be slightly better predictors of [M(org)] than $\{M\}_T$; this improvement is probably due to the fact that weak extractants, despite their inherent problems, do not extract those metal forms (e.g., metals bound in lattice positions in alumino-silicates) that are clearly inaccessible for accumulation by benthic organisms (Luoma, 1989). It is probable that the partial extraction reagents solubilize, but not selectively, trace metal forms that are bioavailable. At present, the predictions of [M(org)], based solely upon extractable metal, remain however unsatisfactory (Luoma, 1989).

The most interesting point to note from the recent field studies mentioned above is that they all have shown that the prediction of [M(org)] is much improved when the trace metal concentration extracted from the sediments is normalized with respect to the iron oxyhydroxide or organic matter content of the sediments. For example, Luoma and Bryan (1978) have shown that the concentration of lead in the estuarine detritus feeder *Scrobicularia plana* was best predicted by the ratio {Pb}/{Fe} extracted with 1 N HCl (Fig. 2; $r^2 = 0.98$); consistently, Langston (1980) found that the ratio {As}/{Fe}, extracted with the same acidic solution, was the best predictor of arsenic concentrations in the same estuarine organism ($r^2 = 0.93$). Langston (1982) demonstrated that the ratio {Hg extracted with HNO_3/organic matter content (%)} best explained the concentration of mercury in the two estuarine species *S. plana* ($r^2 = 0.63$) and *Macoma balthica*. Tessier *et al.* (1983; 1984) have shown that for two freshwater filter-feeding bivalves, *Anodonta grandis* and *Elliptio complanata*, collected from a restricted geographical area downstream from a mining complex, Cu and Pb levels were best related to the ratios {Cu}/{Fe} ($r^2 = 0.95$ for *E. complanata*; $r^2 = 0.87$ for *A. grandis*), {Pb}/{Fe} ($r^2 = 0.83$ for the gills of *E. complanata*; $r^2 = 0.82$ for the gills of *A. grandis*) where {Cu}, {Pb} and {Fe} refer to concentrations extracted with

$NH_2OH \cdot Cl$. In other words, according to these studies, trace metal concentrations in the bivalves, [M(org)], are best expressed as:

$$[M(org)] = k \frac{\{S_n\text{-}OM\}}{\{S_n\text{-}OH\}} + [M(org)]^\circ \qquad (1)$$

where k is a proportionality constant, $\{S_n\text{-}OH\}$ is the

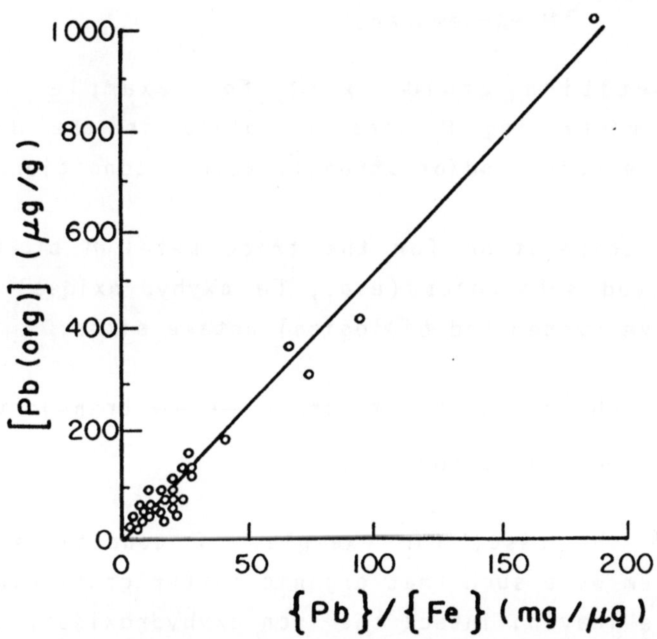

Figure 2: *Relationship between Pb concentration in the tissues of Scrobicularia plana and the ratio {Pb}/{Fe} extracted from sediments with HCl 1N. (After Luoma and Bryan, 1978).*

concentration of a given phase "n" in the sediment (e.g., Fe oxyhydroxides; organic matter), $\{S_n\text{-}OM\}$ is the concentration of the trace metal M associated with this phase, and $[M(org)]^\circ$ is the y-intercept (which is usually close to zero). At least three mechanisms could explain such inverse relationships between metal concentrations in the organisms and the concentration of a potential sink in the sediments (Luoma and Bryan, 1978; Tessier *et al.*, 1984; Campbell and Tessier, 1989):

MECHANISM I. Competition for uptake sites in the digestive system between the trace metal M of interest and a second metal M' (e.g., Fe) that is a major constituent of the substrate (e.g., Fe oxyhydroxides) that binds M:

$$\begin{matrix} M \\ M' \end{matrix} + \text{X-membrane} \begin{cases} \text{M-X-membrane} \longrightarrow \text{transport of M} \\ \text{M'-X-membrane} \end{cases} \quad (2)$$

Such a competition could exist for example if iron oxyhydroxides containing M were dissolved in the digestive system due to reducing and/or strongly acidic conditions.

MECHANISM II. Competition for the trace metal M of interest between unreacted substrates (e.g., Fe oxyhydroxides) present in the digestive system and biological uptake sites:

$$M \begin{cases} + \text{X-membrane} \rightleftarrows \text{M-X-membrane} \longrightarrow \text{transport} \\ + \equiv S_n\text{-OH} \rightleftarrows S_n\text{-OM} \end{cases} \quad (3)$$

This would be the case, for example, if conditions in the digestive system were such that organic matter containing M was digested, while leaving intact the iron oxyhydroxides.

MECHANISM III. Uptake of dissolved M in sorptive equilibrium in the external medium with component n of the sediments. Assuming that [M(org)] is related to $[M^{z+}]$ (see section 2 above) and taking into account sorption in the external medium (see section 3.2.1 below, and particularly equation (10)), one can write:

$$[M(org)] = k\ [M^{z+}] = k \frac{\{S_n\text{-OM}\} \cdot [H^+]^x}{\{S_n\text{-OH}\} \cdot {}^*K_A(n) \cdot N_s(n)} \quad (4)$$

where $^*K_A(n)$ is an overall apparent constant for the sorption of M on the substrate n, $N_s(n)$ is the density of sorption sites (moles of sites per g of substrate), and x is the apparent number of protons involved in the sorption reaction. Equation (4) shows that a proportionnality between [M(org)] and the ratio $\{S_n\text{-OM}\}/\{S_n\text{-OH}\}$, such as that indicated by equation (1), should be observed if the values of $[H^+]$ (or pH), x, and $N_s(n)$ are similar for all the sampling sites that served to establish the empirical relation (1). For the studies with freshwater bivalves (Tessier *et al.*, 1983; 1984), the values of $N_s(n)$ and of the exponent x (which both depend, for a given M, essentially upon the nature of the substrate), and pH were presumably similar for all the sites in the mining area of Rouyn-Noranda, as they were located in a limited area of relatively constant geology. This is probably also true for the sites in estuaries (i.e., studies with *S. plana* and *M. balthica*; Langston,1980; 1982; Luoma and Bryan, 1978). In the cases where $[H^+]$, x and $N_s(n)$ are constant, equations (4) and (1) are equivalent. In support of mechanism III, i.e. the uptake of dissolved rather than particulate metal, it should be mentioned that high levels of trace metals are found to be associated with the gills and mantle of the two freshwater pelecypods (*E. complanata* and *A. grandis*); these two organs contribute a large proportion of the total metal body burden. They are flushed with the large amounts of water necessary for feeding and respiration and have large surface areas.

3.2 Activity-based models

An alternative approach to the use of chemical extractions for predicting [M(org)] is the development of models to estimate the free aquo-ion concentration, $[M^{z+}]$. These models (Fig. 3) assume that the biological effects of the metals are correlated to the free-ion concentration, $[M^{z+}]$ (see section 2). An additional assumption is that the free-ion concentration can be related to sediment characteristics such as the concentrations of some sediment components, and those of trace metals associated with these components. The rationale for the models is that $[M^{z+}]$ can be predicted (and hence the

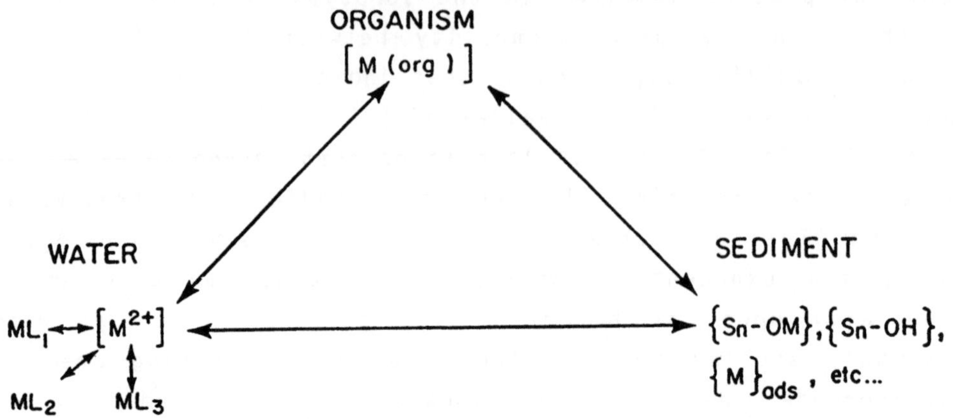

Figure 3: *Schematic representation of the basic relationships assumed between metal concentrations in the aquatic organisms, the water and the sediments in the activity-based models.*

biological effects) from these sediment characteristics, thus circumventing the difficulty of measuring $[M^{z+}]$ in natural waters.

3.2.1 Equilibrium partitioning

The equilibrium partitioning model developed recently (Jenne *et al.*, 1986) is drawn from adsorption models that describe trace metal partitioning among water and various geochemical phases present in aquatic sediments (Oakley *et al.*, 1981; Luoma and Davis, 1983). The adsorption of a trace metal M on a substrate "n" of the sediments can be characterized in a simplified manner by:

$$\equiv S_n\text{-}OH_x + M^{z+} \overset{*K_A(n)}{\rightleftarrows} \equiv S_n\text{-}OM + x\ H^+ \qquad (5)$$

$$*K_A(n) = \frac{\{\equiv S_n\text{-}OM\} \cdot [H^+]^x}{\{\equiv S_n\text{-}OH_x\} \cdot [M^{z+}]} \qquad (6)$$

where charges are omitted for simplicity; $\{\equiv S_n\text{-}OH_x\}$ and $\{\equiv S_n\text{-}OM\}$ are the concentrations of free sites and of sites occupied by the metal respectively for substrate n; $^*K_A(n)$ is an overall apparent equilibrium constant for the adsorption of M on that substrate. The notation "≡" refers to adsorption sites, whereas {} and [] refer to concentrations of the solid and dissolved species respectively. At low adsorption densities, that is when the concentration of occupied adsorption sites is small compared to the total concentration of sites, the condition:

$$\{\equiv S_n\text{-}OH_x\} \simeq \{\equiv S_n\text{-}O\text{-}\}_T \qquad (7)$$

should be fulfilled, where $\{\equiv S_n\text{-}O\text{-}\}_T$ is the total concentration of sites for substrate n. The concentrations of free and occupied sites can be related to the concentration of component n and that of trace metal M associated with it respectively by the following equations:

$$\{\equiv S_n\text{-}O\text{-}\}_T = N_s(n) \cdot \{S_n\text{-}OH\} \qquad (8)$$

$$\{\equiv S_n\text{-}OM\} = \{S_n\text{-}OM\} \qquad (9)$$

where $N_s(n)$ is the density of adsorption sites of component n, $\{S_n\text{-}OH\}$ is the concentration of this component in the sediments, and $\{S_n\text{-}OM\}$ is the concentration of M associated with this component. Equation (9) assumes that M is associated with the adsorption sites of component n in the ratio 1:1. Combining equations (6) to (9) leads to:

$$K_A(n) = \frac{^*K_A(n) \cdot N_s(n)}{[H^+]^x} = \frac{\{S_n\text{-}OM\}}{\{S_n\text{-}OH\} \cdot [M^{z+}]} \qquad (10)$$

where $K_A(n)$ is an overall apparent equilibrium constant which is a function of pH.

The equilibrium partitioning model assumes that sorbed

metals in the sediments are close to equilibrium with the surrounding waters, and that only three sediment components, namely Fe oxyhydroxides, Mn oxyhydroxides, and reactive particulate organic matter (RPOC), are responsible for most of the sorption of trace metals. If the latter assumption is taken as valid, three equations similar to equation (10) can be written where $\{S_nOH\}$ is replaced by {Fe-OH}, {Mn-OH} and {RPOC}, whereas $\{S_n\text{-}OM\}$ is substituted by {Fe-OM}, {Mn-OM} and {RPOC-M}. These three equations, together with the following definition of total sorbed metal concentration, $\{M\}_{ads}$:

$$\{M\}_{ads} = \{Fe\text{-}OM\} + \{Mn\text{-}OM\} + \{RPOC\text{-}M\} \qquad (11)$$

allow calculation of $[M^{z+}]$, provided that $\{M\}_{ads}$, {Fe-OH}, {Mn-OH} and {RPOC} are measured, and the $K_A(n)$ and $N_s(n)$ values are known. Such an approach is presently being evaluated by the US EPA in an attempt to develop sediment quality criteria of general applicability (Shea, 1988). The main problems to be solved, before the model can be adopted, are related to the choice of the input data to the model (particularly $K_A(n)$ and $N_s(n)$) that are valid for natural systems.

The determination of sediment component concentrations and the total sorbed metal concentration, which are necessary inputs to the model, relies upon partial extraction reagents (Jenne, 1987; Crecelius *et al.*, 1987). These reagents are probably more selective for the determination of the component concentrations than for the measurement of $\{M\}_{ads}$. The determination of the latter variable will be subject to problems associated with the use of partial extraction reagents (see section 3.1 above). As for the three sediment components, each of them presents a spectrum of reactivities (e.g., various crystallinity states for the metal oxyhydroxides; great diversity of organic matter); it is difficult at the moment to suggest chemical reagents that will dissolve selectively the fractions of these components that are reactive with respect to sorption of the trace metals.

Site densities can in principle be determined (Luoma

and Davis, 1983) by various methods including fast tritium exchange (Yates and Healy, 1976), acid-base titration (Bourg, 1983; Huang and Stumm, 1973), metal adsorption isotherms at constant pH (Dempsey and Singer, 1980), or calculations based on the specific surface area of the solid and that occupied by the adsorbed metal (James and Healy, 1972). Determination of site densities for single synthetic phases (e.g., amorphous Fe oxyhydroxides; goethite) by various methods is not always consistent (Luoma, 1986). This probably reflects both the complexity of the single substrates, and the limitations of the methods available. More importantly, the techniques developed for single-component systems cannot be used to determine the site density of a particular component within a mixture of components such as a sediment. The only means to have access to the site density value of a component of a natural sediment would be to isolate that component from the sediment matrix while preserving its integrity. Recent experiments where inert (Teflon) surfaces were inserted vertically in lake sediments, and left *in situ* for periods of weeks to months, have shown that it is possible to collect separately on these surfaces Fe and Mn oxyhydroxides that are formed diagenetically in the sediments (Belzile *et al.*, 1989b); this method of isolation with minimal perturbation should allow measurement of site densities of components of natural sediments. For the moment, any calculation of $[M^{z+}]$ with the activity-based model must rely upon site densities obtained in the laboratory for single-component systems; however, there is no guarantee that they will apply to natural systems.

Equilibrium constants (K_A) for the adsorption of trace metals onto various substrates (e.g., iron and manganese oxyhydroxides, clays, humic substances) resembling components of natural sediments have been determined mostly in well-defined media (Davis and Leckie, 1978; Farrah and Pickering, 1979; Loganathan and Burau, 1973; Beveridge and Pickering, 1980). Sophisticated models have been developed (Stumm *et al.*, 1970; Davis and Leckie, 1978) for interpreting the adsorption of metal ions at oxyhydroxide surfaces under various physico-chemical conditions; these models have not however been applied to complex natural systems. Several studies have attempted to estimate binding constants for the adsorption of trace metals on natural Fe oxyhydroxides from *in situ* measurements (Johnson, 1986; Tessier *et al.*, 1985; 1989a; 1989b), based on equation (10). This latter equation can be written as follows if adsorption on Fe oxyhydroxides is considered specifically:

$$K_A(Fe) = \frac{{}^*K_A(Fe) \cdot N_s(Fe)}{[H^+]^x} = \frac{\{Fe\text{-}OM\}}{\{Fe\text{-}OH\} \cdot [M^{z+}]} \qquad (12)$$

where {Fe-OH} is the concentration of iron oxyhydroxides, {Fe-OM} is that of metal associated with the iron oxyhydroxides, whereas the other terms have been defined previously. In these experiments, {Fe-OH} and {Fe-OM} were estimated by partial extraction of suspended sediments rich in iron (Johnson, 1986) or of oxic lake sediments (Tessier *et al.*, 1985, 1989a,b); these values were used in equation (12), together with the estimated ambient free-ion concentrations of trace metals, to calculate $K_A(Fe)$ values derived from natural Fe oxyhydroxides for lakes of various pH values. Equation (12) predicts that a plot of log $K_A(Fe)$ vs pH should yield a straight line of slope x and intercept log ($^*K_A(Fe) \cdot N_s(Fe)$). Figure 4 shows an example of the results obtained for Zn by both Johnson (1986) and Tessier *et al.* (1989a). Similar straight lines of log $K_A(Fe)$ vs pH were obtained for Cd, Cu, Ni, and Pb. Comparison of these field-derived K_A values with those measured in the laboratory indicated significant differences; in general laboratory-derived K_A values are one or more order of magnitude lower than field-derived ones. These differences can be explained by various factors (Tessier *et al.*, 1985). Surface characteristics such as porosity and surface area differ between fresh and aged Fe oxyhydroxides and between those derived from Fe(III) and Fe(II) sources (Crosby *et al.*, 1983); natural iron oxyhydroxides are commonly formed from the oxidation of Fe(II), whereas in the laboratory they are often prepared by precipitation of Fe(III) salts. Ageing times employed for laboratory-prepared Fe oxyhydroxide phases (minutes to a few days; Millward and Moore, 1982; Swallow *et al.*, 1980) are not necessarily representative of those occurring in natural environments. Kinetic studies of the sorption of trace metals on natural sediments have shown that a few days are necessary for the distribution coefficient (ratio $\{M\}_T/[M]_T$) of most metals to remain constant with time (Santschi *et al.*, 1986; Balistrieri and Murray, 1984). Reactions proceed for much longer times in the natural systems than in the laboratory where only minutes (Millward and Moore, 1982) to hours (Benjamin and Leckie, 1981) are usually allowed. Other differences between field and laboratory conditions that may lead to divergence between laboratory- and field-derived binding constants include density of adsorption, dissolved ligand concentrations, ionic strength and temperature.

In summary, the equilibrium partitioning approach has the merit of being based on sound physico-chemical and biological principles, and thus it holds considerable promise as a general model for predicting metal bioaccumulation from the dissolved phase. At the present time, however, it is of limited applicability, since all the data needed for handling the complex sediment-metal-water interactions are not yet available.

3.2.2 Use of a surrogate parameter

An approach simpler than the equilibrium partitioning model consists of focusing on the sorption of M on a single component, in particular Fe oxyhydroxides, rather than on the three components retained by the previous model, to estimate $[M^{z+}]$. As discussed previously, description of the sorption of a trace metal on Fe oxyhydroxides leads to equation (12) which, if combined with the assumption that [M(org)] is correlated with $[M^{z+}]$, leads to:

$$[M(org)] = k\ [M^{z+}] = k\ \frac{\{Fe\text{-}OM\} \cdot [H^+]^x}{\{Fe\text{-}OH\} \cdot {}^*K_A(Fe) \cdot N_s(Fe)} \qquad (13)$$

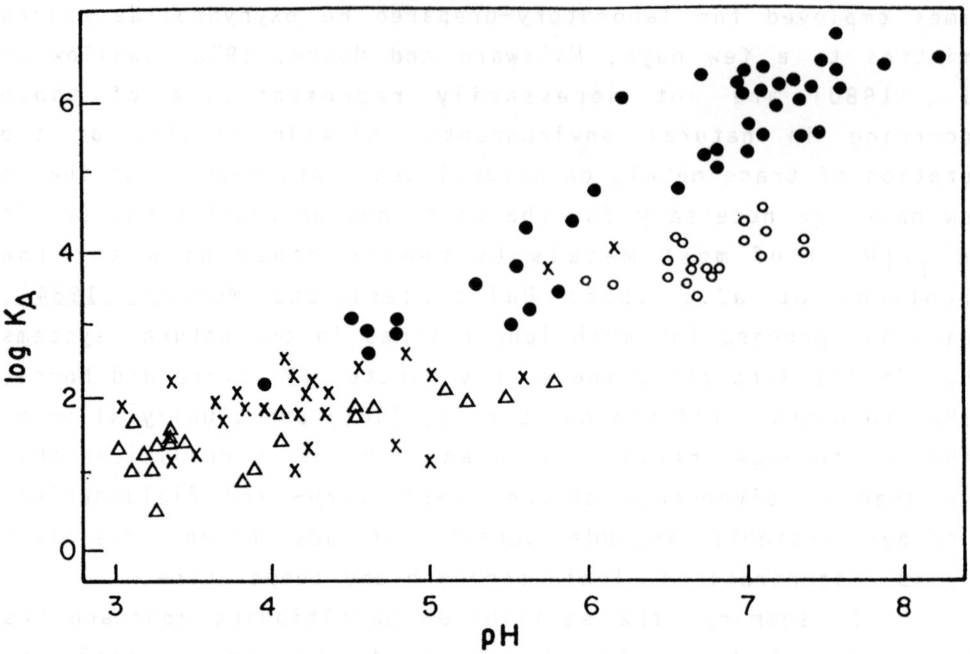

Figure 4: *Apparent overall equilibrium constant for the sorption of Zn on natural iron oxyhydroxides. Data are from various lakes (●; Tessier* et al., *1989) or from the Carnon river system (Johnson, 1986; o uncontaminated drainage waters; x contaminated surface waters; Δ mine waters).*

According to this equation, the measurement of {Fe-OM}, {Fe-OH} and pH should allow the prediction of [M(org)] provided that $*K_A$(Fe) and N_s(Fe) are constant and that the ratio $k/*K_A(Fe)\cdot N_S(Fe)$ has been measured; the ratio on the right hand side of equation (13) represents in fact a surrogate measure of $[M^{z+}]$, a variable that is difficult to determine in natural waters. The capacity of equation (13) to predict [M(org)] is being tested presently (Tessier *et al.*, unpublished results) with the freshwater pelecypod *E. complanata*, for M = Cd in lakes of varying pH values; to this end, {Fe-OH} and {Fe-OCd} are obtained by extracting surficial sediments adjacent to the animals with a reducing reagent, $NH_2OH\cdot HCl$. Preliminary results are shown in Fig. 5.

Comparison of Figures 5a and 5b indicates that equation (13) predicts [Cd(org)] much better than does equation (1). This observation constitutes a strong support for mechanism

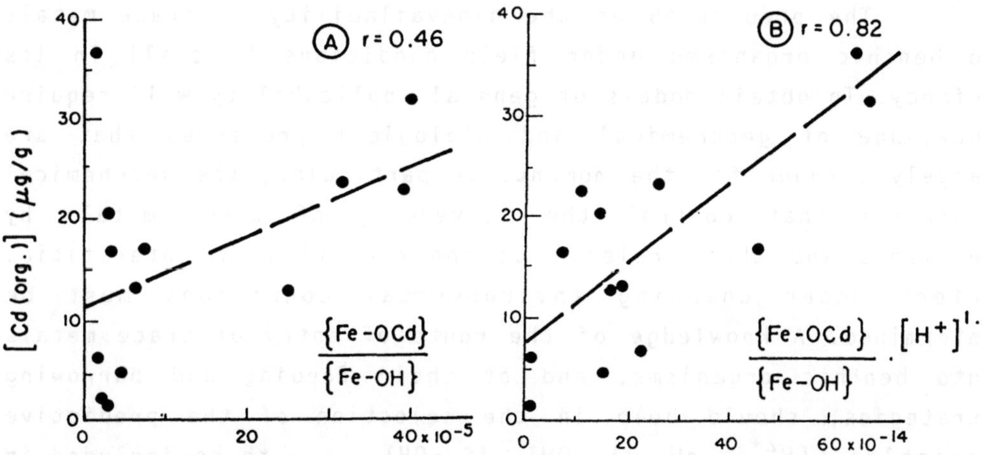

Figure 5: *Relationships between Cd concentration in the tissues of* *Elliptio complanata* *and the ratios* $\{Fe\text{-}OCd\}/\{Fe\text{-}OH\}$ *(A) and* $\{Fe\text{-}OCd\}\cdot\{H^+\}^x/\{Fe\text{-}OH\}$ *(B)* .

III, which involves water as the vector of trace metal uptake, rather than mechanisms I and II which imply ingestion of sediments (see section 3.1 above). It also highlights the importance of the free-ion activity for predicting [M(org)]. The predictive power of equation (13) must, however, be tested for more lakes, other metals and other organisms before its generality can be proclaimed.

4. Conclusions

Trace metals occur in various forms in both water and sediments, and accumulation or toxic effects are related to only certain of these forms. At present, decisions concerning the management of trace metals are based on total concentrations in water or sediments, or on the results of sediment elutriation. Neither of these measurements provides good predictions of metal accumulation in benthic invertebrates or of detrimental effects.

The prediction of the bioavailability of trace metals to benthic organisms under field conditions is still in its infancy. To obtain models of general applicability will require knowledge of geochemical and biological processes that are largely unknown for the moment. In particular, the geochemical processes that control the scavenging of trace metals by sediments and their release to the overlying or interstitial waters under changing environmental conditions must be determined. A knowledge of the route of entry of trace metals into benthic organisms, and of their feeding and burrowing strategies, should help in the selection of the predictive variables ($[M^{z+}]$, pH, $\{S_n\text{-OM}\}$, $\{S_n\text{-OH}\}$, ...) to be included in a generic model. Biological factors (e.g., age, sexual condition, etc.) that affect the accumulation of metals must also be quantified.

Acknowledgements

Financial support from the Québec Fond pour la Formation de Chercheurs et l'Aide à la Recherche, from the Wildlife Toxicology Fund and from the Natural Sciences and Engineering Research Council of Canada is acknowledged.

References

Adams WJ (1987) Bioavailability of neutral lipophilic organic chemicals contained on sediments: a review. In: Dickson KL, Maki AW, Brungs WA (eds) Fate and Effects of Sediment-Bound Chemicals in Aquatic Systems. Pergamon Press, New York, p 219.

Balistrieri LS, Murray JW (1984) Marine scavenging: trace metal adsorption by interfacial sediment from MANOP site H. Geochim Cosmochim Acta 48:921-929.

Belzile N, Lecomte P, Tessier A (1989a) Testing the potential readsorption of trace elements during partial chemical extractions of bottom sediments. Environ Sci Technol 23:1015-1020.

Belzile N, DeVitre R, Tessier A (1989b) *In situ* collection of diagenetic iron and manganese oxyhydroxides from natural sediments. Nature 340:376-377.

Benjamin MM, Leckie JO (1981) Multiple-site adsorption of Cd, Cu, Zn, and Pb on amorphous iron oxyhydroxide. J Colloid Interface Sci 79:209-221.

Beveridge A, Pickering WF (1980) Influence of humate-solute interactions on aqueous heavy metal ion levels. Water Air Soil Pollut 14:171-185.

Bourg A (1983) Modélisation du comportement des métaux traces à l'interface solide-liquide dans les systèmes aquatiques. Rapport du B.R.G.M. No 62.

Campbell PGC, Tessier A (1989) Geochemistry and bioavailability of trace metals in sediments. In: Boudou A, Ribeyre F (eds). Aquatic Ecotoxicology. Fundamental Concepts and Methodologies. CRC Press, Boca Raton, FL, Vol. 1, p 125.

Crecelius EA, Jenne EA, Anthony JS (1987) Sediment quality criteria for metals: optimization of extraction methods for determining the quantity of sorbents and adsorbed metals in sediments. Report prepared for US Environmental Protection Agency, Washington, DC.

Crosby SA, Glasson DR, Cuttler AH, Butler I, Turner DR, Whitfield M, Millward GE (1983) Surface areas and porosities of Fe(III)- and Fe(II)-derived oxyhydroxides. Environ Sci Technol 17:709-713.

Davis JA, Leckie JO (1978) Surface ionization and complexation at the oxide/ water interface. II. Surface properties of amorphous iron oxyhydroxide and adsorption of metal ions. J Colloid Interface Sci 67:90-107.

Dempsey BA, Singer PC (1980) The effects of calcium on the adsorption of zinc by MnO_x(s) and $Fe(OH)_3$(am). In: Baker RA (ed) Contaminants and Sediments. Ann Arbor Science Publisher Inc., Ann Arbor, p 159.

Dickson W (1980) Properties of acidified waters. In: Ecological Impact of Acid Precipitation. Proc Intern Conf Sandefjord, Norway, p 75-83.

Engler RM, Brannon JM, Rose J, Bigham G (1977) A practical selective extraction procedure for sediment characterization. In: Yen TF (ed) Chemistry of Marine Sediments. Ann Arbor Science Publishers Inc., Ann Arbor, p 163.

Farrah H, Pickering WF (1979) pH effects in the adsorption of heavy metal ions by clays. Chem Geol 2:317-326.

Förstner U (1982) Accumulative phases for heavy metals in limnic sediments. Hydrobiologia 91:269-284.

Harvey RW, Luoma SN (1985a) Separation of solute and particulate vectors of heavy metal uptake in controlled suspension-feeding experiments with *Macoma balthica*. Hydrobiologia 121:97-102.

Harvey RW, Luoma SN (1985b) Effect of adherent bacteria and bacterial extracellular polymers upon assimilation by *Macoma balthica* of sediment-bound Cd, Zn and Ag. Mar Ecol Progr Ser 22:281-289.

Huang CP, Stumm W (1973) Specific adsorption of cations on hydrous γ-Al_2O_3. J Colloid Interface Sci 43:409-420.

James RO, Healy TW (1972) Adsorption of hydrolysable metal ions at the oxide-water interface. III. A thermodynamic model of adsorption. J Colloid Interface Sci 40:65-81.

Jenne EA (1987) Sediment quality criteria for metals: II. Review of methods for quantitative determination of important adsorbents and sorbed metals in sediments. Report prepared for US Environmental Protection Agency, Washington, DC.

Jenne EA, DiToro DM, Allen HE, Zarba CS (1986). An activity-based model for developing sediment criteria for metals: Part I. A new approach. In: Lester JN, Perry R, Sterritt RM (eds) Proc Intern Conf on Chemicals in the Environ. Selper Ltd, London, p 560.

Johnson CA (1986) The regulation of trace element concentrations in river and estuarine waters contaminated with acid mine drainage: the adsorption of Cu and Zn on amorphous Fe oxyhydroxides. Geochim Cosmochim Acta 50:2433-2438.

Langston WJ (1980) Arsenic in U.K. estuarine sediments and its availability to deposit-feeding bivalves. J Mar Biol Ass UK 60:869-881.

Langston WJ (1982) Distribution of mercury in British estuarine sediments and its availability to deposit-feeding bivalves. J Mar Biol Ass UK 62:667-684.

Loganathan P, Burau RG (1973) Sorption of heavy metal ions by a hydrous manganese oxide. Geochim Cosmochim Acta 37:1277-1293.

Luoma SN (1983) Bioavailability of trace metals to aquatic organisms - a review. Sci Tot Environ 28:1-22.

Luoma SN (1986) A comparison of two methods for determining copper partitioning in oxidized sediments. Mar Chem 20:45-59.

Luoma SN (1989) Can we determine the biological availability of sediment-bound trace elements? Hydrobiologia 176:379-396.

Luoma SN, Bryan GW (1978) Factors controlling the availability of sediment-bound lead to the estuarine bivalve *Scrobicularia plana*. J Mar Biol Ass UK 58:793-802.

Luoma SN, Davis JA (1983) Requirements for modelling trace metal partitioning in oxidized estuarine sediments. Mar Chem 12:159-181.

Luoma SN, Jenne EA (1976) Factors affecting the availability of sediment-bound cadmium to the estuarine deposit feeding clam, *Macoma balthica*. In: Cushing E (ed) Radioecology and Energy Resources. Dowden, Hutchinson and Ross, Inc, Stroudsberg, p 283.

Luoma SN, Jenne EA (1977) The availability of sediment-bound cobalt, silver, and zinc to a deposit-feeding clam. In: Wildung RE, Drucker H (eds) Biological Implications of Metals in the Environment. NTIS CONF-750920, Springfield, VA, p 213.

Millward GE, Moore RM (1982) The adsorption of Cu, Mn, and Zn by iron oxyhydroxide in model estuarine solutions. Water Res 16:981-985.

Morel FMM (1984) Principles of Aquatic Chemistry. John Wiley & Sons, New York.

Oakley SM, Nelson PO, Williamson KJ (1981) Model of trace-metal partitioning in marine sediments. Environ Sci Technol 15:474-480.

Rendell PS, Batley GE, Cameron AJ (1980) Adsorption as a control of metal concentrations in sediment extracts. Environ Sci Technol 14:414-418.

Robbins JM, Lyle M, Heath GR (1984) A sequential extraction procedure for partitioning elements among co-existing phases in marine sediments. Report 84-3, College of Oceanography, Oregon State University, Corvallis, OR.

Santschi PH, Nyffeler UP, Anderson RF, Schiff SL, O'Hara P, Hesslein RH (1986) Response of radioactive trace metals to acid-base titrations in controlled experimental ecosystems: evaluation of transport parameters for application to whole-lake radiotracer experiments. Can J Fish Aquat Sci 43:60-77.

Shea D (1988) Developing national sediment quality criteria. Environ Sci Technol 22:1258-1261.

Smith AL, Green RH, Lutz A (1975) Uptake of mercury by freshwater clams (Family Unionidae). J Fish Res Board 32:1297-1303.

Stumm W, Huang CP, Jenkins SR (1970) Specific chemical interaction affecting the stability of dispersed systems. Croat Chim Acta 42:223-245.

Sunda WG, Engel DW, Thuotte RM (1978) Effect of chemical speciation on toxicity of cadmium to grass shrimp, *Palaemonetes purgio*: importance of free cadmium ion. Environ Sci Technol 12:409-413.

Swallow KC, Hume DN, Morel FMM (1980) Sorption of copper and lead to hydrous ferric oxide. Environ Sci Technol 14:1326-1331.

Tessier A, Campbell PGC, Bisson M (1979) Sequential extraction procedure for the speciation of particulate trace metals. Anal Chem 51:844-851.

Tessier A, Campbell PGC, Auclair JC (1983) Relationships between trace metal partitioning in sediments and their bioaccumulation in freshwater pelecypods. In: Proc 4th Intern Conf Heavy Metals Environ. CEP Consultants Ltd, Edinburgh, UK, p 1086.

Tessier A, Campbell PGC, Auclair JC, Bisson M (1984) Relationships between the partitioning of trace metals in sediments and their accumulation in the tissues of the freshwater mollusc *Elliptio complanata* in a mining area. Can J Fish Aquat Sci 41:1463-1472.

Tessier A, Rapin F, Carignan R (1985) Trace metals in oxic lake sediments: possible adsorption onto iron oxyhydroxides. Geochim Cosmochim Acta 49:183-194.

Tessier A, Carignan R, Dubreuil B, Rapin F (1989a) Partitioning of zinc between the water column and the oxic sediments in lakes. Geochim Cosmochim Acta 53:1511-1522.

Tessier A, Carignan R, Belzile N (1989b) Reactions of trace elements near the sediment-water interface in lakes. In: DePinto J, Lick W (eds). Transport and Transformation of Contaminants near the Sediment-Water Interface (submitted for publication).

Tipping E, Hetherington NB, Hilton J (1985) Artifacts in use of selective chemical extraction to determine distributions of metals between oxides of manganese and iron. Anal Chem 57:1944-1946.

Yates DE, Healy TW (1976) The structure of the silica/electrolyte interface. J Colloid Interface Sci 55:9-19.

Zamuda CD, Sunda WG (1982) Bioavailability of dissolved copper to the American oyster *Crassostrea virginica*. I. Importance of chemical speciation. Mar Biol 66:77-82.

Environmental Chemistry of the Actinide Elements

Marco S Caceci
IRDI-DERDCA-DRDD-SESD-SCPCS
Commissariat à l'Energie Atomique
B.P. 6, F-92265 Fontenay-aux-Roses CEDEX
France

1. Introduction

Research in the chemistry of the actinide (An) elements has known in recent years a significant shift of focus: from metallurgy and reprocessing in the age of the atom bomb and of the nuclear reactor, to geochemistry, now that the primary issue is environmental protection. The tools remain basically the same, the glove-box and the radiation counter, but the chemistry, in neutral or alkaline media, is quite different from that of nitric acid solutions.

The actinides are the group of 15 elements from ^{89}Ac to ^{103}Lr characterized by the gradual filling-up of the 5f electron shell. Only two actinide elements exist in nature in significant amounts, thorium (^{232}Th, half life $1.4 * 10^{10}$ y) and uranium (^{235}U, $7.04 * 10^{8}$ y and ^{238}U, $4.46 * 10^{9}$ y). The earth's crust contains on average 9.6 ppm Th and 2.7 ppm U. Sea water contains 3 ppb U and 0.05 ppb Th. Vast deposits of minerals of Th (monazite: a phosphate) and U (uraninite: an oxide) are known. World production is estimated at 50000 and 500 tons/year of uranium and thorium respectively. Enrichment plants produce, for use as reactor fuel, uranium with a content of fissile ^{235}U higher than the natural 0.711 %. Depleted uranium metal, used as ballast in ships and planes and in the manufacture of bullets and rocket heads, is an important potential source of environmental contamination. Thorium can be used as fertile source of ^{233}U in breeder reactors, but in spite of the extensive research toward the development of a thorium nuclear fuel cycle, no such large scale applications are presently in sight.

Thorium and uranium are toxic both chemically and radiologically, having a specific activity of 0.11 and 0.34 μCi/g respectively. Alpha and beta decays produce a cascade of daughter nuclides some of which (^{220,222}Rn, ^{226,228}Ra) are of occupational and epidemiological concern.

Of the actinide elements produced in the radioactive decay chain of natural Th and U, ^{231}Pa (half life $3.27 * 10^{4}$ y) from ^{235}U can establish its own chemical paths in the environment. Long-lived heavier actinides exist naturally only in negligeable amounts but are now produced synthetically by sequential neutron activation and spallation of uranium in nuclear reactors: the most important being

NATO ASI Series, Vol. G 23
Metal Speciation in the Environment
Edited by J. A. C. Broekaert, Ş. Güçer, and F. Adams

^{237}Np, $^{238,239,240,242,244}Pu$, $^{241,243}Am$, and $^{242,244}Cm$. Thorium and protactinium are of relatively little concern here because they are virtually absent from nuclear waste and are of limited environmental mobility. Actinium and the elements beyond curium are too short lived to be significant from a geochemical point of view. Their chemistry is expected to mimic that of curium.

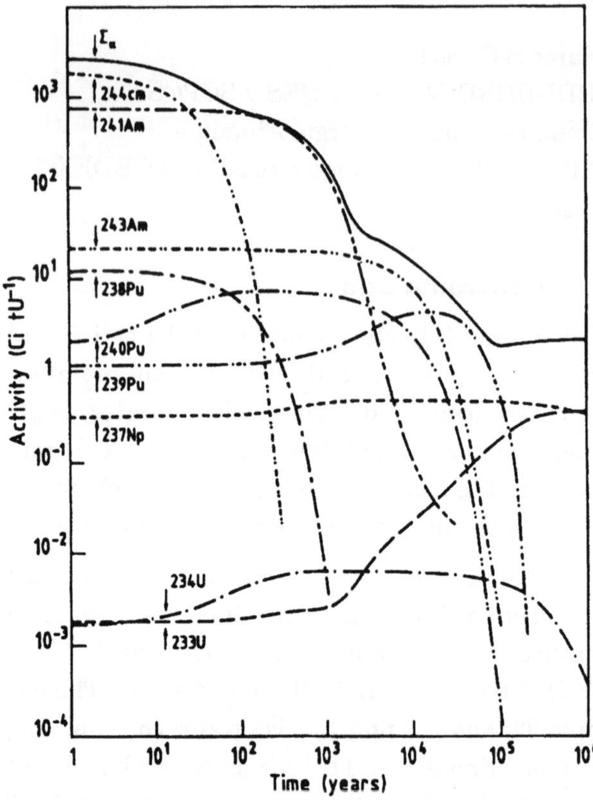

Figure 1. *Decay of alpha activity in typical PWR waste glass.*

The radioactivity of spent nuclear fuel is due to fission products (mainly $^{134,137}Cs$ and ^{90}Sr) and to the actinide elements. Most fission products decay by beta emission and have relatively short half lives, so that an initial (after four years cool-down) activity of 10^6 Ci/tonne U in typical waste repository glass decays to about 20 Ci/tonne U in about a thousand years. The actinide elements in spent fuel, on the other hand, decay mainly by alpha emission. Their extreme radiological toxicity is due to the higher energy and higher damaging effect of alpha particles, as well as to their marked affinity for bone tissue and long residence time in living organisms. Their content in typical waste repository glass is about $3 * 10^3$ Ci/tonne U initially, still more than 200 Ci after a thousand years, and significant even after one hundred thousand years (Fig. 1).

Since the actinides are here to stay, it is of interest to learn how to dispose of them safely and effectively. Of the fraction, however small, that eventually reaches the environment at large, we want to be able to predict the behavior in the geological and biological sphere.

An increasing public concern for environmental protection issues, the May 1986 accident in a nuclear reactor in Chernobyl and consequent dispersal of radioactive materials, and the surfacing of serious contamination and waste disposal problems in a number of military installations in the USA, together with the problem of disposing of the growing mass of spent nuclear fuel, have stimulated continuing research efforts in the years since Watters *el al.* (1983) reviewed actinide behavior in the environment.

Significant progress has been achieved in our understanding of actinide solution chemistry in neutral and alkaline media; a number of issues pertinent to actinide transport (notably, the interaction with humic matter and colloids) are being actively investigated; field investigations, laboratory studies, and computer modeling appear to be converging into a complex but consistent scenario.

2. Solution chemistry

The actinides are all metals; their aqueous chemistry [Katz *et al.*, 1986. Choppin, 1983] is characterized by:

- many possible oxidation states (fig. 2), most of which exibit characteristic narrow absorption bands (fig. 3).

- existence (except for Pa) of unique and stable dioxo "actinyl" species in the formal V and VI oxidation states. The effective charge on the central atom, hence complexation strength, increases in the sequence $AnO_2^+ < An^{3+} <= AnO_2^{2+} < An^{4+}$.

- "hard" acid behavior: strong affinity for hydroxide, carbonate, fluoride, phosphate ions.

Most absorption bands are due to Laporte forbidden f-f transitions, have absorption coefficients about two orders of magnitude larger than in the lanthanides, and lie in the visible or near infrared. These bands are sensitive to the environment (complexation) around the metal ion, and constitute a valuable tool for speciation studies: specially built spectrophotometers [Caceci, 1989a] or thermal lensing apparatus [Beitz and Hessler, 1980] are often required due to relatively low absorption coefficients and narrow bandwidths. The luminescence of UO_2^{2+}, Cm^{3+} and, to a lesser degree, Am^{3+} allows their detection at ppt concentration levels by time resolved fluorimetry [Moulin C *et al.*, 1989].

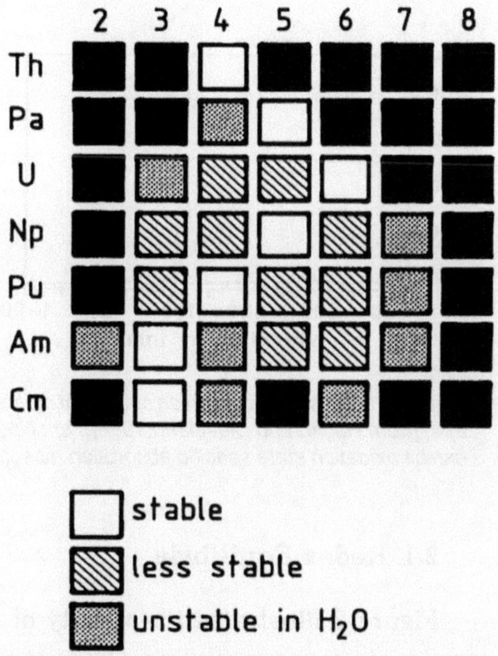

Figure 2. *Oxidation states of the actinide elements in acidic aqueous solution.*

The luminescent uranyl (UO_2^{2+}) ion has a high oxidation potential in its excited state. It was calculated that in the upper layers of the oceans each uranyl ion is excited by sunlight on average every 10 minutes [Joergensen and Reisfeld, 1982]. It is quite possible that uranyl-catalysed photooxidation of organic matter may prove to be significant - besides Fe(III)-photocatalized processes - as a mechanism of degradation of the organic matter transported by rivers to the oceans, thus demonstrating that at least one actinide has a beneficial role in the biological theatre. Radiolysis may play a significant role in the near field. Alpha radiation produces in aqueous solutions a number of radicals which can both oxidize and reduce the actinides. H_2O_2, formed by OH radical recombination, normally acts as a reducing agent. Brine solutions tend to become oxidizing: species such as ClO^- and ClO_2^- may induce formation of mobile Am(V).

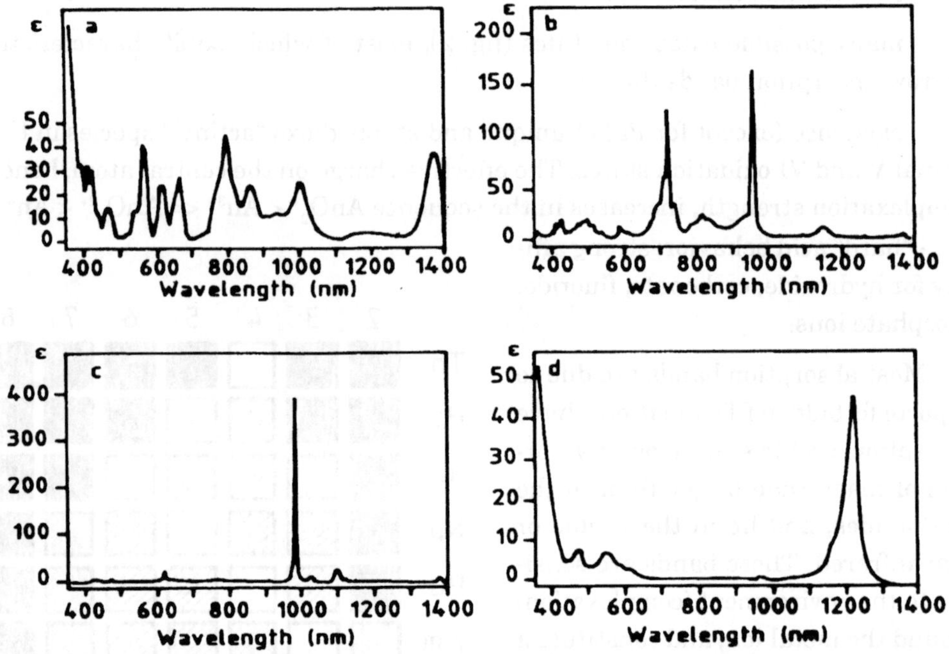

Figure 3. *UV-Vis-NIR absorption spectra of different oxidation states of Np in aqueous solution ($HClO_4$ 2 M) [from Hagan and Cleveland, 1966]. a) : Np(III); b): Np(IV); c): Np(V); d): Np(VI). Most actinides exhibit oxidation state specific absorption spectra useful for analytical purposes.*

2.1. Redox Equilibria

Figure 2 illustrates the variety of oxidation states accessible to the lighter actinides in aqueous solution. Oxidation states indicated as most stable are such in acidic solution. Formal potentials in 1 M perchlorate media (Riglet *et al.*, 1989a and 1989b, Fig 4) change dramatically at higher pH's. Measurements in neutral or alkaline media are made difficult by the poor solubility of the hydroxides, so that redox equilibria under environmentally relevant conditions are probably better computed by extrapolation using hydrolysis constants obtained, for example, from solubility studies. Solubility in carbonate media is relatively high for species other than An(III), and - when equilibria are sufficiently rapid - redox potentials can be measured reliably and used to compute stability constants with carbonate.

Pentavalent, hexavalent, and very probably heptavalent species exist in solution as linear dioxo cations with formal charge +1, +2 and +3 and water or ligand molecules in the equatorial plane. Some indications exist that "non-dioxo" cations of V or VI oxidation states may exist as transients in solution or in solid compounds [Musicas, 1976].

Redox equilibria are established rapidly in the III-IV and the V-VI couples, but are slower in complexing media (carbonate) and whenever metal-oxygen ("-yl") bonds must be formed or broken. It is remarkable that in concentrated carbonate media all four oxidation states of americium can coexist [Bourges *et al.*, 1983].

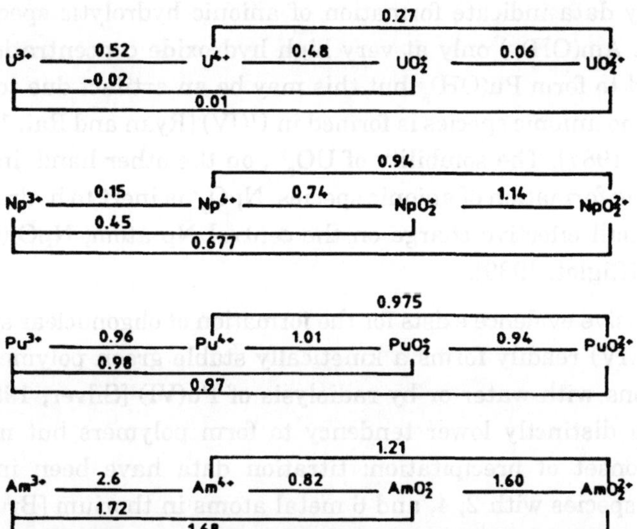

Figure 4. *Formal redox potentials in 1 M acid of U, Np, Pu, Am.*

2.2. Hydrolysis

Extensive studies of actinide hydrolysis [Baes and Messmer, 1976] have been conducted by potentiometric titrations, solubility, solvent extraction and more recently also by spectrophotometry in the case of Np [Riglet, 1989]. Limited thermodynamic data exist as a function of ionic strength [Caceci and Choppin, 1983a] and temperature. Experimental difficulties related to carbonate contamination, unwanted redox reactions, poor solid crystallinity, adsorption losses [Caceci and Choppin, 1983b] and radiolytic problems contribute to scatter in the literature data.

The tendency to hydrolysis grows in the series: An(V) < An(III) $\simeq$ An(VI) < An(IV), while solubility follows an opposite trend. According to Allard (1981), in first approximation the pKa values at room temperature for the first hydrolysis step of an "average" actinide (An) ion (in other words, the pH at which the ratio $[An^{x+}]/[AnOH^{x-1+}]$ is unity) and the negative logarithm of the solubility product (pKso) of the hydroxide solid can be assumed to be:

	pKa	pKso
An^{4+} :	0.4	54
AnO_2^{2+} :	5.2	23
An^{3+} :	6.5	23
AnO_2^{+} :	9.1	9

Solubility data indicate formation of anionic hydrolytic species in trivalent actinides (ex. $Am(OH)_4^-$) only at very high hydroxide concentrations. Pu(IV) has been reported to form $Pu(OH)_5^-$ but this may be an artifact due to CO_3^{2-} contamination, since no anionic species is formed in U(IV) [Ryan and Rai, 1983] nor Th(IV) [Bruno *et al.*, 1987]. The solubility of UO_2^{2+}, on the other hand, increases at high pH's indicating formation of anionic species. NpO_2^+ is inert to hydrolysis up to pH 9 due to the small effective charge on the central Np atom; $NpO_2(OH)_2^-$ is formed above pH 11 [Riglet, 1989].

No conclusive evidence exists for the formation of oligonuclear species in An(III) or An(V). Pu(IV) readily forms a kinetically stable green polymer on dilution of acidic solutions with water or by radiolysis of Pu(VI) [Silver, 1988]. Th(IV) and U(IV) have a distinctly lower tendency to form polymers but may oligomerize around the onset of precipitation: titration data have been interpreted with formation of species with 2, 4, and 6 metal atoms in thorium [Bruno *et al.*, 1987] and a similar behavior is observed with U(IV).

In the hydrolysis of UO_2^{2+}, $(UO_2)_3(OH)_5^+$, a triangular array of uranyl groups linked by hydroxyl groups, plus a single oxygen atom in the middle of this triangle or hexagon, has such a high stability constant (3 M + 5 OH -> M3OH5: log beta: 54.4: Baes and Mesmer, 1976) that it is expected to dominate speciation even at extreme dilutions. Other polymeric U(VI)/hydroxide moieties are fairly well established, namely the 2:2 and the anionic 3:7 species. Np(VI) and Pu(VI) have a similar but weaker tendency to form oligonuclear species.

2.3. Complexation with inorganic ligands

Except for An(III) all oxidation states are fairly soluble in carbonate or bicarbonate media. Interest in actinide chemistry in carbonate media has led to continuing spectroscopic and electrochemical studies on uranium [Bruno *et al.*, 1989], neptunium [Varlashkin *et al.*, 1984], plutonium [Wester and Sullivan, 1983] and americium [Berger *et al.*, 1988].

In complexation with the carbonate ion the trend in strength of the complexes is An(V) < An(III) < An(VI) < An(IV) with log beta 1:1 values of about 4.3 (Np(V)), 8.2 (Am(III)), 9.8 (U(VI)) and 13 (Pu(IV)) respectively. The limiting species are of stoichiometry 1:3 for trivalent, pentavalent and hexavalent ions, and 1:5 for the tetravalent species.

There is considerable uncertainty in the numerical values of An(IV) carbonate complexation: only the stability constant of the 1:5 complex is well known in U(IV) (log beta $\simeq$ 40: Ciavatta *et al.*, 1983). A reported value of 47.1 $\simeq$ 3 for the 1:1 Pu(IV) carbonate complex [Kim *et al.*, 1983], while in excellent accord with older results, is completely unreasonable (Newton and Sullivan, 1985). A value about 34 orders of magnitude lower (log beta 1:1 $\simeq$ 13: Silva, 1985) should be preferred. A predictable consequence of the absence of reliable data on Pu(IV) carbonate complexation has been the omission of these complexes from the data bases used in many speciation modeling programs.

Trimers of composition $(MO_2)_3(CO_3)_6^{6-}$ (M = U, Np, Pu) have been characterized [Grenthe *et al.*, 1986a] having a stability constant of about 53.5 (U(VI)). Formation of bicarbonate complexes was claimed for Am(III) [Bidoglio, 1982] but is now thought unlikely at ordinary pressure [Nitsche *et al.*, 1989]. The evidence for $Am(OH)_x(CO_3)_y$ species is fairly strong, and there are indications of similar mixed species formation in Pu(IV). Mixed OH^- - CO_3^{2-} complexes have been observed spectrophotometrically in Np(V) [Riglet, 1989] in highly alkaline solutions. Of the many mixed hydroxo-carbonate UO_2^{2+} species proposed in the literature, oligo-nuclear $(UO_2)_{11}(OH)_{12}(CO_3)_6^{2-}$ is fairly popular. An(IV) mixed hydroxo-carbonate species have been suggested: thorium solubility data indicate formation of a 6:6:6 Th-OH-CO_3 species [Bruno *et al.*, 1987].

Due to the high negative charge of carbonate complexes, their stability constants and the solubility of the solid phases are strongly influenced by ionic strength and temperature. Grenthe *et al.*, (1986b) among others applied the Guggenheim-Scatchard S.I.T. (specific ion interaction) theory to fit satisfactorily changes in complexation constants in solutions up to 3.5 M in ionic strength.

Fluoride, phosphate and sulphate ions, among the many inorganic ligands found in natural waters, can be relevant in actinide environmental chemistry, since they form both relatively strong complexes and sparingly soluble solids, particularly with trivalent and tetravalent ions (AnF_3 pKso: 10; AnF_4 pKso: 23; $AnPO_4$ pKso: 23 [Allard, 1981]). Trivalent rare earths and tetravalent thorium in nature are found in fluorocarbonates (bastnasite) and phosphates (monazite). It is surprising that few thermodynamic data are available for these and other minerals. In this regard it has been pointed out [Nash *et al.*, 1988] that mixed complexes (possibly fluorocarbonates) may account for the higher than expected solubility of Am(III) in a number of groundwaters.

2.4. Interaction with organic ligands

A large fraction of soil organic carbon and of water dissolved organic carbon (DOC) is constituted by humic (HA) and fulvic acid (FA) - collectively known as humic substances. Humic acid is soluble in bases and insoluble in acids, while fulvic acid is soluble in both [Aiken *et al.*, 1985]. The borderline between these two mixtures of compounds is somewhat blurred but it is related to a stronger tendency of humic acid to complex metal ions and to flocculate.

Humic substances contain carboxylic, hydroxilic and phenolic groups which are mainly responsible for complexation of "hard" metal ions [Schnitzer and Khan, 1972]. They are generally considered to play an important role in the transport of actinide elements in the geosphere, particularly in soil and surface fresh waters, where they are present in the largest concentrations. The interaction with metals is complex [Falck, 1988]: a simple stability constant approach is only a rough approximation and it is often more convenient to consider the existence of a continuous distribution of sites [Altmann and Buffle, 1988]. Partly irreversible kinetic effects [Cacheris and Choppin, 1987], the precipitation of metal humates past definite metal and electrolyte concentrations and pH [Caceci, 1989b], and the reducing properties of humic substances [Nash *et al.*, 1981] do not help to simplify experiments and models.

In their study of the interaction of HA with Am(III), Torres and Choppin (1985) interpreted solvent extraction data in terms of formation of 1:1 and 1:2 complexes. The first stability constant, with humic acid concentration expressed in grams per liter, was found to be $10^{6.5 \pm 0.2}$ l g^{-1}. Later work by ion exchange at higher pH substantially confirmed these results, although a 1:1 model was deemed sufficient [De Brabandere, 1989]. On the other hand ultrafiltration [Caceci, 1985], spectrophotometry [Moulin *et al.*, 1987] and size exclusion ion chomatography [Moulin,

1986] yielded stability constants about two orders of magnitude lower. An explanation for this inconsistency has not been found. In any case, Am(III) will be present in most surface waters (with the possible important exception of seawater) mostly as a humate complex, according to calculations [Choppin and Allard, 1985, Moulin *et al.*, 1988] and experimental evidence [Nakayama and Nelson, 1988].

In an experiment in which the load of Eu on humic acid was investigated as a function of pH [Caceci, 1985], it was found that complexation could be modeled as resulting from two classes af sites in similar amounts and of widely different strength (log beta about 4.5 and 2 l g^{-1} respectively). Varying the pH affected the relative amounts of these two sites but not their sum, indicating that a common chemical substrate binds in both cases. It appears now that the transition from "strong" to "weak" complexation may correspond to the onset of flocculation.

In the system Th(IV)-HA or Pu(IV)-HA stability constants are extremely high [Choppin and Allard, 1985]. Field studies indicate that DOC reduces the adsorption of plutonium on sediments in rivers and lakes already at the 1 ppm level [Nelson *et al.*, 1985].

NpO_2^+, as expected, interacts weakly with HA. A spectrophotometric investigation of the shift of the 980 nm band on complexation presents a clear isobestic point, but a simple 1:1 model fails to fit the data; ultrafiltration results indicate a continuous distribution of sites yielding a linear plot with a slope of -0.5 in a log-log scale of the ratio metal bound / metal free vs total metal [Caceci, 1985, unpublished results].

HA interaction with UO_2^{2+} is about 2 orders of magnitude weaker than with Am(III) [Choppin and Allard, 1985]. Giesy *et al.* (1986), using a Gaussian-Scatchard model and program GEOCHEM, predicted that only about 28 % of uranyl ion in the humate rich surface waters investigated was bound to humic substances.

Humic acid has reducing properties. It reduced Np(VI) to Np(V) almost instantaneously, Pu(VI) to Pu(IV) slowly, but failed to reduce U(VI) in two months [Nash *et al.*, 1981]. There is evidence, on the other hand, that important bodies of uranium ore may have originated from the reduction of uranyl by humic matter in anoxic conditions.

The interaction of the actinides with many organic ligands has been extensively investigated in the literature. A number of synthetic chelating agents, notably EDTA, are now ubiquitous in the environment and may be associated with enhanced actinide mobility in soil.

3. Speciation in water

Possibly the reference "standard" state in nature is the ocean: I = 0.7, pH = 8.2, $pCO_2 \simeq 3.5$, T = 277 K. In waters, soils and the underground, "normal" conditions span a range of pH between 4 and 9 [Baas Becking *et al.*, 1960], and redox potentials lie in the limits set by atmospheric oxygen in the surface and by sulfide ion and Fe(II) in the underground. Concrete, metals, and other chemicals injected in the environment by human activities have created some extreme environmental habitats. The solubility and the redox state in solution of the actinides can be computed at any given set of values for pH, Eh and chemical composition from thermodynamic data. The redox state of a nuclide will determine its mobility and its biological uptake.

Pourbaix diagrams represent the areas of predominance of the different oxidation states of a nuclide as a function of Eh and pH [Brookins, 1988]. The diagrams in figures 5 (following Langmuir, 1978), 6, 7 and 8 (according to Robouch, 1989) were computed combining room temperature redox, hydrolysis and carbonate complexation data assuming normal partial CO_2 pressure and neglecting complexation by humic matter or other anions. The results may surprise the chemist accustomed to handling actinides in acidic solutions.

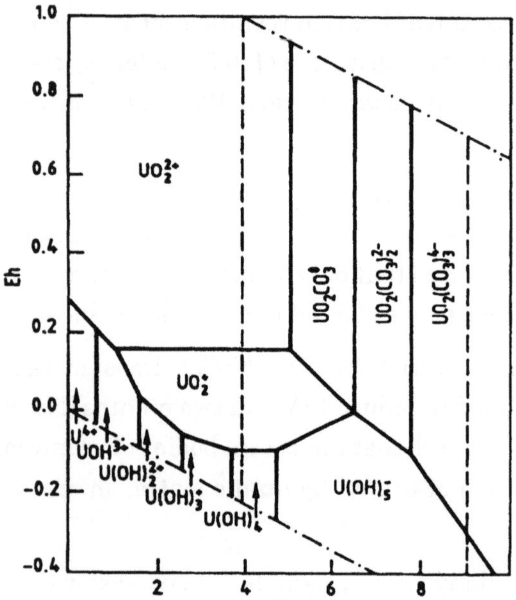

Figure 5. *pH-Eh diagram for tracer uranium in aqueous solution. $[CO_2]$: $10^{-3.5}$ atm [from Kniewald and Branica, 1988 and Langmuir, 1978].*

Uranium (fig. 5) in seawater is present mainly as the tricarbonate uranyl(VI) complex. In a region at low pH and slightly reducing conditions pentavalent UO_2^+ will dominate in solution over solid UO_2 [Langmuir, 1978]. The existence of U(V) in significant amounts in natural systems remains to be proved, but its presence in open waters as a transient species at the oxic-anoxic boundary is supported by experiment [Kniewald and Branica, 1988]. The solubility and mobility of uranium will be favored by a high redox potential and by a high carbonate ion concentration.

Neptunium is the only actini-

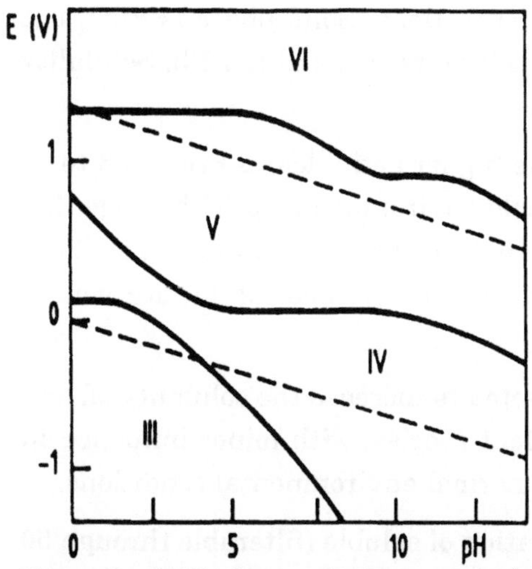

Figure 6. *pH-Eh diagram for tracer neptunium in aqueous solution. [CO_2]: $10^{-3.5}$ atm [Robouch,1989]*

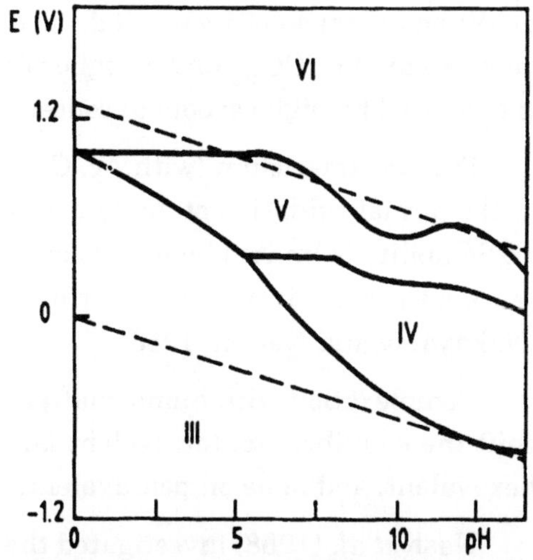

Figure 7. *pH-Eh diagram for tracer plutonium in aqueous solution. [CO_2]:$10^{-3.5}$ atm [Robouch,1989]*

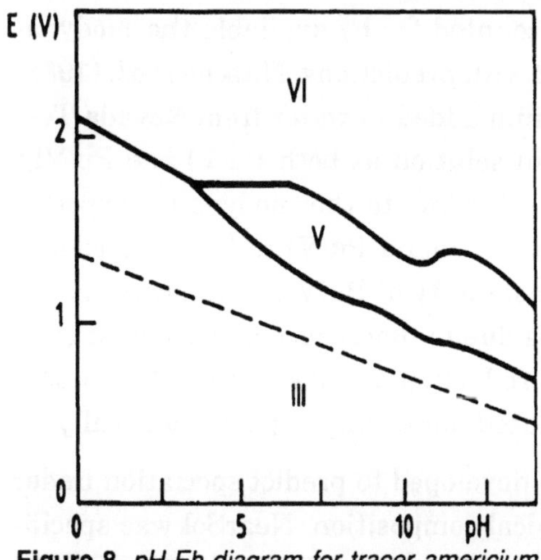

Figure 8. *pH-Eh diagram for tracer americium in aqueous solution. [CO_2]:$10^{-3.5}$ atm [Robouch, 1989]*

de element for which the pentavalent, monopositive species is the most stable in acidic solution. Its speciation diagram (fig. 6) indicates that Np(V) (as hydroxo and/or carbonate complex) will predominate under oxidizing conditions. Under reducing conditions, neptunium will be highly insoluble and present in solution as Np(IV). This prediction was experimentally verified by Pryke and Rees (1986) in simulated respository waters of controlled redox potential. In any case the solid phase will be highly insoluble NpO_2.

The chemistry of plutonium is probably the most complex for a metal in the periodic table. All four oxidation states have a region of predominance in solution in the environmental pH-Eh range (fig. 7). The most striking feature is the predicted instability of Pu(IV) at pH values between 0 and 5-6. In seawater Pu(VI) (as the triscarbonato complex) should be most stable, but in fresh waters (pH 4 - 7) Pu(V) prevails under oxic conditions. Lowering the redox potential shifts the equilibria towards very insoluble Pu(IV) in alkaline regions and Pu(III) in acidic regions. It must be remarked that calculated speciation patterns could be significantly affected by even small changes in the values of the stability constants. The mobility of plutonium will be enhanced by oxidizing, but also by very reducing conditions because of increasing Pu(III) concentrations. Solid phase is PuO_2 but may become $Pu_2(CO_3)_3$ under higher CO_2 partial pressures and low Eh. Solubility is enhanced by high carbonate concentrations.

Finally, americium (with Ac, Cm and the higher actinides) is predicted to be in the formal oxidation state 3+ over the whole environmental pH-Eh range (fig. 8). Solubility is limited by formation of $AmOHCO_3$ or $Am_2(CO_3)_3$. Indeed, no difference was found in the behavior of Am and Cm in a number of surface waters [Nakayama and Nelson, 1988].

Complexation with humic matter is expected to increase the solubility of, and shift the equilibria to, tetravalent and trivalent species, with minor influence on hexavalent, and none on pentavalent, under normal environmental conditions.

Nash *et al.* (1988) investigated the speciation of soluble (filterable through 50 nm pores) Np, Pu and Am added to a number of natural waters, including seawater. Correlations between carbonate and fluoride content and Pu(IV) concentration were observed that could not be accounted for by available thermodynamic data, but in general a fair accord existed with predictions. Nitsche *et al.* (1988) studied the oxidation state of tracer plutonium added to water from Nevada Test Site and found that plutonium was found in solution as both Pu(V) and Pu(VI). The ratio Pu(VI)/Pu(V) increased with pH according to thermodynamic expectations. Morse and Choppin (1986) among others found Pu(V) to be the predominant form of plutonium in seawater. The instability of Pu(VI) in seawater is not predicted by thermodynamic data: it may be due to inaccuracies in the data or to the existence of a dynamic system in which reduction of Pu(VI) to Pu(V) is provoked by traces of humic acid or by H_2O_2 produced, for example, photochemically.

A number of computer codes have been developed to predict speciation under given conditions of pH, Eh, pCO_2 and chemical composition. NearSol was specifically developed to assess near field solubilities [Leach and Prike, 1986], EQU3/6, PHREEQE and WATEQF [Plummer, 1976] model comprehensively far field conditions, GEOCHEM [Sposito and Mattigod, 1979] accounts for humic acid complexation. The need of validating the consistency of thermodynamic databases used by different modelling programs has led to intercomparison efforts, notably project CHEMVAL [Chandratillake *et al.*, 1988]. When predicting numerically speciation in natural environments differences in the databases account for many but in no way all of the differences in the computed results: according to a recent informal comparison test, of a dozen programs using the same database on the same set of problems only about one half converged to identical and consistent results.

4. Sorption and coprecipitation on minerals and colloids

The interaction of metal ions with solid surfaces (bottom sediments, soils and minerals) is a complex phenomenom both from the theoretical and the experimental perspective. Investigations are still of a somewhat qualitative nature and modeling at an exploratory stage.

Solubility calculations based on thermodynamic databases are unnecessarily conservative - or useless - for far-field mobility prediction purposes, since in real natural systems the actinide elements will be highly diluted and present as a minority or trace component dissolved in other mineral phases: their «effective» solubility products may be many orders of magnitude lower [Bruno *et al.*, 1985]. Kinetics of precipitation and of dissolution must also be included in a comprehensive model: progress is being made in geochemical modeling [Delany *et al.*, 1986] and, besides the vast literature on common mineral phases, some data already exist concerning nuclear glass and UO_2(fuel) kinetics of dissolution.

In the study of the interaction of dissolved species with solid surfaces, an ion exchange or surface adsorption representation results in a constant value, at trace concentrations, of the ratio $[M]_{solid\ phase}$ (in moles/gram) / $[M]_{solution}$ (in moles/liter), the distribution coefficient K_D. Excessive confidence in the use of K_D values has received wide criticism [Sholkovitz, 1983]: in particular, it appears inappropriate to enter the mass of solid phase in an equilibrium calculation, when only the surface is assumed to take part in the sorption of metal ions. On the other hand, K_D "constants" can be of practical use to obtain a qualitative picture of the extent of metal uptake on solids. In any case, measurements of distribution coefficients are plagued by poor reproducibility [Benes *et al.*, 1988], only in part due to losses on walls and incomplete phase separations, or slow kinetics [Jannasch *et al.*, 1988].

Desorption of ^{137}Cs from sediment gave a non-monotonic variation of K_D vs. time [Benes *et al.*, 1988]. The apparent distribution coefficient first decreased due to release of ^{137}Cs into the solution, but then increased again. A likely explanation can be found in an irreversible diffusion of the tracer in the bulk of the solid phase. It is reasonable to assume that, on the geological time scale, the actinides will penetrate (some of) the minerals with which they come in contact, either by direct diffusion, or by coprecipitation mechanisms.

Bruno and Sandino (1988) drew attention to the importance of coprecipation in the geochemical time-scale, and demonstrated in the U(Th,La,Ba)O_2 system a sequential transition from surface adsorption (modeled by Langmuir isotherms) to surface coprecipitation (Freundlich isotherm) to coprecipitation (controlled by Berthelot-Nernst or Hoskins distribution laws).

Trivalent and tetravalent species display very high distribution coefficients on the common mineral surfaces: iron oxides - which have received special attention [Music and Ristic, 1989] because of their use in waste decontamination - silica, alumina, clays.

The tendency of NpO_2^+ to be sorbed on minerals increases in the order: oxidized clay < montmorillonite < MnO_2 < kaolinite < FeOOH << carbonate sediment < calcite < aragonite [Keeney-Kennicutt and Morse, 1984] while PuO_2^+ adsorption grows in the order: montmorillonite < MnO_2 < oxidized clay < kaolinite << calcite < FeOOH <= aragonite < carbonate sediment [Keeney-Kennicutt and Morse, 1985]. The behavior of plutonium can be attributed to catalytic reduction to Pu(IV) on the goethite surface [Sanchez *et al.*, 1985]. Keeney-Kennicutt and Morse (1985) indicated that both surface oxidation and reduction can occour on FeOOH, and observed that this process could be driven by light.

In open waters the role of particles in the scavenging of metal ions has been recognized early [Krauskopf, 1954] and has led to the development of consistent numerical models [Nyffeler *et al.*, 1986]. It has been observed that increasing particle concentration induces a decrease in apparent K_D, which may be an artifact related to the formation of filterable colloids [Morel and Gschwend, 1987]. Honeymann *et al.* (1988) integrated particle concentration effects into an empirical scavenging model for thorium in seawater capable of reproducing both experimental sorption rates and K_D's.

Incorporation of Pu(VI) (or Pu(V)) in $CaCO_3$ may explain the observed vertical distribution of plutonium in the oceans: the significantly higher concentrations found close to the Pacific ocean bottom at 5.9 km depth and possibly in the 3 km layer [Nakanishi *et al*., 1984] may be due to release from sinking, dissolving aragonite.

Colloidal particles have received increasing attention as possible vectors for the actinides in transport mediated by water flow [Avogadro and De Marsily, 1984, Lieser *et al.*, 1986]. Colloids can be defined as particles of diameter between 1 nm and 1 μm. The distinction from solid (larger) particles is somewhat arbitrary: from a practical point of view, colloids are those particles than can pass through 0.45 um filters and do not settle in a gravitational field, but can be isolated by ultrafiltration and ultracentrifugation.

Colloids can adsorb (by physical sorption, chemisorption, ion exchange or coprecipitation) radionuclides from the surrounding solution to form "pseudocolloids" (in German "Fremdkolloiden" [Lieser *et al.*, 1986]). The migration of these colloidal particles will be controlled by filtration and sedimentation, as well as by their tendency to adhere to one another or to mineral surfaces, which is controlled in turn by their surface (zeta) potential.

Even if colloids are not thermodynamically stable [Vold and Vold, 1983] towards growth [Heicklen, 1976] and coagulation [Stumm and Morgan, 1981], significant amounts exist at least in surface waters [Hahn, 1983], as a result of mechanical stress or of precipitation from highly supersaturated solutions. It was indicated that during the dissolution of nuclear glass those elements for which both the dissolution rate and the maximum solubility are lower than that of the silicon dioxide matrix will be essentially released in the form of colloids [Avogadro and De Marsily, 1984].

Plutonium and americium added to water samples even at 10^{-8} - 10^{-10} M levels were found to be present in colloidal form [Nash *et al.*, 1988]. The different nature of Pu(IV) and Am(III) colloids was evidenced by the observation that colloidal americium dissolves promptly in diluted acid while plutonium requires prolonged standing [Tsvetaeva *et al.*, 1986].

In their investigation of water sampled from a borehole drilled in lavas 300 m to the side of a cavity created by a 200-500 kton nuclear explosion ten years before, Buddemeyer and Hunt (1988) found that virtually all of the lanthanide activity was transported by colloids (> 3 nm). While locations so energetically stressed may not be representative of the underground in general, the possibility of colloid transport appears demonstrated.

Humic matter also adsorbs on mineral surfaces [Davis, 1982] and colloids, resulting in either enhancement or decrease of metal adsorption. Few investigations have dealt with such mineral-organic matter-actinide equilibria, undoubtedly because of their complexity. V Moulin *et al.* (1989) report that log K_D for Am(III) on alumina, which varied from 10^1 ml g^{-1} at pH 3 - 5 to 10^3 ml g^{-1} at pH 6 - 10, upon addition af humic acid increased in the lower pH range and decreased in the higher pH range. This behavior appears related to a decrease at high pH of the distribution coefficient of HA on alumina, itself strongly dependent on HA concentration due to surface saturation phenomena. At low pH humic acid is strongly bound to Al_2O_3 and increases Am(III) adsorption. At higher pH, due to charge repulsion, it remains in solution and holds Am(III) there. Being negatively charged in a larger pH range, silica does not bind humic acid as strongly as alumina, and the effect of humic acid on Am(III) sorption is negative at all pH's and proportional to its concentration [Moulin and Stammose, 1989].

5. Conclusions

The fundamental chemistry of the actinides in the environmental pH-Eh range is now fairly well understood in the thermodynamic frame. Lacunae persist, notably concerning speciation at higher temperatures and pressures.

Extensive observations of the behavior of the actinides in a number of environmental systems, primarily the sea and fresh waters, are in fair accord with theoretical models. Chemistry in the underground (migration in rock) is still at an exploratory stage, undoubtedly because of the very complexity of the issue. Water-rock systems must be seen in their dynamic evolution: colloid transport can only be modeled with the help - beside chemistry - of physics and mathematics.

A solid understanding of actinide chemistry and environmental migration mechanisms must be at the foundation of nuclear waste disposal and radiation protection policies, and a necessary component of environmental impact, risk characterization, and accident management strategies.

6. Aknowledgements

Thanks are due to Alain Billon, Jordi Bruno, Valerie Moulin, and Pierre Vitorge for helpful criticism.

7. References

Aiken R, McKnight DM, Wershaw RL, McCarthy P (1985) Humic substances in soil, sediment and water. John Wiley & Sons, New York

Allard B (1981) Solubilities of actinides in neutral or basic solutions. In: Edelstein NM (ed.) Actinides in perspective (Proc. of the Actinides 1981 Conference. Pacific Grove, California, USA, 10-15 September 1981). Pergamon, Oxford, p 553

Altmann RS, Buffle J (1988) The use of differential equilibrium functions for interpretation of metal binding in complex ligand systems: its relation to site occupancy and site affinity distributions. Geochim Cosmochim Acta 52:1505-1519

Avogadro A, De Marsily G (1984) The role of colloids in nuclear waste disposal. Mat Res Soc Symp Proc 21:497-505

Baas Becking LGM, Kaplan IR, Moore D (1960) Limits of the natural environment in terms of pH and oxidation-reduction potentials. J Geol 68:243-284

Baes CF, Mesmer RF (1976) The hydrolysis of cations. John Wiley and Sons, New York.

Beitz JV, Hessler JP (1980) Oxidation state specific detection of transuranic ions in solution. Nuclear Technology 51:169-177

Benes P, Kuncova M, Slovak J, Lam Ramos P (1988) Analysis of the interaction of radionuclides with solid phase in surface waters using laboratory model experiments: methodical problems. J Radioanal Nucl Chem Articles 125:2:295-315

Berger P, Blanc P, Bourges J (1988) Electrochemical generation and stability of americium(IV) and americium(V) in potassium bicarbonate and carbonate solution. Radiochimica Acta 43:217:222

Bidoglio G (1982) Characterization of Am(III) complexes with bicarbonate and carbonate ions at groundwater concentration levels. Radiochem Radioanal Letters 53:1:45-60

Bondietti EA, Reynolds SA, Shanks MH (1975) Interaction of plutonium with complexing substances in soils and natural waters. In: Transuranium Nuclides in the Environment - Proceedings of a Symposium in San Francisco, 178-21 November 1975. IAEA-SM-199/52, IAEA Vienna p 273

Bourges JY, Guillaume B, Koehly G, Hobart DE, Peterson JR (1983) Coexistence of americium in four oxidation states in sodium carbonate - sodium bicarbonate medium. Inorg Chem 22:1179-1184.

Brookins DG (1988) Eh-pH diagrams for geochemistry. Springer-Verlag, Berlin

Bruno J, Grenthe I, Munoz M (1985)Studies of radionuclide coprecipitation-solid solution formation. The UO_2(s)-La$(OH)_3$(s) coprecipitation as an analogue for the UO_2(s)-Pu$(OH)_3$(s) system. Mat Res Soc Symp Proc 50:717:726

Bruno J, Casas I, Grenthe I, Lagerman B (1987) Studies of metal carbonate equilibria. 19. Complex formation in the Th(IV)-H_2O-CO_2(g) system. Inorg Chim Acta 140:299-301

Bruno J, Sandino A (1988) The thermodynamics and kinetics of coprecipitation and its effects on radionuclide solubility. Radiochimica Acta 44/45:17-21

Bruno J, Grenthe I, Robouch P (1989) Studies of metal carbonate equilibria. 20. Formation of tetra(carbonato)uranium(IV) ion, $U(CO_3)_4^{4-}$, in hydrogen carbonate solutions. Inorg Chim Acta 158:221-226

Buddemeyer RT, Hunt JR (1988) Transport of colloidal contaminants in groundwater: radionuclide migration at Nevada Test Site. Applied Geochem 3:535-548

Caceci MS, Choppin GR (1983a) The first hydrolysis constant of U(VI). Radiochimica Acta 33:207-212

Caceci MS, Choppin GR (1983b) An improved technique to minimize cation absorption in neutral solutions. Radiochimica Acta 33:113-114

Caceci MS (1985) The interaction of humic acid with Eu(III): complexation strength as a function of load and pH. Radiochimica Acta 39:51-56

Caceci MS (1989a) A simple, performant diode array spectrophotometer. Computers Chemistry 13:1:33-38

Caceci MS (1989b) Evidence for large organic scatterers in humic acid samples. Org Geochem (in press)

Cacheris WP, Choppin GR (1987) Dissociation kinetics of thorium-humate complex. Radiochimica Acta 42:185-190

Chandratillake MR, Newton GWA, Robinson VJ (1988) Chemval project. Comparison of thermodynamic databases used in geochemical modelling. Topical Report EUR 11891 EN, ECSC-EEC-EAEC, Brussels

Choppin GR (1983) Solution chemistry of the actinides. Radiochimica Acta 32:43-53

Choppin GR, Allard B (1985) Complexes of actinides with naturally occurring organic compounds. In: Freeman AJ and Keller C (eds) Handbook of the physics and chemistry of the actinides. Elsevier Science, Amsterdam, p 407

Ciavatta L, Ferri D, Grenthe I, Salvatore F, Spahiu K (1983) Studies of metal carbonate equilibria. 4. Reduction of the tris(carbonato)dioxouranate(VI) ion, $UO_2(CO_3)_3^{4-}$, in hydrogen carbonate solution. Inorg Chem 22:2088-2092

Davis JA (1982) Adsorption of natural dissolved organic matter at the oxide/water interface. Geochim Cosmochim Acta 46:2381-2393

De Brabandere J (1989) Humic acid complexation of europium in Boom clay. Doctoral thesis Nr. 177, Faculty of Agriculture, Catholic Univ Leuven (Belgium)

Delany JM, Puigdomenech I, Wolery TY (1986) Precipitiation kinetics option for for the EQ6 geochemical reaction path code. LLNL, Report UCRL-53642

Falck WE (1988) A review of modelling the interaction between natural organic matter and metal cations. British Geological Survey Technical Report WE/88/49

Giesy JP, Geiger RA, Kevern NR (1986) UO_2^{2+}- humate interactions in soft, acid, humate-rich waters. J Environ Radioactivity 4:39-64

Grenthe I, Riglet Ch, Vitorge P (1986a) Studies of metal-carbonate complexes. 14. Composition and equilibria of trinuclear neptunium(VI) and plutonium(VI) carbonate complexes. Inorg Chem 25:1679-1684

Grenthe I, Robouch P, Vitorge P (1986b) Chemical equilibria in actinide carbonate systems. J Less-Common Metals 122:225-231

Hagan PG, Clevelend JM (1966) The absorption spectra of neptunium ions in perchloric acid solution. J Inorg Nucl Chem 28:2905-2909

Hahn HH (1983) Kolloidale Wasserinhaltsstoffe in natuerlichen Gewaessern. Wasserwisrtschaft 73:11:434-441

Heicklen J (1976) Colloid formation and growth. Academic Press, New York

Honeyman BD, Balestrieri LS, Murray JW (1988) Oceanic trace metal scavenging: the importance of particle concentration. Deep Sea Research 35:2:227-246

Jannasch H W, Honeymann BD, Balestrieri LS, Murray JW (1988) Kinetics of trace element uptake by marine particles. Geochim Cosmochim Acta 52:567-577

Joergensen CK, Reisfeld R (1982) Uranyl Photophysics. Structure and Bonding 50:122-171

Katz JJ, Seaborg GT, Morss LR (1986) (Eds) The chemistry of the actinide elements (2nd edition). Chapman and Hall, London

Keeney-Kennicutt WL, Morse JW (1984) The interaction of $Np(V)O_2^+$ with common mineral surfaces in dilute aqueous solutions and seawater. Marine Chem 15:133-150

Keeney-Kennicutt WL and Morse JW (1985) The redox chemistry of $Pu(V)O_2^+$ interaction with common mineral surfaces in dilute solutions and seawater. Geochim Cosmochim Acta 43:781-787

Kim JI, Lierse Ch, Baumgartner F (1983) Complexation of the plutonium(IV) ion in carbonate-bicarbonate solutions. In: Carnall WT, Choppin GR (Eds) ACS Symp Ser No 216, American Chemical Society, Washington, p 317

Kniewald G, Branica M (1988) Role of uranium(V) in marine sedimentary environments: a geochemical possibility. Marine Chem 24:1-12

Krauskopf KB (1954) Factors controlling the concentration of thirteen trace metals in seawater. Geochim Cosmochimm Acta 12:331-344

Langmuir D (1978) Uranium solution - mineral equilibria at low temperatures with applications to sedimentary ore deposits. Geochim Cosmochim Acta 42:547-569

Leach SJ, Pryke DC (1986) NearSol, a simple program to model actinide speciation and solubility under waste disposal conditions: a users guide. AERE Harwell Report R-12066

Lieser KH, Gleismann B, Peschke S, Steinkopff TH (1986) Colloid formation and sorption of radionuclides in natural systems. Radiochimica Acta 40:39-47

Maes A, Cremers A (1985) Radionuclide sorption in soils and sediments: oxide-organic matter competition. In: Bulman RA, Cooper JR (Eds) Speciation of fission and activation products in the environment. Proceedings of Speciation-85 Seminar, Christ Church, Oxford, UK, 16-19 April 1985. Elsevier Applied Science, London, New York, p 93

Morel FFM, Gschwend PM (1987) The role of colloids in the partitioning of solutes in natural waters. In: Stumm W (Ed) Aquatic surface chemistry: chemical processes at the particle-water interface. John Wiley, New York, p 405

Morse JW, Choppin GR (1986) Laboratory sudies of plutonium in marine systems. Marine Chem 20:73-89

Moulin C, Decambox P, Mauchien P (1989) Direct trace determination of Cm by laser-induced time-resolved spectrofluorimetry. Radiochimica Acta (in press)

Moulin V (1986) Les acides humiques et leurs interactions avec les elements metalliques Cu(II), Eu(III), Th(IV), U(VI): apport d'une methode de chromatographie par exclusion sterique et recherche de modeles de complexation. Report CEA-R-5354

Moulin V, Robouch P, Vitorge P, Allard B (1987) Spectrophotometric study of the interaction between americium(III) and humic materials. Inorg Chem Acta 140:303-306

Moulin V, Robouch P, Vitorge P, Allard B (1988) Environmental behavior of americium(III) in natural waters. Radiochimica Acta 44/45:33-37

Moulin V, Stammose D, and Mahn Tran T (1989) Chemical processes at the mineral oxide-water interface: system Al_2O_3-organic matter- Am(III). In: Miles (Ed) Water-Rock Interaction. Balkema, Rotterdam, p 505

Moulin V, Stammose D (1989) Effect of humic substances on Am(III) retention onto silica. Mat Res Soc Symp Proc 127:723-727

Music S, Ristic M (1988) Adsorption of trace elements or radionuclides on hydrous iron oxides. J Radioanal Nucl Chem 120:289-304

Musicas C (1976) Contribution to the study of U(V) ions and pentavalent transuranics in aqueous solution. J Inorg Nucl Chem Supplement 1976:171-177

Nakanishi T, Yajima M, Senaga M, Takei M, Ishikawa A, Sakamoto K, Sakanoue M (1984) Determination of ^{239,240}Pu in seawater. Nucl Inst Methods Phys Res 223:239-242

Nakayama S, Nelson DM (1988) Comparison of distribution coefficients for americium and curium: effects of pH and naturally occourring colloids. J Environ Radioactivity 8:173-181

Nash KL, Fried S, Friedman AM, Sullivan JC (1981) Redox behavior, complexing, and adsorption of hexavalent actinides by humic acid and selected clays. Environ Sci Technol 15:7:834-837

Nash KL, Cleveland JM, Rees TF (1988) Speciation pattern of actinides in natural waters: a laboratory investigation. J Environ Radioactivity 7:131-157

Newton TW, Sullivan JC (1985) Actinide carbonate complexes in aqueous solution. In: Freeman AJ, Keller C (Eds) Handbook on the physics and chemistry of the actinides. Elsevier Science, Amsterdam, p 387

Nelson DM, Penrose WR, Karttunen JO, Mehlhaff P (1985) Effects of dissolved organic carbon on the adsorption properties of plutonium in natural waters. Environ Sci Technol 19:127-131

Nitsche H, Lee SC, Gatti RC (1988) Determination of plutonium oxidation state at trace levels pertinent to nuclear waste disposal. J Radioanal Nucl Chem Articles 124:1:171-185

Nitsche H, Standifer EM, Silva RJ (1989) Americium(III) carbonate complexation in aqueous perchlorate solution. Radiochimica Acta 46:185-189

Nyffeler UP, Santschi PS, Li Y-H (1986) The relevance of scavenging kinetics to modeling of sediment-water interactions in natural systesm. Limnology and Oceanography 31:277-292

Plummer LN, Jones BF, Truesdell AH (1976) WATEQF-a Fortran(IV) version of WATEQ, a computer program for calculating chemical equilibium in natural waters. US Geological Survey Water Resources Investigation 76-13

Pryke DC, Rees JH (1986) Understanding the behavior of the actinides under disposal conditions: a comparison between calculated and experimental solubilities. Radiochimica Acta 40:27-32

Riglet Ch (1989) Chime du neptunium et autres actinides en milieu carbonate. Doctoral Thesis, University Paris 6, March 17

Riglet Ch, Vitorge P, Grenthe I (1989a) Standard potentials of the (MO_2^{2+}/MO_2^{+}) systems for uranium and other actinides. Inorg Chim Acta 133:323-329

Riglet Ch, Robouch P, Vitorge P (1989b) Standard potentials of the (MO_2^{2+}/MO_2^{+}) and (M^{4+}/M^{3+}) redox systems for neptunium and plutonium. Radiochimica Acta 46:85-94

Robouch P (1989) Contribution a la prevision du comportement de l'americium, du plutonium et du neptunium dans la geosphere: donnees chimiques. Report CEA-R-5473

Ryan JL, Rai D (1983) The solubility of uranium(IV) hydrous oxide in sodium hydroxide solution under reducing conditions. Polyhedron 2:947-952

Sanchez AL, Murray JW, Sibley TH (1985) The adsorption of plutonium on goethite. Geochim Cosmochim Acta 49:2297-2307

Schnitzer M, Khan SU (1972) Humic substances in the environment. Marcel Dekker, New York

Sholkovitz ER (1983) The geochemistry of plutonium in fresh and marine water environments. Earth Sci Rev 19:95-161

Silva RJ (1985) Carbonate complexation of Pu(IV) in aqueous solution. Paper 44, Div Nucl Chem Tech, 189th ACS Meeting, Miami Beach, FL April 28-May 3

Silver GL (1988) Formation of plutonium(IV) colloid from plutonyl cations. J Radional Nucl Chem Letters 126:6:429-433

Sposito G, Mattigod SV (1979) GEOCHEM: a computer program for the calculation of chemical equilibria in soil solutions and other natural water systems. Kearney Foundation of Soil Science, University of California, Riverside

Stumm W, Morgan J (1981) Aquatic Chemistry. John Wiley and Sons, New York

Torres RA, Choppin GR (1984) Europium(III) and americium(III) stability constants with humic acid. Radiochimica Acta 35:143-147

Tsvetaeva NE, Filin VM, Ragimov TK, Shcherbakov BYa, Rudaya LYa, Shapiro KYu (1986) Comparative behavior of americium and plutonium in wastewater. Radiokhimiya 28:1:129-133

Varlashkin PG, Hobart DE, Begun GM, Peterson JR (1984) Electrochemical and spectroscopic studies of neptunium in concentrated aqueous carbonate and carbonate-hydroxide solutions. Radiochimica Acta 35:91-96

Vold RD, Vold MJ (1983) Colloid and interface chemistry. Addison-Wesley, Reading, Massachussets

Watters RL, Hakonson TE, Lane LJ (1983) The behavior of actinides in the environment. Radiochimica Acta 32:89-95

Wester DW, Sullivan JC (1983) The absorption spectra of Pu(VI), -(V) and -(IV) produced electrochemically in carbonate-bicarbonate media. Radiochem Radioanal Letters 57:1:35-42

ACCUMULATION AND DISTRIBUTION OF ^{99}Tc IN THREE BEAN PLANT VARIETIES

Bennàssar, A., Cabot, C., Vázquez, M. D.*, Poschenrieder, Ch.*, Barceló J.*
Dep. Biologia i C.S. Institut d'Estudis Avançats.
Universitat de les Illes Balears.
Car. Valldemossa, Km 7.5.
07071 Palma de Mallorca, Spain.

1. Introduction

In nature technetium-99 (^{99}Tc) exists in very small amounts as a product of the spontaneous fission of ^{235}U (1). The potencial entrances of ^{99}Tc into the environment are principally arise from the nuclear fuel cycle, which yields high amounts of this isotope by slow neutron fission of ^{235}U and ^{239}Pu (6.3 and 6.1 % respectively) (2), and from nuclear weapons, hospitals and research institutions.

In soils, ^{99}Tc may be present as soluble pertechnetate ion ($^{99}TcO_4^-$), which is highly mobile within trophic chains. Thus ^{99}Tc is a potentially dangerous ambient pollutant if for some reason its concentration in the environment were to increase. Phytotoxic effects of ^{99}Tc have already been described in experiments with algae and higher plants (3). Toxic effects of ^{99}Tc occur almost immediately and are due to chemical rather than radiological damage. Nevertheless radiation effects cannot be discounted when high ^{99}Tc concentrations are present (4)

Technetium-99 has been reported to be rapidly absorbed by plant species such as soybean, corn, oat, barley, radish and wheat (4), and very high transfer factors for this isotope have been found. Within plants, ^{99}Tc is principally accumulated in leaves and apical meristems (3).

The aims of this study were to determine if there were any differences in the accumulation and distribution of ^{99}Tc among three bush bean varieties and to study the effects of ^{99}Tc on cell ultrastructure.

---

*Dep. Fisiologia Vegetal. Facultad de Ciències. Universitat Autonoma de Barcelona. 08193 Bellaterra, Spain.

Supported by DGICYT (Spain project PB88-0234)

NATO ASI Series, Vol. G 23
Metal Speciation in the Environment
Edited by J. A. C. Broekaert, Ş. Güçer, and F. Adams

2. Material and Methods

Three bush bean (*Phaseolus vulgaris* L.) varieties, Contender, Garrafal Nana and Superviolette were grown in greenhouse in 25% Hoagland and Arnon nutrient solution (5) with 0, 0.075, 0.15, 0.3 and 0.6 mg/l ^{99}Tc supplied as NH_4TcO_4, at pH levels of 5 and 6.8. Choice of these ^{99}Tc concentrations was made according to previous studies on ^{99}Tc toxicity levels for bush bean plants (6). The maximum established environmental concentrations from fallout are substancially lower (i.e. 1.0 pg/g ^{99}Tc in soils) (7).

After 12 days, six plants per treatment were harvested and fresh and dry weights (after 24h at 70 ºC) of roots, stems, primary and trifoliolate leaves were recorded. Technetium-99 content of the different plant organs was analyzed by liquid scintillation spectrometry following a special digestion procedure for volatile samples (8).

For cell ultrastructure studies, bean plants var. Contender were grown in 10% Hoagland and Arnon nutrient solution (5) with 0 and 0.12 mg/l ^{99}Tc supplied as NH_4TcO_4 at pH 5.0. After 6 days plants were harvested, growth parameters recorded and ^{99}Tc content analyzed following the same procedure presented above. Samples from primary leaves and root tips were taken for TEM and SEM studies. Sample preparation was performed as previously described (9).

Results

Plant productivity was fairly affected by ^{99}Tc supply at the short period of time considered (12 days). Figure 1 shows the relationship between ^{99}Tc supply and total plant dry weights at pH 5 (A) and 6.8 (B).

Whereas Contender and Superviolette showed a homogeneous response to ^{99}Tc supply at both pH levels tested, Garrafal Nana exhibited an irregular response at pH 5.0 being the most inhibited variety at pH 6.8. Technetium-99 distribution was the same in the three varieties increasing in the order shoot < root < trifoliolate leaf < primary leaf (Fig. 2).

Technetium-99 accumulation level changed depending on pH and variety (Fig 2). Contender and Garrafal Nana showed higher ^{99}Tc content at pH 5 whereas Superviolette exhibited higher ^{99}Tc values at pH 6.8. Contender accumulated the highest ^{99}Tc tissue concentrations. Primary leaves hold values up to those found for essential micronutrientes such as Fe. Electron microscopy studies considered the effects of ^{99}Tc at different levels in roots and primary leaves of 6 days old

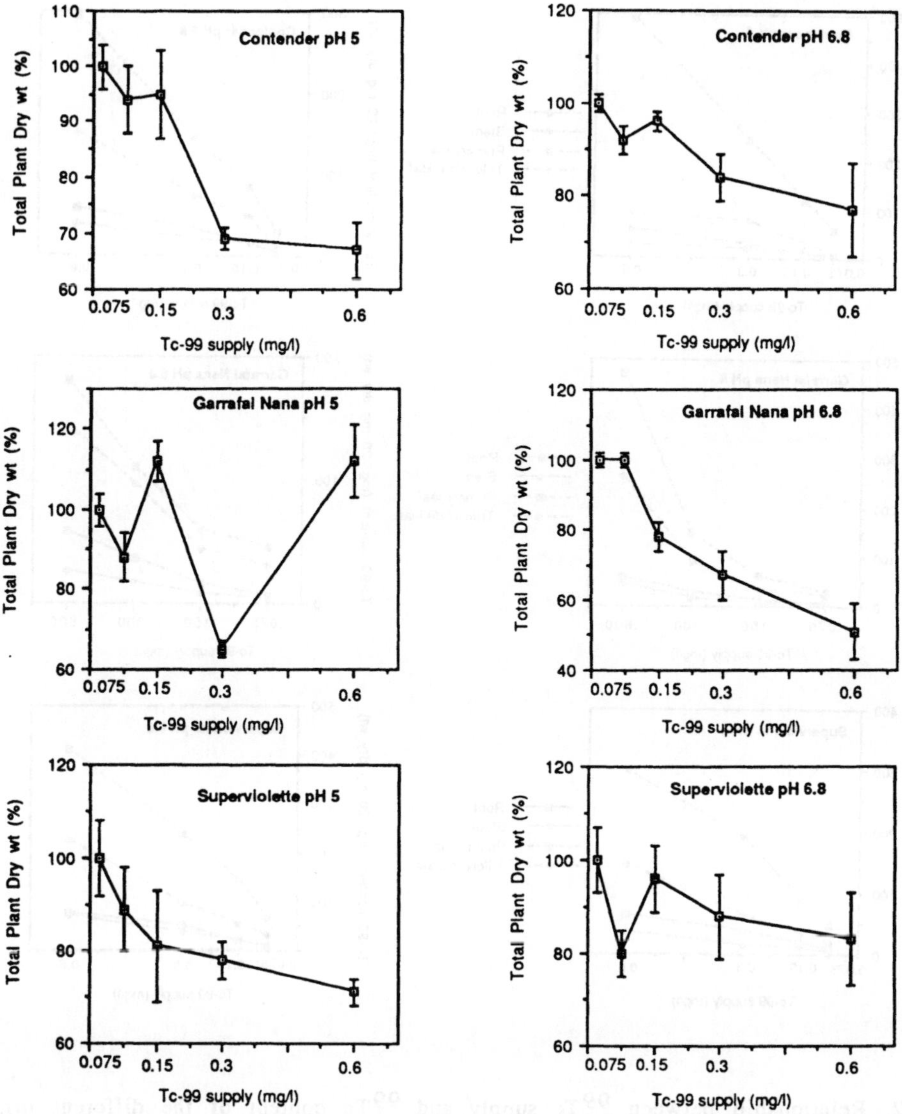

Fig 1. Relationship between ^{99}Tc (mg/l) and total plant dry weight (g) of the three bush bean varieties Contender, Garrafal Nana and Superviolette at pH 5 and pH 6.8.

^{99}Tc-treated plants. TEM micrographs from primary leaves revealed the accumulation of electron-dense material within the vacuoles of epidermal cells. The tonoplast of these cells was heavily stained (Fig. 4). Chloroplasts from ^{99}Tc-treated plants frequently were abnormally swollen and presented distorted

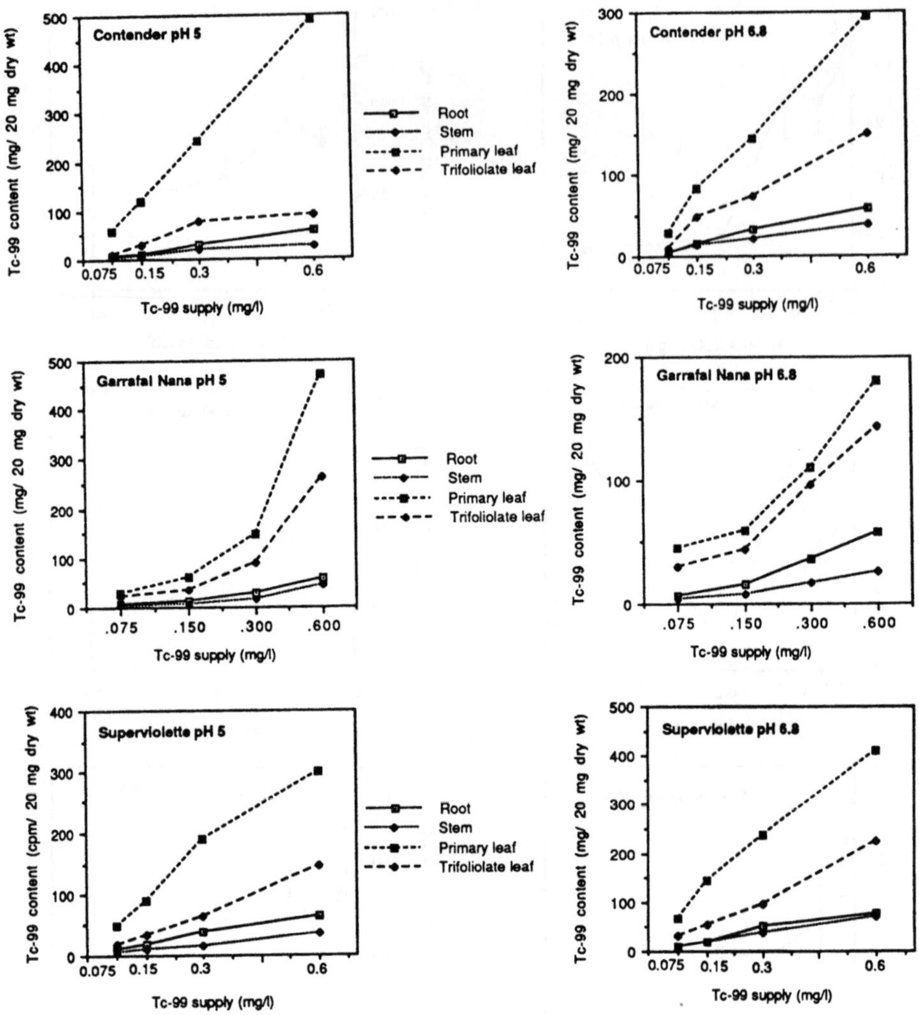

Fig 2. Relationship between ^{99}Tc supply and ^{99}Tc content of the different organs at pH 5 and pH 6.8 in *Phaseolus vulgaris* var Superviolette , Contender and Garrafal Nana.

thylakoid membranes. Starch was only found within a few chloroplasts of ^{99}Tc-treated plants. The absence of starch within most of the chloroplasts (Fig. 5) from leaves affected by ^{99}Tc toxicity contrasted with the high starch amounts found in control plants (Fig. 3). In ^{99}Tc-treated plants, vacuoles from leaf parenchyma cells frequently contained myelin-like whorls, vesicles and membraneous fragments. SEM micrographs from root tips of ^{99}Tc-treated plants revealed an increased development of the root cap (Fig. 7) in comparison to controls (Fig. 6). Vacuoles of root cortex cells from ^{99}Tc-treated plants contained osmiophilic material with the appearance of protein bodies (Fig. 8).

Discussion

In the short period of time considered (12 days) the three bush bean varieties tested showed low sensitivity to ^{99}Tc supply.

Our results on ^{99}Tc accumulation and distribution in *Phaseolus vulgaris* are similar to those previously described by others (3,4,9,10) for different higher plant species. The analysis of ^{99}Tc concentrations within different plant organs showed that ^{99}Tc is highly mobile within plants and, unlike other heavy metals, ^{99}Tc is not accumulated within roots but leaves. Thus, in contrast to other polluting trace elements ^{99}Tc is rather retained within the root by the endodermis barrier but easily attains the long-distance transport way, i.e. the xylem.

Although all bean plant varieties tested showed the same distribution pattern of ^{99}Tc within their different organs, there were varietal differences concerning the amount of ^{99}Tc taken up. Furthermore, ^{99}Tc uptake was significantly influenced by pH. The fact that pH influenced ^{99}Tc uptake in different ways in the diverse varieties suggests that the pH more significantly affected the growth and metabolism of the different bean varieties than the availability or ionic form of ^{99}Tc.

According to the distribution pattern of ^{99}Tc within plants, the ultrastructure of leaves was much more affected than that of roots. Within epidermal cells, a high accumulation of ^{99}Tc transported by the transpiration stream may be responsible for the electron-dense precipitates observed. Autoradiographic studies were undertaken to confirm this hypothesis. Technetium-99 injured cell membranes as indicated by the appearance of myelin whorls within vacuoles and the distortion of thylakoids. The absence of starch in chloroplasts from ^{99}Tc-treated plants accords with previous reports on ^{99}Tc-induced inhibition of photosynthesis (11) which may lead to carbon starvation and growth reduction.

Moreover ^{99}Tc seems to interfere with the use of storage protein, as indicated by the high amount of protein bodies found in root cortex cells. The increased root cap size in ^{99}Tc-treated plants resulted from an inhibition of root cell slough off. This may be related to ^{99}Tc effects on mucilage production in the root cap.

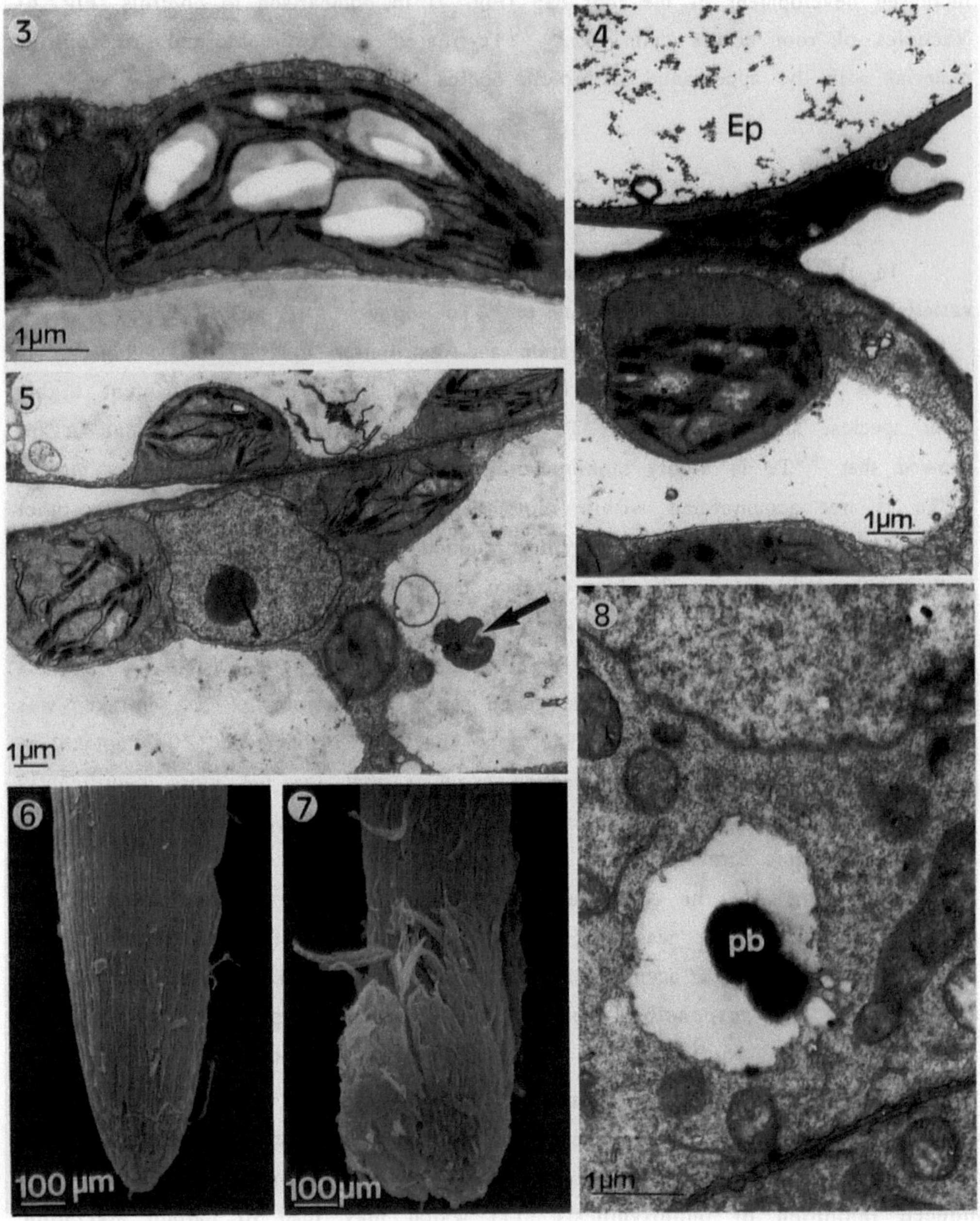
3
1µm
4
Ep
1µm
5
1µm
8
pb
1µm
6
100 µm
7
100µm

Fig 3. LM micrograph from a leaf parenchyma cell of a bush bean plant grown in control nutrient solution. Note the well organized thylakoid membrane system and the abundance of starch.
Fig 4. TEM micrograph from a primary leaf of a bush bean plant grown in solution containing 0.12 mg ^{99}Tc/l. Note the electron-dense precipitates in the vacuole of the epidermis cell (Ep) and the swollen chloroplast without starch.
Fig 5. TEM micrograph of primary leaf parenchyma cells from a busch bean plant grown in nutrient solution containing 0.12 mg ^{99}Tc/l. Note the disorganization of the thylakoid membrane system and the absence of starch in most of the choroplasts. Myelin-like whorls (arrow) indicate alterations at the membrane level.
Fig 6. SEM micrograph from the root tip of a control plant.
Fig 7. SEM micrograph from the root tip of a bush bean plant exposed to 0.12 mg ^{99}Tc/l in nutrient solution. Note the increased development of the root cap.
Fig 8. Differentiating cell from the root tip of a bush bean plant grown in nutrient solution containing 0.12 mg ^{99}Tc/l. Note the presence of protein bodies (pb).

In conclusion we can state that there are varietal differences for ^{99}Tc uptake in bean plants; however all varieties tested accumulate substantial amounts of the element, which is extremely injurious to the plant's metabolism.

References

1) Wildung RE, Garland TR, Cataldo DA (1977) Accumulation of technetium by plants. Health Phys. 32: 314-317

2) Anderson TJ, Walker RL (1980) Determination of picogram amounts of Technetium-99 by resin bead mass spectrometric isotope dilution. Anal. Chem. 52: 709-713

3) Cataldo DA, Wildung RE, Garland ThD (1983) Root absorption and transport behavior of technetium in soybean. Plant Physiol. 73: 849-852

4) Berlyn GP, Dhillon DD, Koslow EE (1980) Technetium: A toxic waste product of the nuclear fuel cycle: effects on soybean, growth and development. Environ. Manag. 4:149-156

5) Hoagland DR, Arnon DI (1950) Agricultural experimental station, Circular 347

6) Bennàssar A, (1988) Tecneci-99 en *Phaseolus vulgaris*, Tesi Doctoral. Universitat de les Illes Balears

7.) Routson RC, Cataldo DA (1978) Accumulation of ^{99}Tc by tumbleweed and cheatgrass grown on arid soils. Health Phys 34: 685-690

8) Bittell R (1980) Radioprotection 15: 141-146

9) Barceló J, Vázquez MD, Poschenrieder Ch (1988) Cadmium-induced structural and ultrastructural changes in the vascular system of bush bean stems. Botanica Acta 101: 254-261

10) Sheppard MI, Vandergraaf TT, Thibault DH, Keith Reid JA (1983) Thechnetium and uranium: sorption by and plant uptake from peat and sand. Health Phys. 44: 635-643

11) Gearing P, Van Baalen C, Parker PL (1975) Biochemical effects of technetium-99 pertechnetate on microorganisms. Plant Physiol. 55: 240-246

Heavy metals in sediments of Turkish river systems - Natural background and anthropogenic effects

F.-M. Kindler and H.E. Sevim
Institut für Sedimentforschung
Universität Heidelberg
Im Neuenheimer Feld 236
6900 Heidelberg
Federal Republic of Germany

Introduction

During the months of July-October 1988, first samples of sediment were taken from the 7 main rivers in Turkey (Fig. 1). To ensure an accurate picture, a total of 250 samples were collected at 10- 20 km intervals (Table 1).

Fig. 1. Map of Turkey - with the investigated Rivers Sakarya, Kizilirmak, Yesilirmak, Seyhan, Ceyhan, Menderes and Gediz

NATO ASI Series, Vol. G 23
Metal Speciation in the Environment
Edited by J. A. C. Broekaert, Ş. Güçer, and F. Adams

River	Length (km)	Number of Samples	Catchment Area (km^2)
Sakarya	824	41	55321
Kizilirmak	1355	76	75120
Yesilirmak	519	30	35958
Ceyhan	509	27	20466
Seyhan	560	32	17496
Menderes	530	23	23889
Gediz	320	19	15616

Table 1. Length of rivers, number of samples, and catchment area of rivers examined

The study was undertaken with the aim of obtaining data on the geochemical background of heavy metals such as Cd, Co, Cr, Cu, Fe, Mn, Ni, Pb, and Zn in areas which have hardly been exposed to anthropogenic influence so far. Given that extensive parts of Turkey consist of magmatic rock including many ultrabasites (serpentinits), the geogenic base load was estimated to be high, particularly with regard to Co, Cr, and Ni. Furthermore, the impact of the accelerating industrialization on river sediments was investigated. The quality of flowing water is vital in view of its exploitation as drinking water and industrial water, as also for irrigation purposes.

Materials and methods

We succeeded in taking samples mainly of the finest sediment available, i.e. in the flood plain area. Samples of non-oxidized sediment were disregarded in this presentation. The following procedures were applied to all samples:

a) Grain size fractioning at the intervals: <0.002 - 0.0063 - 0.020 - 0.063 - >0.063 (mm) (wet sieving and Atterberg sedimentation-tube).
b) Measurement of the heavy metals Cd, Co, Cr, Cu, Fe, Mn, Ni, Pb, and Zn - in the bulk sample as well as in the fraction <0.002 mm (AAS /GFAAS).
c) Six extraction steps based on a scheme given by Tessier *et al.* (1979) with modifications by Shuman (1985).

Extraction scheme:

I exchangeable 1 M $NH_4OAc.$ + $C_2H_4O_2$, pH 7, shaking 2h, (1 : 25)

II carbonatic 1 M NaOAc. + $C_2H_4O_2$, pH 5, shaking 12h, (1 : 25)

III easy/mod. reducible (amorphous Mn/Fe-oxides), 0.2 M $(NH_4)_2C_2O_4$ + 0.2 M HO_2CCO_2H pH 3, shaking (dark) 24h, (1 : 100)

IV mod./strong reducible (cristalline Fe-oxides), 0.2 M $(NH_4)_2C_2O_4$ + 0.2 M HO_2CCO_2H + 0.1 M $C_6H_8O_6$ pH 3, water bath 97° C 1h, (1 : 100)

V oxidizable (organic + sulfidic), 30% H_2O_2 + 0.02 M HNO_3 pH 2, water bath 85° C 4h + 3.2 M $NH_4OAc.$/20% HNO_3 30min, (1 : 25)

VI residual HNO_3 conc., 150° C 3h, (1 : 25)

Results

Owing to temporary high water flows (strong seasonal fluctuations, currents caused by dams), most samples, the majority of which were taken from the flood plain, were extremely sandy (Fig. 2). At the time the sample was collected (i.e.during low water), only 1-2 mm at the surface contained fine sediment.

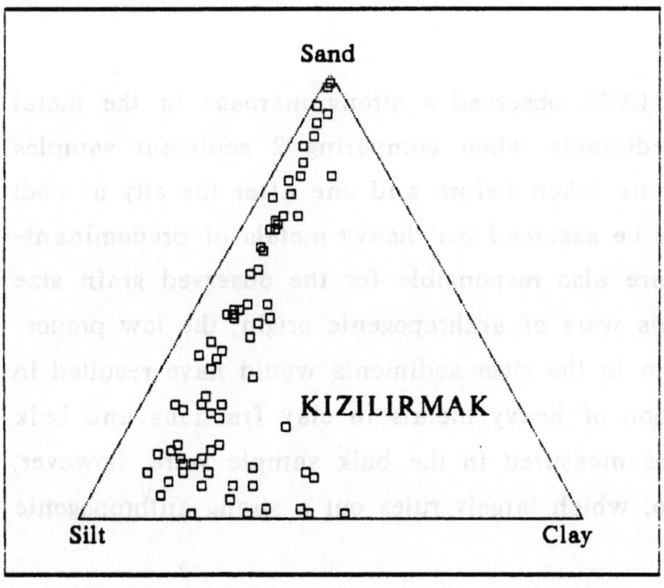

Fig. 2. Example of particle-size distribution in the River Kizilirmak sediments (76 Samples)

The concentration of heavy metals (Cu, Fe, Mn, Pb, and Zn) within the individual grain classes increases considerably as the grain size decreases (Fig. 3). A plausible explanation for this might be surface dependent binding forms (such as Mn-/Fe-oxides), since the relation of surface/volume increases noticeably with decreasing grain size (Brook and Moore 1988). In the case of Co and Cr a tendency toward medium grain sizes was observed, as also in the case of Ni to a lesser degree. This might be due to a concentration of mineral components in coarser grain classes.

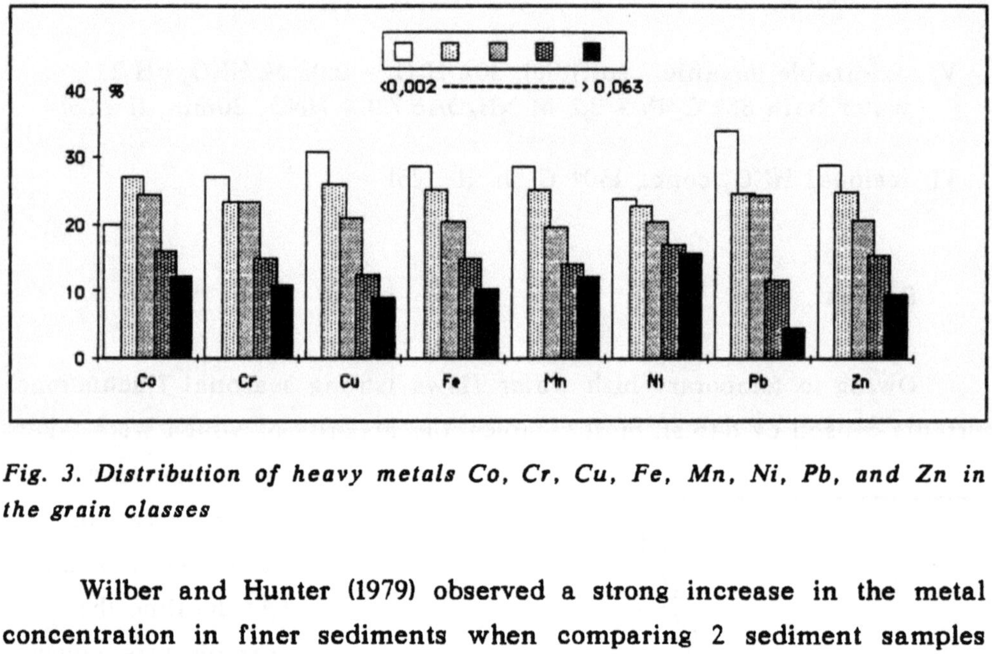

Fig. 3. Distribution of heavy metals Co, Cr, Cu, Fe, Mn, Ni, Pb, and Zn in the grain classes

Wilber and Hunter (1979) observed a strong increase in the metal concentration in finer sediments when comparing 2 sediment samples from the Saddle River - one taken before and one after the city of Lodi (New Jersey). Thus it may be assumed that heavy metals of predominantly anthropogenic source are also responsible for the observed grain size effect. If the heavy metals were of anthropogenic origin, the low proportion of the fraction < 2 μm in the river sediments would have resulted in a more varied concentration of heavy metals in clay fractions and bulk sample. The actual values measured in the bulk sample were, however, only 50% lower in average, which largely rules out a strong anthropogenic influence.

A comparison of the mean values in the fraction < 2 μm with the data on pre-civilizatory shale, presented by Turekian and Wedepohl (1961) reveals a 6-fold value for Ni (Seyhan). Also the values of Co, Cr, Pb, and Zn are 2-3 times as high as in the compared data, which supplies background data in the FRG and in other countries (Fig. 4).

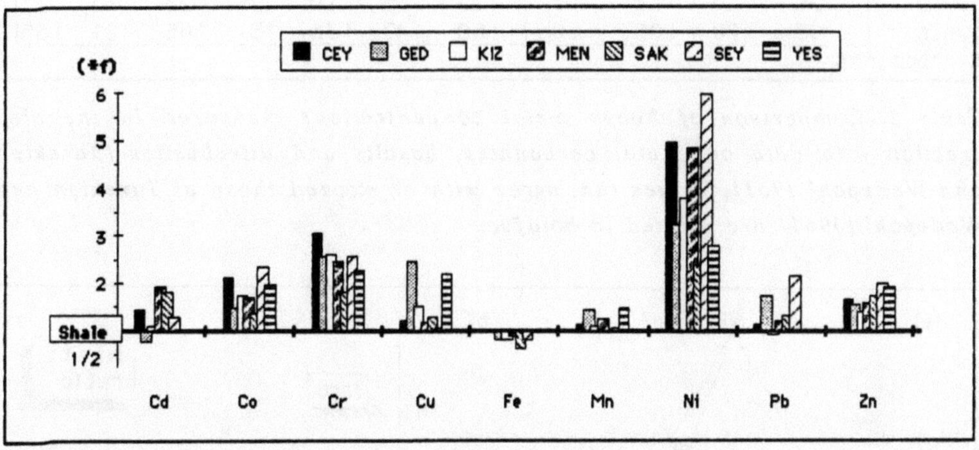

Fig. 4. Comparison between heavy metal concentrations in the clay fraction of all rivers and the data presented by Turekian and Wedepohl (1961) on unpolluted shale (no civilizatory influence). The Ni-concentration in the clay fraction of the River Seyhan is 6 times as high as in the compared data

On the other hand, there is good agreement between various characteristic lithogenic data mainly for basalts and ultrabasites (Table 2). The ratio of Co:Cr:Ni additionally confirms the predominantly geogenic origin of the metals (Fig. 5).

A similarly evident geogenic enrichment of Co, Ni, and Cr by factors of ten and more was found by Förstner (1977) in recent lake deposits of a greenstone-belt in Western Australia.

In the examined samples from Turkish rivers there is a lesser proportion of residual fraction (Fig. 6) than Salomons and Förstner (1980) found in their study on rivers world-wide that are hardly exposed to civilizatory influence (Table 3).

ppm	Shale	Carb.	Bas.	ultrab.	CEY	GED	KIZ	MEN	SAK	SEY	YES
Cd*	**300**	35	**220**	<100	**434**	236	**329**	580	**540**	390	**311**
Co	**19**	< 1	**48**	150	**40**	28	**33**	33	**26**	44	**38**
Cr	90	11	**170**	**1600**	**275**	232	**237**	223	**163**	233	**205**
Cu	**45**	4	**87**	10	**56**	110	**69**	53	**58**	48	**99**
Fe**	**4,7**	0,4	**8,7**	9,4	**4,9**	3,9	**3,9**	4,1	**2,9**	3,9	**4,5**
Mn	**850**	**1100**	**1500**	**1620**	**964**	1222	**955**	1066	**855**	911	**1273**
Ni	68	20	**130**	**2000**	**338**	222	**259**	331	**162**	407	**190**
Pb	**20**	9	6	1	**23**	35	**19**	25	**26**	43	**21**
Zn	**95**	20	**105**	50	**160**	147	**136**	151	**165**	191	**185**

*ppb **%

Table 2. Comparison of heavy metal concentrations measured in the clay fraction with data on shale, carbonates, basalts and ultrabasites (Turekian and Wedepohl 1961). Values that agree with or exceed those of Turekian and Wedepohl (1961) are printed in boldface

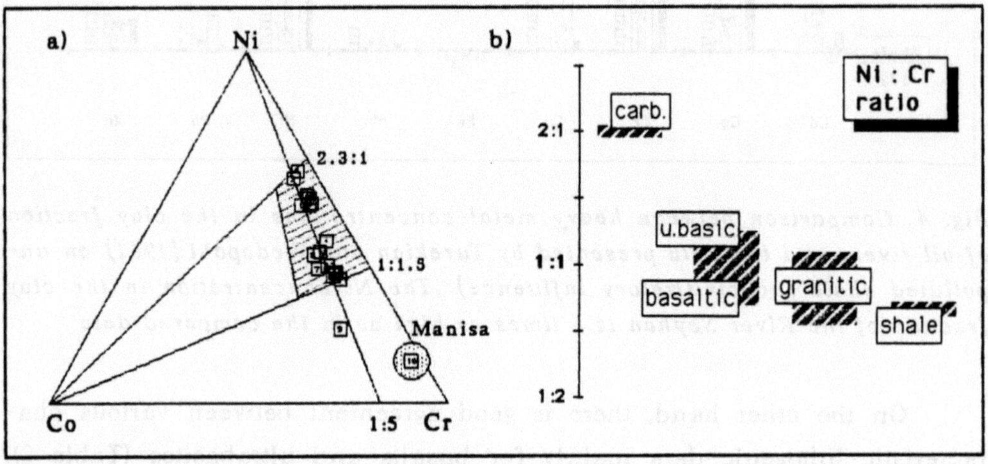

Fig. 5. The ratio Co: (Cr/Ni) (a) appears to be quite constant (1:5) due to the low Co concentration. The ratio Ni:Cr fluctuates between 2.3:1 and 1:1.5 which corresponds to that of various rocks (Turekian and Wedepohl 1961)(b)

Zn in %	Step	CEY	GED	KIZ	MEN	SAK	SEY	YES	Rhein	Zaire	Ottawa R.
	6	24	25	30	24	25	34	20	17	53	83
	6+4	39	45	47	39	36	64	31			

Table 3. Proportion of residual fraction (step 6) in various international rivers with different civilizatory influence (Salomons and Förstner 1980), and in the investigated Turkish rivers

The climate favours weathering residue such as crystalline Fe- oxides, in particular haematite (α-Fe_2O_3). Red suspension material gave the River Kizilirmak its colouring and its name (Kizilirmak = red river). In the applied extraction procedure by Shuman (1985) which is extremely selective, especially for crystalline Fe-oxides, these oxides are recorded in a separate step (IV). Investigations by Rapin and Föstner (1983) show that the extraction method used by Salomons *et al.* (1980), i.e. by determining the reducible fraction with the help of NH_2OH HCl, does not lead to complete extraction of the crystalline oxides. This is only achieved in the last step (residual fraction). The H_2O_2 step used does not make a division into the two important phases possible: sulfides (diagenetic, detrigenic), and organic substance. It does, however, have a high selectivity for PbS (Rapin and Förstner 1983). This explains the high percentage (70%) of lead extracted in this step from sediment from the Kizilirmak (Fig. 6). The only metals worth mentioning, that are bound in carbonate and exchangeable form are Cd, Mn, and Pb (Fig. 6). A significant decrease in residue-bound Cr can generally be observed about 1-2 samples downstream after big cities (20-30 km). This additional metal content is apparently fixed in the sediment in an unstable form only (Fig. 7).

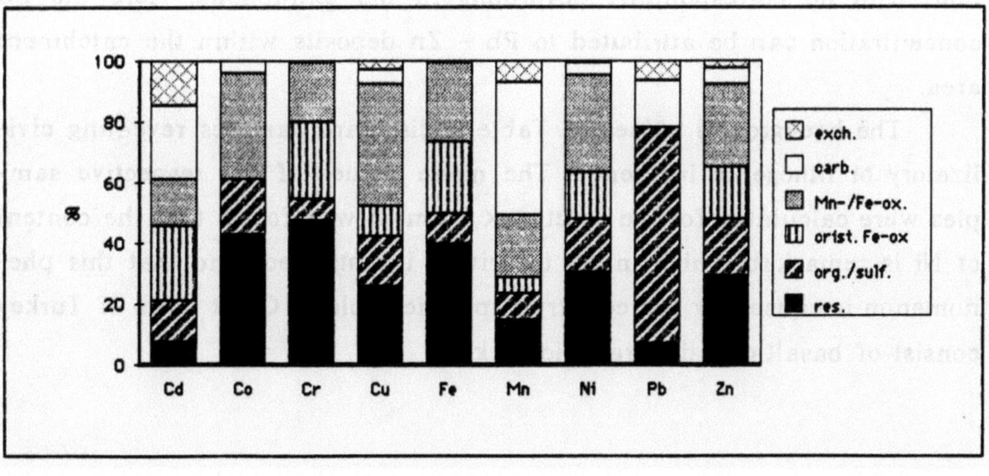

Fig. 6. Chemical partition of heavy metals Cd, Co, Cr, Cu, Fe, Mn, Ni, Pb and Zn in Kizilirmak sediments - the same has been observed in all Turkish rivers

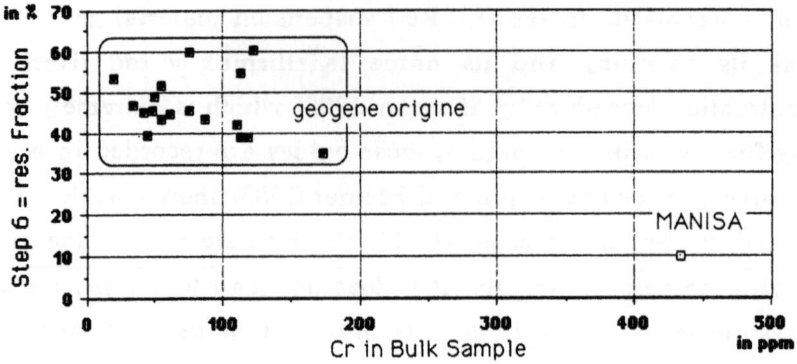

Fig. 7. The sample taken after the town of Manisa in western Turkey showed a 4-5-fold content of Cr. The residue-bound proportion makes up only 10%, as opposed to 40-60% in the remaining samples from the Gediz

With the help of Mn-54 isotopes, Förstner and Schoer (1984) were able to prove that heavy metal pollution of, in particular, anthropogenic origin is - unlike the base load - only loosely bound in the sediment.

In the lower course of the Sakarya a noticeable increase in Zn content (5-10 times above the mean value) was recorded in the clay fraction, with no corresponding deviations in the extractions. Here the Zn concentration can be attributed to Pb - Zn deposits within the catchment area.

The background values in Table 4 disregard samples revealing civilizatory or lithogenic influence. The mean values of the respective samples were calculated for the fraction < 2 μm. It was found that the content of Ni is remarkably high in all the rivers investigated, and that this phenomenon is caused by the country's specific geology. Great parts of Turkey consist of basaltic and ultrabasic rocks.

1)	Shale	2)	CEY	GED	KIZ	MEN	SAK	SEY	YES
in ppm									
Cd*	300		430	224	250	580	540	364	300
Co	19		37	27	33	31	26	44	37
Cr	90		274	197	234	217	156	231	204
Cu	45		55	57	65	53	51	45	98
Fe**	4,7		4,9	3,6	3,9	4,1	2,9	3,9	4,4
Mn	850		963	1040	946	1066	793	889	1273
Ni	68		299	211	256	313	162	396	189
Pb	20		23	30	14	25	25	35	18
Zn	95		160	119	124	151	129	171	185

*ppb **%

Table 4. Background Values - Mean values: Without samples which have significant maximum values in the extraction steps I and II as well as in the fraction < 2 μm

In order to classify local anthropogenic heavy metal pollution I_{geo} index was calculated and classification in I_{geo}- classes was carried out (after Müller 1979) (Table 5).

$$I_{geo} = \log_2 C_n / 1.5 B_n$$

C_n is the measured concentration of the element 'n' in the fraction < 2 μm and B_n is the geochemical background value which is given for the Turkish rivers in Table 4. The factor 1.5 is used because of the possible variations of the background data.

Figure 8 shows the resulting Igeo classification in the immediate catchment area of several big cities in Turkey.

pollution	*sed. acc.* I_{geo}	I_{geo}*-class*
very strong	*» 5*	*6*
strong to very strong	*»4-5*	*5*
strong	*»3-4*	*4*
moderate to strong	*»2-3*	*3*
moderate	*»1-2*	*2*
none to moderate	*»0-1*	*1*
practically none	*« 0*	*0*

Table 5. Index of geoaccumulation (Igeo) of trace metals in sediments and their classification (after Müller, 1979) The heavy metal concentration double with every class

	BILEÇIK	ESKIŞEHIR	KAYSERI	KIRIKKALE	MANISA	SIVAS
Cd	1	0	1	4	0	1
Cr	1	1	0	0	2	0
Cu	2	1	0	2	3	0
Pb	3	1	1	2	2	4
Zn	2	1	2	2	2	0

Fig. 8 Heavy metal pollution in rivers inverstigated near big cities - I_{geo} classification (after Müller, 1979)

Acknowledgement

We want to thank Otto-Benecke-Stiftung, LGF of the University of Heidelberg and DAAD for their support.

References

BROOK E J, MOORE JN (1988) Particle-Size and Chemical Control of As, Cd, Cu, Fe, Mn, Ni, Pb and Zn in Bed Sediment from the Clark Fork River, Montana (USA). Sci Total Envir 76: 247-266

FÖRSTNER U (1983) Assessment of Metal Pollution in Rivers and Estuaries. In: Thornton I (ed) Applied Environmental Geochemistry, 1983: 395-423. Academic Press, London

FÖRSTNER U, SCHOER J (1984) Some typical examples of the importance of the role of sediments in the propagation and accumulation of pollutants. In: Sediments and Pollution in Waterways - General Considerations. IAEA-TecDoc - 302: 137-158. Vienna: Internatinal Atomic Energy Agency

MÜLLER G (1979) Schwermetalle in den Sedimenten des Rheins - Veränderungen seit 1971. Umschau 79: 778-783

RAPIN F, FÖRSTNER U (1983) Sequential leaching techniques for particulate metal speciation: The selectivity of various extractants. In: Müller G (ed) Heavy metals in the environment, Int Conf Heidelberg, Sept. 1983, Vol 2, Edinburg CEP 1074-1077

SALOMONS W, FÖRSTNER U (1980) Trace metal analysis on pollutted sediments. II. Evaluation of environmental impact. Environ Technol Lett 1: 506-517

SHUMAN LM (1985) Fractionation method for soil microelements. Soil Sci 140: 11-22

TESSIER A, CAMPBELL PGC, BISSON M (1979) Sequential extraction procedure for the speciation of particulate trace metals. Anal Chem 51: 844-851

TUREKIAN KK, WEDEPOHL LH (1961) Distribution of elements in some major units of the earth's crust. Bull Geol Soc Amer 72: 175

WILBER WG, HUNTER J V (1979) The impact of urbanization on distribution of heavy metals in bottom sediments of the Saddle River. Water Resour Bull 15: 790-800

CHROMIUM TREATMENT OF WASTEWATERS BY CHEMICAL METHODS

F. Şengül, A. Türkman
Dokuz Eylül University
Faculty of Engineering and Architecture
Department of Environmental Enginering
Bornova İzmir Turkey

1. Introduction

Chromium is a relatively common element, but the chromium ions in water bodies mainly originate from industrial wastewater discharges. Chromium has several oxidation states among which the mostly encountered ones are Cr^{+3} and Cr^{+6}. In industrial wastes, chromium is present primarily in the hexavalent form as chromate ($CrO_4^{=}$) and dichromate ($Cr_2O_7^{=}$).

Thermodynamic data show that Cr^{+3} and Cr^{+6} species are stable in water and that solubility of chromium is less than 10^{-8} moles per liter (equivalent to 0.5 µg/l of Cr) only between pH 8.0 and 9.5 and in solutions having a little lower E_h than is normally imparted by saturation with atmospheric oxygen. The anionic species dichromate and chromate are relatively stable in ground-water environments. In rather strongly reducing environments below a pH 8.0, the predominant form is the complex cation $CrOH^{+2}$ (Hem, 1971).

The chloride, nitrate and sulfate forms of the trivalent chromic salts are readily soluble in water but the hydroxide and carbonate are quite insoluble. Of the hexavalent chromate salts, only sodium, potassium and ammonium chromates are soluble. The corresponding dichromates are also quite soluble.

Chromium compounds are added to cooling water to inhibit corrosion. They are employed in the manufacture of inks and industrial dyes and paint pigments, as well as in chrome tanning, aluminum anodizing and

NATO ASI Series, Vol. G 23
Metal Speciation in the Environment
Edited by J. A. C. Broekaert, Ş. Güçer, and F. Adams

other metal cleaning, preparing and electroplating operations. The major source of waste chromium is the chromic acid bath and rinsewater used in metal plating operations (Patterson, 1985).

All chromium compounds tend strongly to be oxidized to the chromic condition. Hexavalent chromium can be reduced to the trivalent form by heat, by organic matter or by reducing agents.

On the basis of many experimental studies conducted on the toxicity of chromium compounds, it appears that the following concentrations of chromium, trivalent or hexavalent will not interfere with the specified beneficial uses (McKee, 1978):

a. Domestic water supply	0.05 mg/l
b. Stock and wildlife watering	5.0 mg/l
c. Fish life	1.0 mg/l
d. Other aquatic life	0.05 mg/l

2. Treatment Technology for Hexavalent Chromium

Reduction of hexavalent chromium from a valence state of plus six to plus three, and subsequent hydroxide precipitation of the trivalent chromic ion, is the most common method of hexavalent chromium treatment.

The standard reduction treatment technique is to lower the pH of the wastewater to 2.0-3.0 with sulfuric acid, and convert the hexavalent chromium to trivalent chromium with a chemical reducing agent such as sulfur dioxide, sodium bisulfite, sodium metabisulfite, sodium hydrosulfite or ferrous sulfate. The trivalent chromium is then removed, usually by hydroxide precipitation (Patterson, 1985). The reactions are given below:

$$Cr_2O_7^{=} + 6\,Fe^{+2} + 14\,H^+ \longrightarrow 2Cr^{+3} + 6Fe^{+3} + 7H_2O$$

$$2Cr_2O_7^{=} + 3S_2O_5^{=} + 10H^+ \longrightarrow 4Cr^{+3} + 6SO_4^{=} + 5H_2O$$

$$2Cr^{+6} + 3SO_2 + 6H_2O \longrightarrow 2Cr^{+3} + 3SO_4^{=} + 12H^+$$

After the completion of reduction from Cr^{+6} to Cr^{+3}, chromium ion is precipitated in the form of hydroxide:

$$2Cr^{+3} + 3Ca(OH)_2 \longrightarrow 2Cr(OH)_3 + 3Ca^{++}$$

Sulfur dioxide is the most popular reducing agent used in treatment of chromium wastes, primarily because it is relatively inexpensive.

The use of metabisulfite as a reducing agent has been reported for several hexavalent chromium treatment plants. Reduction is carried out at pH 2.0, which is achieved by addition of sulfuric acid.

The one-step Sulfex TM process (Permutit) is carried out at pH 8-9 with solid ferrous sulfide in excess, eliminating pH reduction required for conventional hexavalent chromium conversion. Performance is claimed equal to or better than chemical reduction/hydroxide precipitation treatment (Patterson, 1985).

The ion exchange process is reported to be economical for chromium recovery and elimination of waste discharge. Cation exchange can be applied to remove trivalent chromium, and anion exchange is employed to remove chromate and dichromate (Jakobsen and Laska, 1977).

Evaporative recovery of concentrated chromate and chromic acid wastes has also proved technically and economically feasible. Plating-waste rinsewater containing only a few mg/l chromic acid can be concentrated by evaporative treatment to above 900 mg/l (Culotta and Swanton, 1970).

Some success has been reported from pilot plant work on chromate removal from dilute wastewaters by activated carbon. Reverse osmosis is also reported to have application for hexavalent chromium treatment although experience with the process is limited. Ion exchange and evaporative recovery have the advantage that the chromium containing process water may be recovered for reuse. If waste flow rates and chromium concentrations are low (less than 50 mg/l), the total water volume for reuse and the amount of chromium may not warrant reclamation (Patterson, 1985).

3. Chromium Treatment Studies

A jar test apparatus with six beakers was used during this study for the treatment of chromium containing wastewaters. pH adjustment was made with the help of a pH meter and chromium concentrations were measured with an Atomic Absorption Spectrophotometer (Varian AA175). The experiments were conducted in two series, with synthetic and real wastewater samples. The blank solutions are prepared by taking into account other constituents in the solution.

Synthetic Samples

Potassium dichromate ($K_2Cr_2O_7$) and tap water were used to prepare synthetic wastewater samples. Reductants have been added to the 1 liter beakers in amounts more than the theoretical. Prior to the 5 minutes of rapid mixing of reductants, the pH value was adjusted to 2.0. Rapid mixing is followed by 30 minutes of sedimentation. After this stage, the pH of the water was increased to 9.0 with the help of lime or caustic soda. Slow mixing was applied for 30 minutes after which 3 hours of sedimentation is allowed.

Sludge volumes were measured and chromium determinations were made on the supernatant samples. For the preservation of 100 ml samples, 5 ml of concentrated nitric acid was added to each, filtered and kept in plastic containers. Chromium removal efficiency was calculated by determining chromium ion concentrations in raw and treated wastewater samples.

The chemicals used in treatability studies were concentrated sulfuric acid and lime for pH adjustment, iron sulfate to reduce chromium (+6) to chromium (+3) and lime to precipitate chromium (+3). Batch treatability studies on potassium dichromate wastewaters with a concentration range of 50 to 1060 mg/l chromium (+6) yielded the results presented in Table 1. All the results were obtained at the experimental conditions of pH= 2.0 for reducing chromium (+6) to chromium (+3) and pH = 9.0 for chemical precipitation.

Another series of experiments was accomplished to study the effect of pH and reducing agent on chemical precipitation. In this series sodium metabisulfite ($Na_2S_2O_5$) was used as reducing agent and pH of the medium was adjusted to pH = 12.0 during chemical precipitation. In order to achieve a complete chromium reduction, 1.5 times the stoichiometric amount of $Na_2S_2O_5$ was used at pH 2.0. Chromium concentrations studied varied between 2.5 mg/l and 600 mg/l. After the completion of the reduction step, pH was increased to 12.0 and 30 minutes of slow mixing was applied, then the water is allowed to settle for 2 hours. Chromium determinations were made on the supernatant samples by using atomic absorption spectrometry. The results of this series of experiments are summarized in Table 2.

The third series of experiments were performed with the real wastewater samples. Wastewater characteristics of the tannery studied are given below:

pH: 3.3
Total suspended solids: 115 mg/l
Chemical oxygen demand: 7200 mg/l
Settleable solids: 10 ml/l
Total chromium: 3968 mg/l

$FeSO_4 \cdot 7H_2O$ or $Na_2S_2O_5$ was used as reducing agent for the treatment of tannery wastewaters. Reduction and chemical precipitation were achieved at pH= 2.0 and pH= 9.0 respectively. The results of experiments are presented in Table 3.

Most of the tanneries in İzmir are located at Yeşildere area and have their collective wastewater treatment plant. Influent wastewater samples are also studied in this context to determine the degree of chromium removal that can be achieved at the treatment plant. Table 4 indicates the experimental results which are obtained by using NaOH for chemical precipitaiton and $Na_2S_2O_5$ for reduction of chromium.

Test performances at different pH values were plotted on a graph to see the effect of pH on treatment efficiency. Figure 1 shows that a better, but not very different efficiency is obtained at pH=9.

Table 1. Experimental results of chromium (+6) treatment of a synthetic wastewater at pH= 9.0

Sample number	Inital Chromium conc. (g/l)	$FeSO_4.7H_2O$ dose (g/l)	$Ca(OH)_2$ dose (g/l)	Chromium conc.in eff.(mg/l)	Chromium removal eff. (%)	Sludge volume (l/m^3)
1	1.06	5.0	3.8	78.0	92.6	130
2	1.06	10.0	3.3	70.4	93.4	150
3	1.06	14.0	3.5	48.3	95.4	190
4	1.06	18.0	3.8	21.2	98.0	195
5	1.06	18.0	4.0	10.7	99.0	200
1	1.00	8.0	10.0	101	90.0	200
2	1.00	10.0	10.0	43	96.0	200
3	1.00	120	10.0	1.73	99.8	205
4	1.00	16.0	10.0	0.96	99.9	210
5	1.00	20.0	10.0	0.87	99.9	215
1	0.50	2.0	6.0	161	68.0	220
2	0.50	6.0	6.0	96.0	73.6	230
3	0.50	8.0	6.0	96.0	81.0	245
4	0.50	10.0	6.0	43.0	91.4	260
5	0.50	12.0	6.0	31.0	94.0	280
6	0.50	16.0	6.0	15.0	97.0	290
7	0.50	18.0	6.0	11.0	98.0	295
8	0.50	20.0	6.0	8.0	98.4	300
9	0.50	22.0	6.0	2.0	99.6	310
1	0.25	2.5	4.0	0.72	99.7	160
2	0.175	2.0	4.0	0.58	99.7	160
3	0.10	2.0	4.0	0.42	99.6	160
4	0.10	2.0	6.0	0.87	99.1	215
5	0.05	1.0	4.0	0.40	99.0	160

Table 2. Experimental results of chromium (+6) treatment of a synthetic wastewater at pH= 12

Sample number	Initial chromium conc.(mg/l)	$Na_2S_2O_5$ dose (g/l)	$Ca(OH)_2$ dose (g/l)	Chromium conc.in eff.(mg/l)	Chromium removal eff.(%)	Sludge volume (l/m^3)
1	600	2.48	4.5	4.99	99.2	870
2	500	2.06	4.5	1.00	99.8	790
3	400	1.65	4.4	0.67	99.8	470
4	300	1.24	4.3	1.04	99.6	410
5	200	0.83	4.2	0.37	99.8	320
6	100	0.41	3.6	0.63	99.4	150
7	80	0.33	3.6	0.70	99.1	42
8	60	0.25	3.2	1.03	98.0	38
9	50	0.21	3.1	0.89	98.2	38
10	40	0.17	3.0	1.00	97.5	34
11	20	0.14	3.4	0.22	98.9	37
12	10	0.10	3.2	0.15	98.5	27
13	5	0.02	3.0	0.10	98.0	33
14	2.5	0.01	3.2	0.05	98.0	30

Table 3. Experimental results of chromium (+6) treatment of a tannery wastewater at pH= 9

Sample number	Initial chromium conc.(mg/l)	$FaSO_4.7H_2O$ dose (g/l)	NaOH dose (g/l)	Chromium conc.in eff.(mg/l)	Chromium removal eff.(%)	Sludge volume (l/m^3)
1	3968	0.5	6.4	4.6	99.90	950
2	3968	1.0	7.0	4.3	99.90	945
3	3968	2.0	7.0	3.3	99.92	940
4	3968	5.0	7.4	3.1	99.92	930
5	3968	1.0	7.8	2.9	99.93	900
6	3968	15.0	8.0	2.5	99.94	870

Table 4. Experimental results for chromium treatment carried out at pH= 12.0 with the influent of Yeşildere collective wastewater treatment plant of tanneries

Sample number	Initial chromium conc. (mg/l)	$Na_2S_2O_5$ dose (g/l)	$Ca(OH)_2$ dose (g/l)	Chromium conc. in eff. (mg/l)	Chromium removal eff. (%)	Sludge volume (l/m^3)
1	99.0	0	5	22.5	74.2	400
2	99.0	4	5	2.1	97.9	400
3	99.0	16	5	1.5	98.5	410
4	99.0	20	5	0.15	99.8	420
5	99.0	40	5	0.10	99.9	430
1	115	0	5	24.0	79.1	380
2	115	2	5	5.6	95.1	380
3	115	4	5	3.2	97.2	400
4	115	6	5	2.8	97.6	400
5	115	8	5	1.3	98.9	400
6	115	10	5	1.1	99.0	400

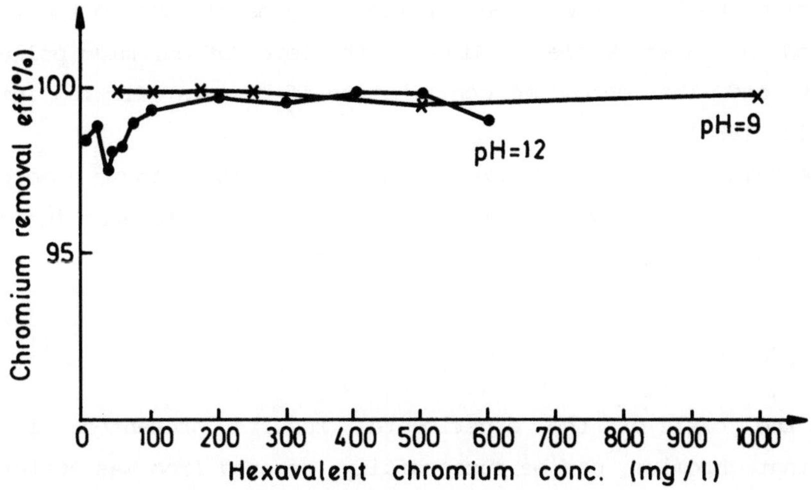

Figure 1. Chromium removal efficiency versus hexavalent chromium concentration at different pH values

Chromium removal at Yeşildere collective wastewater treatment plant is very important from an environmental point of view. Yeşildere tanneries wastes are discharged to Melez Creek which flows into the İzmir

Table 5. Chromium concentrations measured at water and sediment samples taken from İzmir Bay and the Creeks flowing into it

Study Area	Cr, µg/l in water	Cr, mg/kg in sediment
Melez Creek	15-150	150-1600
Arap Creek	12-43.5	50-410
Manda Creek	4-70	35-700
Bornova Creek	2-7.6	50-80
İzmir Bay (Outer)	4.6	28.5 (avg.)
İzmir Bay (Inner)	4.5-27.5	111.0 (avg.)

Bay. Chromium build-up took place at İzmir Bay mainly due to the tanneries located in İzmir. Table 5 indicates the level of chromium pollution at the Bay and the chromium contribution of creeks flowing into it (Müezzinoğlu, 1989).

The values in Table 5 were obtained by flame atomic absorption spectrophotometry for river water samples and graphite furnace injection system for seawater and sediment samples (APHA, 1981).

4. Discussion

On the basis of the experimental results presented, it was concluded that chromium can be successfully removed from wastewaters by the chemical method described. According to the "Water Pollution Control Regulation" of Turkey, chromium concentration at seawater should not exceed 0.1 mg/l (Official Gazette, 1988). When discharged directly into the receiving waters, chromium standard for industrial wastewaters is 2 mg/l for 2 hours composite sample and 1 mg/l for 24 hours composite sample. These values show the importance and necessity of chromium treatment in wastes.

Chromium removal rate of 99% was achieved at best for the synthetic wastewater samples containing 1.06 g/l of chromium. Although very high removal efficiencies were obtained in the range of 93 % to 99 % in this case effluent water quality was still far from meeting the discharge quality requirements. The reason for the insufficient chromium removal may be related to the inadequate doses of chemicals. In fact, when higher doses were applied for 100 mg/l, chromium containing wastewaters, an effluent quality of 0.87 mg/l of chromium was achieved. For this case the best removal efficiency was 99.9 %.

An obvious relation between the chemical dosage and chromium removal efficiencies can be observed from Table 1 for 500 mg/l chromium containing wastewaters. As the reducing agent dose was increased gradually, chromium concentration at the effluent was lowered and the sludge volume increased.

Comparison of tes performances at pH = 12.0 and pH = 9.0 reveals that the chemical precipitation at pH=9.0 is slightly more efficient for chromium removal. Representative results are given in Tables 1 and 2. Chromium removal efficiencies at pH = 12.0 are also very high ranging between 97.5 % and 99.8 %. Water Pollution Control Regulation discharge limit is exceeded only for 600 mg/l chromium containing wastewaters. For this case an additional chromium removal will be necessary.

Besides synthetic wastewater samples, real wastewater samples were also tested in order to evaluate chromium treatability. Table 3 and Table 4 show these results. Reduction of chromium content from 3968 mg/l to 2.5 mg/l indicates a very good chromium removal from tannery wastewaters. Use of $FeSO_4.7H_2O$ or $Na_2S_2O_5$ as reducing chemical does not change the removal efficiencies considerably.

From the data presented in Table 4, it may be estimated that chromium removal can be applied successfully in the collective wastewater treatment plant. For this case, influent chromium concentrations varied between 99 mg/l and 115 mg/l, and the lowest chromium concentration achieved at the effluent was 0.10 mg/l (99.99 % removal).

This study has shown that chromium may easily be removed from wastewaters by chemical treatment. Although the exact doses of chemicals may be determined at full scale plant studies, this study gives an idea about the amount of chemicals required for such a treatment. Also effects of various parameters on actual treatment conditions such as pH and the type of chemical to be used may be determined by the use of the data obtained here.

Use of lime may be proposed instead of sodium hydroxide as the precipitating agent. Lime may be preferred because of higher removal efficiencies, however, the amount of suldge is higher in this case. Removal of chromium from sludges is another important topic that should be taken into consideration from point of view of environmental protection.

References

APHA, AWWA, WPCF (1981) Standard Methods for the Examination of Water and Wastewater, Fifteenth Edition, American Public Health Association, Washington.

Culotta JM, Swanton WF (1977) Case Histories of Plating Waste Recovery Systems. 56 th Annual Conference of the American Electroplaters Society. Detroit, MI.

Eckenfelder WW (1980) Principles of Water Quality Management. CBI Publishing Company, Boston.

Faust SD, Aly DM (1983) Chemistry of Water Treatment. Butterworth Publishers, Boston, London.

Gloyna FE, Eckenfelder WW (1970) Water Quality Improvement by Physical and Chemical Processes, University of Texas Press, Austin.

Gould J (1982) The Kinetics of Hexavalent Chromium Reduction by Metallic Iron. Water Res. 16, pp. 871-877.

Hem, JD (1971) Study and Interpretation of the Chemical Chracteristics of Natural Water. Geological Survey Water Supply Paper 1473, United States Government Printing Office, Washington.

Jakobsen K, Laska R (1977) Advanced Treatment Methods for Electroplating Wastes, Poll. Eng.: pp. 42-46.

Müezzinoğlu A, Şengül F (1989) Chromium, Cadmium and Mercury Pollution in İzmir Bay. Plants and Pollutants in Developed and Developing Countries, Ed. by: Dr.Münir Ahmet Öztürk, Ege University, pp. 37-49, İzmir.

McKee JE, Wolf HW (1978) Water Quality Criteria. California State Water Resources Control Board, Publication No: 3-a, California.

Official Gazette (1988) Water Pollution Control Regulation, Gazette No. 19919, pp. 41-42, Ankara.

Patterson JW (1985) Industrial Wastewater Treatment Technology Second Edition, Butterworth Publishers, Boston London Durban.

Şengül F, Türkman A, Filibeli A (1986) A Case Study on Chromium Wastes Treatment. Preprints of the Third Symposium on Environmental Management for Developing Countries, İstanbul.

Bacterial Response to Copper in the Environment: Copper Resistance in *Escherichia coli* as a Model System

B.T.O. Lee, N.L. Brown*, S. Rogers, A. Bergemann, J. Camakaris and D.A. Rouch*
Department of Genetics
The University of Melbourne
Parkville
Victoria 3052
Australia

1. Introduction

In order to study the biological effects of heavy metals, it is necessary to make a distinction between two types of heavy metals (1) metals that are toxic *per se* and (2) metals which are essential to the growth and maintenance of living organisms but which are toxic in excess. Copper is an extremely common heavy metal in the environment. Although it is an essential element, it is toxic at high concentrations. Copper causes cell damage through, for example, modification of the active sites of cellular enzymes and the peroxidation of membranes. So, copper must not be in a free state either in the cell or at the environment/cell interface. It is involved in essential redox reactions and is a component of a number of enzymes, which means that normal copper metabolism must be closely regulated in both low and high copper environments. If any resistance to high copper concentrations is developed in response to environmental changes it must be regulated together with the normal copper utilization systems (Rouch *et al.* 1989a,b). In contrast, mercury salts are highly toxic to all living organisms and have no beneficial function. They are widely distributed, as a result of both geochemical and human activities and this, coupled to the toxicity, has resulted in the evolution and spread of mercury resistance in bacteria (Brown *et al.* 1989; Silver and Misra, 1988). Mercury resistance is the best characterized of all resistances to heavy metals and involves the sequestration of mercuric ions in the periplasmic space, their transport into the cell and their reduction to the non-toxic form, metallic mercury, Hg(0). This makes biological sense in that a highly toxic compound is sequestered in the periplasm and not released to the cellular constituents until it has been detoxified.

There are similarities between the biological requirements to protect against excess Cu or Hg in terms of binding proteins and transport of the metals, as mercury is taken up into the cell to be detoxified and copper is taken up, sequestered and mostly exported. But there will be differences in the regulation, as copper resistance involves a two-way control, that is, the

*School of Biological Sciences, The University of Birmingham, Birmingham B15 2TT, U.K.

NATO ASI Series, Vol. G 23
Metal Speciation in the Environment
Edited by J. A. C. Broekaert, Ş. Güçer, and F. Adams

sensing of both high and low level heavy metal concentration while mercuric ion resistance is only required to deal with toxicity by responding to mercuric ions at any concentration and is therefore a one-way control system,that is, sense and switch on/ off. These extremes then represent different aspects of the same problem, the maintenance of cellular homeostasis, and are probably best studied and modelled in *Escherichia coli* because of the ease with which it can be genetically manipulated.

2. Results and discussion

The normal response to elevated levels of any heavy metal is stress, which can lead to death of the organism or the utilization of some resistance mechanism

A. Response of bacteria to copper stress

The isolation of heavy metal resistant organisms would be expected in those situations where the environmental stress exceeds the capacity of the organism to cope by using the normal metabolic systems. In terms of copper stress, resistant strains of *Escherichia coli* were isolated from faeces of pigs fed on a diet supplemented with copper sulphate (100 p.p.m.) (Tetaz and Luke,1983), and strains of *Pseudomonas syringae* were isolated from plants presumably exposed to copper as antifungal or antimicrobial agents (e.g.Bordeaux mixture) (Cooksey, 1987;Mellano and Cooksey, 1988a,b). In both of these organisms the resistance is plasmid coded and inducible (Rouch *et al.* 1989a,b Mellano and Cooksey, 1988). In *E. coli*, a strain carrying the *pco* (plasmid-borne copper-resistance) determinant previously grown in 0.4 mM added copper shows only a slight lag in growth after a shift to 12 mM added copper.

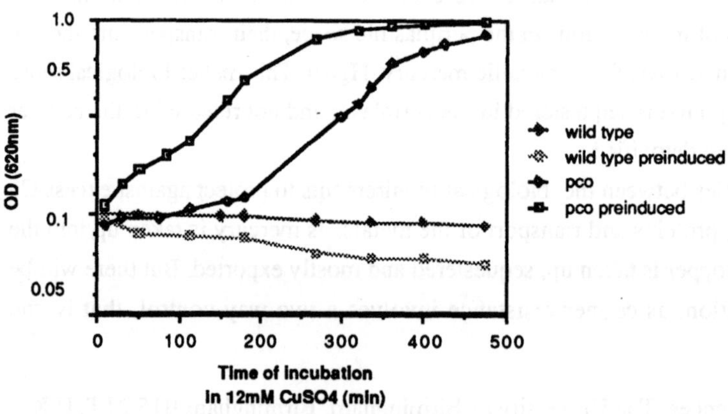

Figure 1. *Growth (OD 620nm) of wild-type* E.coli *and* E.coli(pco), *carrying the* pco *resistance determinant, in 12 mM CuSO4 after pregrowth in broth in the presence of 0.4 mM CuSO4 or its absence.*

On the other hand, the same strain grown without added copper showed a 125 minute lag in growth when the copper concentration was increased.(Figure 1). This contrasts with the response of the same strain without the *pco* determinant, which showed no growth when challenged with 12 mM added copper. Obviously, low level copper concentrations are being managed, while high level copper resistance mechanisms require additional system(s) for management.

In *E.coli* both resistant and wild-type, actively growing strains reach saturating cell densities in broth at all levels of added copper (as $CuSO_4$) up to 8mM in broth. So, exponentially growing wild-type cells can tolerate up to this concentration of copper (data not shown) in contrast to stationary phase cultures which start to die at 0.1mM in broth (Figure 2). Therefore the level of resistance /tolerance is higher during active growth. It is also evident that accumulation of copper correlates with the degree of cell survival.This may be compared to resistant cells which grow and survive at least to 20 mM even in stationary phase (Figure 2).

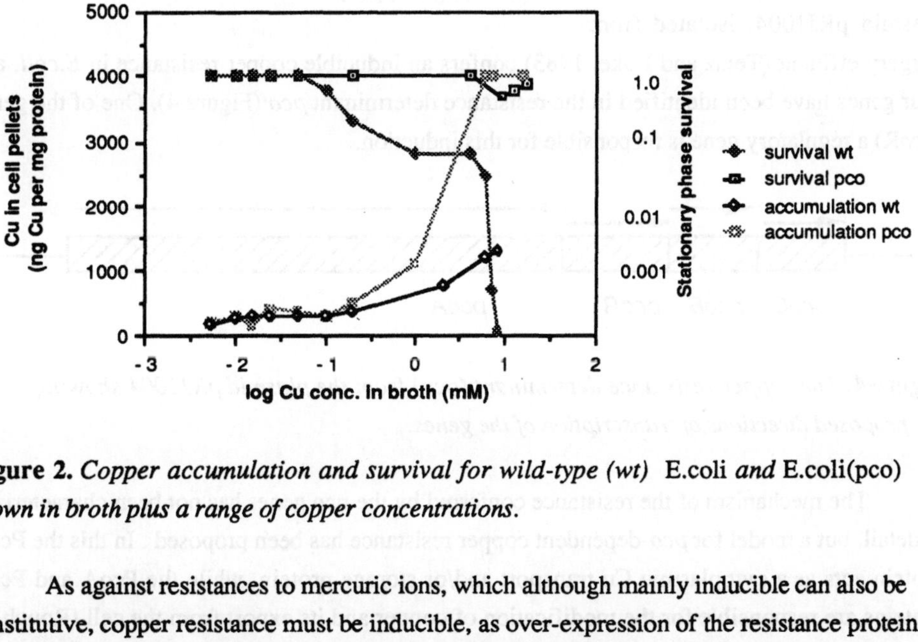

Figure 2. *Copper accumulation and survival for wild-type (wt)* E.coli *and* E.coli(pco) *grown in broth plus a range of copper concentrations.*

As against resistances to mercuric ions, which although mainly inducible can also be constitutive, copper resistance must be inducible, as over-expression of the resistance proteins would be lethal due to a decrease in available cellular copper giving rise to copper starvation. When we investigated the level of expression of either copper or mercury resistance as a function of metal concentration (Figure 3), we found a linear induction response for copper, with increasing levels, whereas a threshold response was found with increasing concentrations of mercury.

The linear response of resistance expression with increasing copper concentration is consistent with the necessity for homeostatic control of this type of resistance, as the level of resistance expression must correlate with the level of environmental copper. A threshold response to mercury is expected, because the homeostatic level of Hg is close to zero, so overexpression of the response is not an immediate problem.

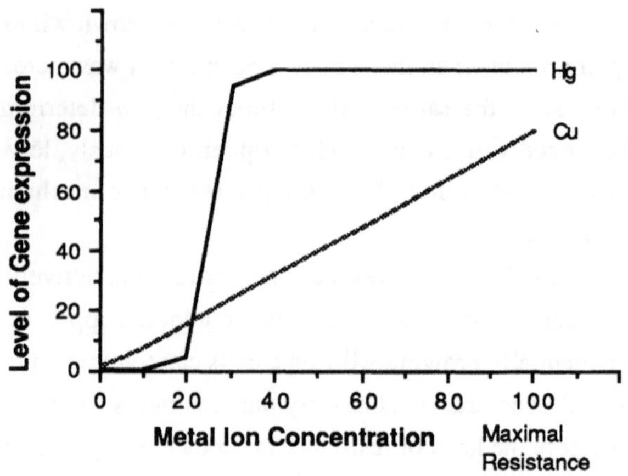

Figure 3. *An idealised representation of induction of the resistance gene(s) vs metal concentration for a toxic metal (e.g. mercury) and an essential but toxic metal (e.g. copper).*

The copper resistance plasmid pRJ1004, isolated from piggery effluent (Tetaz and Luke, 1983) confers an inducible copper resistance in *E.coli*, and four genes have been identified in the resistance determinant *pco* (Figure 4). One of the genes (*pco*R) a regulatory gene is responsible for this induction.

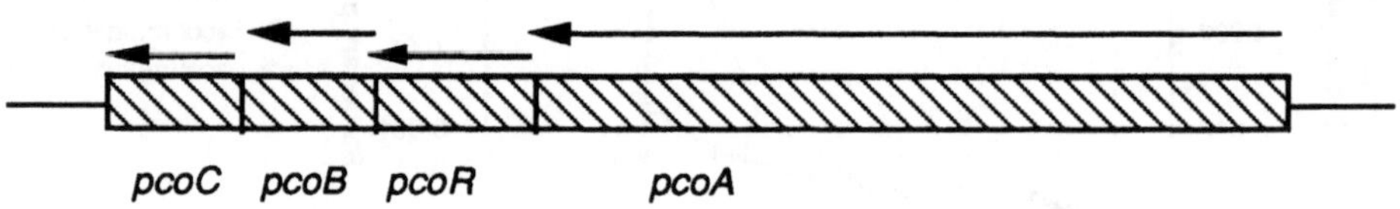

Figure 4. *The copper resistance determinant* (pco) *from the plasmid pRJ1004 showing the proposed directions of transcription of the genes.*

The mechanism of the resistance conferred by the *pco* genes has not been characterized in detail, but a model for *pco*-dependent copper resistance has been proposed . In this the PcoC protein acts as a cytoplasmic Cu transport and/or storage protein, while the PcoA and PcoB proteins are responsible for the modification of copper and its export from the cell (Rouch, et al. 1989a). As it is predicted that copper resistance involves a two-way control, it is expected that an interaction would exist between the plasmid and the chromosomal copper handling systems. Thus, a range of copper-sensitive/copper-dependent mutants were isolated (Rouch *et al.* 1989b; *vide infra*) and tested for their interaction with the expression of plasmid-coded resistance. The mutants fell into two classes; those that allowed the expression of normal, plasmid-coded, resistance and those that gave the mutant (sensitive) phenotype. The fact that this second class of mutants was obtained, indicates that some chromosomally-encoded genes must be involved in the expression of resistance to copper. However, in the comparable one way system of mercury resistance no such interaction with chromosomal genes has been found.

B. Response of bacteria to normal copper levels

As *E.coli* must normally sequester and utilize copper from the environment, a simple model of copper utilization would be: UPTAKE---BINDING---EFFLUX (and REGULATION). If this is so, then it should be possible to isolate mutants in all the metabolic steps involved. The expectation is that such mutants would be copper-sensitive/copper-dependent. Isolation of such mutants has been undertaken (Rouch *et al.* 1989 b, and in preparation), Two copper-uptake systems (*cutA* and *cutB*) have been identified genetically; one is specific for copper, the other transports Zn(II) and may be the normal zinc uptake system. Once inside the cell copper must be prevented from damaging cell constituents, and copper storage and or intracellular transport proteins may be responsible for this. Two storage/transport genes have been identified (*cutE* and *cutF*). Copper excess to the requirement of the cell must be stored or exported and two genes whose products are involved in copper efflux (*cutC* and *cutD*) have been identified.The expression of the genes must be regulated in order to maintain copper homeostasis; another gene *cutR* may partly or completely fill this role.

C. Interaction between the systems for high and low level copper responses

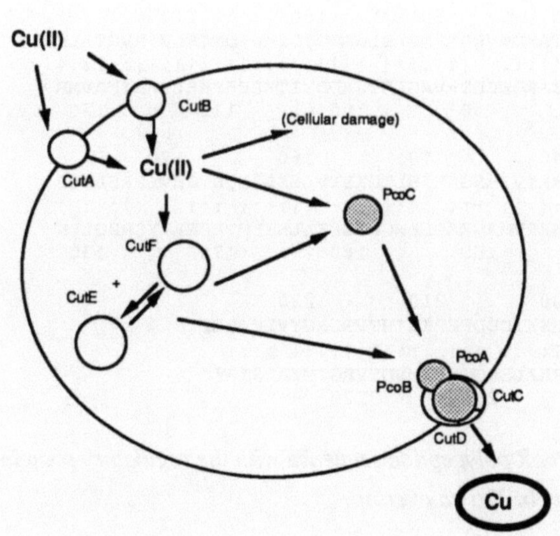

Figure 5. *An outline model of the chromosomal* (cut) *and plasmid* (pco) *genes in plasmid (pRJ1004) mediated copper resistance in* E.coli.

A priori, one expects that the expression of copper-resistance in *E.coli* would involve the genes for normal copper metabolism. Studies in *pco*-containing, wild-type and cut-mutant strains have shown that the *cutA* (uptake) and *cutD* (efflux) genes

(See Figure 5) are required for the expression of copper resistance.

A further interaction between the chromosomal (metabolic) and plasmid (resistance) genes occurs in the regulation of gene expression. Mutants in *pcoR* still show repression of the remaining *pco* genes due to the *cutR* gene on the chromosome. Any reciprocal effect of *pcoR* on the *cut* genes is yet to be determined.

We have some nucleotide sequence data from the *pcoRBC* region of the resistance determinant. We have a complete, but not fully confirmed sequence of the PcoR protein. This partial sequence shows good homology (Figure 6) with a number of prokaryotic regulatory proteins which are regulatory proteins in two component systems (Ronson et al.,1987; Magasanik, 1988). These include OmpR, PhoB and SfrA from *E.coli*, SpoOF from *Bacillus subtilis*, VirG from *Agrobacterium tumifaciens* and NtrC from *Klebsiella pneumoniae*. These proteins and others respond to environmental stimuli; and the second component of each system is a membrane bound (e.g.OmpR/EnvZ system) or intracellular (e.g.NtrB/NtrC system) sensor protein. In the case of the OmpR/EnvZ and PhoB/PhoR systems it is known that the gene products become phosphorylated (Magasanik, 1988).It appears, therefore, that the pcoR protein is a member of this class of DNA-binding proteins that do not contain helix-turn-helix motifs (Pabo and Sauer, 1984), and, if it behaves in a manner similar to OmpR, PhoB etc., it may act as a regulator of gene expression only after it has been modified by a sensor protein which remains to be identified.

```
                 10        20        30        40        50        60
PcoR     MQRILIVEDEQKTGRYLQQGLVEEGYQADLFNNGRDGLGAASKGQYDLIILDVMLPFLDG
          :|||:||||: :   :   | ::|:|:   ::  ::::: ::   |||:|| |||  :|
PhoB    MARRILVVEDEAPIREMVCFVLEQNGFQPVEAEDYDSAVNQLNEPWPDLILLDWMLPGGSG
                10        20        30        40        50        60

                 70          80        90       100       110
PcoR     WQIISALRESG--HEEPVLFLTAKDNVRDKVKGLELGADDYLIKPFDFTELVARVRTLLR
          |:|: |: ::  :: X|::|||:::  |:|:||| |||||::|||: :||||||::::|
PhoB     IQFIKHLKRESMTRDIPVVMLTARGEEEDRVRGLETGADDYITKPFSPKELVARIKAVMR
                70        80        90       100       110       120

        120        130       140       150       160       170
PcoR     RARSQA-ATVCTIADMTVDMVRRTVIRSGKKIHLTGKEYVLLELLLQRTGEVLPRSLISS
         X :: | ::| :::::::| ::: |: ::: :::: :|: ||:::: ::: | :|: : :
PhoB     RISPMAVEEVIEMQGLSLDPTSHRVMAGEEPLEMGPTEFKLLHFFMTHPERVYSREQLLN
               130       140       150       160       170       180

         180       190       200       210       220
PcoR     LVWNMNFDSDTNVIDVAVRRLRSKIDDDFEPKLIHTVRGAGYVLEIREE
          ||: |   :::::|| :||||: :: : : ::::||||:|| :::|
PhoB     HVWGTNVYVEDRTVDVHIRRLRKALEPGGHDRMVQTVRGTGYRFSTRF
               190       200       210       220
```

Figure 6. *Alignment of the provisional PcoR polypeptide sequence with the regulatory protein PhoB from E.coli, showing 31.9% identity in 226 aa overlap.*

The similarities between the low and high level copper responses leads to the speculation that the two systems function by environmental sensing. Presumably, they must monitor internal levels as well, since the internal environment is crucial to the maintenance of homeostasis (Figure 7). That is, it must sense the external copper environment and internally adjust the copper concentration to maintain a steady state. In the case of copper this homeostatic control must be a two way mechanism. Available copper must be regulated so that the cell is able to respond to changes towards both high and low copper concentrations away from the homeostatic level. The wild-type cell can respond to control signals by adjusting chromosomally regulated uptake/export of copper, while the resistant cell can, in addition, adjust expression of the resistance mechanism. In contrast, the comparable mercury resistance system involves a one way control for metal homeostasis, that is, to sense mercuric ions entering the cell and to increase the level of resistance. Although expression of mercuric ion resistance is reduced in the absence of the metal, this response is not due to the requirement for metal homeostasis.

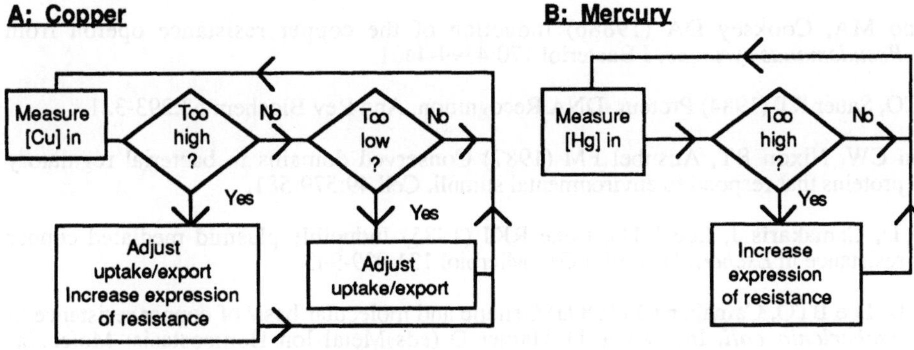

Figure 7. *Proposed circuits of copper (two way) and mercury (one way) regulatory systems for heavy metal homeostasis.*

Thus, the copper resistance system is useful not only in understanding direct responses to environmental stress but has wider application in understanding homeostatic mechanisms in general, and in providing some insight into the genetic factors involved in the homeostasis of heavy metals.

Acknowledgements

This work was supported by grants from the Australian Research Council, the Murdoch Institute and the Royal Society. B.T.O.L is supported by a Royal Society Guest Research

Fellowship and N.L.B. was supported by a Melbourne University Visiting Research Fellowship and by a Royal Society EPA Cephalosporin Senior Research Fellowship. We thank Julian Parkhhill for help with the diagrams and for discussion.

References

Brown NL, Lund PA, Ni Bhriain N (1988) Mercury resistance in bacteria.In: Hopwood DA, Chater KF (eds) The Genetics of Bacterial Diversity. Academic Press, London NewYork, pp175-195

Cooksey DA (1987) Characterization of a copper resistance plasmid conserved in copper-resistant strains of *Pseudomonas syringae* pv. *tomato*. Appl Envir Microbiol 53:454-456.

Magasanik B (1988) Reversible phosphorylation of an enhancer binding protein regulates the transcription of bacterial nitrogen utilization genes. Trends Bioch Sci 13:475-479.

Mellano MA, Cooksey DA (1988a) Nucleotide sequence and organization of copper resistance genes from *Pseudomonas syringae* pv.*tomato*. J Bacteriol 170:2879-2883.

Mellano MA, Cooksey DA (1988b) Induction of the copper resistance operon from *Pseudomonas syringae*. J Bacteriol 170:4399-4401.

Pabo CO, Sauer RT (1984) Protein -DNA Recognition. Ann Rev Biochem 53:293-321.

Ronson CW, Nixon BT, Ausubel FM (1987) Conserved domains in bacterial regulatory proteins that respond to environmental stimuli. Cell 49:579-581.

Rouch D, Camakaris J, Lee BTO, Luke RKJ (1985) Inducible plasmid mediated copper resistance in *Escherichia coli*. J Gen Micrbiol 131:939-943.

Rouch D, Lee BTO, Camakaris J (1989a) Genetic and molecular basis of copper resistance in *Escherichia coli*. In: Winge D, Hamer D (eds)Metal Ion Homeostasis: Molecular Biology and Chemistry. Alan R Liss, New York, p439-446.

Rouch D, Camakaris J, Lee BTO (1989b) Copper transport in Escherichia coli. In: Winge D, Hamer D (eds) Metal Ion Homeostasis: Molecular Biology and Chemistry. Alan R Liss, New York, p 469-477.

Silver S, Misra T (1988) Plasmid-mediated heavy metal resistance. Ann Rev Microbiol 42: 717-743.

Tetaz T, Luke RKJ (1983)Plasmid controlled resistance to copper in *Escherichia coli*. J Bacteriol 154: 1263-1268.

Panel Discussions

In what follows a brief summary is presented on three panel discussions organised during the workshop. The first deals with the dynamic behaviour of metal species and was animated by Förstner, Buffle and Tessier. The second is a discussion on complexation in the environment and was coordinated by Lund, Frimmel and Astruc. Finally, a discussion on analytical techniques for metal speciation was chaired by Adams, Broekaert, Klockow and Irgolic.

1. Dynamic Behaviour of Metal Species in the Environment

A typical 'dynamic' environment is a system where mechanical, biological and chemical phenomena take place with strong gradients of chemical parameters (E_h, pH) and variations in time (e.g. biological activity). This has been shown in the case of contaminated soils, where organic matter may act as chelator, but also as reductor and salt ions may cause a release of metals. It also applies to estuarine sediments where for instance, the sulfide oxidation is sped up by up to 10^6 (1.000.000) times due to the action of bacteria. Here too, changes in salinity may lead to a release of metals, and also the food chain plays or may play a role. For the case of municipal waste disposal, organic acids (pH = 5) may lead to a release of metals in the acidogenic phase, whereas in the post-methanogenic phase, the oxidation of sulfide may occur and at a pH of ~ 6.5 release of Cd and Zn, however, not of Cu and Pb, may follow. It has been emphasized that for drawing up a picture of these phenomena, engineering work must go hand in hand with analytical chemistry. On the other hand, it is not sufficient to measure the species but conclusions should be drawn which allow prognoses and show up how waste can be stored without risks of release.

It should be mentioned that 'macro'-modelling of the environment needs to be shared by modelling on the 'micro' scale, describing the equilibria between particles and colloids, complexants and metals as well as their interaction with cells. In order to describe the related mechanisms, equilibrium constants, dissociation rate constants

NATO ASI Series, Vol. G 23
Metal Speciation in the Environment
Edited by J. A. C. Broekaert, Ş. Güçer, and F. Adams

and diffusion coefficients of all complex species must be known, for which up to now, the determination is hampered by a lack of methodology.

The measurement of size distributions is helpful in this respect; filtration, however, may falsify the real particle size as e.g. by the use of Nuclepore filters, aggregations are easily formed (see Electron Microscope images of vertical transects). For a determination of the bioavailability, however, a discrimination between dissolved and particulate matter is required. Voltammetric analyses are a means to replace filtration for distinguishing between both, as all matter ≥ 40 nm does not contribute to the signal. This might be an alternative to EPMA subsequent to filtration. The dissociation rate constant of complexes is also problematic, and information is often only obtained from titration curves and modelling. However, information is available only at concentration levels ≥ 10^{-5} mol/L. Since in reality, concentrations are as low as 10^{-8}, natural dissociation rates may be in the order of magnitudes smaller than those measured in the laboratory. For some elements (Am), measurements have also been carried out at concentration levels of 10^{-7} mol/L and, indeed, in these cases, the kinetics in reality last for months. The desequilibration in real systems may be very complex as a result of multiple and competitive reactions. Moreover, in most systems, steady state conditions are never obtained due to a continuous introduction of material (as e.g. in lakes) and accordingly, laboratory experiments are of limited value. It was stressed that the work on speciation in the environment is presently based too much on thermodynamics and too litle on kinetics. This may surprise, when comparing to the way followed in atmospheric chemistry. Although in aquatic systems, the interactions are slower, mass transfer should nevertheless be taken into account, especially in fast systems. In the latter case, off-line analyses may then become problematic and sensoring is, in fact, required.

Several comments were made and discussed:

One must investigate with the aid of highly sensitive methods over a long-term period and include the biological uptake, otherwise modelling alone will be of little information.

One should also search for methods which can detect the species and especially the macromolecular complexing agnets present in the environment, and this very frequently at low concentration levels. For anthropogenic substances, this might be easily done, however, for the natural species, it may be much more difficult.

The comparison of laboratory experiments and real systems is indeed a challenge; this applies also to comparisons between batch and column experiments, as the effect of mass transfer is not understood.

It should be pointed out where the state-of-the-art now is compared to 5 years ago; then speciation studies mostly uniquely involved modelling and experimental approaches were limited.

New is that experiments in the field increased much, but also that computer simulations allowed it to improve the modelling work.

Also the knowledge that many reactions involving organics are not reversible and the consideration of the time scale improved (especially for long-term experiments). Also additional extractants have been tested.

The future lies in simulation of the environment on computer; it should become possible then to calculate for a given input of pollutants the environmental impact. To do this, however, much data material will have to be collected.

Indeed, progress in models is required, but its remains questionable, if this can be made use of for real world situations. People measure more in the fild but especially need methodology studies for in situ measurements without disturbing the system.

Many processes involving trace metals are occuring in lakes. For example, recent measurements have shown that settling particles (including phytoplankton) are important for transporting trace metals to the bottom sediments of certain lakes; in other lakes (acid lakes) diffusional transport across the sediment water interface may be an important deposition mechanism. Reactions occurring in the upper layers of sediments are important in the overall cycling of trace metals deposited on the sediments. Surficial layers of the sediments are indeed the locus of several diagenetic reactions (formation of iron and manganese oxides or sulfides, degradation of organic matters) that can influence the behaviour of trace metals. In a sediment column, the concentrations of trace metals are generally much lower in the interstitial waters than in the solids. Small changes in the concentrations of the trace metals associated with the solids can thus induce changes in their dissolved concentrations in the adjacent porewates; profiles can be obtained, for example, with porewater peepers (in situ dialysis). These profiles help in identifying the diagenetic processes. As example, profiles of Fe often show a peak a few centimetres below the sediment-water interface due to iron reduction, which indicates that Fe(II) is diffusing both upward and downward. A position of the Fe(II) that diffuses upward from anoxic sediments is oxidized to Fe(III) and retained in the surface sediments as reactive anthigenic Fe oxyhydroxides, whereas another position can diffuse out of the sediments to the overlying water. The downward diffusion of Fe(II) depends on the presence of an Fe(II) sink, presumably sulfide originating from the reduction of sulfate. The formation of these anthigenic components can be portrayed by using Teflon collector sheets inserted vertically into the sediments. The Teflon sheets collect in the following order of depth in the sediment: manganese oxides, iron oxides and sulfides. Profiles of dissolved arsenic concentrations show a peak of this element close to the depth of dissolved Fe peaks, indicating a close association of As with the iron cycle. In acid lakes, dissolved Cu, Ni and Zn concentrations are typically high in the overlying water and decrease sharply below the sediment-water interface. The gradients of concentration of these metals indicate that the sediments of acid lakes act as traps for these trace metals. Use of Fick's first law allows estimation of the fluxes, provided the appropriate diffusion coefficients are known.

These few examples show the dynamic behaviour of trace metals in the water column and in the recent sediment layers.

2. Complexation in the Environment

The panel discussion focused on several major issues of concern to environmental scientists namely, (1) how the details of chemical complexation is important to bio-uptake and toxicity of metals, (2) the need for a more interdisciplinary approach to bio-uptake and bio-toxicity, and (3) the need for a reference humic substance for scientists doing work on chemical interactions with humic substances.

Illucidation of the role of metal complexation in bio-uptake and bio-toxicity awaits more detailed studies by bio-inorganic chemists and requires more interdisciplinary team efforts to link the details of the chemical separation of metal complexes to the mechanisms of metal uptake by biota. Several mechanisms are thought to be responsible for metal uptake by biota, and detailed experimental designs to evaluate the role of complexation requires a more integrated interdisciplinary effort. The term 'bio-availability' was thought to be unspecific as related to questions of chemical form (i.e. complexed, reduced, etc.), and sharpening of the definition awaits more specific mechanistic information on metal transport across biological tissue barriers inside biological material. What role the form of metal species plays prior to, during and after transport across biological membranes needs considerable attention.

The question of need for a reference humic substance material was discussed. Although not unanimous, the general concentus indicated that considerable benefit would accrue from a 'reference humic substance' which would meet the criteria of being (1) well characterized physically and chemically, (2) accessible, (3) affordable, (4) available in large quantities. Questions were raised concerning long-term stability of the material and whether homogeneity could be assured for large quantities. A suggestion was offered which proposed possible synthesis of a reference material using organic synthetic chemistry and building in the features of humic substance believed to be responsible for the chemical behaviour of the humic material.

Thoughts on future areas for work included structural information on metal complexes in natural systems and the stability and lability of the complexes under natural conditions. In addition, it was suggested that chemical speciation of metal ions in bio-uptake and bio-toxicity studies should get greater attention.

3. Analytical Techniques of Metal Speciation

Ideally analytical methodology for speciation purposes should be such that the measurement occurs on the undisturbed material in its natural environment. Sampling is liable to induce changes which then result in a different signal generation. Metal speciation on the samples in their undisturbed dynamic habitat will remain wishful thinking for a long time to come. The sophisticated techniques suitable to analyse e.g.

an air particulate matter while travelling through the atmosphere, a colloid floating in a river, ... have not been developed yet. Sensors may be considered a modest attempt in this direction.

In the meantime in-situ techniques on specimens which are taken out of their natural environment into laboratory circumstances are supposedly the best approach to minimalize the alterations brought along by the sampling act. It became clear from the many discussions that sampling plays the more crucial role in metal speciation work.

Optical techniques are expected to play the leading role in this field because they are far less interfering than most other techniques such as electrochemical methods, extraction procedures, precipitation techniques, mass spectrometry, a.o. However, all the latter techniques result in a particular type of speciation only, defining species according to the technique applied. This "shortcoming" will stay as it is, it being inherent to the nature of the different methodologies applied.

In metal speciation there appears an outspoken different line of thoughts between researchers working on the bench and those who are developing models to describe the whole system and try to find definitive solutions to the speciation work by applying chemometrics.

The methodology for speciation work is very much dependant on the nature of the material to be speciated: air particulate matter, water samples, colloidal systems, soil, living systems, ... all require their own procedures.

As far as the element detection is concerned there is the choice between an array of multi-element techniques, such as ICP-AES, ICP-MS, XRF, MS, versus mono-element techniques, such as the classical AAS and electrochemical techniques. The development of hyphenated techniques is expected to become more and more popular in the near future. It will go hand in hand with commercial interests.

Researchers doing metal speciation work may find inspiration for developing new techniques by adopting existing knowledge from other scientific disciplines, such as organic chemistry and bio-chemistry.

The need for standard reference materials certified for metal species is evident. They are a prerequisite to harmonize the analytical results and to provide a sound system of quality assurance.

Due to the multidisciplinary nature of metal speciation, the analytical chemist is engaged in the whole problematic of this endeavour. He has to take into account why these data are needed, assist to define the problems and adapt to a problem oriented approach. The search not only goes for the identification of the molecular species and their properties but also for the prediction of future implications of existing species in the ecosystem. The state of the art appears that only a rather immature knowledge of speciation has been built up, yet it will be gradually extended.

Sampling has already been mentioned as a major difficulty. When this aspect has been solved to a satisfactory degree, then separation strategies have to be developed. These are of cause dependent on the signal detection. Use of high-power lasers and measurement of the fragmentation products with mass spectrometry, and other methods not yet thought of, will enable totally new developments. Biosensors were also mentioned as a new approach to elucidate complex metal species.

The panel concluded that the following topics are most important:

- the strategy of reliable sampling, including preservation of the species is fundamental to the success of metal speciation work
- the emphasis which should be put on the development of hyphenated techniques, preserving the molecular identity and including a suitable detection
- the progress needed in the field of in-situ techniques; a number of existing ones have been cited but there is a demand for many more.

speciation speciation speciation speciaton speciation speciation speciation

WORDS! Lots of them have been made about them. Words are very important for communication. Proper communication depends on the well defined meaning of words. Many - if not most - problems between two people, among experts, among members of different scientific disciplines, even among moles of chemists, and certainly among the various forms of speciators are traceable to misunderstandings that arose through misuse of words or through misinterpretation of the meaning of words. In a moment of deep insight an ancient Chinese philosopher replied when asked which one thing he would change to improve relations among peoples: **the language!** Language consists of words. To improve the language, its building blocks must be properly shaped and kept in shape. Scientists always strived to adhere to the recommendable principle of **"one word-one meaning."** To keep the language of science pure, international committees are in watchful sessions and exert all their expertise to prevent the accumulation of ambiguous terms in the vocabulary of science and to eradicate those confusing terms that sneaked into general use in spite of their strenuous efforts. During the past decade
such a term "**speciation**" and related words assaulted the one-word-one-meaning principle. Even a cursory reading of papers dealing with trace element compounds will reveal, that **speciation** certainly is a word-chameleon. **Speciation** is used in at least three meanings. It stands for the assembly of trace element compounds in a sample, for the processes that convert trace element compounds into other compounds, and for the analytical operations that result in the identification or quantification of trace element compounds. Such mayhem with scientific terms cannot be tolerated. A committee of IUPAC came to the rescue of this untenable and unscientific situation and suggested the following definition:

SPECIATION IS THE PROCESS YIELDING EVIDENCE OF THE ATOMIC OR MOLECULAR FORM OF AN ANALYTE.

For the sake of clarity all speciators are encouraged to adhere to this definition or not use this term at all. Severe unknown penalties await the non-conformists.

However, the situation is too serious to be solved by a committee alone. The NATO Advanced Study Institute on **Metal Speciation in the Environment** was the ideal forum to confuse the situation even further. Heated discussion raged, opponents and proponents had a difficult time to keep their scientific cool, and important activities such as swimming in the Mediterranian were sacrificed to the pursuit of clarity. To channel these clarification efforts into productive directions, a contest was organized for all those who had a burning interest in the meaning of **speciation.** The entries were numerous and showed deep insight into the difficult problem. A distillation of the submissions is presented here to keep the results of hours of informed contemplation from sinking into the hades of oblivion. A word of caution to the readers: don't take most of the definitions too serious. Accept them rather as the best that could be achieved under the circumstances, particularly in view of the prizes offered to the winners, which and who shall remain unknown.

Speciation is the tittilating effort to **speciate** among the different definitions of **speciation** to identify the correct meaning of the term **speciation.**

According to the Turkish Minister for Tourism **Speciation** is the acronym for:
Seeing Pamukale, Ephesus, Cesme In Ancient Turkey Is Obviously Necessary.

According to IUPAC V_3 **Metal Speciation** stands for: Search for Problems in the Environment by Chemical Irritation of Atoms and Ions.

Metal Speciation is the acronym for: Modelling Exciting , Tantalizing, and Labile-Sensual Practices Encouraging Contacts Inside a Transient Institute of NATO.

Spec(z)i in German means "good friend." Consequently, **speciation** is the process of making friends.

The determination of species in environmental samples is like standing in a dark tunnel, noticing a faint light (the perceived solution to the problem), proceeding toward the light to find the end of the tunnel (the final solution to the problem), and suddenly being run over by a train.

Catchung **species** is quite an art.
When you try to catch them, some fall apart.
You must use the tool "very-quick-see"
to identify and determine such a **specie.**
Your fame will rise to celestial height
should your identification turn out to be right.

METAL SPECIATION is one of a series of strategies in science that allows to cover up the fact that one is not able to determine the total concentration of a metal, although other investigators claim to succeed admirably in this endeavor.

In **METAL SPECIATION** a variable number of molecular **SPECIES** are added or left out as needed to make the sum of their concentrations agree with the independently determined total concentration of a metal within rather large experimental errors.

METAL SPECIATION in the environment deals with assemblies of atoms, many of which are unstable to sunlight and fall apart just as assemblies of scientists attending the NATO Advanced Study Institute at Cesme disperse into the aqueous environment , when the sun shines invitingly warm in the afternoon.

METAL SPECIATION is like fishing: too many almost got it; few catch really big fish; when someone is successful, he always claims to have caught the best and biggest fish; success is always the result of extraordinary inspiration, perseverance, insight, and cleverness.

SPECIATION is the term used to raise funds from NATO AUTHORITIES for the purpose of proving that metals are distributed in the environment in forms different from bullets and missiles.

METAL SPECIATION from the analytical chemist's point of view is the extraordinarily painful, tedious, and operator-dependent process that yields specific, selective, real, dubious, or useless information on the chemical, physical, metaphysical, astrological, or psychic condition of metals and metalloids in their natural, environmental, physiological, galactic, or intergalactic state. Because of likely unknown spectral, atomic, molecular, complex-formational, physical, or contaminational artifacts, all numerical results must be viewed with extreme caution, with a jaundiced eye, and with a mole of sodium chloride.

METAL SPECIATION is

- a Turkish delight.
- the analytical chemist's answer for eternal employment.
- a religious cult dedicated to the preservation of analytical chemists.
- a society for the prevention of cruelty to humic acids and fulvic acids.
- a good reason to organize a NATO Advanced Study Institute in Cesme.
- a welcome opportunity to meet scientists whose names are familiar from the literature.

THE METAL ION

A **metal ion** hard of nature
sits in soil, is bound to major
donor sites such as C=O
and vibrates mildly to and fro.

Such vibrations bring much pleasure
and are something worth to treasure.
However, it is also great
when changed is the vibration rate

The metal ion spoke: "I like variety!
Therefore, I go next door to see
the soft-touch sulfur in the ring,
to which I like quite well to cling.

I know the rule, that hard an ion
should not use softies for a tie-on.
However, when I spot a sulfur-blond,
I can't resist, I have to bond.

Should there be some inquiries
by people using ICPs
about the whereabouts of me,
just tell them that I am not free.

I certainly do not find cool
the hard-with-hard and soft-soft rule.
My dislike's bad for speciations,
but gives wonderful vibrations.

Should this cause problems to you chaps,
dont blame me! Perhaps,
your Mister Pearson
was not right-on".

Kurt J. Irgolic
NATO Advanced Study Institute on
"Metal Speciation in the Environment"
October 9 -20, 1989
Golden Dolphin, Cesme. Turkey.

NATO ASI Series G

Vol. 1: **Numerical Taxonomy.** Edited by J. Felsenstein. 644 pages. 1983. (out of print)

Vol. 2: **Immunotoxicology.** Edited by P.W. Mullen. 161 pages. 1984.

Vol. 3: **In Vitro Effects of Mineral Dusts.** Edited by E.G. Beck and J. Bignon. 548 pages. 1985.

Vol. 4: **Environmental Impact Assessment, Technology Assessment, and Risk Analysis.** Edited by V.T. Covello, J.L. Mumpower, P.J.M. Stallen, and V.R.R. Uppuluri. 1068 pages. 1985.

Vol. 5: **Genetic Differentiation and Dispersal in Plants.** Edited by P. Jacquard, G. Heim, and J. Antonovics. 452 pages. 1985.

Vol. 6: **Chemistry of Multiphase Atmospheric Systems.** Edited by W. Jaeschke. 773 pages. 1986.

Vol. 7: **The Role of Freshwater Outflow in Coastal Marine Ecosystems.** Edited by S. Skreslet. 453 pages. 1986.

Vol. 8: **Stratospheric Ozone Reduction, Solar Ultraviolet Radiation and Plant Life.** Edited by R.C. Worrest and M.M. Caldwell. 374 pages. 1986.

Vol. 9: **Strategies and Advanced Techniques for Marine Pollution Studies: Mediterranean Sea.** Edited by C.S. Giam and H.J.-M. Dou. 475 pages. 1986.

Vol. 10: **Urban Runoff Pollution.** Edited by H.C. Torno, J. Marsalek, and M. Desbordes. 893 pages. 1986.

Vol. 11: **Pest Control: Operations and Systems Analysis in Fruit Fly Management.** Edited by M. Mangel, J.R. Carey, and R.E. Plant. 465 pages. 1986.

Vol. 12: **Mediterranean Marine Avifauna: Population Studies and Conservation.** Edited by MEDMARAVIS and X. Monbailliu. 535 pages. 1986.

Vol. 13: **Taxonomy of Porifera from the N.E. Atlantic and Mediterranean Sea.** Edited by J. Vacelet and N. Boury-Esnault. 332 pages. 1987.

Vol. 14: **Developments in Numerical Ecology.** Edited by P. Legendre and L. Legendre. 585 pages. 1987.

Vol. 15: **Plant Response to Stress. Functional Analysis in Mediterranean Ecosystems.** Edited by J.D. Tenhunen, F.M. Catarino, O.L. Lange, and W.C. Oechel. 668 pages. 1987.

Vol. 16: **Effects of Atmospheric Pollutants on Forests, Wetlands and Agricultural Ecosystems.** Edited by T.C. Hutchinson and K.M. Meema. 652 pages. 1987.

Vol. 17: **Intelligence and Evolutionary Biology.** Edited by H.J. Jerison and I. Jerison. 481 pages. 1988.

Vol. 18: **Safety Assurance for Environmental Introductions of Genetically-Engineered Organisms.** Edited by J. Fiksel and V.T. Covello. 282 pages. 1988.

Vol. 19: **Environmental Stress in Plants. Biochemical and Physiological Mechanisms.** Edited by J.H. Cherry. 369 pages. 1989.

Vol. 20: **Behavioural Mechanisms of Food Selection.** Edited by R.N. Hughes. 886 pages. 1990.

Vol. 21: **Health Related Effects of Phyllosilicates.** Edited by J. Bignon. 462 pages. 1990.

NATO ASI Series G

Vol. 22: **Evolutionary Biogeography of the Marine Algae of the North Atlantic.** Edited by D. J. Garbary and G. R. South. 441 pages. 1990.

Vol. 23: **Metal Speciation in the Environment.** Edited by J. A. C. Broekaert, Ş. Güçer, and F. Adams. 655 pages. 1990.

MIX
Papier aus verantwortungsvollen Quellen
Paper from responsible sources
FSC® C105338

If you have any concerns about our products,
you can contact us on
ProductSafety@springernature.com

In case Publisher is established outside the EU,
the EU authorized representative is:
Springer Nature Customer Service Center GmbH
Europaplatz 3, 69115 Heidelberg, Germany

Printed by Libri Plureos GmbH
in Hamburg, Germany